Stock Market
Volatility

T0225472

CHAPMAN & HALL/CRC FINANCE SERIES

Series Editor

Michael K. Ong

Stuart School of Business
Illinois State of Technology
Chicago, Illinois, U. S. A.

Aims and Scopes

As the vast field of finance continues to rapidly expand, it becomes increasingly important to present the latest research and applications to academics, practitioners, and students in the field.

An active and timely forum for both traditional and modern developments in the financial sector, this finance series aims to promote the whole spectrum of traditional and classic disciplines in banking and money, general finance and investments (economics, econometrics, corporate finance and valuation, treasury management, and asset and liability management), mergers and acquisitions, hinsurance, tax and accounting, and compliance and regulatory issues. The series also captures new and modern developments in risk management (market risk, credit risk, operational risk, capital attribution, and liquidity risk), behavioral finance, trading and financial markets innovations, financial engineering, alternative investments and the hedge funds industry, and financial crisis management.

The series will consider a broad range of textbooks, reference works, and handbooks that appeal to academics, practitioners, and students. The inclusion of numerical code and concrete real-world case studies is highly encouraged.

Published Titles

Introduction to Financial Models for Management and Planning, **James R. Morris and John P. Daley**

Stock Market Volatility, **Greg N. Gregoriou**

Forthcoming Titles

Decision Options: The Art and Science of Making Decisions, **Gill Eapen**

Emerging Markets: Performance, Analysis, and Innovation, **Greg N. Gregoriou**

Portfolio Optimization, **Michael J. Best**

Proposals for the series should be submitted to the series editor above or directly to:
CRC Press, Taylor & Francis Group
4th, Floor, Albert House
1-4 Singer Street
London EC2A 4BQ
UK

CHAPMAN & HALL/CRC FINANCE SERIES

Stock Market Volatility

Edited by

Greg N. Gregoriou

SUNY
Plattsburgh, New York, U. S. A.

CRC Press
Taylor & Francis Group
Boca Raton   London   New York

CRC Press is an imprint of the
Taylor & Francis Group, an **informa** business

A CHAPMAN & HALL BOOK

Chapman & Hall/CRC
Taylor & Francis Group
6000 Broken Sound Parkway NW, Suite 300
Boca Raton, FL 33487-2742

First issued in paperback 2017

© 2009 by Taylor & Francis Group, LLC
Chapman & Hall/CRC is an imprint of Taylor & Francis Group, an Informa business

No claim to original U.S. Government works

ISBN-13: 978-1-4200-9954-6 (hbk)
ISBN-13: 978-1-138-11516-3 (pbk)

Library of Congress Cataloging-in-Publication Data

Stock market volatility / Greg N. Gregoriou.
     p. cm. -- (Chapman & Hall/CRC Finance series ; 2)
   Includes bibliographical references and index.
   ISBN 978-1-4200-9954-6 (acid-free paper) 1. Stock exchanges. 2.
   Stocks--Prices. I. Gregoriou, Greg N. II. Title. III. Series.

HG4551.S825 2009
332.63'222--dc22   2008048116

Visit the Taylor & Francis Web site at
http://www.taylorandfrancis.com

and the CRC Press Web site at
http://www.crcpress.com

Table of Contents

Acknowledgments

I would like to thank the editor, Dr. Sunil Nair, for his valuable suggestions and comments on the manuscript, Jessica Vakili, project coordinator, Sarah Morris, editorial assistant, and finally Jay Margolis, project editor. It is these wonderful people at Chapman-Hall/Taylor & Francis Group that makes working with book publishers a wonderful experience. In addition, I also thank numerous anonymous referees that were part of the review and selection process.

The Editor

Greg N. Gregoriou is professor of finance in the School of Business and Economics at State University of New York (Plattsburgh). A native of Montreal, Professor Gregoriou obtained his joint PhD in finance at the University of Quebec at Montreal, which merges the resources of Montreal's major universities: McGill, Concordia, and HEC. Professor Gregoriou's interests focus on hedge funds and managed futures. In addition to his university studies, Greg has completed several specialized courses from the Canadian Securities Institute. Greg has published more than fifty academic articles in more than a dozen peer-reviewed journals, such as the *Journal of Portfolio Management, Journal of Futures Markets, European Journal of Operational Research, Annals of Operations Research, Computers and Operations Research, Journal of Asset Management, Journal of Alternative Investments, European Journal of Finance,* and *Journal of Wealth Management,* as well as more than twenty book chapters. Greg is hedge fund editor and editorial board member for *Journal of Derivatives and Hedge Funds,* a London, UK–based academic journal, and also editorial board member of the *Journal of Wealth Management* and the *Journal of Risk and Financial Institutions.* He has published twenty-eight books with Chapman-Hall/CRC, John Wiley & Sons, McGraw-Hill, Bloomberg Press, Elsevier Butterworth-Heinemann, Palgrave-MacMillan, and Risk books.

The Contributors

David E. Allen is professor of finance at Edith Cowan University, Perth, Western Australia. He is the author of three monographs and over seventy refereed publications on a diverse range of topics covering corporate financial policy decisions, asset pricing, business economics, funds management and performance bench marking, volatility modeling and hedging, and market microstructure and liquidity.

Turan G. Bali received his PhD from the Graduate School and University Center of the City University of New York in 1999. He is the David Krell Chair Professor of Finance at Baruch College and the Graduate School and University Center of the City University of New York. His fields of specialization are asset pricing, risk management, fixed-income securities, interest rate derivatives, and dynamic asset allocation. He has published about fifty articles in leading journals in economics and finance, including the *Journal of Finance, Journal of Financial Economics, Management Science, Journal of Business, Journal of Financial and Quantitative Analysis, Journal of Economic Dynamics and Control, Journal of Money, Credit, and Banking, Journal of Empirical Finance, Journal of Banking and Finance, Financial Management, Journal of International Money and Finance, Risk,* and many others. He is an associate editor of the *Journal of Banking and Finance,* the *Journal of Futures Markets,* and the *Journal of Risk.*

Sumon Kumar Bhaumik is a senior lecturer at Brunel University and an associate editor of *Emerging Markets Finance and Trade.* He is also a research fellow at the William Davidson Institute at the University of Michigan, Ann Arbor, and IZA—Institute for the Study of Labor in Bonn. He has a PhD in economics from the University of Southern California and has worked at a number of organizations, including ICRA Limited (the Indian associate of Moody's Investors Service), London Business School, and Queen's University Belfast.

Laurent Bodson is PhD candidate in finance and FNRS Research fellow at the HEC, Business School of the University of Liège. His areas of expertise include portfolio and risk management, as both a practitioner and researcher. He is also specialized in investment analysis, derivatives, style analysis, stock market price behavior and integration of higher order moments.

Suchismita Bose is an economist at ICRA Limited (the Indian associate of Moody's Investors Service). She has a PhD from the Indian Statistical Institute at Calcutta.

Wolfgang Breuer was born in 1966. Since March 2000, he has been a full professor of finance at the RWTH Aachen University, Germany's leading technical university. From October 1995 to February 2000 he was a full professor of finance at the University of Bonn. He earned his PhD degree in February 1993 and his habilitation degree in July 1995, both at the University of Cologne. After his diploma in 1989 he worked 1 year in Frankfurt as a consultant at McKinsey & Co., Inc., before he continued his academic career. Wolfgang Breuer has written about a dozen books, sixty book chapters, and forty peer-reviewed journal articles (among others, in the *Journal of Banking and Finance*, *Journal of Futures Markets*, *Journal of Institutional and Theoretical Economics*, and *European Journal of Finance*) comprising a great variety of topics in the field of finance. His current research interests focus on portfolio management, international financial management, and corporate finance.

Charlie X. Cai is a lecturer at the University of Leeds, United Kingdom. He has published a number of papers in international peer-reviewed journals. His research specialties are in asset pricing, volatility, and market microstructure.

Christian Calmès holds a PhD in economics from UQAM. He also holds a MSc in economics from Laval University. He is associate professor of finance at the Department of Administrative Sciences of the University of Quebec, Outaouais (UQO). He was previously a senior economist at the Bank of Canada. Professor Calmès is a permanent member of the Laboratory for Research in Statistics and Probability (LRSP). He is presently on the editorial board of review for *L'Actualité économique, revue d'analyse* *économique*. He is also on the editorial board of New Economics Papers, which includes eighty specialized periodicals. He is editor of two specialized electronic publications: New Economic Papers in Business Economics and New Economic Papers in Regulation. His research interests focus on the following areas: self-enforcing labor contracts in macroeconomics, macroeconomic dynamics, corporate finance, and the management of financial institutions. Professor Calmès has published books in economics and published articles in the following journals: *International Advances in Economic Research, Journal of International Financial Markets, Institutions and Money, Journal of Financial Transformation, Swiss Journal of Economics and Statistics*, and *L'Actualité économique, Revue d'analyse économique*.

Rachael Carroll obtained her BSc and MSc degrees in statistics from University College Dublin. She is now research associate in the Institute for International Integration Studies at Trinity College Dublin. Rachael's prime research interests include GARCH modeling of volatility in equity markets, and volume and volatility in the international markets for corporate control.

Thomas C. Chiang is the Marshall M. Austin Professor of Finance at Drexel University. He is the author of numerous articles in refereed journals and two books. His recent research interests have included financial contagion, international finance, asset pricing, and financial econometrics. His articles have appeared in the *Journal of Money, Credit and Banking, Journal of International Money and Finance, Applied Financial Economics, Weltwirtschaftliches Archiv*, and *Physical Review E*, among others. Dr. Chiang received his PhD from Pennsylvania State University, with a concentration in financial economics and econometrics.

Rosa Cocozza, who has her MA in banking and finance and PhD in business administration, is professor of financial risk management at the Faculty of Economics of the Università di Napoli Federico II. Member of American Risk and Insurance Association (ARIA), European Association of University Teachers of Banking and Finance (Wolpertinger Club), and Associazione dei Docenti di Economia dei Mercati e degli Intermediari Finanziari (ADEIMF), she is on the editorial board of the ADEIMF Working Paper Series. Her research focuses on risk management processes and techniques within financial institutions. Author of more than thirty papers on quantitative management modeling for financial intermediaries, she has also published two monographs: one on credit pricing and the other on interest rate risk management for life insurers.

Yves Crama is professor of operations research and production management, and director general of HEC—Management School of the University of Liège in Belgium. He holds a PhD in operations research from Rutgers University. He is interested in the theory and applications of optimization, mathematical modeling, and algorithms, with an emphasis on applications arising in supply chain management and finance. He has published over sixty papers on these topics in leading international journals, is the coauthor of a monograph on production planning, and is the associate editor of several scientific journals.

Nazmi Demir is an assistant professor and the chairman of the Department of Banking and Finance of Bilkent University in Ankara, Turkey. Previously he was an assistant professor in the Department of Economics in Bilkent University. He worked as director general for research and then deputy undersecretary in the Ministry of Agriculture, Forestry and Rural Affairs. He was board member for the international research institutes of the World Bank, ICARDA, and IBPGR for 6 years in each, and a 6-year representative of the Middle East agricultural research for the IBRD. Additionally, he was visiting scientist for 1 year in CIMMYT, Mexico City. He holds MSc and PhD degrees, both from the University of California, Davis in agricultural economics. His main interest is banking and finance as well as agricultural economics as related to the environment. His papers have been published in *Economic Letters, Canadian Journal of Agricultural Economics, Developing Economics, Indian Economic Review*, and others.

K. Ozgur Demirtas is an associate professor of finance at Zicklin School of Business of Baruch College in New York. Professor Demirtas received his PhD in finance from Boston College in 2003. Professor Demirtas has a vast base of research and publications and has won many awards, grants, and fellowships for his diverse and successful research record. His work has been published in top academic journals such as *Journal of Financial and Quantitative Analysis, Journal of Banking and Finance, Journal of Futures Markets, Finance Letters*, and *International Journal of Revenue Management*, as well as practitioner-oriented journals such as *Journal of Portfolio Management*. He also has a distinguished record for teaching. He received the Donald J. White Teaching Excellence Award during his studies at Boston College. He has been elected as the best teacher within the finance department of Baruch College. In 2004, Professor Demirtas received the Zicklin School of Business Teaching Excellence Award. Finally, in 2007, he became the youngest faculty to receive the prestigious Presidential Excellence Award for Distinguished Teaching.

Robert B. Durand is associate professor of finance at the University of Western Australia. His research interests include asset pricing, portfolio theory, and behavioral finance. He has published a number of papers on asset pricing in Australia and the integration of the Australian and world markets.

Cumhur Ekinci is an assistant professor at Istanbul Technical University. He holds a BA in economics from Bogazici University, an MA in finance from the University of Paris I Pantheon-Sorbonne, and a PhD in finance from the University of Aix-Marseille III. Dr. Ekinci worked in a school trading room at CNAM in Paris and gives courses about financial markets at CNAM, University of Aix-Marseille II, and ENPC. His research topics include market microstructure, high-frequency data, competition among market venues, hedge funds business, and algorithmic trading.

Craig Ellis is an associate professor of finance at the University of Western Sydney, Australia. His primary research interests include topics relating to financial asset return distributions and the statistical and economic implications of nonrandom behavior for financial asset pricing. Craig has published and refereed numerous articles in journals including *Chaos Solitons and Fractals, Economics Letters, International Review of Financial Analysis*, and *Physica A*.

Robert Faff holds a chair at Monash University, Australia. He has been ranked as one of the most prolific researchers in the world by a number of academic surveys. To date, he has published over 120 articles in leading finance journals.

Dean Fantazzini is a lecturer in econometrics and finance at the Moscow School of Economics, Moscow State University. He graduated with honors from the Department of Economics at the University of Bologna (Italy) in 1999. He obtained the Master in Financial and Insurance Investments at the Department of Statistics, University of Bologna (Italy), in 2000 and the PhD in economics in 2006 at the Department of Economics and Quantitative Methods, University of Pavia (Italy). Before joining the Moscow School of Economics, he was research fellow at the Chair for

Economics and Econometrics, University of Konstanz (Germany), and at the Department of Statistics and Applied Economics, University of Pavia (Italy). He is a specialist in time-series analysis, financial econometrics, and multivariate dependence in finance and economics. He has to his credit more than twenty publications, including three monographs.

Giampaolo Gabbi is full professor of financial markets and risk management at the University of Siena, Italy, and professor at SDA Bocconi, Milan, where he coordinates several executive courses on financial forecasting and risk management. He coordinates the MSc in finance at the University of Siena and is also head of the financial areas of masters in economics in the same university. Professor Gabbi holds a PhD in banking and corporate management. He has published many books and articles in refereed journals, including *Managerial Finance*, the *European Journal of Finance*, and the *Journal of Economic Dynamics and Control*.

Bartosz Gebka is lecturer in finance at the Newcastle University Business School, UK. His research interests are in the efficiency of emerging markets, with a focus on the role of trading volume in asset pricing, the differences between institutional and individual investors, and corporate governance. He has published in the *Journal of International Financial Markets, Institutions and Money, Global Finance Journal, International Review of Financial Analysis*, and others, and acts as a referee for numerous academic journals and publishers. He obtained his PhD at the European University in Frankfurt (Oder), Germany, and was also a visiting research fellow at the MMU in Manchester, UK, and a lecturer at the ITESM, Cuernavaca, Mexico.

Massimo Guidolin (PhD, 2000, University of California) is a chair professor of finance at Manchester Business School. He also served as an assistant vice president and senior policy consultant (financial markets) within the U.S. Federal Reserve system (St. Louis FED), where he still covers advising roles. From December 2007 he has been co-director of the Center for Analysis of Investment Risk, at Manchester Business School. His research focuses on predictability and nonlinear dynamics in financial returns, with applications to portfolio management, and sources and dynamics of volatility and higher-order moments in equilibrium asset pricing models. He has published papers in the *American Economic Review, Review of Financial Studies, Journal of Business, Journal of Econometrics*, and *Economic Journal*, among others.

Jodie Gunzberg joined the Marco Consulting Group in 2007 as director of research/manager search, where she is responsible for investment research and manager selection. She has several years of investment experience across asset classes including equities, fixed income, real estate, hedge funds, and commodities. Prior to joining MCG, Jodie held various analyst, portfolio management, and risk management positions where she built security selection models, risk management systems, and engineered new strategies. She also has publications on commodity and hedge fund investing. Jodie holds her MBA from the University of Chicago and her BS in mathematics from Emory University. She is a CFA charterholder and member of the CFA Institute and the CFA Society of Chicago.

Marc Gürtler has been since 2002 a full professor of finance at the Technical University of Braunschweig, Germany. Before coming to Braunschweig he was an assistant professor of finance at the RWTH Aachen University. He earned his PhD degree in 1997 at the University of Bonn and his habilitation degree in 2002 at the RWTH Aachen University. From 1993 to 1994 he worked as a risk manager in the department of asset management of AXA Colonia Insurance Company, Cologne. His research interests include, in particular, portfolio management, credit risk management, and international financial management. He has written several books and peer-reviewed journal articles, and contributed to other books.

Sam Hakim is an adjunct professor of finance at Pepperdine University in Malibu, California. He is concurrently a vice president of risk management at Energetix LLP, an energy company in Los Angeles. Previously Dr. Hakim was director of risk control at Williams, an oil and gas company in Houston. Dr. Hakim was also financial economist at Federal Home Loan Bank in Washington, DC. Between 1989 and 1998 Dr. Hakim was an associate professor of finance and banking at the University of Nebraska at Omaha. Dr. Hakim is the author of more than forty articles and publications. He is an Ayres fellow with the American Bankers Association in Washington, DC, and a fellow with the Economic Research Forum. He holds a PhD in economics from the University of Southern California.

David Hillier is centenary professor and Ziff Chair in Financial Markets at University of Leeds, United Kingdom. His research interests are in financial markets, corporate finance, and corporate governance.

Jason C. Hsu oversees the research and investment management areas at Research Affiliates. He manages $35 billion in the firm's subadvisory and hedge fund businesses as well as direct researches on asset allocation models that drive the firm's global macro and GTAA products and equity strategies that underpin RA's Fundamental Index* concept. Jason is an adjunct professor in finance at the UCLA Anderson Business School and served as visiting professor at the UC Irvine Paul Merage School of Management and the School of Commerce at Taiwan National Chengchi University. Jason graduated summa cum laude from California Institute of Technology and earned his PhD in finance from the University of California, Los Angeles.

Georges Hübner (PhD, INSEAD) is the Deloitte Professor of Financial Management and is co-chair of the Finance Department at HEC— Management School of the University of Liège. He is an associate professor of finance at Maastricht University and academic expert at the Luxembourg School of Finance, University of Luxembourg. He is affiliate professor of finance at EDHEC and Solvay Business School. He is also the founder and CEO of Gambit Financial Solutions, a financial software spin-off company of the University of Liège. Georges Hübner has taught at the executive and postgraduate levels in several countries in Europe, North America, Africa, and Asia. He regularly provides executive training seminars for the preparation of the financial risk manager (FRM) and chartered alternative investment analyst (CAIA) certifications. His research articles have been published in leading scientific journals, including *Journal of Banking and Finance*, *Journal of Empirical Finance*, *Review of Finance*, *Financial Management*, and *Journal of Portfolio Management*. Georges Hübner was the recipient of the prestigious 2002 Iddo Sarnat Award for the best paper published in the *Journal of Banking and Finance* in 2001.

Zeynep İltüzer is a teaching and research assistant at Istanbul Technical University (ITU). She has a BS in mathematical engineering from Yıldız Technical University and an MS in management engineering from ITU. She is preparing a PhD at ITU and works in the field of risk management.

Elena Kalotychou (BA/MA in Cantab mathematics, MSc in operational research, PhD in finance) is currently lecturer in finance at the Faculty of

Finance, Cass Business School, City University, London. Her research interests are in international credit risk, financial econometrics, and forecasting. She has published in the *International Journal of Forecasting, Computational Statistics and Data Analysis, Financial Markets, Institutions and Instruments, Journal of Multinational Financial Management, Journal of International Financial Markets, Institutions and Money,* and *Applied Economics.*

Colm Kearney is professor of international business in the Business School and research associate of the Institute for International Integration Studies at Trinity College Dublin. Prior positions include professor of finance and economics at the University of Technology Sydney and senior consultant to the Australian federal treasurer and finance ministers. Colm's research focuses on international business, international finance, and modeling the determination and transmission of volatility.

Claudia Klüppelberg holds the chair of mathematical statistics at the Center for Mathematical Sciences of the Munich University of Technology. She has held positions at the University of Mannheim and in the Insurance Mathematics group of the Department Mathematik at ETH Zurich. Her research interests combine applied probability and statistics with special application to finance and insurance risk processes. She is an elected fellow of the Institute of Mathematical Statistics, a member of the editorial board of the Springer Finance book series, and associate editor of several scientific journals. Besides numerous publications in scientific journals, Claudia Klüppelberg co-authored the book *Modelling Extremal Events for Insurance and Finance* (Springer, 1997) with P. Embrechts and T. Mikosch, published in New York.

Raphael W. Lam is economist at the International Monetary Fund and holds a PhD in economics from the University of California, Los Angeles.

Peter Lerner received his undergraduate and graduate education in physics at Moscow Institute for Physics and Technology and Lebedev Institute for Physical Sciences. He conducted research with Los Alamos National Laboratory and Penn State University. During this time, he authored more than fifty papers and book contributions in optics, atomic physics, and materials science. In 1998, Peter graduated from Katz School of Business (University of Pittsburgh) with an MBA and worked 2 years as a risk quant in energy trading. He received his PhD in finance from Syracuse University in 2006 at the ripe old age of 48.

Suntharee Lhaopadchan holds a lectureship at Faculty of Management Sciences, Kasetsart University, Thailand. At present, she holds a research position at Amsterdam Business School, Universiteit van Amsterdam, The Netherlands. Her doctoral thesis was on financial market volatility.

Feifei Li is responsible for quantitative research on the Research Affiliates' Fundamental Index' along with other Research Affiliates equity products and strategies. She utilizes advanced econometric/statistical models to analyze investment strategies for their performance over time and by industry. Prior to joining Research Affiliates, Feifei was employed as a credit analyst with the Bank of China and with PricewaterhouseCoopers, both in Beijing. Feifei completed her PhD in finance at the University of California, Los Angeles, where she conducted empirical research on corporate finance and event-driven investment strategies. She earned a bachelor of arts from Tsinghua University's School of Management and Economics in Beijing, China. Feifei is a certified financial risk manager.

Andrea Limone graduated in financial markets economics from the University of Siena. She obtained her master's degree in financial management of insurance companies from the University of Roma "La Sapienza." She is a research grant holder at the Law and Economics Department, University of Siena, and a lecturer in financial management of life products. She is a risk management and insurance teaching assistant.

Ross Maller is professor of mathematical finance at the Australian National University and is the author or co-author of over a hundred papers and two books in probability, statistics, and mathematical finance. He has held visiting positions at various institutions, including Cornell University and the University of Manchester, and is currently associate editor of the *Journal for Theoretical Probability*. His research interests include derivatives pricing using Levy process models, and portfolio analysis.

Gernot Müller is postdoc at the chair of mathematical statistics at the Center for Mathematical Sciences of the Munich University of Technology. His research covers various topics, including computational and asymptotic statistics for discrete- and continuous-time models with applications to financial markets.

Simon Neaime is professor and chair of the Department of Economics, American University of Beirut, Lebanon. He has numerous academic journal articles published in top-ranking economic and finance journals, and has also published several graduate textbooks in financial, monetary, and international economics. His most recent articles appeared in the *Review of Middle East Economics and Finance, Journal of Economic Integration, North American Journal of Economics and Finance*, and *International Economic Journal.*

Ignacio Olmeda (PhD in finance) is an associate professor of economics and computer science at the University of Alcalá. He has been a Fulbright Visiting Scholar at several institutions in the United States and Asia. He is the director of the masters of finance at CIFF (Bank of Santander—UAH) and the director of the SUN Microsystems Laboratory of Computational Finance.

Mehmet Orhan is an associate professor at the Economics Department of Fatih University, Istanbul. At the same time, he is the director of the Social Sciences Institute, which is responsible for the coordination of graduate programs. He has his PhD from Bilkent University, Ankara, and has graduated from the Industrial Engineering Department of the same university. He had a full scholarship until he got his PhD. His main interest is econometrics, both theoretical and applied. He has published articles in *Economics Letters, International Journal of Business, Applied Economics*, and *Journal of Economic and Social Research*. His theoretical research interests include HCCME estimation, robust estimation techniques, and Bayesian inference. He is working on IPO performances, hedge fund returns, tax revenue estimation, and international economic cooperations as part of his applied research studies.

Jack Penm is currently at academic level D at the ANU. He has an excellent research record in the two disciplines in which he earned his two PhDs, one in electrical engineering from the University of Pittsburgh and the other in finance from ANU. He is an author/co-author of more than eighty papers published in various internationally respectful journals.

Valerio Potì is a finance lecturer at Dublin City University. After graduating from Bocconi University of Milan, Valerio worked for many years as a derivatives trader. He later taught international finance at Queen's University Belfast and obtained his PhD at Trinity College in Dublin.

He has publications in international peer-reviewed journals and is the author of numerous book chapters on the volatility and co-dependency of asset returns, on asset pricing, and on performance attribution. His consulting experience includes advising a number of high-standing banking institutions on capital allocation and risk management. He is now working on the pricing of nonlinear strategies and alternative investments performance evaluation.

Roberto Renò obtained his PhD at Scuola Normale Superiore, Pisa, Italy, and associate professor of financial mathematics at University of Siena, Italy. His research interests cover financial mathematics, financial econometrics, and empirical finance. He is author of several research papers published in international scholar journals.

Marco Rossi is senior economist at the International Monetary Fund and holds a PhD in quantitative economics from the Catholic University of Louvain and the London School of Economics.

Marcel Scharth is a master's graduate in economics from the Department of Economics, Pontifical Catholic University of Rio de Janeiro (PUC-Rio), and is currently undertaking his PhD studies under a scholarship at the Tinbergen Research Institute, the Netherlands.

Martin Scheicher is a senior economist in the European Central Bank. His research interests are risk management and empirical finance. His most recent work has focused on the market pricing of credit derivatives. His work includes papers published in the *Journal of Banking and Finance* and the *Journal of Futures Markets*. He was educated at the University of Vienna and London School of Economics.

Michael Schyns obtained a PhD in portfolio optimization at the University of Liege. He is currently professor of information systems at HEC—Management School of the University of Liège, where he heads the Operations Department. His main research interest is combinatorial optimization applied to management problems and statistics.

Rituparna Sen is assistant professor of statistics at the University of California at Davis. She obtained her PhD in statistics from the University of Chicago in 2004. Her main research interests are in applications of statistics in finance, specifically asset pricing and hedging in the incomplete market, stochastic volatility, and co-volatility in the presence of microstructure noise, jumps, and asynchronicity. In statistical methodology she is interested in convergence of stochastic processes, Bayesian filtering, asymptotic inference, functional data analysis, and hidden Markov models.

Rudra Sensarma teaches and researches finance at the University of Hertfordshire Business School. He has also worked at the University of Birmingham, the Indian Institute of Management, and the Reserve Bank of India. His interests are in the areas of money, banking, and financial markets.

M. Nihat Solakoglu is an assistant professor in the Banking and Finance Department of Bilkent University in Ankara, Turkey. Previously he was an assistant professor in the department of management at Fatih University. Before joining Fatih University, he worked for American Express in the United States in the international risk management, international information management, information and analysis, and fee services marketing departments. He received his PhD in economics and master's degree in statistics from North Carolina State University. His main interests are applied finance and international finance. His papers have been published in *Applied Economics, Applied Economics Letters, Journal of International Financial Markets, Institutions and Money, Journal of Economic and Social Research*, and others.

Daniel Sotelsek (PhD in economics) is an associate professor of economics at the University of Alcalá. Before joining the University of Alcalá he was a professor of several institutions in Argentina and occupied managing positions in the financial sector. He is the director of the masters of banking at CIFF (Bank of Santander–UAH) and the director of the Institute of Latin American Studies.

Sotiris K. Staikouras is associate professor of financial institutions management at Cass Business School, City University, London. His major research interests are risk analysis and management of financial institutions, asset pricing, and financial modeling. Dr. Staikouras has worked as a research advisor and analyst at London Clearing House and other institutions. He holds a PhD degree in finance from Cass Business School, and his research has been published in U.S. and European journals.

Olaf Stotz earned his PhD degree at RWTH Aachen University in 2003, where he currently is assistant professor of finance. His research has been published in several academic journals and has been awarded with several prizes. Before his position at RWTH Aachen University, he worked for several years as a quantitative researcher in the investment industry.

Maike Sundmacher is a lecturer in finance at the School of Economics and Finance, University of Western Sydney. She teaches in corporate finance, bank management, and credit risk management. Maike is enrolled in a PhD degree at the Macquarie Graduate School of Management and researches in the areas of capital markets and risk management in financial institutions.

Oktay Taş is an associate professor and chair of accounting and finance at Istanbul Technical University (ITU). After a BA and an MA in accounting and finance at Marmara University, he completed his PhD at the Technical University of Berlin. Professor Tas teaches financial management, portfolio management, and investment. His areas of interest are corporate finance, financial analysis and auditing, and financial derivatives.

R. D. Terrell is a financial econometrician, and officer in the general division of the Order of Australia. He served as vice chancellor of the ANU from 1994 to 2000. He has also held visiting appointments at the London School of Economics, the Wharton School, University of Pennsylvania, and the Econometrics Program, Princeton University. He has published a number of books and research monographs and around eighty research papers in leading journals.

Raymond Théoret holds a PhD in economics (financial economics) issued by the University of Montreal. He is professor of finance at l'École des Sciences de la Gestion (ESG) of the University of Quebec, Montreal (UQAM). He was previously professor in financial economics at l'Institut d'Économie Appliquée, located at HEC Montreal. He was an economic and financial consultant at various financial institutions in Quebec and secretary of Campeau Commission on the improvement of the situation of financial institutions in Montreal, which gives way to the foundation of Institut de Finance Mathématique de Montréal. Professor Théoret has published many articles and many books on financial engineering, especially in the fields of numerical methods, computational finance, asset pricing, and banking. His articles appear in the following journals: *Journal of Wealth Management*, *Journal of Derivatives and Hedge Funds*, *International Advances in Economic Research*, *L'Actualité Économique*, *Journal of Theoretical Accounting Research*, *Luxembourg Economic Papers*, and *Journal of Risk and Insurance*. He is member of the Corporate Reporting Chair located at ESG-UQAM.

Audrey Wang is manager of alternatives and senior investment manager analyst, joined the Marco Consulting Group in 2001 as an investment manager analyst, where she is responsible for due diligence on investment managers. Since 2004, she has been focusing on hedge funds and alternative strategies. Prior to MCG, she worked at Morningstar and Merrill Lynch. She earned her undergraduate degree from Loyola University, Chicago, and is currently pursuing her MBA at DePaul Kellstadt.

Chunchi Wu received his PhD from the University of Illinois at Urbana-Champaign in 1982. For many years he worked at the Whitman School of Management, Syracuse University, as an assistant professor, associate professor, professor, and the head of its PhD program (1997–2005). In 2005–2007, he headed finance direction at Singapore Management University. Professor Wu has published extensively in a number of leading finance journals, including *Journal of Business*, *JFQA*, and *JFE*. Currently, he is Jeffrey E. Smith Missouri Professor in Finance at University of Missouri-Columbia.

Vladimir Zdorovtsov is a vice president of State Street Global Advisors. He heads up the firm's Advanced Research Center in London and is also responsible for European Active Equity research. His current research activities include applications of natural language processing, market microstructure, and behavioral finance to quantitative security selection models. Previously, Vladimir led SSGA's Absolute Return Strategy and the Portfolio Construction and Transaction Cost Analysis research teams. Prior to joining the company in 2004, Vladimir worked as a senior economist at Law and Economics Consulting Group and as director of trading at Short Term Capital Management LLC, which he co-founded. Vladimir earned a PhD in finance from the University of South Carolina and an MBA from the University of Arkansas. He also received a BS with a concentration in international economics from Sumy State University.

I

Modeling Stock
Market Volatility

CHAPTER 1

An Overview of the Issues Surrounding Stock Market Volatility

Elena Kalotychou and Sotiris K. Staikouras[*]

CONTENTS

1.1 INTRODUCTION

Financial markets can move quite dramatically, and stock prices may appear too volatile to be justified by changes in fundamentals. Such observable facts have been under scrutiny over the years and are still being studied vigorously (LeRoy and Porter, 1981; Shiller, 1981; Zhong et al., 2003).

Volatility as a phenomenon as well as a concept remains central to modern financial markets and academic research. The link between volatility and risk has been to some extent elusive, but stock market volatility

---

[*] The authors thank Gang Zhao and Kjell Horn for excellent research and editorial assistance, respectively. The usual disclaimer applies.

is not necessarily a bad thing. In fact, fundamentally justified volatility can form the basis for efficient price discovery, while volatility dependence implies predictability, which is welcomed by traders and medium-term investors.

The importance of volatility is widespread in the area of financial economics. Equilibrium prices, obtained from asset pricing models, are affected by changes in volatility, investment management lies upon the mean-variance theory, while derivatives valuation hinges upon reliable volatility forecasts. Portfolio managers, risk arbitrageurs, and corporate treasurers closely watch volatility trends, as changes in prices could have a major impact on their investment and risk management decisions.

The current chapter provides an overarching review of the equity market volatility, covering areas that have caught the attention of practitioners and academics alike.* It aims to enlighten financiers and anyone interested in equity markets about the theories underlying stock market volatility, the historical trends and debates in the field, as well as the empirical findings at the forefront of academic research.

In what follows, Section 1.2 discusses the link between volatility and speculative action, Section 1.3 looks at the interface among information, volume, and volatility, and Section 1.4 explores the impact of derivatives trading on the underlying asset's volatility. Section 1.5 considers some stylized facts related to equity volatility, Section 1.6 compares rival volatility forecasts, and Section 1.7 focuses on volatility trading. Section 1.8 concludes the chapter.

1.2 SPECULATION AND VOLATILITY

Speculators are usually seen with some sort of resentment by the wider community.† From the early days, scholars have either supported that speculators stabilize prices (Smith, 1776; Mill, 1871; Friedman, 1953) or argued that speculators make money at the expense of others, which in turn produces a net loss and results in unnecessary price fluctuations (Kaldor, 1960; Stein, 1961; Hart, 1977). In any event, large institutional

---

* Because of the vast amount of research available, and to keep the task manageable, this chapter has no intention to lessen the importance of any studies excluded. Further and more specialized reading can be obtained from the references provided in the papers cited in this chapter. For market microstructure issues see O'Hara (1997).

† Carpenter (1866, p. 84) quotes Abraham Lincoln: "For my part, I wish every one of them [speculators] had his devilish head shot off." The role of speculators has also been discussed by Walras (1896) and Keynes (1936).

investors should be able to insure against excess fluctuations (at least in the short run), while small agents may have to bear the consequences. Under these circumstances, greater instability leads to real economic costs.

Analysts often argue that there is a link between speculation and volatility, while some even commit themselves to the *post hoc ergo propter hoc* fallacy. It is crucial, however, to distinguish between the order of events and the factors that rule out any connection between the two episodes, i.e., understand the concept of coincidental correlation, or more formally separate the notion of correlation and causation. In essence, a case could be made that speculators act as momentum traders by identifying peaks and troughs in retrospect, which in turn accelerates upward/downward movements or even increases the amplitude and frequency of fluctuations. What determines the level of disruption in the cash market is the speculators' (poor) forecasting ability and lack of information (Baumol, 1957; Seiders, 1981).[*] But from a practical point of view, how do speculators inject excess volatility (if any) in financial markets?

Volatility is an inevitable market experience mirroring (1) fundamentals, (2) information, and (3) market expectations. Interestingly, these three elements are closely associated and interact with each other. Adjustments in equity prices (should) echo changes in various aspects of our society such as economic, political, monetary, and so forth. That is, corporate profitability, product quality, business strategy, political stability, interest rates, etc., should have a role to play in shaping the intensity of price fluctuations, as the market moves from one equilibrium to another.[†] At the same time, information about changes in fundamentals should spark market activity changing the landscape of future prices. In fact, the process can be viewed as a "game" where the sequence becomes one of changes in fundamentals, information arrival, and new expectations (hence new trading positions), which in turn results in an endless cycle where these events embrace each other in a series of lagged responses.

The point here is what kind of information speculators[‡] possess, which raises a few interesting questions. First, do speculators have superior access

---

[*] Early research has produced mixed results (Telser, 1959; Kemp, 1963; Farrell, 1966; Hart and Kreps, 1986). Noise traders can also be held responsible for fueling price instability. These investors irrationally trade on information immaterial to equity values (Black, 1986).

[†] Under market efficiency (Fama, 1970, 1991) any changes should be reflected in prices instantaneously.

[‡] We refer here to institutional speculators such as hedge funds, investment houses, etc., which aim to profit from changes in market conditions.

to information? Speculators devote more resources to follow the markets and, because of their size, are able to reduce any associated expenses. Second, do speculators, by means of expertise/knowledge, better interpret the same set of information than others? Theoretically, sophisticated speculators should be one step ahead. On the other hand, historical cases (Metallgesellschaft (1993), Orange County (1994), Daiwa Bank (1995), Chase Manhattan (1997), UBS (1998), LTCM (1998), etc.) do not endorse such a claim, which places more emphasis on the roots of excessive volatility and market instability. Third, on the basis of information received/interpreted, do speculators behave in a proactive rather than inactive way? This actually leads us, indirectly, to the concept of herding behavior. Market analysts sometimes pin down the origins of volatility to either uninformed trading or collective irrationality—possibly resulting from herding behavior. Such an approach reinforces the view that speculation can lead to unjustified price variability.

The debate over speculation and excess volatility has become more of a two-handed lawyer problem. If speculators indeed lead the market, then we shall observe faster price adjustments on the basis of their actions. It would also be hard to blame them for acting quicker than others, or hold them responsible for long-term excess volatility. Besides, such volatility should fade away rather quickly in an efficient market. On the other hand, if speculators simply follow the market or possess the same information set—interpreted in the same way as by the rest of the market—then their actions would lack the material information* required to justify price changes or even excess volatility.

Nonetheless, professional market players measure their performance against their peers' (Lakonishok et al., 1992a), while some tend to "rationally" herd (Lakonishok et al., 1992b; Wermers, 1999; Grinblatt et al., 1995; Welch, 2000). The contagion advocated by the second group of studies preserves reputation since the failure/loss is shared with the market peers.† This issue has been the subject of analysis (Devenow and Welch, 1996; Calvo and Mendoza, 2000), but bear in mind that markets closely watch those who tend to lose as a result of taking decisions different from

---

* Information is material if it has an impact on securities prices when it becomes publicly available for the first time. If it has no impact on prices, it is largely irrelevant, although it may cause portfolio adjustments that leave prices unchanged.

† Actually, this is not a new approach, as Keynes, (1936, p. 158) states, "It is better for reputation to fail conventionally than to succeed unconventionally."

their peer group. Finally, note that the approaches discussed are based on the fact that legitimate information (about future demand) governs the actions of speculators. Yet, price manipulation[*] is a market reality. It is certainly possible that through price manipulation excess profits can be earned (Allen and Gorton, 1992; Allen and Gale, 1992; Jarrow, 1992, 1994; Cooper and Donaldson, 1998), but all largely depends upon the underlying model assumptions, such as risk aversion, information, etc.

1.3 INFORMATION, LIQUIDITY, AND VOLATILITY

Volatility is a natural consequence of trading, which occurs through the news arrival and the ensuing response of traders. The chain reaction of market participants[†] will force equity prices to reach a postinformation equilibrium level. Revision of expectations and subsequent actions will be reflected in the liquidity of the particular market and specifically on the amount of stocks traded. If we place the above process in a continuous time of revising expectations, and since the underlying prime mover is common, i.e., flow of information, then it is expected that information, liquidity, and volatility are related.

The relation among information, volume (liquidity), and volatility is consistent with four competing propositions: the mixture of distributions hypothesis (MDH) (Clark, 1973; Epps and Epps, 1976; Harris, 1986, 1987), the sequential information hypothesis (Copeland, 1976; Morse, 1980; Jennings et al., 1981; Jennings and Barry, 1983), the dispersion of beliefs approach (Harris and Raviv, 1993; Shalen, 1993), and the information trading volume model of Blume et al. (1994).

The motivation behind the MDH is drawn by the apparent leptokurtosis exhibited in daily price changes attributed to the random events of importance to the pricing of stocks. The MDH postulates that volume and volatility are contemporaneously and positively correlated, while jointly driven by

---

[*] In March 2008, the SEC launched an inquiry suspecting short selling and market manipulation surrounding Lehman Brothers (shares fell 40%). The SEC also investigated the trading of Bear Stearns shares prior to its purchase by JP Morgan (March 2008). During the same period, in the UK, the Bank of England and the FSA investigated allegations of market abuse by traders spreading false rumors to profit from short selling. Shares of Halifax, UK's largest mortgage lender, fell 20% amid speculation of stock shorting by hedge funds as well as claims of emergency funding.

[†] Obviously their expectation about future prices will determine their trading activity. Hedgers will mainly respond in order to secure their future income, while speculators will take advantage should their expectations about future volatility come true.

a stochastic variable defined as the information flow.[*] The question of how noncorrelated news can change these variables in a simultaneous fashion, prompted Andersen (1996) to argue for a modified version of MDH, where information is serially correlated implying that current volume and volatilities are affected by their past values. The MDH is subject to one limitation: it fails to consider the precision or quality of information.

Under the sequential information hypothesis, information is absorbed by traders on a group-by-group basis who then trade upon the arrival of news. The implication of this model is that the volume-volatility relation is sequential, not contemporaneous. A number of incomplete equilibria are observed before a final equilibrium is attained—when all traders observe the same information set. The sequential response to the arrival of information implies that price volatility is forecastable, based on the knowledge of trading volume. Yet, the model is not flawless as (1) it does not account for the fact that traders learn from the market price as other traders become informed, and (2) it implies that volume is greatest when all investors agree on the meaning of the information.

The dispersion of beliefs model posits that the greater the dispersion of beliefs among traders, the higher the volatility/volume relative to their equilibrium values. The approach engulfs both informed and uninformed segments of financial markets, with uninformed traders reacting to changes in volume/prices as if these changes reflect new information. On the other hand, knowledgeable investors make their trades on price reflecting fair values, as they possess homogeneous beliefs. It is therefore expected that uninformed investors will shake prices and increase price volatility.

Finally, the information trading volume approach is based on the notion that volume plays an informationally important role in an environment where traders receive pricing signals of different quality. Of paramount importance is the assumption that the equilibrium price is nonrevealing given that pricing signals alone do not provide sufficient information to ascertain the underlying value. Trading volume is treated as containing information regarding the quality of signals received by traders, whereas prices alone do not. This in turn leads to the formulation of a link among trading volume, the quality of information flow, and volatility. It is also argued that traders who use information contained in market statistics do better than those who do not.

---

[*] The mixture of distributions model does not explicitly preclude a lead-lag relation between volume and volatility.

Over the recent years, scholars have made noteworthy advances in equity volatility modeling by taking into account features of returns not previously considered. One of the assumptions underlying time-series models is that time intervals over which price variations are observed are fixed. Price changes and news arrival, however, can take place in irregular time intervals. Empirical evidence using high-frequency data indicates that adjusting volume and volatility for the duration between trades provides time-consistent parameter estimators in microstructure models, while allowing for proper integration of the information—proxied by trade intensity—into the regression model (Engle and Russell, 1998; Dufour and Engle, 2000; Engle, 2000). Recent research shows that volatility and volume are persistent and highly autocorrelated, while shorter time duration between trades implies higher probability of news arrival and higher volatility (Xu et al., 2006). The findings suggest that there is an inverse relation between price impact of trades and duration between trades. A similar relationship is documented for the speed of price adjustment to trade-related information and the time interval between transactions.

The issue of information asymmetry is also important. Agents with different information sets take different trade positions, while their actions flag signals and cause a persistent impact on equity prices. As trading actions spread, news is conveyed into the market and stock prices adjust to reflect expectations based on previous trades and all available information. Empirical research (Glosten and Harris, 1988; Hasbrouck, 1988, 1991a, 1991b) has put forward models to understand the equity pricing function by integrating the news arrival process into equity prices. The results suggest that past price changes as well as signed trades have a persistent impact on current price changes, thereby being important in determining the intrinsic value of stocks.

Within a noisy rational expectations framework Wang (1994) and Blume et al. (1994) unveil a positive association between volume and price changes. McKenzie and Faff (2003) take into account liquidity disparities for equities, as they exert a significant impact on individual stocks but not on indices. They show that conditional autocorrelation in equity returns is highly dependent on trading volume for individual stocks but not for indices. Elsewhere, Li and Wu (2006) find that by controlling for the effect of informed trading, return volatility is negatively correlated with volume. This is consistent with the contention that liquidity increases market depth and reduces price volatility.

In general, empirical research supports a positive correlation between equity price changes and volume.* Nonetheless, the difficulty in evaluating such a relationship stems from the ambiguity regarding the information content of volume. We would rather suggest that volume provides insights on the dispersion and quality of information signals, rather than representing the information signal per se.

1.4   DERIVATIVES TRADING AND VOLATILITY

The general belief that futures trading triggers excess speculation, and possible price instability, has been a fertile research terrain for many scholars (Damodaran and Subrahmanyam, 1992). The implications for policy makers and those responsible for regulating futures trading have also been noted. The debate became more vivid after "Black Monday," which has led to much interest in examining volatility in modern financial markets.

It is not yet clearly established whether derivatives induce excess volatility in the cash market and thus destabilize equity prices. Financial bubbles along with the existence of speculators have been addressed (Edwards, 1988a, 1988b; Harris, 1989; Stein, 1987, 1989) as other potential sources of excess price variability.† It is also true that closer to the expiration day, traders attempt to settle their contracts, close their trading positions, and aggressively arbitrage on price differences. Miller (1993) finds that futures trading has raised volatility in the Japanese market, possibly attributed to low-cost speculative opportunities. These arguments along with the discussion in Section 1.2 underline the role of derivatives trading in destabilizing financial markets.

On the other hand, there is a consensus that derivatives trading contributes to stabilizing the underlying equity market. The very nature of derivatives is risk reducing, being a platform for competitive price discovery, and acting as a hedging device for buyers and sellers. Derivatives also increase market liquidity and expand the investment opportunity

---

* For more evidence see Crouch (1970), Rogalski (1978), Smirlock and Starks (1985), Wood et al. (1985), Richardson et al. (1986), Gallant et al. (1992), Richardson and Smith (1994), Kandel and Pearson (1995), Chan and Fong (2000), Chordia and Subrahmanyam (2001), Chen et al. (2001), and Llorente et al. (2002). The following studies use a GARCH framework: Lamoureux and Lastrapes (1990), Najand and Yung (1991), Sharma et al. (1996), Brooks (1998), and Kalotychou and Staikouras (2006). The volume-volatility theory and a survey of the literature can be found in O'Hara (1997) and Karpoff (1986, 1987).

† Financial fads, investment trends, and social norms can also notably contribute to equity price changes (Shiller, 1984). See Shiller (1989) for other aspects of market volatility.

set at lower transaction costs and margin requirements. Exchange traded derivatives are more centralized, enabling participants to trade and communicate their information more effectively. Assuming that derivatives do attract rational traders, then equity prices should move closer to their fundamentals and markets should become less volatile. Based on intraday data, Schwert (1990) shows that the equity cash market is 40% less volatile than its counterpart futures arena, while Merton (1995) argues that the volatility's asymmetric response to the arrival of news is reduced in the presence of futures markets.

Yet, anecdotal evidence both supports and refutes the aforesaid hypotheses. Moreover, tightening any regulatory framework in the derivatives market is not empirically endorsed. With the lack of a clear-cut theoretical background that justifies market realities, the question becomes an empirical one. At times, when fluctuations are large, they can easily call into question the collective rationality of the market. The issue is whether volatility is a sign of collective irrationality or is consistent with the kind of fluctuations expected to arise naturally from the actions of less informed investors.

Early evidence (Bessembinder and Seguin, 1992) points out that futures trading improves liquidity and depth in the cash equity market, which is corroborated by more recent studies (Board et al., 2001). Analysis of the FTSE100, S&P500, and DJIA indices (Robinson, 1994; Pericli and Koutmos, 1997; Rahman, 2001) reveals either a volatility reduction in the postfutures phase or no change in the conditional volatility over the two periods. Elsewhere, findings indicate that twenty-three international stock indices exhibit either a reduction or no change in volatility during the postfutures period, while the opposite applies for the U.S. and Japanese equity markets (Gulen and Mayhew, 2000). Recently, Dawson and Staikouras (2008) investigated whether the newly cultivated platform of derivatives volatility trading has altered the variability of the S&P500 index. They document that the onset of the CBOE volatility futures trading has lowered the equity cash market volatility, and reduced the impact of shocks to volatility. The results also indicate that volatility is mean reverting, while market data support the impact of information asymmetries on conditional volatility. Finally, comparisons with the UK and Japanese indices, which have no volatility derivatives listed, show that these indices exhibit higher variability than the S&P500.

The dynamic interaction between derivatives and cash equity markets engulfs the issue of volatility's asymmetric response to the arrival of news

(Engle and Ng, 1993). In other words, do market participants react differently upon the arrival of bad and good news? The information transmission mechanism, from futures to spot market, is yet unclear. The role of asymmetries in the futures market[*] will have implications for the effectiveness of policy frameworks at both an institutional and a state level. Early evidence unveils that bad news in the futures market increases volatility in the cash markets more than good news (Koutmos and Tucker, 1996; Antoniou et al., 1998), while postfutures asymmetries are significantly lower for major economies, except the United States and United Kingdom. When both spot and futures markets are examined, it seems that asymmetries run from the spot to the futures market. The leverage hypothesis[†] is not the only force behind asymmetries, as market interactions, noise trading, and irrational behavior may well contribute to the rise of asymmetries.

Analysts and traders use techniques such as portfolio insurance, sentiment, and other technical indicators, as well as extrapolative expectations that are in line with the positive feedback trading approach. The latter calls for tracking market movements in retrospect of a trend change. On that basis, as futures do attract a diverse number of participants, then some form of market destabilization may take place. Recent evidence (Antoniou et al., 2005; Chau et al., 2008) indicates that feedback trading is either reduced or not attributed, at least in large part, to the existence of futures markets. When feedback trading does take place, both rational and any other investors/speculators tend to join the trading game, which in the short run may drive prices away from fundamentals.[‡] On the other hand, in efficient markets and under rational expectations, the effect of feedback trading might be limited as speculators will ultimately start liquidating their positions, driving equity prices closer to their intrinsic values.

Finally, research has concentrated on stock indices rather than individual shares. It is a fact, however, that individual share futures (ISFs) are traded in modern markets, and their analysis sheds light on financial markets' behavior (McKenzie et al., 2001; Chau et al., 2008). It is true that equity indices capture wide-market forces, but when it comes to identifying the

---

[*] In that respect, Staikouras (2006) provides some evidence for the UK interest rate market, while Dawson and Staikouras (2008) offer findings for the S&P500.

[†] Negative equity returns imply higher leverage, through the reduced firm's market value, which in turn increases the firm's perceived riskiness and leads to higher volatility.

[‡] Asset values with long-term swings away from fundamentals could be translated to predictable stock returns, in the long run, which in turn broaches the idea of market inefficiency.

origins of a phenomenon, the large number of constituent stocks poses an obstacle. Liquidity is another motive behind such an analysis, as indices are more liquid than individual stocks, amplifying any possible impact of stock index futures on the underlying asset. At the same time, the underlying asset on stock index futures is not traded as opposed to ISFs, making the latter an apt alternative for investigation. In a multiaspect examination, McKenzie et al. (2001) study the systematic risk, asymmetries, and volatility of ISFs. Their stock-specific empirical findings add to the mixed results of the ongoing literature. They detect a clear reduction in beta risk and unconditional volatility, during the post-IFS listing, and offer some mixed evidence regarding the change in conditional volatility, while asymmetric response is not consistent across all stocks.

1.5 STYLIZED FACTS OF VOLATILITY MODELING

It is well established by now that equity volatility is time varying and tends to display patterns, thereby rendering the stock returns' empirical distribution nonnormal. Several historical time-series models have been proposed to account for such features. The simplest class of historical volatility models lies on the premise that past standard deviations of returns can be estimated. The most naive historical volatility model is the *random walk*, where the best forecast of today's volatility is yesterday's realized value, i.e., $\hat{\sigma}_t^2 = \sigma_{t-1}^2$. Another approach is the *historical average* (HA), which amounts to a long-term average of past standard deviations. Whereas the HA uses all past standard deviations, the *moving average* (MA) discards older information by deploying a rolling window of fixed length (N), typically 20 to 60 trading days. The MA volatility forecast is

$$\hat{\sigma}_t^2 = (1/N) \sum_{i=1}^{N} \hat{\sigma}_{t-i}^2 = (1/N) \sum_{i=1}^{N} r_{t-i}^2$$

where r_t is the observed return on day t, with squared returns typically used as an estimate of the ex-post daily variance. A drawback of the MA is that all past observations carry the same weight, while the so-called ghosting feature* should not be ignored.

---

* The volatility forecast increases as a direct result of including a particular high observation. After N days this observation is dropped out of the estimation window, causing a sudden fall in volatility, *ceteris paribus*.

A more refined approach is the *RiskMetrics* model (JP Morgan, 1996), which uses an *exponentially weighted moving average* (EWMA) to forecast volatility and gives greater importance to more recent volatility estimates. The EWMA variance forecast is formulated as

$$\hat{\sigma}_t^2 = (1-\lambda) \sum_{i=1}^{N} \lambda^{i-1} r_{t-i}^2$$

where the decay parameter is set at $\lambda = 0.94$ for daily and $\lambda = 0.97$ for monthly forecasts, and a window of $N = 75$ days is typically used. The EWMA posits geometrically declining weights on past observations, giving greater emphasis to new information. The smaller the λ, the higher the impact of recent news is and the faster the decay in weights for old news.

Volatility clustering is a characteristic of equity returns and mirrors the leptokurtosis (fat tails) in the returns' distribution. Volatility clustering refers to large/small price changes being followed by large/small changes in either direction. It has been attributed to the quality of information reaching the market in clusters (Gallant et al., 1991), as well as to the time-varying rate of information arrival and news processing by the market (Engle et al., 1990). One of the major breakthroughs in financial economics is the modeling of nonconstant variances (*conditional heteroskedasticity*) and volatility clustering in equity returns.[*] The GARCH framework builds on the notion of volatility dependence to measure the impact of last period's forecast error and volatility in determining current volatility. The simplest GARCH specification is formulated as

$$r_t = \mu + \varepsilon_t \qquad \varepsilon_t \sim N\left(0, \sigma_t^2\right)$$
$$\sigma_t^2 = \alpha_0 + \alpha_1 \varepsilon_{t-1}^2 + \beta \sigma_{t-1}^2$$

where the ARCH term α_1 measures the extent to which a volatility shock today feeds into tomorrow's volatility and represents the short-run persistence of shocks on return variance. The GARCH term β is the contribution of older shocks to the long-run persistence. Akgiray (1989) finds that a GARCH(1,1) model is sufficient to capture all volatility clustering.

---

[*] See Engle (1982), Bollerslev (1986, 1987), and Engle and Bollerslev (1986). For academic surveys see Bollerslev et al. (1992, 1994), and for a practical review of these models see Engle (1993).

It is also shown that the GARCH estimator performs better than the ARCH, EWMA, and HA for predicting monthly U.S. stock index volatility. Interestingly, ARCH effects are more pronounced in daily and weekly data and tend to dampen as the frequency of the data decreases.

The degree of volatility persistence ($\alpha_1 + \beta$) measures the rate at which the volatility feedback effect decays over time.[*] High persistence ($\alpha_1 + \beta$ close to 1) means volatility shocks are felt further into the future, albeit at a progressively smaller extent. That is, mean reversion toward the long run variance will take several days, although shocks decay rather quickly in longer horizons over a month. The fall in persistence, when using monthly data, indicates that volatility predictability based on current information weakens. Christoffersen and Diebold (2000) suggest that there is little value in forecasting volatility for more than 10 days ahead. Volatility persistence has two implications: predictability of future economic variables[†] and predictability of changes in the risk-return trade-off over business cycles.

In the limiting case of $\alpha_1 + \beta = 1$, a process known as integrated GARCH (IGARCH), the shock will have a permanent effect on the variance process; i.e., after the shock the variance will rise and will remain at that level. In fact, the EWMA variance estimate, which can be expressed as $\sigma_t^2 = (1-\lambda) r_{t-1}^2 + \lambda \sigma_{t-1}^2$, is an IGARCH process. Therefore, the one-day-ahead EWMA volatility forecasts are very close to those of the GARCH. Nevertheless, be aware that longer-horizon forecasts are markedly different, as the EWMA is not mean reverting. At the same time, mean reversion in monthly volatility is well established. RiskMetrics accounts for this by proxying the latest variance innovation using the 25-day MA, rather than the latest squared return. Thus, the month τ variance is expressed as

$$\sigma_\tau^2 = (1-\lambda) \sum_{i=1}^{25} r_{t-i}^2 + \lambda \sigma_{\tau-1}^2$$

Asymmetry, long memory, and spillover effects are by now stylized facts that characterize the behavior of global equity market volatility. From an empirical point of view, volatility is higher in bearish markets than it is in bullish markets (asymmetry), indicating a negative correlation between

---

[*] Typical financial time series may have GARCH persistence of 0.90 to 0.99 for daily data.

[†] Economic variables and stock market volatility are related (Schwert, 1989; Campbell et al., 2001).

future conditional volatility and current stock returns (Black, 1976). Apart from the leverage effect, mentioned in the previous section, volatility feedback has been put forward as another justification of volatility asymmetries (Campbell and Hentschel, 1992). According to this hypothesis, the causality runs from volatility to prices; i.e., positive shocks to volatility increase future risk premium, and if dividends remain the same, then the stock price should fall. In emerging markets, asymmetric volatility has been identified at the aggregate market level (Chiang and Doong, 2001), but there is no evidence of how this asymmetric volatility occurs at sector and firm levels. Yet, volatility feedback does not preclude the presence of leverage effects. Christie (1982) tests Black's hypothesis by analyzing a cross section of firms. Although he finds that there is a strong correlation between asymmetry and leverage, the leverage itself is not sufficient to explain the asymmetric effects.

A number of extensions to the standard GARCH model have been suggested, such as the exponential GARCH (EGARCH) (Nelson, 1991) and threshold GARCH (TGARCH) (Glosten et al., 1993; Zakoian, 1994), to accommodate the asymmetric nature of volatility. Pagan and Schwert (1990) notice that EGARCH yields somewhat better predictions for monthly U.S. stock index volatility than GARCH, whereas Franses and van Dijk (1996) argue that asymmetric models fare no better than simple GARCH for forecasting the weekly volatility of European stock market indices. Brailsford and Faff (1996) support the TGARCH for the Australian stock market, albeit only slightly better in performance than the simple random walk, HA, MA, and EWMA. A GARCH-in-mean parameterization is also proposed (Engle et al., 1987) to formalize the idea that risk is priced by the market and risk premia vary with volatility.

Looking at high-frequency data, volatility changes slowly and shocks tend to take a long time to decay (long memory). Fractionally integrated GARCH (FIGARCH) specifications have been developed (Baillie et al., 1996) as a way of modeling the hyperbolic rather than exponential decay of shocks. Financial globalization has eased the transmission of price fluctuations from one market to others, or among assets in the same market (Hamao et al., 1990; Conrad et al., 1991; Koutmos and Booth, 1995; Karolyi and Stulz, 1996). Such contagion effects increase during periods of high-equity market volatility. Volatility co-movements or spillovers endorse a multivariate framework for forecasting the variance-covariance structure of asset returns. To this end, multivariate GARCH (MGARCH) specifications have been suggested for modeling asset interdependence as well as the dynamics of volatility and covariance/correlation. The BEKK estimator, introduced by Engle and Kroner (1995), directly characterizes volatilities and covariances. In a multiasset setting, however, conditional correlation models are more appealing, and among them the dynamic conditional correlation (DCC) model (Engle, 2002) is becoming all the more popular (Kalotychou et al., 2008).

1.6 THE RIVAL VOLATILITY FORECASTS

An alternative to time-series approaches is market-based volatility forecasts, which are derived from the Balck-Scholes European option pricing formula. Volatility is a crucial factor in determining the value of an option, and using numerical methods, the option's implied volatility can be obtained. Nowadays, traders quote options in terms of volatility rather than price, since an option's implied volatility is a more useful measure of its relative value. This is because the price of an option is associated with the price of its underlying asset, and if an option is held as part of a delta-neutral portfolio, then the next most important factor in determining the value of the option is its implied volatility.

There has been a burgeoning literature on the forecasting ability of GARCH models and their relative merits over alternative approaches, such as implied volatility. When it comes to forecasting future volatility of financial assets, simple historical volatility and GARCH models are roughly comparable, but lag behind their main rival of implied volatility. There is some support for the EWMA approach (Dimson and Marsh, 1990; Tse and Tung, 1992; Figlewski, 1997), pointing out that GARCH models are not superior to their simpler rivals. Looking at the S&P indices, implied volatility has often been found to outperform time-series approaches (Fleming, 1998; Blair et al., 2001; Hol and Koopman, 2002). Nonetheless, implied-based volatility forecasts face some shortcomings, as they are not available for all assets and are prone to volatility smile effects, namely, different strike prices yield different volatilities. Donaldson and Kamstra (2004) suggest that market conditions influence the relative performance of historical and market-based measures of volatility. ARCH-type models are found to be better in periods of low trading volume when information is stale, while implied volatility leads in periods of intense trading with rich information flow. Vasilellis and Meade (1996) advocate combining implied and GARCH volatility forecasts. As earlier discussed, the role of trading volume in improving the predictability of market volatility has also been attested (Lamoureux and Lastrapes, 1990; Brooks, 1998), but with limited success.

Studies by Akgiary (1989), Heynen and Kat (1994), Franses and van Dijk (1995), Brailsford and Faff (1996), Figlewski (1997), and Brooks (1998) find that a regression of realized volatility on out-of-sample GARCH volatility predictions yields a low coefficient of determination (R^2 often <10%). Nevertheless, substantial gains can be achieved by improving the measure of actual or population volatility, which is a latent process. A very common, albeit crude proxy for the unobserved daily price variation is the daily squared return. Andersen and Bollerslev (1998) emphasize how the evaluation of volatility forecasts depends on the underlying measure of ex-post volatility. They show that daily squared returns are an unbiased but very noisy estimator of the conditional variance undermining the forecasting performance of GARCH models. When cumulated squared intraday returns proxy the true volatility, the accuracy of GARCH forecasts increases tenfold. This study has been the impetus for the growing interest in exploiting intraday prices to better characterize the all-important volatility of financial assets. The availability of intraday data has created new grounds for unravelling volatility dynamics with more precision. Fuertes et al. (2008) compare the relative merits of various high-frequency volatility estimators under different market conditions. The realized power variation (Barndorff-Nielsen and Shephard, 2004) provides the most accurate one-day-ahead forecasts, while forecast combination of different measures is also fruitful. The use of intraday price information to forecast volatility is most beneficial when trading volume is low or when the stock is on a bullish trend.

1.7 VOLATILITY TRADING

As foreign exchange investors think they can foretell the currency trends, or bond traders believe they know something about the future path of interest rates, laying a bet on the level of expected volatility surfaces as another trading vehicle. *Volatility trading* has become the new buzzword in financial markets and established itself as a new asset class. A range of structured derivatives products, either over the counter or, more recently, exchange traded, are available to investment banks, hedge funds, and traders to accommodate their bets on stock market volatility. CBOE and EUREX have launched a range of volatility derivatives (on S&P500, DJIA, SMI, NASDAQ-100, DAX, etc.) to house the demand for these instruments.

Among the various CBOE products, VIX* has been a popular volatility index and has been regarded as the world's premier barometer of investors' sentiment and market volatility. It is worth noting that while variance futures are based on realized volatility, implied volatility†—as measured by VIX—reflects expectations, hence its market name as "the fear gauge." In general, VIX starts to rise during times of financial stress and lessens as investors become complacent. Thus, the greater the fear, the higher the volatility index would be. Remember, during the collapse of the fixed-income arbitrage hedge fund LTCM and the Russian debt crisis, the S&P500 plunged to 957 (August 1998) and 970 (October 1998) while the VIX soared above 45 in each trough.

But what is it about these instruments that make them particularly attractive to some investors? First, hedging volatility in equity markets with options instead of volatility derivatives, as was the case until recently, is not ideal. Remember, options hedge against price risk but delta-hedging is inaccurate, since many of the Black-Scholes assumptions are violated in real world. Second, volatility is mean reverting, which makes volatility products appealing to statistical arbitrage funds and other market-neutral players.‡ Third, the existence of stochastic volatility and stochastic jumps makes financial markets incomplete and acts as a deterrent in achieving an optimum payoff. Finally, empirical evidence supports a negative/low correlation between volatility and equities (Dawson and Staikouras, 2008). Volatility contracts could, therefore, reduce risk or increase portfolio diversification instead of using other assets, such as commodities, precious metals, property, etc.

At the same time it is worth mentioning some issues pertinent to volatility trading. Many commentators use the VIX to represent the overall sentiment for equity options. Yet this relationship could be easily overstated, as these two are often unrelated since different forces drive the volatility

---

* For a detailed description of the VIX contract see http://cfe.cboe.com/Products/Products_VIX.aspx.

† Note that realized volatility tends to be lower than implied, mirroring investors' preference to avoid short options.

‡ Statistical arbitrage involves data mining and statistical methods, as well as automated trading systems. It is not without risk, as it depends heavily on market prices returning to a historical or predicted "normal." Market neutral strategies (or relative value trading) involve buying one asset (undervalued) and simultaneously selling another (overvalued) while avoiding systematic risk. In theory, the portfolio would make money on the increase or decrease in the spread between the two positions, and would be unaffected by the absolute level of the assets.

of index options compared to that of equity options. This becomes clearer when volatility is considered at an industry level or compared to the time span of VIX and nonindex equity options. Volatility is usually assumed to be high in certain sectors (IT) and low in others (utility), implying that a single volatility figure for all equities may be a crude proxy. Also, VIX reflects expectations over a 30-day period, while most nonindex equity options exhibit high liquidity within 2- to 6-month maturities.

Given the various features of the volatility trading platform, a number of market professionals would be interested in these instruments. Participants in the credit market who anticipate volatility increases (spreads widen) can hedge their credit exposure with volatility index futures. Portfolio managers exposed to correlation risk can go short on volatility, while volatility index futures can be used for directional bets on expected volatility. Traders selling (buying) options can do so when the volatility index trades high (low) and buy (sell) the option back (later) at a lower (higher) price when volatility decreases (falls). Moreover, a number of professions, such as long-equity funds, are implicitly short volatility due to the negative correlation between equity indices and volatility.[*] The hedge fund industry is actively involved in absolute return strategies[†] (merger arbitrage/event-driven hedge) by shorting the stock of a bidding firm and being long on the target firm. If the deal goes through, they earn the spread, but if volatility rises, the spread widens and the deal may collapse. Other short-volatility categories could involve portfolio managers, where increased volatility will inflate their tracking errors, or active portfolio rebalancing could result in greater transaction costs.

Research shows that there is an economic motivation behind variance contracts (Branger and Schlag, 2006). They argue that investors who are willing to trade variance risk are likely to prefer such contracts as opposed to standard options. Their main argument is that variance contracts are superior to dynamic and semistatic replication strategies due to discrete trading, incorrect parameter estimation, and model risk in the sense of misspecification problems. Christoffersen and Diebold (2006) find that

---

[*] This point reinforces the use of volatility derivatives, because global correlations have increased, rendering geographic diversification as an ineffective portfolio hedging strategy.

[†] The return is associated with the risk of certain corporate transactions, and is established independently of the market return. Under a merger arbitrage, risk arbitrageurs bet on the rise (decline) of the target (bidding) firm's equity, while under event-driven hedge the philosophy is exactly the same, focusing on important business events (divestments, recapitalization, restructuring, etc.) that have an impact on corporate equity.

volatility fluctuations prompt variations in the sign of equity returns, supporting the idea that forecastability (dependence) in stock market volatility can assist in generating directional change forecasts. For instance, in the case of a positive (negative) return-volatility relationship a predicted rise in volatility for the next day will indicate a buy (sell) signal. Christoffersen et al. (2007) propose models that use conditional moments (mean, variance, skewness, and kurtosis) to predict the market direction of change, and more specifically the probability of positive returns.

Finally, active portfolio managers can take advantage of dynamic volatility timing strategies by allocating assets according to recursively updated variance-covariance estimates. The dynamic nature of volatility and correlations among assets has put into question the simple static investment strategies. Multivariate and rolling GARCH estimators with increasing levels of sophistication have been developed to accommodate this. But do their statistical advantages translate into economic value? Fleming et al. (2001) show that switching from a static portfolio approach to a dynamic volatility timing strategy yields portfolios with improved performance. In a follow-up article, Fleming et al. (2003) demonstrate that this improved portfolio selection strategy can be enhanced even more by using intraday instead of daily price data. The incremental gains to a risk-averse investor who adopts dynamic volatility timing, based on intraday information, can be up to 200 basis points a year.

1.8 CONCLUSION

Volatility has enjoyed the attention of a wide audience, ranging from research laboratories to institutional investors. The mere fact that the stock market is estimated at about $51 trillion, as well as its fundamental role for raising corporate capital, has placed it at the forefront of economic and policy debates. The current survey highlights the diverse themes that equity volatility touches upon, and offers some insights into empirical findings across international stock markets.

The current discussion delineates the controversial role of speculators, and recognizes the impact of price manipulation and herding behavior on equity markets. Research on the volume-volatility relationship shows that there is a positive correlation between the two, but it is not clear as yet whether volume can represent the information signal per se. Regarding the role of derivatives, evidence both supports and refutes the argument of market stabilization, while new research areas, such as individual share futures, feedback trading, and volatility futures, are explored. Moreover, a number of stylized facts are considered, including volatility clustering, asymmetries, long memory and persistence, volatility co-movements, along with the pertinent advances on the modeling front. When it comes to forecasting volatility, implied volatility models are favored over historical and conditional volatility frameworks. Finally, *volatility derivatives* seem to be the new buzzword, and to that extent the chapter illustrates some aspects of this new trading platform.

Looking ahead, one should be very open-minded when analyzing stock market data, since it is difficult to identify and model all the factors responsible for price swings. Most of the time, the interaction among drivers of volatility makes it difficult to isolate their precise impact, while economic theory could remain silent as to why markets have moved toward a certain direction. From a methodological perspective, modeling and forecasting equity volatility warrants even further investigation. What is complicated but at the same time appealing is to design an empirical framework that attenuates the intricate characteristics of a stochastic global environment.

REFERENCES

Akgiray, V. (1989). Conditional heteroscedasticity in time series of stock returns: Evidence and forecasts. *Journal of Business* 62:55–80.

Allen, F., and D. Gale. (1992). Stock price manipulation. *Review of Financial Studies* 5:503–29.

Allen, F., and G. Gorton. (1992). Stock price manipulation, market microstructure and asymmetric information. *European Economic Review* 36:624–30.

Andersen, T. G. (1996). Return volatility and trading volume: An information flow interpretation of stochastic volatility. *Journal of Finance* 51:169–204.

Andersen, T., and T. Bollerslev. (1998). Answering the sceptics: Yes, standard volatility models do provide accurate forecasts. *International Economic Review* 39:885–906.

Antoniou, A., P. Holmes, and R. Priestley. (1998). The effects of stock index futures trading on stock volatility: An analysis of the asymmetric response of volatility to news. *Journal of Futures Markets* 18:151–66.

Antoniou, A., G. Koutmos, and A. Pericli. (2005). Index futures and positive feedback trading: Evidence from major stock exchanges. *Journal of Empirical Finance* 12:219–38.

Baillie, R., T. Bollerslev, and H. Mikkelsen. (1996). Fractionally integrated generalized autoregressive conditional heteroscedasticity. *Journal of Econometrics* 74:3–30.

Barndorff-Nielsen, O. E., and N. Shephard. (2004). Power and bipower variation with stochastic volatility and jumps. *Journal of Financial Econometrics* 2:1–48.

Baumol, W. J. (1957). Speculation, profitability and stability. *Review of Economics and Statistics* 39:263–71.

Bessembinder, H., and P. J. Seguin. (1992). Futures trading activity and stock price volatility. *Journal of Finance* 47:2015–34.

Black, F. (1976). Studies of stock price volatility changes. In *Proceedings of the Business and Economics Statistics Section*, 177–81. American Statistical Association, Boston.

Black, F. (1986). Noise. *Journal of Finance* 41:529–43.

Blair, J. B., S.-H. Poon, and S. J. Taylor. (2001). Forecasting S&P100 volatility: The incremental information content of implied volatilities and high frequency index returns. *Journal of Econometrics* 105:5–26.

Blume, L., D. Easley, and M. O'Hara. (1994). Market statistics and technical analysis: The role of volume. *Journal of Finance* 49:153–81.

Board, J., G. Sandmann, and C. Sutcliffe. (2001). The effect of futures market volume on spot market volatility. *Journal of Business Finance and Accounting* 28:799–819.

Bollerslev, T. (1986). Generalized autoregressive conditional heteroscedasticity. *Journal of Econometrics* 31:307–27.

Bollerslev, T. (1987). A conditionally heteroscedastic time series model for speculative prices and rates of return. *Review of Economics and Statistics* 69:542–47.

Bollerslev, T., R. Y. Chou, and K. F. Kroner. (1992). ARCH modeling in finance: A review of the theory and empirical evidence. *Journal of Econometrics* 52:5–59.

Bollerslev, T., R. F. Engle, and D. Nelson. (1994). ARCH models. In *Handbook of Econometrics*, ed. R. F. Engle and D. L. McFadden, 4:2959–3038, Amsterdam: Elsevier.

Brailsford, T. J., and R. W. Faff. (1996). An evaluation of volatility forecasting techniques. *Journal of Banking and Finance* 20:419–38.

Branger, N., and C. Schlag. (2006). An economic motivation for variance contracts. Paper presented at the American Finance Association 2006 Meetings, Boston.

Brooks, C. (1998). Predicting stock index volatility: Can market volume help? *Journal of Forecasting* 17:57–80.

Calvo, G. A., and E. G. Mendoza. (2000). Rational contagion and the globalization of security markets. *Journal of International Economics* 51:79–113.

Campbell, J., M. Lettau, B. Malkiel, and Y. Xu. (2001). Have individual stocks become more volatile? An empirical exploration of idiosyncratic risk. *Journal of Finance* 56:1–43.

Campbell, Y., and L. Hentschel. (1992). No news is good news: An asymmetric model of changing volatility in stock returns. *Journal of Financial Economics* 31:281–318.

Carpenter, F. (1866). *Six months at the White House with Abraham Lincoln*. New York: Hurd.

Chan, K., and W. Fong. (2000). Trade size, order imbalance, and the volatility-volume relation. *Journal of Financial Economics* 57:247–73.

Chau, F., P. Holmes, and K. Paudyal. (2008). The impact of universal stock futures on feedback trading and volatility dynamics. *Journal of Business Finance and Accounting* 35:227–49.

Chen, G., M. Firth, and O. M. Rui. (2001). The dynamic relation between stock returns, trading volume, and volatility. *Financial Review* 36:153–74.

Chiang, T. C., and S. C. Doong. (2001). Empirical analysis of stock returns and volatility: Evidence from seven Asian stock markets based on TAR-GARCH model. *Review of Quantitative Finance and Accounting* 17:301–18.

Chordia, T., R. Roll, and A. Subrahmanyam. (2001). Market liquidity and trading activity. *Journal of Finance* 56:501–30.

Christie, A. A. (1982). The stochastic behaviour of common stock variances: Value, leverage and interest rates effects. *Journal of Financial Economics* 10:407–32.

Christoffersen, P., and F. Diebold. (2000). How relevant is volatility forecasting for financial risk management? *Review of Economics and Statistics* 82:12–22.

Christoffersen, P., and F. Diebold. (2006). Financial asset returns, direction-of-change forecasting, and volatility dynamics. *Management Science* 52:1273–87.

Christoffersen, P., F. X. Diebold, R. Mariano, A. Tay, and Y. Tse. (2007). Direction-of-change forecasts based on conditional variance, skewness and kurtosis dynamics: International evidence. *Journal of Financial Forecasting* 1:3–24.

Clark, P. K. (1973). A subordinated stochastic process model with finite variance for speculative prices. *Econometrica* 41:135–55.

Conrad, J., M. Gultekin, and G. Kaul. (1991). Asymmetric predictability of conditional variances. *Review of Financial Studies* 4:597–622.

Cooper, D. J., and R. G. Donaldson. (1998). A strategic analysis of corners and squeezes. *Journal of Financial and Quantitative Analysis* 33:117–37.

Copeland, T. E. (1976). A model of asset trading under the assumption of sequential information arrival. *Journal of Finance* 31:1149–67.

Crouch, R. L. (1970). The volume of transactions and price changes on the New York Stock Exchange. *Financial Analysts Journal* 26:104–9.

Damodaran, A., and M. Subrahmanyam. (1992). The effects of derivative securities on the markets for the underlying assets in the United States: A survey. *Financial Markets, Institutions and Instruments* 1:1–21.

Dawson, P., and S. K. Staikouras. (2008). The CBOE volatility futures trading and the S&P500 cash market. Paper presented at the 2008 Midwest Finance Association Annual Meeting, San Antonio, TX.

Devenow, A., and I. Welch. (1996). Rational herding in financial economics. *European Economic Review* 40:603–15.

Dimson, E., and P. Marsh. (1990). Volatility forecasting without data-snooping. *Journal of Banking and Finance* 14:399–421.

Donaldson, G., and M. Kamstra. (2004). *Volatility forecasts, trading volume and the ARCH versus option-implied volatility trade-off*. Federal Reserve Bank of Atlanta WP 04-06.

Dufour, A., and R. F. Engle. (2000). Time and the price impact of a trade. *Journal of Finance* 55:2467–98.

Edwards, E. R. (1988a). Does futures trading increase stock market volatility? *Financial Analysts Journal* 44:63–69.

Edwards, E. R. (1988b). Futures trading and cash market volatility: Stock index and interest rate futures. *Journal of Futures Markets* 8:421–39.

Engle, R. F. (1982). Autoregressive conditional heteroscedasticity with estimates of the variance of UK inflation. *Econometrica* 50:987–1007.

Engle, R. F. (1993). Statistical models of financial volatility. *Financial Analysts Journal* 49:72–78.

Engle, R. F. (2000). The economics of ultra-high-frequency data. *Econometrica* 68:1–23.

Engle, R. F. (2002). Dynamic conditional correlation: A simple class of multivariate generalized autoregressive conditional heteroscedasticity models. *Journal of Business and Economic Statistics* 20:339–50.

Engle, R. F., and T. Bollerslev. (1986). Modeling the persistence of conditional variances. *Econometric Reviews* 5:1–50.

Engle, R. F., T. Ito, and W. Lin. (1990). Meteor showers or heat waves? Heteroscedastic intra-daily volatility in the foreign exchange market. *Econometrica* 58:525–42.

Engle, R. F., and K. Kroner. (1995). Multivariate simultaneous GARCH. *Econometric Theory* 11:122–50.

Engle, R. F., D. M. Lilien, and R. P. Robins. (1987). Estimating time-varying risk premia in the term structure: The ARCH-M model. *Econometrica* 55:391–408.

Engle, R. F., and V. K. Ng. (1993). Measuring and testing the impact of news on volatility. *Journal of Finance* 48:1749–78.

Engle, R. F., and J. R. Russell. (1998). Autoregressive conditional duration: A new model for irregularly spaced transaction data. *Econometrica* 66:1127–62.

Epps, T. W., and M. L. Epps. (1976). The stochastic dependence of security price changes and transaction volumes: Implication for the mixture of distributions hypothesis. *Econometrica* 44:305–21.

Fama, E. F. (1970). Efficient capital markets: A review of theory and empirical work. *Journal of Finance* 25:383–417.

Fama, E. F. (1991). Efficient capital markets: II. *Journal of Finance* 46:1575–617.

Farrell, M. J. (1966). Profitable speculation. *Economica* 33:183–93.

Figlewski, S. (1997). Forecasting volatility. *Financial Markets, Institutions and Instruments* 6:1–88.

Fleming, J. (1998). The quality of market volatility forecasts implied by S&P 100 index option prices. *Journal of Empirical Finance* 5:317–45.

Fleming, J., C. Kirby, and B. Ostdiek. (2001). The economic value of volatility timing. *Journal of Finance* 56:329–52.

Fleming, J., C. Kirby, and B. Ostdiek. (2003). The economic value of volatility timing using "realized" volatility. *Journal of Financial Economics* 67:473–509.

Franses, P. H., and D. van Dijk. (1996). Forecasting stock market volatility using non-linear GARCH models. *Journal of Forecasting* 15:229–35.

Friedman, M. (1953). *The case for flexible exchange rates: Essays in positive economics.* Chicago: Chicago University Press.

Fuertes, A., M. Izzeldin, and E. Kalotychou. (2008). On forecasting daily stock volatility: The role of intraday information and market conditions. Working paper, Cass Business School, London.

Gallant, A. R., D. A. Hsieh, and G. E. Tauchen. (1991). On fitting a recalcitrant series: The pound/dollar exchange rate 1974–1983. In *Nonparametric and semiparametric methods in econometrics and statistics*, ed. W. A. Barnett, J. Powell, and G. E. Tauchen. Cambridge, UK: Cambridge University Press, 199–240.

Gallant, A. R., P. E. Rossini, and G. Tauchen. (1992). Stock prices and volume. *Review of Financial Studies* 5:199–242.

Glosten, L., and L. Harris. (1988). Estimating the components of bid–ask spread. *Journal of Financial Economics* 21:123–42.

Glosten, L., R. Jagannathan, and D. Runkle. (1993). On the relation between the expected value and the volatility of the nominal excess return on stocks. *Journal of Finance* 48:1779–802.

Grinblatt, M., S. Titman, and R. Wermers. (1995). Momentum investment strategies, portfolio performance, and herding: A study of mutual fund behaviour. *American Economic Review* 85:1088–105.

Gulen, H., and S. Mayhew. (2000). Stock index futures trading and volatility in international equity markets. Working paper, Purdue University, West Lafayette, IN.

Hamao, Y., R. Masulis, and V. Ng. (1990). Correlations in price changes and volatility across international stock markets. *Review of Financial Studies* 3:281–307.

Harris, L. (1986). Cross-security tests of the mixture of distribution hypothesis. *Journal of Financial and Quantitative Analysis* 21:39–46.

Harris, L. (1987). Transaction data tests of the mixture of distributions hypothesis. *Journal of Financial and Quantitative Analysis* 22:127–41.

Harris, L. (1989). S&P 500 cash stock price volatilities. *Journal of Finance* 44:1155–76.

Harris, M., and A. Raviv. (1993). Differences of opinion make a horse race. *Review of Financial Studies* 6:473–506.

Hart, O. D. (1977). On the profitability of speculation. *Quarterly Journal of Economics* 91:579–97.

Hart, O. D., and D. M. Kreps. (1986). Price destabilizing speculation. *Journal of Political Economy* 94:927–52.

Hasbrouck, J. (1988). Trades, quotes, inventories and information. *Journal of Financial Economics* 22:229–52.

Hasbrouck, J. (1991a). Measuring the information content of stock trades. *Journal of Finance* 46:179–207.

Hasbrouck, J. (1991b). The summary of informativeness of stock trades: An econometric analysis. *Review of Financial Studies* 4:571–95.

Heynen, R., and H. Kat. (1994). Volatility prediction: A comparison of stochastic volatility, GARCH(1,1) and EGARCH(1,1) models. *Journal of Derivatives* 1994:50–65.

Hol, E., and S. J. Koopman. (2002). *Stock index volatility forecasting with high frequency data*. Tinbergen Institute Discussion Paper 2002-068/4.

Jarrow, R. A. (1992). Market manipulation, bubbles, corners, and short squeezes. *Journal of Financial and Quantitative Analysis* 27:311–36.

Jarrow, R. A. (1994). Derivative securities markets, market manipulation, and option pricing theory. *Journal of Financial and Quantitative Analysis* 29:241–61.

Jennings, R. H., and C. Barry. (1983). Information dissemination and portfolio choice. *Journal of Financial and Quantitative Analysis* 18:1–19.

Jennings, R. H., L. T. Starks, and J. C. Fellingham. (1981). An equilibrium model of asset trading with sequential information arrival. *Journal of Finance* 36:143–61.

JP Morgan. (1996). *RiskMetrics™—Technical document*. 4th ed. New York: JP Morgan.

Kaldor, N. (1960). *Essays on economic stability and growth*. London: Gerald Duckworth.

Kalotychou, E., and S. K. Staikouras. (2006). Volatility and trading activity in short sterling futures. *Applied Economics* 38:997–1005.

Kalotychou, E., S. K. Staikouras, and G. Zhao. (2008). Taking advantage of global diversification: A multivariate GARCH approach. Paper presented at the 28th Symposium of Forecasters, June, Nice, France.

Kandel, E., and N. D. Pearson. (1995). Differential interpretation of public signals and trade in speculative markets. *Journal of Political Economy* 103:831–72.

Karolyi, G., and R. Stulz. (1996). Why do markets move together? An investigation of US-Japan stock return comovements. *Journal of Finance* 51:951–86.

Karpoff, J. M. (1986). A theory of trading volume. *Journal of Finance* 41:1069–87.

Karpoff, J. M. (1987). The relation between price changes and trading volume: A survey. *Journal of Financial and Quantitative Analysis* 22:109–26.

Kemp, M. C. (1963). Speculation, profitability, and price stability. *Review of Economics and Statistics* 45:185–89.

Keynes, J. M. (1936). *The general theory of employment, interest and money*. London: Macmillan.

Koutmos, G., and G. Booth. (1995). Asymmetric volatility transmission in international stock markets. *Journal of International Money and Finance* 14:747–62.

Koutmos, G., and M. Tucker. (1996). Temporal relationships and dynamic interactions between spot and futures stock markets. *Journal of Futures Markets* 16:55–69.

Lakonishok, J., A. Shleifer, and R. W. Vishny. (1992a). The structure and performance of the money management industry. *Brookings Papers on Economic Activity, Microeconomics* 1992:339–91.

Lakonishok, J., A. Shleifer, and R. W. Vishny. (1992b). The impact of institutional trading on stock prices. *Journal of Financial Economics* 32:23–44.

Lamoureux, C. G., and W. D Lastrapes. (1990). Heteroscedasticity in stock return data: Volume versus GARCH effects. *Journal of Finance* 45:221–29.

LeRoy, S., and R. Porter. (1981). The present value relation: Tests based on variance bounds. *Econometrica* 49:555–74.

Li, J., and C. Wu. (2006). Daily return volatility, bid-ask spreads, and information flow: Analyzing the information content of volume. *Journal of Business* 79:2697–740.

Llorente, G., R. Michaely, G. Saar, and J. Wang. (2002). Dynamic volume-return relation of individual stocks. *Review of Financial Studies* 15:1005–47.

McKenzie, M. D., T. Brailsford, and R. W. Faff. (2001). New insights into the impact of the introduction of futures trading on stock price volatility. *Journal of Futures Markets* 21:237–55.

McKenzie, M. D., and R. W. Faff. (2003). The determinants of conditional autocorrelation in stock returns. *Journal of Financial Research* 26:259–74.

Merton, R. C. (1995). Financial innovation and the management and regulation of financial institutions. *Journal of Banking and Finance* 19:461–81.

Mill, J. S. (1871). *Principles of political economy II.* 7th ed. Longmans, Green, Reader & Dyer, London.

Miller, M. H. (1993). The economics and politics of index arbitrage in the US and Japan. *Pacific-Basin Finance Journal* 1:3–11.

Morse, D. (1980). Asymmetrical information in securities market and trading volume. *Journal of Financial and Quantitative Analysis* 15:1129–46.

Najand, M., and K. Yung. (1991). A GARCH examination of the relationship between volume and price variability in futures markets. *Journal of Futures Markets* 11:613–21.

Nelson D. (1991). Conditional heteroscedasticity in asset returns: A new approach. *Econometrica* 59:347–70.

O'Hara, M. (1997). *Market microstructure theory.* Oxford: Blackwell.

Pagan, A. R., and G. W. Schwert. (1990). Alternative models for conditional stock volatility. *Journal of Econometrics* 45:267–90.

Pericli, A., and G. Koutmos. (1997). Index futures and options and stock market volatility. *Journal of Futures Markets* 17:957–74.

Rahman, S. (2001). The introduction of derivatives on the Dow Jones Industrial Average and their impact on the volatility of component stocks. *Journal of Futures Markets* 21:633–53.

Richardson, G., S. E. Sefcik, and R. Thompson. (1986). A test of dividend irrelevance using volume reaction to a change in dividend policy. *Journal of Financial Economics* 17:313–33.

Richardson, M., and T. Smith. (1994). A direct test of the mixture of distributions hypothesis: Measuring the daily flow of information. *Journal of Financial and Quantitative Analysis* 29:101–16.

Robinson, G. (1994). The effects of futures trading on cash market volatility: Evidence from the London Stock Exchange. *Review of Futures Markets* 13:429–52.

Rogalski, R. J. (1978). The dependence of prices and volume. *Review of Economics and Statistics* 60:268–74.

Schwert, G. W. (1989). Why does stock market volatility change over time? *Journal of Finance* 44:1115–53.

Schwert, G. W. (1990). Stock market volatility. *Financial Analysts Journal* 46:23–34.

Seiders, D. F. (1981). GNMA pass through securities: Discussion. *Journal of Finance* 36:484–86.

Shalen, K. T. (1993). Volume, volatility and the dispersion of beliefs. *Review of Financial Studies* 6:405–34.

Sharma, J. L., M. Mougoue, and R. Kamath. (1996). Heteroscedasticity in stock market indicator return data: Volume versus GARCH effects. *Applied Financial Economics* 6:337–42.

Shiller, R. J. (1981). Do stock prices move too much to be justified by subsequent changes in dividends? *American Economic Review* 71:421–36.

Shiller, R. J. (1984). Stock price and social dynamics. *Brookings Papers on Economic Activity* 1984:457–510.

Shiller, R. J. (1989). *Market volatility*. Cambridge, MA: MIT Press.

Smirlock, M., and L. Starks. (1985). A further examination of stock price changes and transaction volume. *Journal of Financial Research* 8:217–25.

Smith, A. (1776). *An inquiry into the nature and causes of the wealth of nations*. London: Adam Smith Institute. Available at www.adamsmith.org/smith/won-index.htm.

Staikouras, S. K. (2006). Testing the stabilisation hypothesis in the UK short-term interest rates: Evidence from a GARCH-X model. *Quarterly Review of Economics and Finance* 46:169–89.

Stein, J. (1987). Informational externalities and welfare-reducing speculation. *Journal of Political Economy* 95:1123–45.

Stein, J. (1989). Overreaction in options markets. *Journal of Finance* 44:1011–23.

Stein, J. L. (1961). Destabilizing speculative activity can be profitable. *Review of Economics and Statistics* 43:301–2.

Telser, L. G. (1959). A theory of speculation relating profitability and stability. *Review of Economics and Statistics* 41:295–301.

Tse, Y. K., and S. H. Tung. (1992). Forecasting volatility in the Singapore stock market. *Asia Pacific Journal of Management* 9:1–13.

Vasilellis, G., and N. Meade. (1996). Forecasting volatility for portfolio selection. *Journal of Business Finance and Accounting* 23:125–43.

Walras, L. (1896). *Studies in social economics*, Lausanne: F. Rouge & Co.

Wang, J. (1994). A model of competitive stock trading volume. *Journal of Political Economy* 102:127–68.

Welch, I. (2000). Herding among security analysts. *Journal of Financial Economics* 58:369–96.

Wermers, R. (1999). Mutual fund herding and the impact on stock prices. *Journal of Finance* 54:581–622.

Wood, R. A., T. H. McInish, and J. K. Ord. (1985). An investigation of transactions data for NYSE stocks. *Journal of Finance* 40:723–39.

Xu, X. E., P. Chen, and C. Wu. (2006). Time and dynamic volume–volatility relation. *Journal of Banking and Finance* 30:1535–58.

Zakoian, J. M. (1994). Threshold heteroscedastic models. *Journal of Economic Dynamics and Control* 18:931–55.

Zhong, M., A. Darrat, and D. Anderson. (2003). Do U.S. stock prices deviate from their fundamental values? Some new evidence. *Journal of Banking and Finance* 27:673–97.

CHAPTER 2

Analysis of Stock Market Volatility by Continuous-Time GARCH Models

Gernot Müller, Robert B. Durand,
Ross Maller, and Claudia Klüppelberg

CONTENTS

2.1 INTRODUCTION

2.1.1 Modeling Market Volatility and the Risk-Return Trade-off

Understanding the volatility of a market is critical to our understanding of finance. The *returns* of an equity market as a whole—where the market's returns may be proxied, for example, by the returns on an index such as the S&P500 index—are frequently modeled as a function of investors' expectations of the market's *volatility* (see, for example, Merton, 1980; French et al., 1987; Abel, 1988; Barsky, 1989). Ang et al. (2006) present evidence that the volatility of the market is a candidate for inclusion as an additional factor augmenting standard multifactor models of the cross section of stock returns (Fama and French, 1993; Carhart, 1997). In arguing that total risk is priced, Ang et al. (2006) present a considerable challenge to paradigms that argue that only diversifiable, or systematic, risk is required to capture the cross section of expected equity returns (Sharpe, 1964).

Nevertheless, while the theorized relationship of the risk of a market to its return has attracted considerable attention, empirical support for the relationship has been mixed and disappointing. To quantify such a relationship, the market's volatility must be estimated, in some way, from the market's returns. Analyses such as those of Ang et al. and Durand et al. (2007) take an indirect route by using a proxy for expected volatility—the Chicago Board of Options Exchange Volatility Index (the VIX)—which is exogenous to the market. Although the use of a proxy is one way of dealing with the issue, still the question of estimating risk directly from returns is appealing, and conditional volatility models represent a natural choice with which to model it.[*]

Hsieh (1991) represents an early attempt to model the S&P500 using high-frequency data (Hsieh uses 15-minute rather than 5-minute observation intervals), and finds support for an EGARCH (4,4) model. Later, Anderson et al. (2002) and Eraker et al. (2003) used models admitting stochastic volatility together with jumps both in returns and in volatility, for the S&P500. Lundblad (2007) estimates daily and monthly volatility of returns on the U.S. market using data stretching from the early 1800s to the 1990s (but the S&P500 index was not available over such a long period). While he finds some evidence of a relationship between volatility and return, he is agnostic as to which of the models he

---

[*] We note that such models have recently attracted some criticism (Ghysels et al., 2005; Durham, 2007).

uses—GARCH (1,1), TARCH (1,1), QGARCH (1,1), or EGARCH (1,1)—
best fits the data.

In the present chapter we will also model the volatility of the U.S. market using the S&P500 index to represent the returns of the market. The S&P500 index is perhaps the most widely followed measure of the U.S. equities market, being made up of 500 of the largest stocks in the market (representing around three-quarters of all U.S. equities).[*] We are not concerned, however, with running a "horse race" to find a best model. Rather, we demonstrate the applicability of a certain continuous-time GARCH model (the COGARCH model of Klüppelberg et al. (2004)) for modeling the time-varying volatility of the S&P500. Maller et al. (2008) have recently shown how to apply this kind of methodology to describe the volatility of the Australian stock market, using it to analyze 10 years of daily data, mostly equally spaced in time, for the ASX200 index. This analysis does not, however, demonstrate the full potential applicability of the COGARCH. Rather than using daily data, the analysis in the present paper will use very high-frequency observations—observations taken at 5-minute intervals—to better approximate the underlying continuous-time framework the methodology is designed to capture. The use of COGARCH enables the analysis of irregularly spaced data without recourse to approximations involving missing values estimation, and in high-frequency data such as studied in this chapter, irregular spacing—here due to a transformation to a business time scale—is a dominant feature of the data.

2.1.2 ARCH/GARCH and COGARCH Modeling

The ARCH/GARCH model paradigm introduced by Engle (1995) and Bollerslev et al. (1995) has been enormously influential and successful in capturing some of the most important empirical features of financial data, and is therefore widely used in finance research and applications.

Empirical studies commonly show that volatility changes randomly in time, has distributions with heavy or semiheavy tails, and clusters on high levels. These stylized features are well modeled by the GARCH family, as has been shown in many studies. For a recent discussion concerning the GARCH(1,1) process, see Mikosch and Stărică (2000).

---

[*] See www.indices.standardandpoors.com for further information on the index.

Up until quite recently, stochastic volatility models have been investigated mostly in discrete time, but with modern easy access to voluminous quantities of very high-frequency data, a demand for continuous-time models that allow, for instance, a more natural analysis of possibly irregularly spaced data, has arisen (for a general overview on stochastic volatility see Shephard, 2005). A first attempt to create a continuous-time GARCH model dates back to Nelson (1990), who, by taking a limit of a discrete-time GARCH process, derived a bivariate diffusion driven by two independent Brownian motions. By contrast, the discrete-time GARCH model incorporates only one source of uncertainty. Consequently, Nelson's continuous-time limit process does not possess the feedback mechanism whereby a large innovation in the mean process produces a burst of higher volatility, which is a distinctive feature of the discrete-time GARCH process. Moreover, the diffusion limit no longer has the heavy tailed distribution of returns needed for realistic modeling of returns in high-frequency financial data.

To overcome these problems, Klüppelberg et al. (2004) suggested an extension of the GARCH concept to continuous-time processes. Their COGARCH (continuous-time GARCH) model is based on a single background driving (continuous-time) Lévy process, which preserves the essential features of discrete-time GARCH processes and is amenable to further analysis, possessing useful Markovian and stationarity properties.

The aim of this chapter is to illustrate the advantages of using the continuous-time COGARCH model for the analysis of very high-frequency, unequally spaced financial data. We proceed by summarizing, in the following section, the currently available literature on the COGARCH, taking a detailed look at the model definition, and also outlining briefly the pseudo-maximum-likelihood method we will use to fit the model to data. Section 2.3 gives an illustrative data analysis, applying the COGARCH model to high-frequency data from the S&P500 stock market index, over the years 1998 to 2007. The model proves to be remarkably stable and informative. We discuss some extensions of the model and other issues in Section 2.4, and conclude with Section 2.5.

2.2 COGARCH: A SUMMARY OF THE CURRENT LITERATURE

Since its introduction in 2004, many aspects of the COGARCH model have been studied. One field of research covered Markovian and stationarity properties, as well as extremal behavior of the model, and its relations to other models. Model extensions, such as a COGARCH(p,q), and a multivariate COGARCH model, have been developed. Regarding data analysis, so far a couple of different approaches have been suggested for estimation of parameters.

In this section we summarize the literature on these topics, and briefly sketch the ideas. Throughout, by COGARCH we will usually mean the COGARCH(1,1) model, as introduced in Section 2.1, because this is the most widely applied version. Since there exists, analogously to the GARCH(1,1) and GARCH(p,q) models, a generalization to COGARCH(p,q), we will always point this out by adding the complexity tupel (p,q) when referring to this kind of model extension.

2.2.1 The COGARCH Model and Its Theoretical Properties

We first recall the definition of the COGARCH process as introduced in Klüppelberg et al. (2004). On a filtered probability space $(\Omega, \mathcal{F}, \mathbb{P}, (\mathcal{F}_t)_{t \geq 0})$ satisfying the "usual hypothesis" (see Protter, 2005), one is given a *background driving Lévy process* $L = (L(t))_{t \geq 0}$. See Applebaum (2004), Bertoin (1996), and Sato (1999) for detailed results concerning Lévy processes. Throughout it is assumed that $\mathbb{E}L(1) = 0$ and $\mathbb{E}L^2(1) = 1$.

Given parameters (β, η, φ), with $\beta > 0, \eta > 0, \varphi \geq 0$, and a square integrable random variable (*rv*) $\sigma(0)$, independent of L, the COGARCH *variance process* $\sigma^2 = (\sigma^2(t))_{t \geq 0}$ is defined as the almost surely unique solution of the stochastic differential equation

$$d\sigma^2(t) = \beta dt - \eta \sigma^2(t-)dt + \varphi \sigma^2(t-)d[L,L](t), \quad t > 0, \qquad (2.1)$$

where $[L,L]$ is the bracket process (quadratic variation) of L (Protter, 2005). Then one defines the *integrated COGARCH process* $G = (G(t))_{t \geq 0}$ in terms of L and σ as

$$G(t) = \int_0^t \sigma(s-)dL(s), \quad t \geq 0. \qquad (2.2)$$

As has been shown in Klüppelberg et al. (2004), Corollary 3.1, the bivariate process $(\sigma(t), G(t))_{t \geq 0}$ is Markovian. Moreover, under a certain integrability condition, and for the right choice of $\sigma^2(0)$, the process $\sigma^2(t)$ is strictly stationary (Klüppelberg et al., 2004, Theorem 3.2). As a consequence, G has stationary increments. Furthermore, Klüppelberg et al. (2004) also provide explicit expressions for the moments and autocorrelation functions of the variance process σ^2

and for the increments $G^{(r)}(t) := G(t+r) - G(t)$ of G. These can be used to estimate the COGARCH parameters by the method of moments, cf. Section 2.2.3.

Fasen et al. (2005) show that the COGARCH model, in general, exhibits regularly varying (heavy) tails, volatility jumps upwards, and clusters on high levels. More precisely, it can be shown that both the tail of the distribution of the stationary volatility and the tail of the distribution of $G(t)$ are Pareto-like under weak assumptions (cf. Klüppelberg et al., 2006). For more details on the theoretical properties of G and σ^2, we refer to Klüppelberg et al. (2004), Fasen et al. (2005), and Klüppelberg et al. (2006).

2.2.2 The Relation between GARCH and COGARCH

The discrete-time GARCH(1,1) model is specified by the mean and variance equations

$$Y_i = \sigma_{i-1}\varepsilon_i$$

and

$$\sigma_i^2 = a + b\sigma_{i-1}^2\varepsilon_{i-1}^2 + c\sigma_{i-1}^2,$$

for $i = 1, 2, \ldots, n$, with σ_0^2 given, and a, b, c as parameters. The ε_i are independent, identically distributed (i.i.d.) random variables with mean 0 and unit variance. By writing a discretized version of Equation (2.2) in the form

$$G_i - G_{i-1} = \sigma_{i-1}\varepsilon_i,$$

where we replace the increment $dL(t)$ with one of an i.i.d. sequence $\varepsilon_1, \ldots, \varepsilon_n$, and similarly, a discretized version of Equation (2.1) in the form

$$\sigma_i^2 = \beta\Delta t_i + \varphi\sigma_{i-1}^2\varepsilon_i^2 + (1-\eta)\sigma_{i-1}^2\Delta t_i,$$

we can see a direct analogy between the discrete- and continuous-time models. In fact, taking $\Delta t_i = 1$ in the continuous-time equations (corresponding to equally spaced data), we see that the models differ only by a nonessential alteration in parameterization.

More generally, in the continuous-time model, we do not need to assume equally spaced data. We can proceed as follows. Starting with a finite interval

$[0,T]$, $T > 0$, take deterministic sequences $(N_n)_{n \geq 1}$ with $\lim_{n \to \infty} N_n = \infty$ and $0 = t_0(n) < t_1(n) < \cdots < t_{N_n}(n) = T$, and for each $n = 1, 2, \ldots$, divide $[0,T]$ into N_n subintervals of length $\Delta t_i(n) := t_i(n) - t_{i-1}(n)$, for $i = 1, 2, \ldots, N_n$. Define, for each $n = 1, 2, \ldots$, a discrete-time process $(G_{i,n})_{i=1,\ldots,N_n}$ satisfying

$$G_{i,n} = G_{i-1,n} + \sigma_{i-1,n} \sqrt{\Delta t_i(n)} \, \varepsilon_{i,n} \quad i = 1, 2, \ldots, N_n \tag{2.3}$$

where $G_{0,n} = G(0) = 0$, and the variance $\sigma_{i,n}^2$ follows the recursion

$$\sigma_{i,n}^2 = \beta \Delta t_i(n) + \left(1 + \varphi \Delta t_i(n) \varepsilon_{i,n}^2\right) e^{-\eta \Delta t_i(n)} \sigma_{i-1,n}^2, \quad i = 1, 2, \ldots, N_n. \tag{2.4}$$

Here the innovations $(\varepsilon_{i,n})_{i=1,\ldots,N_n}$, $n = 1, 2, \ldots$, are constructed using a "first jump" approximation to the Lévy process developed by Szimayer and Maller (2007), which divides a compact interval into an increasing number of subintervals and for each subinterval takes the first jump exceeding a certain threshold. Finally, embed the discrete-time processes $G_{\cdot,n}$ and $\sigma_{\cdot,n}^2$ into continuous-time versions G_n and σ_n^2 defined by

$$G_n(t) := G_{i,n} \text{ and } \sigma_n^2(t) := \sigma_{i,n}^2, \text{ when } t \in [t_{i-1}(n), t_i(n)], \ 0 \leq t \leq T \tag{2.5}$$

with $G_n(0) = 0$. The processes G_n and σ_n are in $\mathbb{D}[0,T]$, the space of càd-làg real-valued stochastic processes on $[0,T]$.

Assume $\Delta t(n) := \max_{i=1,\ldots,N_n} \Delta t_i(n) \to 0$ as $n \to \infty$. As one main result of their paper, Maller et al. (2008) showed then that the discretized, piecewise constant processes $(G_n, \sigma_n^2)_{n \geq 1}$ defined by Equation (2.5) converge in distribution as $n \to \infty$ to the continuous-time processes (G, σ^2) defined by Equations (2.1) and (2.2). Further, this result was used by Maller et al. (2008) to develop a pseudo-maximum-likelihood estimation procedure for the parameters in the COGARCH. We sketch this method in Section 2.2.3.

2.2.3 Estimation Procedures

Haug et al. (2007) suggested moment estimators for the parameters of the COGARCH process based on equally spaced observations. Using the fact that the increments of the COGARCH process are strongly mixing with exponential rate, they showed that under moment conditions the resulting estimators are consistent and asymptotically normal. The paper by

Müller (2007) shows it is also possible to use Bayesian methods to estimate the COGARCH model. At the time of writing, however, this method is restricted to the case where COGARCH is driven by a compound Poisson process. More generally, Maller et al. (2008) describe a straightforward and intuitive pseudo-maximum-likelihood (PML) method based on the GARCH approximation to COGARCH. This estimation procedure is well adapted for the analysis of unequally spaced data, as we will illustrate in the next section. By applying it, we can easily account also for a transformation to a business time rather than calendar time scale. The general strategy is as follows.

Suppose we are given observations $G(t_i)$, $0 = t_0 < t_1 < \cdots < t_N = T$, on the integrated COGARCH as defined and parameterized in Equations (2.1) and (2.2), assumed to be in its stationary regime. The $\{t_i\}$ are assumed fixed (nonrandom) time points; set $\Delta t_i := t_i - t_{i-1}$. Let $Y_i = G(t_i) - G(t_{i-1})$, $i = 1, \ldots, N$, and denote the observed returns. We wish to estimate the parameters (β, η, φ). From Equation (2.2) we can write

$$ Y_i = \int_{t_{i-1}}^{t_i} \sigma(s-)dL(s), $$

and because σ is Markovian (Klüppelberg et al., 2004, Theorem 3.2), Y_i is conditionally independent of $Y_{i-1}, Y_{i-2}, \ldots$, given the natural filtration of the Lévy process L, with conditional expectation 0, and a conditional variance given by Equation (3.2) of Maller et al. (2008). To ensure stationarity, we take $\mathbb{E}\sigma^2(0) = \beta/(\eta - \varphi)$, with $\eta > \varphi$.

We can then apply the PML method, as in Maller et al. (2008), assuming at first that the Y_i are conditionally $N(0, \rho_i^2)$, and using recursive conditioning and a GARCH-type recursion for the variance process to write a pseudo-log-likelihood function for $Y_1, Y_2, \ldots, Y_N$. Taking as starting value for $\sigma^2(0)$ the stationary value $\beta/(\eta - \varphi)$, one can maximize the function $\mathcal{L}_N$ in Equation (3.3) of Maller et al. (2008) to get PML estimates of (β, η, φ).

2.3   DATA ANALYSIS

In this section we illustrate how to apply the COGARCH model to some market data. After a brief description of the raw data on the S&P500 index, we discuss data cleaning and preprocessing. Then we report on results of fitting the COGARCH model, using the PML method described in Section 2.2.3, after rescaling calendar time to a business time scale.

2.3.1 Description of Raw Data

Intraday data on the value of the S&P500 index was obtained from the TAQTIC database maintained by the Securities Industry Research Centre of Asia-Pacific (SIRCA). Based on this tick-by-tick index data, we compute and analyze 5-minute log-returns from the index, separately for the years 1998 to 2007. Table 2.1 reports, for each year, the number of trading days and the total number of observations, i.e., the number of log-returns. On average, we have 19,169 log-returns per year. However, even for years with the same number of trading days, the values can differ quite significantly. The first reason is that on a couple of trading days each year the NYSE opens later or closes earlier for annual memorials and holidays. For example, in Table 2.1, Christmas Eve, as long as it does not coincide with Saturday or Sunday, is counted as a full trading day, although the NYSE closes at 1 p.m. on December 24. The second reason for irregularities is special events. Here we just note, as an example, one of many which occurred within our time frame since 1998: on September 11, 2002, the NYSE did not open until 11 a.m. due to the memorial events commemorating the 1-year anniversary of the attack on the World Trade Center. However, in Table 2.1 this day is again counted as a full trading day. For a complete list of trading hours exceptions at NYSE since 1885 see http://www.nyse.com/pdfs/closings.pdf. The third reason is that each year a few observations are missing, usually since some data are obviously erroneous and have to be removed from the data set However, taking advantage of our continuous-time approach, we will not use interpolation to fill in the missing values, but instead take the time difference to the previous available observation into account.

TABLE 2.1 Number of Trading Days at NYSE and Number of Log-Returns Based on 5-Minute Index Data from the S&P500, for Years 1998 to 2007

Year	Trading Days	Observations	Year	Trading Days	Observations
1998	245	18,934	2003	251	19,260
1999	250	19,359	2004	249	19,373
2000	251	19,311	2005	252	19,617
2001	245	18,468	2006	251	19,460
2002	251	19,028	2007	251	19,458

2.3.2 Data Cleaning: Local Trends and Local Volatility Weights

Before applying the COGARCH model to the data, one has to think carefully about which features of the data are to be captured by the COGARCH, and which are not. The COGARCH is mainly designed to describe the behavior of the volatility in the data. Moreover, since we assume the COGARCH parameters β, η, and φ to be constant over time, we must check first whether the data indeed show a stationary volatility pattern over the whole time frame. In our data, this is definitely not the case. See, for instance, the 2002 returns (c.f. Figure 2.1, first and second rows). Therefore, we first preprocess the data by estimating local trends and local volatility weights. This procedure is necessary since we have around 20,000 observations per year, so that we cannot expect the COGARCH parameters to be constant over the whole time frame. On the other hand, we must be careful not to destroy the volatility structure that we want to describe by the COGARCH model. Therefore, we aim at a standardization procedure, which uses trends and volatility weights over longer

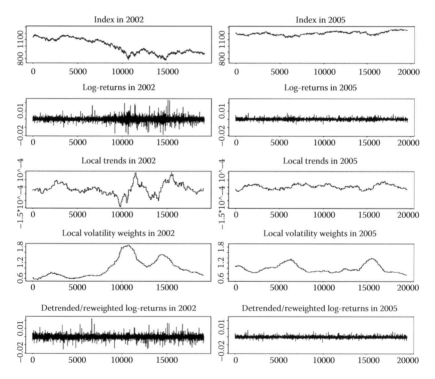

FIGURE 2.1   Five-minutely observations from S&P500 index data: log-returns, local trends, local volatility weights, and detrended and reweighted log-returns for years 2002 and 2005.

periods, such as 1 month, which corresponds to around 1,600 observations in our setup.

In the following we denote the observed log-returns by y_i, $i=1,\ldots,T$, whereas D denotes the number of trading days in the year under consideration (e.g., in 2002 we have $T=19,208$ and $D=251$). Next we have to introduce some functions, to be able to cover all aspects of the data within our subsequent formulas. First, let $d:\{1,\ldots,T\}\to\{1,\ldots,D\}$, $i\mapsto d(i)$, denote a function that returns the trading day for observation i, $A:\{1,\ldots,D\}\to\mathbb{N}$, $d\mapsto A(d)$ denote a function that returns the number of available observations on trading day d, and $N:\{1,\ldots,D\}\to\mathbb{N}$, $d\mapsto N(d)$, a function that returns the number of missing observations on trading day d. On a regular trading day we usually have $A(d)=78$ and $N(d)=0$, since we take 5-minute data between 9:30 a.m. and 4:00 p.m. Note that $N(\cdot)$ can also be 0, when the NYSE opened later or closed earlier, since $N(\cdot)$ counts only missing values when trading really took place at that time. Although only very few observations are missing overall (less than 0.5%), we introduce, to act very precisely, the function $I:\{1,\ldots,T\}\to\mathbb{N}$, $i\mapsto I(i)$, which returns the number of 5-minute intervals elapsed before observation i. Usually $I(i)=1$, and if, e.g., one observation is missing, $I(i)=2$, and so on. Later, we will also need the more precise functions $I_k:\{1,\ldots,T\}\to\mathbb{N}$, $i\mapsto I_k(i)$, for $k\in\{9,10,\ldots,15\}$, which specify how many 5-minute intervals during trading hour k elapsed before observation i. For example, if we have an observation at 9:55 a.m. and the value at 10:00 a.m. was deleted, so that the next observation i is from 10:05 a.m., we have $I(i)=2$, $I_9(i)=I_{10}(i)=1$, and $I_k(i)=0$ for $k=11,\ldots,15$.

We now assume that the log-returns follow the model

$$y_i = m_{d(i)} + v_{d(i)}x_i, \ i=1,\ldots,T$$

where $m_{d(i)}$ represents a *local trend*, $v_{d(i)}$ a *local volatility weight*, and the x_i are detrended and locally reweighted log-returns. This approach takes irregularities of the stock prices and index data into account, which cannot be captured by the COGARCH model. Both local trends and volatility weights are assumed to be constant over trading days, and are estimated as follows. Using a fixed $M\in\mathbb{N}$, the local trends $m_{d(i)}$ are estimated as moving averages over $2M+1$ trading days:

$$\hat{m}_{d(i)} = \left[\sum_{d=d(i)-M}^{d(i)+M}(A(d)+N(d))\right]^{-1}\sum_{d=d(i)-M}^{d(i)+M}\sum_{\{j|d(j)=d\}}y_j, \ M<d(i)\le D-M$$

For the cases $d(i) \leq M$ and $d(i) > D - M$ we can again employ this formula by using data from the previous and following years, respectively, or else we can shrink the time frame of estimation. Note that the addition $A(d) + N(d)$ makes sense, since log-returns are additive.

Similarly, the volatility weights are estimated, for some $V \in \mathbb{N}$, by computing preliminary weights

$$\hat{v}^*_{d(i)} = \left[\sum_{d=d(i)-V}^{d(i)+V} (A(d) + N(d)) \right]^{-1} \sum_{d=d(i)-V}^{d(i)+V} \sum_{\{j|d(j)=d\}} |y_j - \hat{m}_{d(j)}|, V < d(i) \leq D - V$$

and then by reweighting these according to

$$\hat{v}_{d(i)} = \frac{\sum_{j=1}^{T} |y_j - \hat{m}_{d(j)}| / \hat{v}^*_{d(j)}}{\sum_{j=1}^{T} |y_j - \hat{m}_{d(j)}|} \hat{v}^*_{d(i)}$$

This implies that for $\hat{x}_i := (y_i - \hat{m}_{d(i)}) / \hat{v}_{d(i)}$ we have $\sum_{i=1}^{T} |\hat{x}_i| = \sum_{i=1}^{T} |y_i - \hat{m}_{d(i)}|$, so that the magnitude of the values is preserved.

In our analysis, we set $M = V = 10$, so that we use a time frame of 21 trading days to determine the local trend and volatility weight. This seems to be a reasonable choice since we usually get larger standard errors for the COGARCH parameter estimates for very large or very small values of M and V. In Figure 2.1, as an example, the third and fourth rows show the estimated local trends and estimated local volatility weights for the years 2002 and 2005. The fifth row shows the detrended and reweighted log-returns. Note that, for each row, the left figure for 2002 has the same scale as the right figure for 2005. We chose the years 2002 and 2005 for illustration since they apparently show quite different patterns; however, the cleaning and preprocessing procedure is the same for all years. We emphasize once more that both $\hat{m}_{d(i)}$ and $\hat{v}_{d(i)}$ depend only on the $d(i)$, so that all observations of the same day are reweighted by the same weight. This way we do not lose information about the dependence of the volatility on the exact trading time during the day. This dependence is accounted for in the following subsection.

2.3.3 Accounting for Trading Time by Transformation of Time

To take the possible impact of trading time on the volatility into account, we first conducted a standard regression analysis for the squared log-returns to check for explanatory variables having an influence on the volatility. We used indicator variables for the month January, for all weekdays from Monday to Friday, and for all trading hours (9, 10, 11,…, 15). For reasons of identifiability we had to remove one weekday and one trading hour (we chose Friday and hour 15), so that this sums up to a collection of 11 variables. The four most significant indicators always turned out to be hours 9, 12, 13, and 10, in this order. All these had p-values of less than 1%.

An easy way to account for these explanatory variables within the COGARCH context is to apply the COGARCH to a fictive business time axis. That means that we do not insert the explanatory variables into the COGARCH model itself, but use the PML method both to rescale the physical time axis and to estimate the COGARCH parameters, simultaneously. The idea is to replace Δt_i in Equations (2.3) and (2.4) by

$$\Delta\tau_i := \Delta t_i I(i) + h_9 I_9(i) + h_{10} I_{10}(i) + h_{12} I_{12}(i) + h_{13} I_{13}(i),$$

where h_9, h_{10}, h_{12}, and h_{13} are unknown parameters. Since we account for missing values within the functions I and I_k, respectively, Δt_i does not depend on i in our setup and serves just as the basic time unit. To get estimates nearly on an annual basis, we choose Δt_i as $1/(\Sigma I(j))$, since $\Sigma I(j) = T + \Sigma N(d) = \Sigma A(d) + N(d)$. However, since the time axis is transformed during the estimation procedure, values such as the exact annualized volatilities in each year have to be computed separately after the estimation procedure.

Of particular interest are the quantities $f_k := (h_k + \Delta t_i)/\Delta t$ for $k = 9, 10, 12, 13$. These report the factors by which the basic time unit has to be rescaled during a certain trading hour to get business time.

2.3.4 Results and Interpretation

Table 2.2 reports the PML estimates together with their corresponding standard errors. We first note that our estimates satisfy the stationarity condition $\eta > \varphi$. Moreover, the estimates for η and φ are very similar for all 10 years, with values around 0.20 and 0.10, respectively. Such stability is very reassuring as to the applicability of the model. As a consequence, the

TABLE 2.2   PML Estimates and Corresponding Approximated Standard Errors
of the Parameters β, η, φ, h_9, h_{10}, h_{12}, and h_{13}

	β		η		φ	
1998	0.00215457	(0.00009185)	0.2035	(0.0099)	0.1002	(0.0075)
1999	0.00212128	(0.00007431)	0.2022	(0.0040)	0.1028	(0.0023)
2000	0.00312924	(0.00015303)	0.2032	(0.0052)	0.1019	(0.0032)
2001	0.00224318	(0.00010266)	0.1988	(0.0047)	0.1035	(0.0019)
2002	0.00385930	(0.00013885)	0.2067	(0.0068)	0.0918	(0.0054)
2003	0.00152757	(0.00011513)	0.2011	(0.0112)	0.1024	(0.0051)
2004	0.00076915	(0.00002222)	0.1979	(0.0037)	0.1048	(0.0025)
2005	0.00071658	(0.00002205)	0.2018	(0.0046)	0.1027	(0.0038)
2006	0.00069020	(0.00002447)	0.2039	(0.0054)	0.1009	(0.0049)
2007	0.00145146	(0.00008295)	0.2017	(0.0071)	0.0999	(0.0053)

	$h_9 \cdot 10^4$		$h_{10} \cdot 10^4$		$h_{12} \cdot 10^4$		$h_{13} \cdot 10^4$	
1998	2.2616	(0.1230)	0.1225	(0.0194)	−0.2587	(0.0081)	−0.2320	(0.0090)
1999	2.0057	(0.1038)	0.2080	(0.0213)	−0.2534	(0.0077)	−0.2290	(0.0085)
2000	1.9162	(0.1035)	0.2232	(0.0219)	−0.2420	(0.0082)	−0.1957	(0.0096)
2001	2.7368	(0.1571)	0.5309	(0.0349)	−0.2257	(0.0096)	−0.1637	(0.0113)
2002	2.7753	(0.1317)	0.4186	(0.0267)	−0.1412	(0.0106)	−0.1381	(0.0107)
2003	3.0822	(0.1526)	0.5732	(0.0329)	−0.1694	(0.0103)	−0.1970	(0.0096)
2004	2.5946	(0.1318)	0.2317	(0.0223)	−0.2085	(0.0091)	−0.1944	(0.0095)
2005	1.8003	(0.0988)	0.1923	(0.0198)	−0.1979	(0.0088)	−0.1504	(0.0106)
2006	2.1767	(0.1170)	0.2547	(0.0228)	−0.1667	(0.0103)	−0.1346	(0.0112)
2007	2.6795	(0.1377)	0.1301	(0.0193)	−0.1965	(0.0093)	0.1766	(0.0099)

Note: For notational convenience, the estimates and standard errors of h_9, h_{10}, h_{12}, h_{13}
have been multiplied by 10,000.

parameter β directly reflects the long-run volatility in each year, since the
mean of the COGARCH variance equation can be expressed as $\beta / (\eta - \varphi)$,
with $\eta - \varphi$ almost constant in our case. For example, from the plots in
Figure 2.1 one can immediately see that in 2002 the overall mean volatility
was much higher than in 2005. This conjecture is now confirmed by the
estimates of β, which are around 0.0039 in 2002 and only 0.0007 in 2005.

Table 2.3 contains the factors that have to be applied to the physical time
axis to get business time. For example, the value 5.3057 of the estimate for
f_9 in 1998 means that during this year business time was running at around
5.3 times faster than physical time between 9:30 a.m. and 10:00 a.m.,
which reflects the high activity in the market after the opening of the
exchange. Between 10:00 a.m. and 11:00 a.m. the activity decreased, but
is still higher than on average. In general, during lunchtime, between

TABLE 2.3   Factors That Have to Be Applied to Get from Physical to
Business Time Scale, for Trading Hours 9, 10, 12, and 13, Respectively

	f_9	f_{10}	f_{12}	f_{13}
1998	5.3057	1.2332	0.5076	0.5583
1999	4.8966	1.4040	0.5077	0.5551
2000	4.7377	1.4353	0.5280	0.6183
2001	6.2005	2.0089	0.5712	0.6889
2002	6.4119	1.8164	0.7246	0.7307
2003	6.9899	2.1140	0.6709	0.6172
2004	6.0300	1.4491	0.5958	0.6231
2005	4.5319	1.3772	0.6117	0.7049
2006	5.2458	1.4967	0.6748	0.7374
2007	6.2170	1.2533	0.6174	0.6561

12:00 p.m. and 2:00 p.m., business time runs slower than physical time; usually the activity is only 50 to 70% of the average observed between 11:00 a.m. and 12:00 p.m. as well as after 2:00 p.m.

2.4   DISCUSSION

2.4.1   Extensions: COGARCH(*p,q*), ECOGARCH, Multivariate Models

There exist a couple of model extensions to the COGARCH model, which we briefly mention in this section. For more details refer to the corresponding chapters.

Motivated by the generalization of the GARCH(1,1) to the GARCH(*p,q*) model, Brockwell et al. (2006) introduced the COGARCH(*p,q*) model. Here the volatility follows a continuous-time ARMA (CARMA) process, which is again driven by a Lévy process. As in the discrete-time case, this model displays a broader range of autocorrelation structures than those of the COGARCH(1,1) process.

Haug and Czado (2007) introduce an exponential continuous-time GARCH (ECOGARCH) process as analog to the EGARCH(*p,q*) models. They investigate stationarity and moments and show an instantaneous leverage effect for the ECOGARCH(*p,p*) model. In a subsequent paper, Czado and Haug (2008) derive a quasi-maximum-likelihood estimation procedure for the ECOGARCH(1,1) model, in the case when it is driven by a compound Poisson process, assuming normally distributed jumps.

In Stelzer (2008a) multivariate COGARCH(1,1) processes are introduced constituting a dynamical extension of normal mixture models and covering again such features as dependence of returns (but without autocorrelation), jumps, heavy tailed distributions, etc. As in the univariate case, the model has only one source of randomness, a single multivariate Lévy process. The time-varying covariance matrix is modeled as a stochastic process in the class of positive semidefinite matrices. The paper analyzes the probabilistic properties and gives a sufficient condition for the existence of a stationary distribution for the stochastic covariance matrix process, and criteria ensuring the finiteness of moments.

As for the univariate COGARCH, the multivariate COGARCH can be extended to a multivariate ECOGARCH model, as is done in Haug and Stelzer (2008).

Analogously to the papers by Szimayer and Maller (2007) and Maller et al. (2008), Stelzer (2008b) generalizes the first jump approximation of a pure jump Lévy process, which converges to the Lévy process in the Skorokhod topology in probability, to a multivariate setting and an infinite time horizon. Applying this result to multivariate ECOGARCH(1,1) processes, he shows that there exists a sequence of piecewise constant processes determined by multivariate EGARCH(1,1) processes in discrete time that converge in probability in the Skorokhod topology to the continuous-time process.

2.4.2 Other Theory

We pointed out in Section 2.2.2 the striking similarity between the theoretical formulations of the discrete-time GARCH and the COGARCH volatility equations. In relation to this, Kallsen and Vesenmayer (2008) derive the infinitesimal generator of the bivariate Markov process representation of the COGARCH model and show that any COGARCH process can be represented as the limit in law of a sequence of GARCH(1,1) processes. The result of Maller et al. (2008) is even stronger. They approximate the COGARCH with an embedded sequence of discrete-time GARCH(1,1) models that converges to the continuous-time model in a strong sense (in probability, in the Skorokhod metric), as the discrete-approximating grid grows finer. Whereas the diffusion limit in law established by Nelson (1990) occurs from GARCH by aggregating its innovations, the COGARCH limit arising in Kallsen and Vesenmayer (2008) and Maller et al. (2008) occurs when the innovations are randomly thinned. We sketched briefly the basic idea of the Maller et al. (2008) approximation in Section 2.3.

The question also arises as to how strong this similarity may be from a statistical point of view, and how this similarity might be measured mathematically. A sophisticated approach to measuring the similarity between two statistical models is Le Cam's framework of *statistical equivalence*. As was shown by Wang (2002), the diffusion limit in law of the GARCH(1,1) established by Nelson (1990) is *not* statistically equivalent to the approximating series of GARCH models. A recent paper by Buchmann and Müller (2008) investigates in detail the possibility of statistical equivalence between the GARCH and COGARCH models. They show that if full information about the volatility processes is available, the limiting COGARCH experiment is in fact equivalent to the approximating sequence of GARCH models. If, however, the corresponding volatilities of the COGARCH process are unobservable, the limit experiment is again not equivalent to GARCH in deficiency.

2.5 CONCLUSION

In Section 2.2 we summarized the literature on the theoretical properties of the COGARCH, estimation procedures, and some model extensions that have been proposed. In Section 2.3 we used the COGARCH model to analyze high-frequency data on the S&P500.

We emphasize once more that the COGARCH model incorporates the most important stylized features of financial data. Since it is a continuous-time model, it provides great flexibility in modeling different aspects of the data. While we are accustomed to seeing financial time series at relatively low, or coarser, frequencies—such as the closing value of the index reported in nightly news broadcasts—it is relatively easy to forget that such values are merely one point on the high-frequency activity function. High-frequency data readily lend themselves to analysis within a continuous-time framework. Further, the COGARCH methodology demonstrated in this chapter is also appropriate to model the discontinuities that are a natural, but often ignored, feature of the data.

However, we must be careful only to apply the COGARCH model to a stationary time series. This may require preprocessing the data as we did in Section 3.2, before the COGARCH model can be used. The PML method that we employed for fitting the COGARCH to the S&P500 data is applicable in general to irregularly spaced data. In this way we can very easily deal with missing data, or transform to another time scale as needed, to reflect some special properties of the data. In our case, we were able to account for the dependence of market activity on trading time. We have chosen to focus on modeling the volatility of the S&P500 due to its importance as a widely followed measure of the market. Its component stocks are large and liquid, and as a consequence, it is reasonable to believe that the behavior of the S&P500 will not be greatly affected by nosynchronous trading, or nontrading, of its constituent securities. Nonetheless, the premises of COGARCH suggest that it appropriately accounts for such microstructure issues. Analysis of COGARCH in microstructure research for stocks where discontinuities arise due to nontrading offers interesting and potentially insightful research opportunities.

Although many aspects of the COGARCH have already been investigated, it remains an area of active research. We have demonstrated its implementation and applicability to an important financial time series. While we have used high-frequency data to demonstrate the potential of COGARCH, we do not wish to suggest that its only application is in microstructure research. We are confident that the technique will become an important and well-known tool for research in financial economics.

REFERENCES

Abel, A. B. (1988). Stock prices under time-varying dividend risk. An exact solution in an infinite-horizon general equilibrium model. *Journal of Monetary Economics* 22:375–93.

Anderson, T. G., L. Benzoni, and J. Lund. (2002). An empirical investigation of continuous-time equity return models. *Journal of Finance* 57:1239–84.

Ang, A., R. J. Hodrick, Y. Xing, and X. Zhang. (2006). The cross-section of volatility and expected returns. *Journal of Finance* 61:259–99.

Applebaum, D. (2004). *Lévy processes and stochastic calculus.* Cambridge Studies in Advanced Mathematics 93. Cambridge, MA: Cambridge University Press.

Barsky, R. B. (1989). Why don't the prices of stocks and bonds move together? *American Economic Review* 79:1132–45.

Bertoin, J. (1996). *Lévy processes.* Cambridge, MA: Cambridge University Press.

Bollerslev, T., R. F. Engle, and D. B. Nelson. (1995). ARCH models. In *The handbook of econometrics*, ed. R. F. Engle and D. McFadden. Vol. 4. Amsterdam: North Holland.

Brockwell, P. J., E. Chadraa, and A. Lindner. (2006). Continuous time GARCH processes. *Annals of Applied Probability* 16:790–826.

Buchmann, B., and G. Müller. (2008). On the limit experiments of randomly thinned GARCH(1,1) in deficiency. Working paper, Monash University, Melbourne, Australia, and Technische Universität München, Munich, Germany.

Carhart, M. M. (1997). On persistence in mutual fund performance. *Journal of Finance* 52:57–82.

Czado, C., and S. Haug. (2008). Quasi maximum likelihood estimation and prediction in the compound Poisson ECOGARCH(1,1) model. Working paper, Technische Universität München, Munich, Germany.

Durand, R. B., D. Lim, and J. K. Zumwalt. (2007). Fear and the Fama-French factors. Working paper. Available at SSRN: http://ssrn.com/abstract=965587.

Durham, G. B. (2007). SV mixture models with application to S&P500 index returns. *Journal of Financial Economics* 85:822–56.

Engle, R. F. (1995). *ARCH: Selected readings*. Oxford: Oxford University Press.

Eraker, B., M. Johannes, and N. Polson. (2003). The impact of jumps in volatility and returns. *Journal of Finance* 58:1269–300.

Fama, E. F., and K. R. French. (1993). Common risk factors in the returns on stocks and bonds. *Journal of Financial Economics* 33:3–56.

Fasen, V., C. Klüppelberg, and A. Lindner. (2005). Extremal behavior of stochastic volatility models. In *Stochastic finance*, ed. A. N. Shiryaev, M. d. R. Grossinho, P. E. Oliviera, and M. L. Esquivel, 107–55. New York: Springer.

French, K., G. W. Schwert, and R. Stambaugh. (1987). Expected stock returns and volatility. *Journal of Financial Economics* 19:3–29.

Ghysels, E., P. Santa-Clara, and R. Valkanov. (2005). There is a risk-return trade-off after all. *Journal of Financial Economics* 76:509–48.

Haug, S., and C. Czado. (2007). An exponential continuous time GARCH process. *Journal of Applied Probability* 44:960–76.

Haug, S., C. Klüppelberg, A. Lindner, and M. Zapp. (2007). Method of moments estimation in the COGARCH(1,1) model. *Econometrics Journal* 10:320–41.

Haug, S., and R. Stelzer. (2008). Multivariate ECOGARCH processes—A multivariate stochastic volatility model with the leverage effect. Working paper, Technische Universität München, Munich, Germany.

Hsieh, D. A. (1991). Chaos and nonlinear dynamics: Application to financial markets. *Journal of Finance* 46:1839–77.

Kallsen, J., and B. Vesenmayer. (2009). COGARCH as a continuous time limit of GARCH(1,1). *Stochastic Processes and Their Applications*, 119:74–98.

Klüppelberg, C., A. Lindner, and R. A. Maller. (2004). A continuous time GARCH process driven by a Lévy process: Stationarity and second order behaviour. *Journal of Applied Probability* 41:601–22.

Klüppelberg, C., A. Lindner, and R. A. Maller. (2006). Continuous time volatility modelling: COGARCH versus Ornstein-Uhlenbeck models. In *From stochastic calculus to mathematical finance. The Shiryaev festschrift*, ed. Y. Kabanov, R. Lipster, and J. Stoyanov. Berlin: Springer.

Lundblad, C. (2007). The risk return tradeoff in the long run: 1836–2003. *Journal of Financial Economics* 85:123–50.

Maller, R. A., G. Müller, and A. Szimayer. (2008). GARCH modelling in continuous time for irregularly spaced time series data. *Bernoulli* 14:519–42.

Merton, R. (1980). On estimating the expected return on the market: An exploratory investigation. *Journal of Financial Economics* 84:323–61.

Mikosch, Th., and C. Stărică. (2000). Limit theory for the sample autocorrelations and extremes of a GARCH(1,1) process. *Annals of Statistics* 28:1427–51.

Müller, G. (2007). MCMC estimation of the COGARCH(1,1) model. Working paper, Technische Universität München, Munich, Germany.

Nelson, D. B. (1990). ARCH models as diffusion approximations. *Journal of Econometrics* 45:7–38.

Protter, P. (2005). *Stochastic integration and differential equations.* 2nd ed. Heidelberg: Springer.

Sato, K. (1999). *Lévy processes and infinitely divisible distributions.* Cambridge, UK: Cambridge University Press.

Sharpe, W. (1964). Capital asset prices: A theory of market equilibrium under conditions of risk. *Journal of Finance* 19:425–42.

Shephard, N. (2005). *Stochastic volatility: Selected readings.* Oxford: Oxford University Press.

Stelzer, R. (2008a). Multivariate continuous time Lévy-Driven GARCH(1,1) processes. Working paper, Technische Universität München, Munich, Germany.

Stelzer, R. (2008b). First jump approximation of a multivariate Lévy driven SDE and an application to ECOGARCH processes. Working paper, Technische Universität München, Munich, Germany.

Szimayer, A., and R. A. Maller. (2007). Finite approximation schemes for Lévy processes, and their application to optimal stopping problems. *Stochastic Processes and Their Applications,* 117:1422–47.

Wang, Y. (2002). Asymptotic nonequivalence of GARCH models and diffusions. *Annals of Statistics* 30:754–83.

CHAPTER 3

Price Volatility in the Context of Market Microstructure

Peter Lerner and Chunchi Wu[*]

CONTENTS

3.1 INTRODUCTION

Microstructure theories provide an answer to the question of why security prices change in the absence of new information announcements and why they can differ from the volatility of the underlying asset even for the 100% equity firm. Two well-known approaches, which analyze the asymmetric information trading problem are Kyle (1985) and Glosten and Milgrom (1985), are dissimilar in mathematical formalism, yet provide similar answers to this question.

---

[*] We thank discussants and chairs Irv Morgan, J. Randolph Norsworthy (RPI), Elin Tully, and Brian Lucey (Dublin) for valuable suggestions, Mark Miller (Syracuse University) and Natasha Trofimenko (Kiel) for the help with statistics, and Kathy Yuan (University of Michigan) for the emerging market database. All errors are our own.

Microstructure theories regard the market as interplay of heterogeneous agents, each with peculiar beliefs about the security's intrinsic value. Typically, agents have no information about one another's expectations, but they can adjust their preferences by observing the prices and trading volumes. For instance, a random fall in stock price can attract value investors, thus creating a bounce-back. Revision of preferences may contribute to price volatility even in the absence of economic events.

Informed trading theory has been applied and tested mostly for the stock and futures markets (Strother et al., 2002; Stoll, 2003; Hasbrouck, 2007). However, corporate or emerging market bonds seem to be more suitable targets of empirical study for the following reasons. Unlike stocks, most risky bonds are traded in large blocks between relatively sophisticated traders (large investment banks, pension funds, and insurance companies), bid-offer spreads are large, and the impact of a single trade is noticeable.

In this chapter, we explain the dependence of price volatility on market frictions based on a market microstructure theory. Conventional features of securities—prices, returns, and volatilities—are linked to asset volatility by the trading parameters, such as trader participation rate and granularity (decimal) of quotations. Because the participation rate of insiders depends on market frictions, such as a market tick, we expect volatility changes to be associated with changes in tick size.

3.2 HEURISTIC EXPLANATION OF THE EFFECT OF FRICTIONS ON VOLATILITY

All informed trading models explain market movements by the interaction of at least three types of agents: informed traders who have inside information on the true price of an asset; liquidity or noise traders, who trade for reasons unrelated to trading returns; and finally, a market maker, who cannot distinguish between informed and uninformed traders and who sets the clearing price according to the orders she receives from the previous two groups of agents. This market maker can be an individual (NYSE) or virtual (electronic trading platform). The only essential feature of the market maker is that she allows for market clearing according to the law of supply and demand.

Below, we list typical features of microstructure models. First, in the long run, only informed traders contribute to volatility because uninformed traders always have an opportunity to trade at the historical average price. In the absence of intervening economic events, uninformed (liquidity) traders will eventually converge on a true price by the law of large numbers. Therefore, asymptotic volatility for pure liquidity trading is always zero. Zero volatility in the end of each trading session (Kyle's $\Sigma(t) \to 0$, $t \to 1$) was first observed by Kyle (1985).

Second, in the absence of frictions, informed traders always have an incentive to execute the trade: to sell if they feel the asset is overpriced and to buy if they feel it is underpriced. If frictions are present, insiders must predict not only the direction of the price movement but also its relative magnitude. Even if the price moves in the trader's direction (with respect to her net inventory), she must be assured that the movement will be large enough to cover the friction-mediated losses.

Market tick is a specific form of market friction. Bid and ask prices can move only in discrete increments of the tick. When the tick is small, prices adjust almost continuously. When the tick is large, there is a discernible jump. When the bid and ask prices are observed with noise, observable bid-ask spreads increase as well.

Third, the average (i.e., long-term) security price changes only in transactions between informed and uninformed traders. Upward and downward bids by liquidity traders on the average offset one another. Informed traders know the true price, and their bids must be equal to offers unless they have information about one another's inventory.

Heuristic equilibrium supply and demand curves, which follow from this picture of informed trading theory, are displayed in Figures 3.1 and 3.2. All our diagrams presume that an insider is a buyer and uninformed traders are sellers.[*] The informed trader always bids at the true price of an asset. In Figure 3.1, the preferences of the liquidity traders have a stochastic distribution, which is schematically presented as an envelope of numerous supply curves.

In Figure 3.1, we show the influence of frictions. In our stylized two-dimensional static setting, informed trader executes outside of the market maker window at slight variance with the original Kyle model. Through observation of our heuristic supply-demand curves, one might notice that the price range of execution increases in the case of diminishing frictions and decreases otherwise.

Static displays of Figures 3.1 and 3.2 are based on the presumption of only one insider who always bids for an optimum quantity. There is no true

---

[*] The diagrams for the opposite case are quite similar.

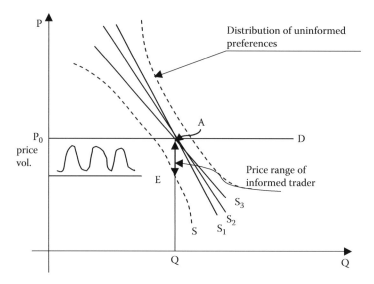

FIGURE 3.1 Equilibrium supply and demand curves for informed and uninformed traders in the case of a frictionless market. The true asset price is P_0. The informed trader is the buyer. The double arrow shows an acceptable range of execution from an informed trader's standpoint, who always bids at or below the true price. Lines S_1, S_2, and S_3 schematically represent fluctuating beliefs of liquidity traders. Point A indicates the average execution price. Point E indicates the limit order price for the most pessimistic liquidity traders. Market price bounces between points A and E.

price volatility in such a model even for a volatile asset because all the bids converge on a uniform true price p_0. To obtain the relationship between price and asset volatility, we need to derive the equilibrium arrival rate of informed traders for an exogenously given spread.

3.3 PARTICIPATION OF INFORMED TRADERS AND SPREADS

Market equilibrium in an ordinary sense in informed trading models is generally achievable only for a very special set of parameters. However, one can define the equilibrium rate of market arrival for informed and uninformed traders conditional on an imperfection.

Insiders know the true value of an asset, and if expected profit does not exceed the bid-ask spread, they will abstain from trading. Liquidity

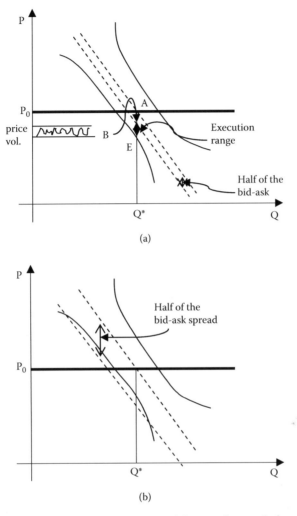

FIGURE 3.2   Supply and demand curves of the market with frictions given an informed buyer, an uninformed seller, and exogenously given bid-ask spread. (a) The acceptable execution range BE for an informed trader shrinks because the market maker reduces the payout to liquidity traders. Volatility follows suit. (b) If the bid-ask spread becomes too large, the informed trader cannot buy the optimal quantity Q^* at all because even the most pessimistic uninformed trader puts higher valuation on the security than $P_0 - \frac{1}{2} \times$ (bid-ask spread).

traders, however, participate in the market for reasons unrelated to adverse selection because the market maker is quasi-monopolist. Participation of insiders does not depend on the relative size of an order; the trade is either profitable or not given the bid-ask spread and the market price. To describe the insider dropout, we present the equation for the proportion of insider trades, which we call the *participation ratio*. The participation ratio can be determined from a particular pricing model through the Proposition 1 (for the proof, see Lerner and Wu, 2005).

Proposition 1

The equilibrium participation ratio $\rho_t = \mu_t/\varepsilon_t + \mu_t$, *where* μ *is the number of insider trades and* ε *is the number of liquidity trades, obeys the equation*

$$\rho_t = \frac{E_t[\max(p - p_0 - \Xi, 0)]}{E_t[\max(p - p_0, 0)]}$$

where p is the market price of a security, p_0 *is the fundamental price, and* Ξ *is the bid-ask spread.*

The equation above can be interpreted as follows. If the insider trader believes that a security in her possession will appreciate, she might exploit this superior information by selling a call option on an asset with a strike price p_0—the natural price of the asset—and replace it with a (cheaper) call option on the same asset with a strike price $p \geq (p_0 + \Xi)$. The share of insiders in the trading is ρ, which is the probability of this trade. When we replace a current price with its expected value at the time *t*, we obtain the above result.

We can compute the participation ratio from the linear equation for ρ if we choose a particular pricing model. The graph of ρ as a function of parameter $E[\Xi^2]/\sigma^2$ (σ = volatility) is given in Figure 3.3 for the Vasiček model (Hull, 1997).

Even in this simplest case, an analytic expression for parameter ρ is cumbersome and we do not provide it here. However, in principle, the equation for ρ can be evaluated for any model for which a closed-form solution for the asset price process is available. We observe that the larger the spread, the lower the number of informed traders. Therefore, insiders lower their participation in batch auctions in response to increased market frictions. Because they are the only purveyors of information in the informed trading format, asymptotic volatility must decrease.

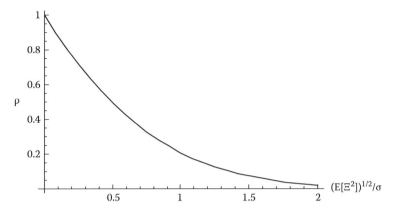

FIGURE 3.3 The participation ratio of informed traders ρ as a function of the spread Ξ in units of volatility σ, $E[\Xi^2]/\sigma^2$ for the Vasiček model.

3.4 MARKET VOLATILITY IN THE CONTEXT OF INFORMED TRADING THEORY

The system of Equations (3.1) (Easley et al., 1996) determines the dynamics of the bids and offers, if the processes for update of prices and probabilities are known. For the period $t + 1$, we get

$$a_{t+1} = E(V_{t+1} \,|\, t) + \frac{\mu P_{g,t+1}}{\varepsilon + \mu P_{g,t+1}} (\bar{V}_{t+1} - E(V_{t+1} \,|\, t))$$

$$b_{t+1} = E(V_{t+1} \,|\, t) + \frac{\mu P_{b,t+1}}{\varepsilon + \mu P_{b,t+1}} (\underline{V}_{t+1} - E(V_{t+1} \,|\, t))$$

(3.1)

where a_t, b_t are ask and bid quotes, respectively, $P_{b,t}$, $P_{g,t}$ are the respective probabilities of bad and good news, and ε and μ are the rates of arrival of liquidity and insider traders on the market. $\bar{V}_t$ and $\underline{V}_t$ are the prices of an asset contingent on good and bad news, and $E(V_{t+1} \,|\, t)$ is an expected asset price given all the information at time t.

In such a general setting one cannot deduce any closed-form expressions for the bid and ask prices. For instance, we did not specify any asset process. However, if we assume that we can neglect moments higher than the second (Gaussian approximation), and that good and bad news arrive with a similar frequency, we can characterize the dynamics in rather simple terms. The following two propositions summarize the results.

Proposition 2

Mid-price $p_t = (a_t + b_t)/2$ in the two-moment approximation evolves as a conditional AR(1) process:

$$\Delta p_{t+1} = \beta \Delta p_t + \sigma \varepsilon_{t+1} \tag{3.2}$$

where

$$\sigma^2 = \frac{1}{T} \sum_{t=1}^{T} \left(\frac{a_{t+1} + b_{t+1}}{2} - \frac{\bar{V}_t + \underline{V}_t}{2} \right)^2 \tag{3.3}$$

β is some constant, and $\Delta p_t = p_t - E[p_t | t-1]$.

Proposition 3

Price volatility σ^2 asymptotically obeys the following equation:

$$\sigma^2 = \frac{\mu^4 \left(\eta_b^4 P_b + \eta_g^4 P_g \right) \sigma_0^2}{4 \left(1 - \frac{\mu^2}{8} \left(\lambda_b^2 + \lambda_g^2 \right) \right)} \tag{3.4}$$

where $\lambda_{b,g} = P_{b,g}/\varepsilon + \mu P_{b,g}$, $\eta_{b,g} = (\mu^{-1} - \lambda_{b,g})$, $\sigma_0^2 = \frac{1}{T}\sum_{t=1}^{T}(\Delta\bar{V}_t)^2 = \frac{1}{T}\sum_{t=1}^{T}(\Delta\underline{V}_t)^2$, is the square of the asset volatility, and $\Delta\bar{V}_t = (\bar{V}_t - E(V_t | t-1))$, $-\underline{V}_t = (E(V_t | t-1) - \underline{V}_t)$ are the updates of beliefs at the step t.

Proofs of Propositions 2 and 3 can be viewed in the web version of the working paper (Lerner and Wu, 2005).

Equation (3.4) expresses asymptotic market price volatility σ through asset volatility σ_0, and the *number* of microstructure parameters, such as average frequency of bad and good news (P_b, P_a) and the rate of arrival of informed (μ) and uninformed (ε) traders on the market. We observe that asymptotic volatility has a highly nonlinear dependence on the ratio between informed and uninformed traders in the market.

Equation (3.4) cannot be tested directly because the rates of arrival of informed and uninformed traders are not observable. However, we can infer that if the market friction suddenly increases, the participation ratio of informed traders will decline according to Proposition 1 and volatility will also decrease with it. This relation can be seen more clearly if we further simplify the formula in Equation (3.4) by assuming that $\varepsilon \gg \mu$, i.e., that the numbers of informed traders are much smaller than the liquidity traders, there is no neutral news, and good and bad news are equally probable. It can be shown that under these conditions $\sigma^2 \approx \varepsilon^2\rho^4\sigma_0^2/2(1-\rho^2/32)$. Thus, the higher the participation ratio of informed traders, the higher is the price volatility.

3.5 SIMULATED MODEL FOR THE TIME EVOLUTION OF PRICES

Despite its seeming simplicity, the EKHP model of Equation (3.1) is very general. In particular, the strategy for liquidity traders can be arbitrary. In our simplified model, we assume that liquidity traders always bid at the historical average price. Surprisingly, our model displays all the important qualitative features of EKHP with only two arbitrary parameters. We specify our model as follows.

There are two types of traders: informed and uninformed. Informed traders always bid at the true price, which changes according to the rule

$$y_t = p_0 + u_t \tag{3.5}$$

where $u_t \sim N(0,\Sigma_0)$. Uninformed investors, however, bid at the historical average price, with normally distributed error terms:

$$\bar{p}_T = \frac{1}{T}\sum_{t=1}^{T} p_t + \varepsilon_t \tag{3.6}$$

$\varepsilon_t \sim N(0,\sigma_u^2)$. Naturally, we assume that $\sigma_u^2 \gg \Sigma_0$. The market clearing price is a linear combination of the price bids by the informed and uninformed traders,

$$p_t = \beta y_t + (1-\beta)\bar{p}_t \tag{3.7}$$

There is no ask price, but it can be incorporated into this framework at the expense of transparency. In Figure 3.4, we compare the results of our simulation to the much more sophisticated model of Bayesian learning of a Vasicek price process (Lerner and Wu, 2005).

Bid prices converge to the true price of an asset; i.e., uninformed investors learn from history. The higher is the correlation between their guesses and the true price (β), the faster the learning process for the investors.

However, quicker learning is accompanied by higher volatility (Figure 3.5). Despite its simplicity, the model of Equations (3.5) to (3.7)

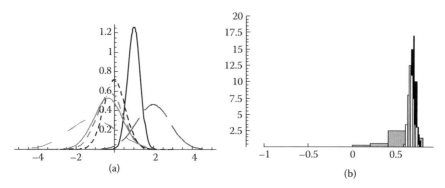

FIGURE 3.4 Distribution of the bids by uninformed investors. (a) Initial bid price is $p_0 = 0$, true price of the asset is $p_0 = 1$. Learning coefficient is equal to $\beta = 0.1$. The volatility of an asset is $\Sigma_0 = 0.1$, the volatility of beliefs of outside investors is $\sigma_0 = 5 \Sigma_0$. The curves represent evolution of the normalized learning distribution as a function of $Y = e^{-\gamma t}$: $Y = \{1$ (black, short dash), 0.3 (gray, long dash), 0.25 (gray, solid), 0.15 (gray, short dash), 0.05 (gray, long dash), and 0 (black, solid)$\}$. (b) The histogram of price distributions from the numerical model of Section 3.5 at $T = 100$ (gray), $T = 1,100$ (white), $T = 3,700$ (black).

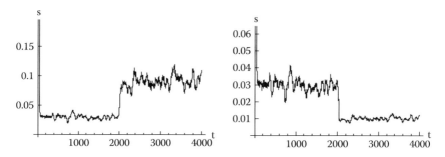

FIGURE 3.5 Evolution of the volatility according to the model in Section 3.5. Parameter β changes at $T = 2,000$ (midpoint). Fundamental price ($p = 1$) and intrinsic volatility of an asset remain unchanged. Higher β, despite producing higher volatility, also leads to more accurate convergence to the true price. Change of volatility on parameter increases. σ is computed as a standard deviation of fifty past returns. (a) Parameter β increases from $\beta = 0.3$ to $\beta = 0.9$. (b) Parameter β decreases from $\beta = 0.3$ to $\beta = 0.1$. Fundamental price ($p = 1$) and intrinsic volatility of an asset remain unchanged.

provides a remarkably good approximation to true behavior according to EKHP Equation (3.1).

3.6 EMPIRICAL EVIDENCE OF VOLATILITY DEPENDENCE ON THE TICK SIZE

An important implication of Propositions 1 to 3 is that price volatility is negatively related to the magnitude of tick size, as well as other frictions that are positively correlated with observed bid-ask spreads. Positive correlation of a tick size with an observed bid-ask spread can be proven in several analytical contexts and is, in fact, a consequence of general signal processing considerations (Caraiscos and Liu, 1984).[*]

The schematic dependence of the spread on the mismatch σ_d between an exact and digitized quote is shown in Figure 3.6. For small σ_d it grows like a square root of σ_d. For large ticks it is approximately proportional to σ_d, but overall dependence of the observed spread on the market tick is highly nonlinear.

Therefore, as tick size increases, price volatility will decrease. On the contrary, if tick size becomes smaller, this will attract more frequent participation of informed traders and volatility will increase.

In the aftermath of the 2000 decimation of the U.S. markets, institutional investors and portfolio managers were indeed expressing these worries. Their large trades were often facing considerable price uncertainty at execution. Because the market depth was lowered after decimation, large market participants had to break their orders into several small orders to complete their trades. When price volatility is generally high, large traders incur significant price risk. Therefore, it appears that the lower bid-ask spread came at the expense of higher price uncertainty after decimation.

The above argument has been articulately rendered by Gibson et al. (2003). They stated, "It is unclear whether the incentives for gathering information will increase, or decrease under decimal pricing. A bid-ask spread that is too large not only imposes a greater fixed cost on informed traders but also increases the probability of the spread straddling the efficient price, thus reducing traders' incentives for obtaining information. Hence, any decrease in the spread swing to decrease market maker rents under decimal pricing ought to increase informed trading." An even more

---

[*] We provide these proofs in Lerner (2007). For the present empirical status of the connection of the tick size and the bid-ask spread, see Ronen and Weaver (2001), Gibson et al. (2003), and Bollen et al. (2004).

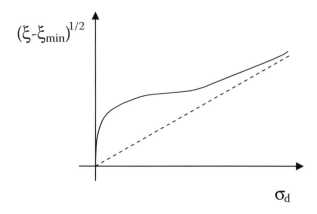

FIGURE 3.6 Symbolic dependence of the signal error induced by digitization according to Caraiscos and Liu (1984). On the horizontal axis, we plot σ_d, a standard deviation of the digitization-induced noise, and on the vertical axis, the square root of the mean square error contributed by digitization. ξ_{min} is the spread under exact quotation.

laconic formulation was produced by L. Harris in his 1997 congressional testimony: "If the tick is too small, front-runners will exploit investors who offer to trade.… Estimates of the benefits to the public from decimalization … do not estimate the increased costs that large traders will pay to avoid front-runners" (Gibson et al., 2003).

Yet, for all the prognostications surrounding decimalization, empirical evidence on the influence of the changeover on American stock markets was decidedly mixed. Bessembinder (1999) and Ronen and Weaver (2001) argue for the diminishing volatility, Chakravarty et al. (2004) for a modest increase, while Gibson et al. (2003) and, recently, Boehmer et al. (2007) find no significant difference. One of the empirical difficulties in uncovering of the effect is that the influence of adverse selection on the bid-ask spread is mixed up with the influence of inventory maintenance costs.

We provide another empirical test for the tick-volatility relation in the situation of a one-stage drastic change in quotation where the inventory maintenance should not have influenced the results because the securities were denominated in U.S. dollars. Our choice of emerging market Eurobonds is prompted by the following considerations. Emerging market bonds in the 1990s had large bid-ask spreads. They were traded infrequently, in relatively large blocks and between sophisticated traders, such as large investment banks and hedge funds. Consequently, we have a better chance to uncover the effects of informed trading in fledgling emerging markets with limited liquidity.[*]

After a period of high inflation (1992–1995), the Russian ruble became substantially devalued. When inflation subsided, there was a decision to redenominate the currency by striking off three zeroes from ruble-denominated money instruments.[†] The replacement of bank notes with three zeroes stricken was widely publicized in advance and was not accompanied by any open-market actions. In our empirical investigation, we examine bond yields that are not sensitive to exchange rates.[‡] However, the effective market tick size changed after redenomination. We can consider the redenomination as equivalent to the changing of the tick by an order of magnitude. Below, we show a domestic currency quotation of an imaginary security with the price in a near-dollar range in the old and new units (1 new = 1,000 old).

Old Quotes			New Quotes		
Bid	Ask	Spread	Bid	Ask	Spread
5,364	5,374	10	5.36	5.37	0.01
5,364	5,379	15	5.36	5.38	0.02
...	...	...	...	...	...
5,378	5,391	13	5.38	5.40	0.02

In other words, a round-off error acted in a similar fashion to the decimalization of the U.S. indexes, only in reverse and on an unparalleled scale. After the changeover, round-off error increased as a percentage ratio to the security price. We presume that the bid-ask spread after the round-off followed the suit.

Our inspection of six Russian Federation bonds with different maturities showed that their yields were roughly comparable (10–11% during the period surrounding New Year's Eve). All yields were dominated at the time by expectations of default, which finally happened on August 17, 1998, and the difference in maturity between the bonds was hardly an important

---

[*] The signaling role of the emerging markets sovereign bonds with respect to other securities of the same domicile have been recently explored by Dittmar and Yuan (2005). Their study suggests an important role of sovereign bond issues as providers of liquidity for the entire national markets.

[†] For an institutional description of Russian Federation debt and the 1990s economic situation in general, see Gaidar (2003). The exchange rate in 1997 was 5,000 to 6,000 rubles per dollar, before the redenomination. After the redenomination the ruble started to be quoted at a rate of roughly 6 RU/$1.

[‡] Russian bonds were originally denominated in U.S. dollars or Deutschemarks.

driver of these spreads. Therefore, we analyzed yield to maturity and the differenced yield as proxies for asset price.

Worthy of note is that bonds were originally denominated in dollars, and the changeover was announced long in advance. In an efficient market, quotation of the ruble exchange rate could not influence the yields because it did not change the fundamentals. However, if we believe that volatility is induced by information dissemination from informed trades, enhanced tick size will reduce their frequency. Insiders will abstain from trading until they can expect profits to compensate them for the increased spreads.

We tested yields of the two series of Russian Federation bonds: Minfin 10% 10-year bond (issued on June 26, 1997) and Minfin 9.25% 5-year bond (issued on November 27, 1996). We examined the period of 90–100 trading days evenly split between 1997 and 1998, the duration being chosen on the basis of bid-ask spread autocorrelations (Lerner and Wu, 2005). Longer samples were likely to be confounded with the August 1998 default, a momentous credit event, which would certainly obscure any microstructure-related changes.

In quantitative terms, we conducted an F-test of variances for two 50-day samples. We performed the F-test with the original yields and with the residuals for the Ordinary Linear Squares (OLS) regression on the calendar dummy, equal to 0 before the New Year and to 1 after the New Year (Table 3.1). The second sample was designed to eliminate effects of unobserved, but

TABLE 3.1   Variance F-test for the Yields of Two Russian Bonds

Sample	Minfin 10%			Minfin 9.25%		
	Average Yield	Standard Deviation	F-statistic	Average Yield	Standard Deviation	F-statistic
Pre-event	11.35%	0.0115	**4.58**	9.78%	0.0100	**2.35**
Post-event	11.24%	0.0054		9.92%	0.0065	
Pre-event	0.00%	0.0115	**4.56**	0.00%	0.0100	**2.35**
Post-event	0.00%	0.0054		0.00%	0.0065	

Note: Ten-year Minfin 10% maturing on June 26, 2007, and 5-year Minfin 9.25% maturing on November 27, 2001. F-statistic was estimated using two 50-trading-day samples (pre-event, beginning on October 20, 1997, and post-event, starting on December 31, 1997, respectively). F-statistic for the OLS residuals of the regression on the dummy, equal to 0 for the period October 27, 1997 to December 31, 1998, and 1 for the period January 1, 1998 to March 16, 1998, is highlighted in gray. Average yields for the de-trended samples are zero. Both tests reject the hypothesis of identical volatility for the pre-event and the post-event sample. Boldfaced is F-statistic rejecting equal variance of samples at 1%.

TABLE 3.2   Chow Breakpoint Statistic for the Yields of Minfin 10% and Minfin 9.25%

Breakpoint (Days)	Minfin 10% Yield Statistic/ Residual Statistic	Probability 1	Minfin 9.25% Yield Statistic/ Residual Statistic	Probability 2
25	0.22/0.61	0.638/0.436	5.38*/3.25	0.023*/0.075
50	0.38/0.01	0.540/0.938	0.77/0.00	0.382/0.987
75	**8.66/6.68***	**0.004/0.011***	**9.71/14.04**	**0.002/0.0003**

Note: Chow statistic for yields and OLS de-trended yields indicates a breakpoint between 50 and 75 days into the 100-day sample. Boldfaced is Chow statistic rejecting the absence of a breakpoint at 1%. Asterisk marks the Chow statistic, which rejects breakpoint at 5%.

possible change in Russian economic fundamentals around the New Year, unrelated to microstructure phenomena. None of the original de-trending OLS regressions on a calendar dummy are significant by themselves ($R^2 \approx 0$), which confirms the absence of fundamental events around the New Year. Furthermore, the sample residual averages differ insignificantly from zero, which indicates successful de-trending. Yet, our results demonstrate a significant (Prob(Null) < 1%) decline of volatility between the pre-event and the post-event samples for both the original and the de-trended sample (Table 3.1).

In another test, we show the Chow statistic (see Greene, 2000) for the breakpoints 25, 50, and 75 days into the 100-day sample, which we perform for both the yields and the first-stage OLS residuals (Table 3.2). The results indicate a statistically significant breakpoint between 50 and 75 days into the sampled period. We view the existence of a structural break as additional confirmation of microstructure-induced regime change. Both above tests confirm that the pre-event and post-event samples correspond to a significant decline in price volatility.

In the case, our OLS de-trending was not sophisticated enough to detect macroeconomic changes unrelated to the change of quote on January 1, 1998; we tested GARCH volatility of the three 46-day samples for the same bonds, chosen as in Figure 3.7. To measure potential change in asymptotic volatility, we use as a proxy the unconditional volatility σ_∞, which is defined by the equation (Tsay, 2002)

$$\sigma_\infty^2 = \alpha_0/(1 - \beta_0 - \gamma_0) \qquad (3.8)$$

In Equation (3.8), α_0, β_0, and γ_0 are the parameters of GARCH(1,1) approximation. Test results are given in Table 3.3. For five out of six

TABLE 3.3 GARCH Statistics

Samples	Yield	$\alpha \times 10^4$	β	γ	Implied σ_∞
Minfin 10% (1)	0.115 (89.5)	0.179 (1.93)	**0.975 (2.82)**	0.00 (−0.00)	6.00%
Minfin 10% (2)	0.117 (203)	**0.074 (2.94)**	0.430 (1.60)	0.00 (−0.00)	1.09%
Minfin 10% (3)	0.112 (197)	0.010 (0.68)	0.671 (1.50)	0.280 (0.72)	1.03%
Minfin 9 ¼% (1)	0.099 (106)	**0.111 (2.06)**	**0.982 (2.79)**	0.00 (−0.00)	5.59%
Minfin 9 ¼% (2)	0.105 (278)	**0.034 (2.02)**	0.833 (1.81)	0.046 (0.29)	1.18%
Minfin 9 ¼% (3)	0.105 (255)	0.025 (1.44)	**0.868 (2.91)**	0.107 (0.39)	2.25%

Note: Parameters for GARCH(1,1) unconditional volatility of the yields on two Russian Minfin bonds (see description to Table 3.1) with Student t in parentheses. The tendency of volatility to decrease after December 31, 1997 serves as an indicator of diminishing volatility as the effect of change in quotation. Nonintercept coefficients significant at 5% are boldfaced.

samples, yields between samples do not change significantly, which makes an explanation of an abrupt change in fundamentals highly unlikely. For both bonds, Table 3.3 shows decline in volatility after the changeover, yet as a rule, unconditional GARCH volatility is not statistically significant.

In Figure 3.7, we show the conditional GARCH variance of the three 46-day samples. The first sample precedes the event by 46 days, the second

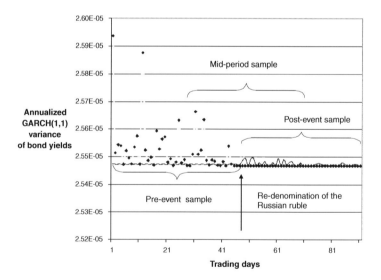

FIGURE 3.7 GARCH(1,1) volatility of the two Russian bonds of the three 46-day samples. Light gray shading indicates pre-event, gray shading indicates mid-term, and dark shading indicates post-event samples, respectively.

includes 23 trading days before the event (mid-period, or mid-term) and 23 trading days after the changeover, and the third follows the event (post-event sample). The change in volatility becomes immediately visible.

3.7 CONCLUSION

Discrete tick size, or any other friction that is positively correlated with the spread, reduces market volatility when compared to the frictionless case. Because price volatility, in the long run, is determined by informed traders, it might change due to regulatory or other noneconomic events, because they accelerate or impede dissemination of insider information through limit orders.

We expressed price volatility through volatility for an underlying asset and microstructure parameters: participation rates of insider and liquidity traders. Equilibrium participation of insiders is governed by our Proposition 1, which can be evaluated analytically for any closed-form solution for the asset price process.

If there were no information asymmetry, a higher-liquidity premium extracted by a market maker (bid-ask spread) would not influence volatility because everyone has the same information concerning orders. In a frictionless market with asymmetry, insiders have an incentive to trade at every opportunity, because they can always use their superior knowledge to benefit at the expense of liquidity traders. Again, they would not change their trading practices because of noneconomic events. In the market with frictions, however, the insiders will trade only if they perceive the momentum in asset prices to be large enough to justify their rents paid to a market maker. Changes in insider participation affect price volatility in the same direction as the changes in bid-ask spread, but in a highly nonlinear fashion.

We observed declining volatility of Russian Federation bonds around the 1,000-fold denomination of the Russian currency. Emerging markets are especially good candidates for testing microstructure events because the market is shallow and is dominated by sophisticated traders. Our results support the contention that an increase in bid-ask spreads after the redenomination of the Russian ruble reduced informed traders' activity and lowered return volatility.

An increase in *observed* bid-ask spread because of quotation tick is an important consequence of the lack of possibility, for the traders, to observe limit orders in their entirety. One can only hypothesize what would happen, if an exchange were to adopt immediate public disclosure of all limit orders. If our last statement is of any guidance, the adverse selection component of the spread will (almost) disappear, but at the expense of drastically increased volatility.

REFERENCES

Bessembinder, H. (1999). Trade execution costs on NASDAQ and the NYSE: A post-reform comparison. *Journal of Financial and Quantitative Analysis* 34:387–407.

Boehmer, E., R. Jennings and L. Wei. (2007). Public disclosure and private decisions: Equity market execution quality and order routing. *Review of Financial Studies* 20:315–58.

Bollen, N., R. Whaley, and T. Smith. (2004). Modeling the bid/ask spread: Measuring the inventory-holding premium. *Journal of Financial Economics* 72:97–141.

Caraiscos, C., and B. Liu. (1984). A round-off error analysis of the LMS adaptive algorithm. *IEEE Transactions on Acoustics, Speech and Signal Processing* 32:34–41.

Chakravarty, S., R. Wood, and R. Van Ness. (2004). Decimals and liquidity: A study of the NYSE. *Journal of Financial Research* 27:75–94.

Dittmar, R., and K. Yuan. (2005). Pricing impact of sovereign bonds. Working paper, University of Michigan, Ann Arbor.

Easley, D., N. Kiefer, M. O'Hara, and J. Paperman. (1996). Liquidity, information, and infrequently traded stocks. *Journal of Finance* 51:1405–36.

Gaidar, E. T., ed. (2003). *The economics of Russian transition*. Cambridge, MA: MIT Press.

Gibson, S., R. Singh, and V. Yeramilli. (2003). The effect of decimalization on the components of the bid-ask spread. *Journal of Financial Intermediation* 12:121–48.

Glosten, L., and P. Milgrom. (1985). Bid, ask and transaction prices in a specialist market with heterogeneously informed traders. *Journal of Financial Economics* 14:71–100.

Greene, W. H. (2000). *Econometric analysis*. Englewood Cliffs, NJ: Prentice Hall.

Hasbrouck, J. (2007). *Empirical market microstructure*. New York: Oxford University Press.

Hull, J. C. (1997). *Options, futures and other derivatives*. Englewood Cliffs, NJ: Prentice Hall.

Kyle, A. S. (1985). Continuous auctions and insider trading. *Econometrica* 53:1315–36.

Lerner, P. (2007). Review of several hypotheses in market microstructure. Working paper, Syracuse University, Syracuse, NY.

Lerner, P., and C. Wu. (2005). Statistical properties of an informed trader model. Working paper, Syracuse University, Syracuse, NY.

Ronen, T., and D. Weaver. (2001). "Teenies" anyone? *Journal of Financial Markets* 4:231–60.

Stoll, H. R. (2003). Market microstructure. In *Handbook of the economics of finance*, ed. G. M. Constantinides, M. Harris, and R. Stultz. Vol. 1A. Amsterdam: Elsevier.

Strother, T. S., J. Wansley, and P. Davis. (2002). The impact of electronic communications networks on the bid-ask spread. Working paper, University of Tennessee, Knoxville, TN.

Tsay, R. (2002). *Analysis of financial time series*. New York: Wiley Interscience Publications.

CHAPTER 4

GARCH Modeling of Stock Market Volatility

Rachael Carroll and Colm Kearney

CONTENTS

4.1 INTRODUCTION

The autoregressive conditional heteroskedastic (ARCH) model was first introduced by Engle (1982) to examine how the variance of inflation evolves over time, and it was quickly extended to the generalized ARCH (GARCH) model by Bollerslev (1986). During the past two decades, the GARCH model has become an important component in the tool kit of business, economic, and financial analysts, researchers, and policy makers. The World Wide Web reveals that over 115,000 articles refer to or use GARCH modeling techniques, over 3,000 articles use the term *GARCH* in their titles, and over 30 articles about GARCH have been cited more than 100 times. Within the finance literature, GARCH modeling is central to asset and derivative pricing, investment analysis, and risk management, and it is now the standard methodology in modeling the causes, transmission, and effects of stock market volatility. The success of the GARCH methodology stems from its parsimonious representation of conditional variance in a manner that is consistent with the stylized facts of many financial time series, such as nonnormality of conditional densities, persistence in variance, and volatility clustering.

Previous authors have surveyed the econometric theory of GARCH (Engle, 1991; Bera and Higgins, 1993; Bollerslev et al., 1994; Bauwens et al., 2006; Teräsvirta, 2006; Silvennoinen and Teräsvirta, 2007) and their applications in financial markets (Bollerslev et al., 1992). In this chapter, we review the most commonly used univariate and multivariate GARCH models for stock market volatility. In so doing, we provide an easy-to-read, notationally consistent, and logically structured description of the most popular variants of the models, to guide financial analysts and researchers through the literature and help with the selection of the most appropriate variants for particular contexts. We begin in Section 4.2 by summarizing the essential characteristics of stock return volatilities that GARCH models should ideally capture. This serves as the benchmark to describe the basic univariate GARCH model and its many extensions. In Section 4.3, we review the important multivariate GARCH models. In Section 4.4, we discuss the most commonly used estimation procedures, and in Section 4.5 we review their applications in modeling stock market volatility. Our summary and conclusions are presented in Section 4.6.

4.2 UNIVARIATE GARCH MODELS

The important characteristics of stock return volatility are described below, many of which have been discussed by Bollerslev et al. (1994) and others. The GARCH models that we describe in this section have been designed to capture particular sets of these characteristics.

1. *Nonnormal returns:* Return distributions tend to be leptokurtic with fat tails and excess peakedness at the means relative to the normal distribution.

2. *Mean reverting volatility:* When volatility is disturbed, it tends to return to its normal level, which may itself vary over time.

3. *Volatility clustering:* Large (small) changes in returns tend to be followed by large (small) changes of either sign.

4. *Co-movements in volatilities:* Volatilities within and across stock markets tend to move together in response to common underlying factors.

5. *Volatility and serial correlation:* Volatility and the serial correlation of returns tend to be negatively correlated.

6. *The leverage effect:* Changes in stock prices tend to be negatively correlated with changes in volatility, so that volatility tends to rise more following large price declines rather than increases of the same magnitude.

7. *The risk premium:* Riskier stocks with greater variance in returns tend to have higher rates of return.

8. *Nontrading periods:* Information accumulates slower when markets are closed than when they are open; it accumulates when they are closed, and tends to be reflected in prices when they reopen. Return variances tend to be greater following weekends and holidays.

9. *Forecastable events:* Anticipated releases of public information and earnings announcements are associated with *exante* volatility.

10. *Macroeconomic variables and volatility:* Macroeconomic uncertainty causes stock market volatility.

GARCH processes have zero mean and are serially uncorrelated with nonconstant variances conditional on the past, but with constant unconditional variances. The dependent variable, y_t, is expressed in terms of ψ_t, the information set available at time t, with h_t denoting the conditional variance. The error or innovation term can be specified as

$$\varepsilon_t = y_t - E\{y_t \mid \psi_{t-1}\} \tag{4.1}$$

where ε_t is a random, unobservable variable with mean and variance conditional on ψ_t. The GARCH model for ε_t has $E\{\varepsilon_t \mid \psi_{t-1}\} = 0$ and $h_t = E\{\varepsilon_t^2 \mid \psi_{t-1}\}$ and is decomposed as

$$\varepsilon_t = z_t h_t^{1/2} \tag{4.2}$$

The sequence $\{z_t\}$ is an independent, identically distributed sequence of random variables with mean zero and unit variance. The GARCH model allows the conditional variance to depend on its own lags and lags of the squared error terms. The GARCH(p,q) model can be written as

$$h_t = \alpha_0 + \sum_{i=1}^{q} \alpha_i \varepsilon_{t-i}^2 + \sum_{j=1}^{p} \beta_j h_{t-j} \tag{4.3}$$

The conditional variance, h_t, is a weighted function of its long-run value (dependent on α_0), information about volatility during previous periods, $\alpha_i \varepsilon_{t-i}^2$, and the fitted variance from previous periods, $\beta_j h_{t-j}$. The model is subject to nonnegativity constraints, $\alpha_0 > 0, \alpha_i \geq 0$ for $i = 1,\ldots, q$ and $\beta_j \geq 0$ for $j = 1,\ldots, p$, to ensure that the variance is strictly positive. The log-likelihood function is

$$L = -\frac{T}{2}\log(2\pi) - \frac{1}{2}\sum_{t=1}^{T}\log(h_t) - \frac{1}{2}\sum_{t=1}^{T}\frac{\varepsilon_t^2}{h_t} \tag{4.4}$$

Engle (1982) derived procedures for maximizing the likelihood function. As α and β are asymptotically independent, the likelihood can be maximized separately.

The basic GARCH model described above has many appealing features that have secured its popularity and usefulness. For example, it can parsimoniously capture leptokurtosis, volatility clustering, nontrading periods,

TABLE 4.1   Univariate GARCH Models

Model	Equation	d	I	α	γ
GARCH	4.3	2	0	Free	0
GJR-GARCH	4.12	2	0	$\alpha_i(1+\gamma)^2$	$-4\alpha_i\gamma_i$
TGARCH	4.14	1	0	Free	0
GTARCH	4.13	1	0	Free	$|\gamma_i| \leq 1$
NGARCH	4.15	Free	0	Free	0
PGARCH	4.16	Free	0	Free	0
APGARCH	4.16	Free	0	Free	$|\gamma_i| \leq 1$
IGARCH	4.10	2	1	Free	0
FIGARCH	4.10	2	$|I| \leq 1$	Free	0

Note: The models in column 1 are described in the text. The "Equation" column gives the equation number, d is the power term, I is the order of integration, and α and γ are parameters.

forecastable events, and the relation between macroeconomic uncertainty and stock market volatility. It has, however, three main limitations. First, it is restricted by nonnegativity constraints that necessitate the imposition of artificial constraints. Second, it cannot capture the leverage effect because the conditional variance depends on the magnitude of lagged residuals and not their signs. Third, it does not allow for feedback from the conditional variance and the conditional mean. The many extensions of the basic GARCH model have been proposed to overcome these limitations. Table 4.1 summarizes the main extensions of the univariate GARCH model that we discuss here.

4.2.1 Nonnormal Conditional Distribution

The unconditional distribution corresponding to the GARCH(p,q) model with conditionally normal errors is leptokurtic, but it is not clear whether the model sufficiently accounts for the observed leptokurtosis in financial time series. Bollerslev (1987) proposed a GARCH(p,q) model with conditionally t-distributed errors and found that the kurtosis of the standardized residuals approximates that of a t-distribution. As the degrees of freedom approach infinity, the t-distribution approaches the normal, but the t-distribution allows for heavier tails. Hansen (1994) proposed the autoregressive conditional density (ARCD) model to allow for both time-varying skewness and kurtosis.

4.2.2 GARCH-in-Mean

Engle et al. (1987) proposed the ARCH-in-mean (ARCH-M) model for estimating time-varying risk premiums with time-varying variances. The GARCH-M version of this model is more commonly used, and is specified as

$$y_t = \mu + \delta\sqrt{h_{t-1}} + \varepsilon_t \tag{4.5}$$

$$h_t = \alpha_0 + \alpha_1\varepsilon_{t-1}^2 + \beta h_{t-1} \tag{4.6}$$

When δ is significantly positive, a higher conditional variance leads to a rise in the mean return. The δ term represents the risk premium, and it captures the stylized fact that stocks with greater variance in their returns tend to have higher mean rates of return.

4.2.3 Exponential GARCH

Nelson (1991) proposed the exponential GARCH (EGARCH) model to allow the conditional variance to depend on both the size and sign of the lagged residuals. Its conditional variance is

$$\ln(h_t) = \alpha_0 + \beta\ln(h_{t-1}) + \gamma\frac{\varepsilon_{t-1}}{\sqrt{h_{t-1}}} + \alpha\left[\frac{\varepsilon_{t-1}}{\sqrt{h_{t-1}}} - \sqrt{\frac{2}{\pi}}\right] \tag{4.7}$$

The logarithmic specification of the EGARCH model ensures the conditional variance is always positive without imposing nonnegativity constraints. It therefore overcomes the first limitation of the basic GARCH model. The β term captures the effect of prior variance terms on the current conditional variance, and the γ term captures the sign of the lagged error term. The EGARCH model also allows asymmetries. If there is a negative relation between returns and volatility, γ will be negative. The absolute value of the standardized error terms, $\frac{\varepsilon_{t-1}}{\sqrt{h_{t-1}}}$, have an expected value $\sqrt{\frac{2}{\pi}}$ assuming the standardized errors are distributed as a $N(0,1)$. If the absolute standardized errors are greater (less) than expected, the conditional variance will rise (fall). Hence, the fourth term in the model captures the magnitude of the lagged error terms.

4.2.4 Integrated GARCH

Engle and Bollerslev (1986) introduced the integrated GARCH (IGARCH) model for cases in which the multistep forecasts of variance do not approach the unconditional variance. Consider the GARCH(p,q) model:

$$h_t = \alpha_0 + \sum_{i=1}^{q} \alpha_i \varepsilon_{t-i}^2 + \sum_{j=1}^{p} \beta_j h_{t-j} \qquad (4.8)$$

where $\alpha_0 > 0, \alpha_i \geq 0$ and $\beta_j \geq 0$ $\forall$ i and j, and the polynomial $1 - \sum_{i=1}^{q} \alpha_i z^i - \sum_{j=1}^{p} \beta_j z^j = 0$ has $I > 0$ unit roots and max$\{p,q\} - I$ roots outside the unit circle. Engle and Bollerslev (1986) describe this as integrated in variance of order d if $\alpha_0 = 0$, and integrated in variance of order I with trend if $\alpha_0 > 0$. For the GARCH(p,q) model to be integrated in variance, the α_i and the β_i values must sum to 1. IGARCH models are persistent in variance because current information remains important for forecasts of the conditional variance for all horizons. Given the IGARCH (1,1) model $h_t = \alpha \varepsilon_t^2 + (1-\alpha)h_{t-1}$, where $0 < \alpha < 1$, it follows that $E_t(h_{t+s}) = h_{t+1}$ and the conditional variance one step ahead is the same as the conditional variance s steps ahead, so today's information remains important and shocks to the system have permanent effects.

The concept of persistence in variance is more complex than persistence in the mean for linear models, because even strictly stationary ARCH models do not always possess finite moments. Chou (1988) showed that temporal aggregation reduces measured persistence in GARCH models. Lamoureux and Lastrapes (1990) suggested that the apparent appearance of high persistence in variance could be due to time-varying parameters, and they proposed a variation of the model that allows for structural shifts in the unconditional variance. Baillie et al. (1996) introduced the fractionally integrated GARCH (FIGARCH) model to encompass the possibility of persistent but not necessarily permanent shocks to volatility. In contrast to the GARCH and IGARCH models, where shocks to the conditional variance either dissipate exponentially or persist indefinitely, the FIGARCH model allows I to take a value between 0 and 1, so the response of the conditional variance to past shocks can decay slowly.

Bollerslev and Mikkelsen (1996) extended this model to a fractionally integrated exponential ARCH (FIEGARCH) model. Writing the GARCH(p,q) model as

$$h_t = \omega + \alpha(L)\varepsilon_t^2 + \beta(L)h_t \qquad (4.9)$$

where L is the backshift operator, Equation (4.9) can be written as an infinite-order ARCH(p) process, $\phi(L)(1-L)\varepsilon_t^2 = \omega + [1-\beta(L)](\varepsilon_t^2 - h_t)$ and $\phi(L) = [1-\alpha(L)-\beta(L)]$. The IGARCH and FIGARCH models can then be written as

$$\phi(L)(1-L)^I \varepsilon_t^2 = \omega + [1-\beta(L)]\left(\varepsilon_t^2 - h_t\right) \tag{4.10}$$

For the GARCH model $I = 0$, for IGARCH $I = 1$, and for FIGARCH $0 < I < 1$.

4.2.5 Other Univariate GARCH Models

The quadratic ARCH (QARCH) model introduced by Sentana (1995) can be interpreted as a second-order Taylor approximation to the conditional variance, or as the quadratic projection of the square innovation on the information set. The conditional variance takes the following form:

$$h_t = \alpha_0 + \alpha_1 \varepsilon_{t-1}^2 + \beta h_{t-1} + \gamma \varepsilon_{t-1} \tag{4.11}$$

The GJR-GARCH model proposed by Glosten et al. (1993) also allows for asymmetry in the GARCH process. The conditional variance of this model is expressed as

$$h_t = \alpha_0 + \alpha_1 \varepsilon_{t-1}^2 + \beta h_{t-1} + \gamma \varepsilon_{t-1}^2 D_{t-1}^- \tag{4.12}$$

where $D_{t-1}^- = 1$ if $\varepsilon_{t-1} < 0$, and $D_{t-1}^- = 0$ if $\varepsilon_{t-1} \geq 0$.

The threshold ARCH (TARCH) model proposed by Zakoian (1994) is similar in structure to the GJR-GARCH, but it models the conditional standard deviation instead of the conditional variance. The conditional standard deviation is given by

$$\sqrt{h_t} = \alpha_0 + \alpha_1 |\varepsilon_{t-1}| + \beta_1 \sqrt{h_{t-1}} + \gamma \varepsilon_{t-1} D_{t-1} \tag{4.13}$$

where $D_{t-1} = 1$ if $\varepsilon_{t-1} < 0$, and $D_{t-1} = 0$ if $\varepsilon_{t-1} \geq 0$.

Taylor (1986) introduced a class of GARCH models that relates the conditional standard deviation of a series to lagged absolute residuals and past standard deviations, where the conditional standard deviation can be specified as

$$\sqrt{h_t} = \alpha_0 + \alpha_1 |\varepsilon_{t-1}|^d + \beta_1 \sqrt{h_{t-1}} \tag{4.14}$$

Higgins and Bera (1992) proposed the nonlinear ARCH (NARCH) model, which can be extended to a nonlinear GARCH model in which $h_t = \sigma^2$, the second moment, can be expressed as

$$\sigma_t^d = \alpha_0 + \alpha_1 |\varepsilon_{t-1}|^d + \beta \sigma_{t-1}^d \tag{4.15}$$

4.2.6 Power GARCH

Ding et al. (1993) proposed the asymmetric power GARCH (APGARCH) model to estimate the optimal power term when the second moment can be specified as

$$\sigma_t^d = \partial_0 + \sum_{i=1}^{q} \alpha_i (|\varepsilon_{t-i}| + \gamma_i \varepsilon_{t-i})^d + \sum_{j=1}^{p} \beta_j \sigma_{t-j}^d \tag{4.16}$$

The power term, d, captures the conditional standard deviation when $d = 1$ and the conditional variance when $d = 2$. Asymmetry is captured by the γ term. The NGARCH model is an APARCH model without the leverage effect. Ding et al. (1993) and Hentschel (1995) show that the APGARCH model nests several other GARCH models by specifying the permissible values for α, β, γ, and d. Standard GARCH models impose a squared term in the second-moment equation. The Taylor (1986) class of GARCH models specifies a power term if $d = 1$. Any positive value can be used to specify the second-moment equation. Brooks et al. (2000) explain that this is because of volatility clustering, and the inclusion of the power term accentuates periods of relative tranquility and volatility by magnifying the outliers. The squared term is particularly suitable when the data are normally distributed, because in this case the distribution can be fully characterized by its first two moments and the squared term reflects the assumption of normality applied to the data. When the data are not normally distributed, higher-order moments need to be considered to adequately describe it, the superiority of the squared term is lost, and other power transformations might be appropriate. The significance of the restrictions required to nest these models can be tested with the likelihood ratio procedure.

TABLE 4.2   Multivariate GARCH Models

Model	Description	Parameters
VEC (p,q)	Stacks lower triangular elements of matrix as a vector	$N(N + 1)(N(N + 1) + 1)2$
DVEC (p,q)	VEC matrices are diagonal	$N(N + 5)/2$
BEKK (p,q,k)	Matrix is positive definite by construction	$N(5N + 1)/2$
Factor GARCH (p,q,k)	Observations are generated by common factors that may be correlated	$N(N + 5)/2$
OGARCH (p,q,m)	Observations are linearly transformed into uncorrelated components by means of orthogonal matrix	$N(N + 5)/2$
CCC	Conditional correlations are constant	$N(N + 5)/2$
DCC	Conditional correlations are time dependent	$(N + 1)(N + 4)/2$
GDC	Captures asymmetry and encompasses DVEC, BEKK, FGARCH, CCC, and DCC	$[N(7N - 1) + 4]/2$

Note: The number of parameters estimated assumes that $p = q = k = m = 1$ and the conditional variances are specified by a GARCH(1,1) for the CCC, DCC, and GDC.

4.3   MULTIVARIATE GARCH MODELS

Financial time series are usually interrelated, and the multivariate GARCH model caters for this by specifying equations for how the covariances and correlations between a number of variables move over time. There are three main types of multivariate GARCH models. The first type models the conditional covariance matrix directly, and includes the vectorized[*] (VEC) and diagonal VEC models of Bollerslev et al. (1988), and the BEKK model of Engle and Kroner (1995). The second type models a parsimonious representation of the covariance matrix and includes the factor GARCH (FGARCH) model of Engle et al. (1990) and the orthogonal GARCH (OGARCH) model of Alexander and Chibumba (1997). The third type models the conditional covariance matrix directly and includes the constant conditional correlation (CCC) model of Bollerslev (1990), the generalized dynamic covariance (GDC) model of Kroner and Ng (1998), and the dynamic conditional correlation (DCC) models of Engle (2002) and Tse and Tsui (2002). Table 4.2 summarizes the main features of these

---

[*] See (Bollerslev, 2008) for a glossary of GARCH

models, which encounter three main challenges. The first is identifying suitable conditions to ensure the variance-covariance matrix, H_t, is positive definite. The second is that identifying conditions for the weak stationarity of the process can be difficult, and the third is that the models' high dimensionality can make them infeasible to estimate. The description here follows the notation used by Bauwens et al. (2006).

4.3.1 VEC-GARCH

If the time series $y_t = (y_{1t}, \ldots, y_{Nt})'$ is an $N \times 1$ vector, it can be expressed as a multivariate GARCH model in the general form

$$y_t \mid \psi \sim N(\mu_t, H_t) \tag{4.17}$$

where μ_t is an $N \times 1$ vector and H_t is an $N \times N$ conditional variance-covariance matrix. The VEC specification of Bollerslev et al. (1988) is applied to the upper or lower triangular elements of a symmetric matrix that stacks each element into a vector with a single column. H_t can be written as

$$vech(H_t) = C + \sum_{i=1}^{q} A_i vech(\varepsilon_{t-i} \varepsilon'_{t-i}) + \sum_{j=1}^{p} B_j vech(H_{t-j}) \tag{4.18}$$

where $\varepsilon \mid \psi_{t-1} \sim N(0, H_t)$, where $\varepsilon = (\varepsilon_{1t}, \ldots, \varepsilon_{Nt})'$ is the $N \times 1$ innovation vector, C is an $N(N + 1)/2 \times 1$ vector, and A_i and B_i are $N(N + 1)/2 \times N(N + 1)/2$ matrices. In the VEC model, each element of the H_t matrix depends on the lagged squared residuals and past variances of all variables in the model as in Equation (4.18). The VEC model is very flexible, but it requires restrictive conditions for H_t to be positive definite for all t, and the number of estimated parameters is large. For example, the simplest bivariate model requires the estimation of twenty-one parameters.

4.3.2 Diagonal VEC GARCH

To economize on the number of parameters requiring estimation in the VEC model, Bollerslev et al. (1988) simplified their VEC model to the diagonal VEC. This reduces the number of parameters by allowing the conditional variance to depend only on its own lagged squared residuals and lagged values. The A_i and B_i matrices become diagonal, and only nine parameters need to be estimated in the bivariate case. With $p = q = 1$, the diagonal VEC is

$$H_t = C^\bullet + A^\bullet \circ (\varepsilon_{t-1} \varepsilon'_{t-1}) + B^\bullet \circ H_{t-1} \tag{4.19}$$

The symbol ∘ denotes the Hadamard product. The variance-covariance matrix has positive numbers on its leading diagonal and is symmetrical around this diagonal. Estimation is less onerous than the VEC model because each equation can be separately estimated. Bera and Higgins (1993) point out that the positive definiteness of the H_t matrix is difficult to impose during estimation and not easy to check, and that no interaction is captured between the different conditional variances and covariances.

4.3.3 BEKK GARCH

The BEKK model of Engle and Kroner (1995) ensures that the variance-covariance matrix is positive definite. It is described as

$$H_t = C^{*'}C^* + \sum_{k=1}^{K}\sum_{i=1}^{q} A_{ik}^{*'} \varepsilon_{t-i}\varepsilon_{t-i}' A_{ik}^* + \sum_{k=1}^{K}\sum_{j=1}^{p} B_{jk}^{*'} H_{t-j} B_{jk}^* \qquad (4.20)$$

where A_{ik}^*, B_{jk}^*, and C^* are $N \times N$ matrices but C^* is upper triangular. The BEKK model is a special case of the VEC model. If $C^{*'}C$ is positive definite, so is the H_t matrix. For the bivariate case, the BEKK model requires the estimation of eleven parameters.

4.3.4 Factor GARCH

Laloux et al. (1999) applied the theory of random matrices to the S&P500 stock returns and showed that since the majority of the eigenvalues cannot be distinguished from the eigenvalues of a random matrix, only a small number of the eigenvalues of the covariance matrix carry information. By imposing appropriate constraints on the matrix entries, the estimation of large covariance matrices becomes less noisy, and techniques such as factor analysis and principal components analysis become useful. The FGARCH model proposed by Engle et al. (1990) assumes that the co-movements of returns are driven by a small number of common factors with GARCH-type structures. H_t is assumed to be generated by K underlying factors. The FGARCH model takes the following form:

$$H_t = C^{*'}C^* + \sum_{k=1}^{K} \lambda_k \lambda_k' \left[\alpha_k (\omega_k' \varepsilon_{t-1})^2 + \beta_k \omega_k' H_{t-1} \omega_k \right] \qquad (4.21)$$

where λ and ω are $N \times 1$ vectors, and α and β are scalars. If $A = \sqrt{\alpha}\omega\lambda'$ and $B = \sqrt{\beta}\omega\lambda'$ the BEKK model is equivalent to the FGARCH model.

4.3.5 Orthogonal GARCH

The OGARCH model proposed by Alexander and Chibumba (1997) assumes that the observations are generated by an orthogonal transformation of N univariate GARCH models. The linear transformation matrix is the orthogonal matrix of eigenvectors of the unconditional covariance matrix of returns. The H_t matrix is generated by $m \leq N$ univariate GARCH models. The OGARCH(1,1,m) model is defined as

$$V^{-1/2}\varepsilon_t = u_t = \Lambda_m f_t \tag{4.22}$$

where $V = diag(v_1, v_2, \ldots, v_N)$, with v_i being the population variance of ε_{it}, Λ_m is the $N \times m$ matrix given by $\Lambda_m = P_m diag(l_1^{1/2} \ldots l_m^{1/2})$, $l_1 \geq \ldots \geq l_m > 0$ are the m largest eigenvalues of the correlation matrix of u_t, and P_m is the $N \times m$ matrix of associated orthogonal eigenvectors. The vector f_t is a random process where the variance of its components can be expressed as a GARCH model. The covariance matrix can be specified as

$$H_t = Var_{t-1}(\varepsilon_t) = V^{1/2} V_t V^{1/2} \tag{4.23}$$

4.3.6 Constant Conditional Correlation Model

Bollerslev (1990) proposed the CCC model with time-varying conditional variances and covariances, but with constant conditional correlations. The variances and covariances can be modeled separately using univariate models to allow different specifications. Based on these conditional variances, the conditional correlation matrix can subsequently be modeled. Assuming constant conditional correlations implies that the conditional covariances are proportional to the product of the corresponding conditional standard deviations, and this reduces the number of parameters to be estimated. The CCC model is defined as

$$H_t = D_t R D_t = \left(\rho_{ij} \sqrt{h_{iit} h_{jjt}} \right) \tag{4.24}$$

where $D_t = diag(h_{11t}^{1/2}, \ldots, h_{NNt}^{1/2})$, h_{iit} can be any univariate GARCH process, and $R = (\rho_{ij})$ is the constant correlation matrix, where $\rho_{ii} = 1, \forall i$. H_t is positive definite if all N conditional variances are well defined and R is positive definite.

4.3.7 Dynamic Conditional Correlation Model

Engle (2002) proposed the DCC model to combine the flexibility of univariate GARCH models with parsimonious parametric models for the correlations. Although nonlinear, they can often be estimated using univariate or two-step methods based on the likelihood function. The DCC model is written as

$$H_t = D_t R_t D_t \qquad (4.25)$$

where $D_t = diag\{\sqrt{h_{i,t}}\}$, $R_t = D_t^{-1} H_t D_t^{-1}$ is the correlation matrix,

$$R_t = diag\left(q_{11,t}^{-1/2} \ldots q_{NN,t}^{-1/2}\right) Q_t diag\left(q_{11,t}^{-1/2} \ldots q_{NN,t}^{-1/2}\right) \qquad (4.26)$$

the $N \times N$ symmetric positive definite matrix $Q_t = (q_{ij,t})$ is given by

$$Q_t = (1 - \alpha - \beta)\bar{Q} + \alpha u_{t-1} u'_{t-1} + \beta Q_{t-1} \qquad (4.27)$$

u_t are the standardized residuals, and $\bar{Q}$ is the $N \times N$ unconditional variance matrix of u_t. Relaxing the constraint of constant correlations is a very significant step forward, but it creates the difficulty that the time-dependent conditional correlation has to be positive definite. The DCC model guarantees this condition is satisfied.

4.3.8 Generalized Dynamic Covariance Model

Kroner and Ng (1998) proposed the GDC model to include asymmetric effects while nesting many other multivariate GARCH models as special cases. The GDC model is written as

$$H_t = D_t R_t D_t + \Phi \circ \Theta_t \qquad (4.28)$$

where $D_t = (d_{ijt})$, $d_{iit} = \sqrt{\theta_{iit}} \forall i$, $d_{ijt} = 0, \forall i \neq j$, and $\Theta = (\theta_{ijt})$, and θ_{ijt} can be specified as a BEKK model. The GDC model has two components. The first term, $D_t R_t D_t$, is similar to the CCC model, but the variance functions are given by the BEKK model. The second term, $\phi \circ \Theta_t$, has zero diagonal elements, but has off-diagonal elements given by the BEKK-type covariance functions, scaled by the ϕ_{ij} parameters. This model encompasses the VEC, BEKK, FGARCH, CCC, and DCC models. The asymmetric dynamic correlation (ADC) matrix model is an extension of the GDC model that permits asymmetric effects in both variances and covariances.

4.4   ESTIMATING GARCH MODELS

GARCH models are usually estimated using numerical procedures to maximize the likelihood function, which produces the most likely values of the parameters given the data. It is important to be aware that the likelihood function can have multiple local maxima, and different algorithms can lead to different parameter estimates and standard errors. Good initial estimates of the parameters are useful to ensure the global maximum is reached. It is also important to be aware that the log-likelihood function can be relatively flat in the region of its maximum value, and in this case different parameter values can lead to similar values of the likelihood function, making it difficult to select an appropriate value.

Most GARCH models are estimated using the Berndt-Hall-Hall-Hausman (BHHH) (1974) algorithm. This algorithm obtains the first derivatives of the likelihood function with respect to the numerically calculated parameters, and approximations to the second derivative are subsequently calculated. Computational speed is increased by not calculating the actual Hessian matrix at each iteration for each time step, but the approximation can be weak when the likelihood function is far from its maximum, thus requiring more iterations to reach the optimum.

The Broyden-Fletcher-Goldfarb-Shanno (BFGS) method solves unconstrained nonlinear optimization problems by calculating the likelihood function gradient in the same way as the BHHH, but it differs in its construction of the Hessian matrix of second derivatives. The BFGS and BHHH are asymptotically equivalent, but can lead to different estimates of the standard errors in small samples. Press et al. (1992) discuss optimization methods in detail. Brooks et al. (2003) review the software packages that are commonly used to estimate GARCH models—EVIEWS, GAUSS-FANPAC, RATS, and SAS—pointing out how different results can be obtained from the alternative packages.

4.5   APPLICATIONS TO STOCK MARKET VOLATILITY

Given the very extensive literature on GARCH modeling of stock market volatility, it is clearly impossible for us to provide a complete review. Instead, we provide a summary review of which models have been used in various contexts. It is appropriate to commence this review by noting that most applications of the GARCH(p,q) model use low orders for the lag lengths p and q, and the GARCH(1,1) is generally found to be the most appropriate for forecasting stock market returns. For example, Corhay and

Rad (1994) fitted various GARCH models to stock returns in European countries and found the GARCH(1,1) model is the most appropriate for forecasting returns. Engle (2004) describes the GARCH(1,1) model as the workhorse of financial applications and claims it can describe the volatility dynamics of stock returns on most developed and emerging markets and most indices of equity returns. In many cases, a slightly better model may be found, but the GARCH(1,1) model usually provides a good starting point.

The GARCH-M model has been used by French et al. (1987) to model the daily S&P index, by Attanasio and Wadhwani (1989) to model monthly and annual returns on UK and U.S. stock indices, and by Friedman and Kuttner (1992) on quarterly U.S. stock indices. They all find positive estimates of the risk aversion parameter with values ranging between 1 and 4.5. More recently, Tsouma (2007) investigated return dynamics in twenty-one mature and twenty emerging markets using an extended AR(1)-GARCH-M model. They show volatility transmission from the leading markets to the others, and allowing for potential structural breaks in mean and variance, they investigated the impact of the October 1997 East Asian financial crisis. The EGARCH model has been used by Nelson (1991) to examine the relation between the level of market risk and returns. In so doing, he showed how positive and negative returns affect conditional variance, how these effects can persist over time, and the implications of thick-tailed conditional distributions of returns. The IGARCH model has been applied to Canadian and Italian returns by Corhay and Rad (1994) and Calvet and Rahman (1995). Bollerslev and Mikkelsen (1996) argue that finding a unit root in variances could reflect restrictive specifications, and they report that U.S. stock market volatility is best modeled by a mean-reverting FIGARCH process.

Brooks et al. (2000) estimated an asymmetric power GARCH model for ten national stock market index returns and a world index. Most of the estimated power terms were between 1.0 and 1.5. They conclude that strong leverage effects are present, and when modeled in a GARCH framework, including a power term is a worthwhile addition to the specification of the model. Ané and Ureche-Rangau (2006) applied a regime switching power GARCH (RS-APGARCH) model to allow heteroskedasticity to vary across regimes together with within-regime volatility persistence. They report that the explosive variance often obtained with GARCH models might result from using single-regime models to capture multiregime processes. This model also allows within-regime asymmetric response to news, and although the leverage effect holds in both regimes, the asymmetric response to news is stronger in the low-volatility regime.

Multivariate GARCH models can be used to evaluate how returns in one stock market influence those in another. Hamao et al. (1990) used the multivariate GARCH model to demonstrate price and volatility spillovers between Japan and the United States, and found that shocks that originate in the United States are larger and more persistent. Karolyi (1995) examined the dynamic relation between Canadian and U.S. returns and volatilities using the BEKK and CCC models. Ramchand and Susmel (1998) used a bivariate switching ARCH (SWARCH) model to test for differences in correlations across variance regimes. They found that correlations between U.S. and other world markets are significantly greater when the U.S. market is in a high-variance rather than a low-variance regime. Ng (2000) constructed a bivariate GARCH(1,1) model that replicates the GDC model originally proposed by Kroner and Ng (1998) and that nests within it the VEC, BEKK, and CCC models. He found that U.S. shocks have larger effects than Japanese shocks on Asian stock markets. Kanas (2000) applied the bivariate EGARCH model to investigate volatility transmission between stock returns and exchange rates in six countries, and found symmetric spillovers from the former to the latter in all but one country. Kearney and Patton (2000) estimated three-, four-, and five-variable GARCH models of exchange rate volatility transmission across the important European Monetary System currencies. They demonstrated that temporal aggregation reduces observed volatility transmission, and that specification robustness checks should be integral to multivariate GARCH modeling. Engle (2002) applied the DCC model to investigate time-varying correlations between the Dow Jones and the NASDAQ and between stocks and bonds. He found that the DCC models are superior to moving average methods and competitive with other multivariate GARCH specifications. Yang (2005) also used the DCC model to examine international stock market correlations between Asian stock markets, showing that the correlations fluctuate over time and rise during periods of high market volatility.

4.6 CONCLUSION

In this chapter, we have summarily reviewed the many variations of univariate and multivariate GARCH models that have been applied to study the evolution of the first two moments of financial and other time series. The discovery that it was possible to formally model many of the stylized facts of stock market behavior constituted a major breakthrough in financial econometrics. The ability of the GARCH model in its many forms to encompass the important stylized facts of equity returns guarantees that it will remain central to modeling the causes and transmission of stock market volatility for the foreseeable future.

REFERENCES

Alexander, C., and Chibumba, A. (1997). Multivariate orthogonal factor GARCH. Mimeograph, University of Sussex.

Ané, T., and Ureche-Rangau, L. (2006). Stock market dynamics in a regime switching asymmetric power GARCH model. *International Review of Financial Analysis* 15:109–29.

Attanasio, O., and Wadhwani, S. (1989). Risk and the predictability of stock market returns. Manuscript, Stanford University, CA.

Baillie, R., Bollerslev, T., and Mikkelsen, H. (1996). Fractionally integrated generalized autoregressive conditional heteroskedasticity. *Journal of Econometrics* 74:3–30.

Bauwens, L., Laurent, S., and Rombouts, J. (2006). Multivariate Garch models: A survey. *Journal of Applied Econometrics* 21:79–109.

Bera, A., and Higgins, M. (1993). ARCH models: Properties, estimation and testing. *Journal of Economic Surveys* 7:305–62.

Berndt, F., Hall, B., Hall, R., and Hausman, J. (1974). Estimation and inference in nonlinear structural models. *Annals of Economic and Social Measurement* 4:653–65.

Bollerslev, T. (1986). Generalised autoregressive conditional heteroskedasticity. *Journal of Econometrics* 31:307–27.

Bollerslev, T. (1987). A conditionally heteroskedastic time series model for speculative prices and rates of return. *Review of Economics and Statistics* 69:542–47.

Bollerslev, T. (1990). Modelling the coherence in short-run nominal exchange rates: A multivariate generalized ARCH model. *Review of Economics and Statistics* 72:498–505.

Bollerslev, T. (2008). Glossary to ARCH (GARCH). Creates research papers, School of Economics and Management, University of Aarhus.

Bollerslev, T., Chou, R., and Kroner, K. (1992). ARCH modeling in finance. *Journal of Econometrics* 52:5–59.

Bollerslev, T., Engle, R., and Nelson, D. (1994). ARCH models. In *Handbook of econometrics*, ed. R. Engle and D. McFadden. Amsterdam: Elsevier Science, 4:2961–3038.

Bollerslev, T., Engle, R., and Wooldridge, J. (1988). A capital asset pricing model with time-varying covariances. *Journal of Political Economy* 96:116–31.

Bollerslev, T., and Mikkelsen, H. (1996). Modeling and pricing long-memory in stock market volatility. *Journal of Econometrics* 73:151–84.

Brooks, C., Burke, S., and Persand, G. (2003). Multivariate GARCH models: Software choice and estimation issues. *Journal of Applied Econometrics* 18:725–34.

Brooks, R., Faff, R., McKenzie, M., and Mitchell, H. (2000). A multi-country study of power ARCH models and national stock market returns. *Journal of International Money and Finance* 19:377–97.

Calvet, L., and Rahman, A. (1995). Persistence of stock return volatility in Canada. *Canadian Journal of Administrative Sciences* 12:224–37.

Chou, R. (1988). Persistent volatility and stock returns—Some empirical evidence using ARCH. *Journal of Applied Econometrics* 3:279–94.

Corhay, A., and Rad, A. (1994). Statistical properties of daily returns: Evidence from European stock markets. *Journal of Business Finance and Accounting* 21:271–82.

Ding, Z., Granger, C., and Engle, R. (1993). A long memory property of stock market returns and a new model. *Journal of Empirical Finance* 1:83–106.

Engle, R. (1982). Autoregressive conditional heteroscedasticity with estimates of the variance of United Kingdom inflation. *Econometrica* 50:987–1007.

Engle, R. (1991). Statistical models for financial volatility. Working paper, University of California, San Diego.

Engle, R. (2002). Dynamic conditional correlation: A simple class of multivariate generalized autoregressive conditional heteroskedasticity models. *Journal of Business and Economic Statistics* 20:339–50.

Engle, R. (2004). Risk and volatility: Econometric models and financial practice. *American Economic Review* 94:405–20.

Engle, R., and Bollerslev, T. (1986). Modelling the persistence of conditional variances. *Econometric Reviews* 5:1–50.

Engle, R., and Kroner, K. (1995). Multivariate simultaneous GARCH. *Econometric Theory* 11:122–50.

Engle, R., Lilien, D., and Robins, R. (1987). Estimating time varying risk premia in the term structure: THE ARCH-M model. *Econometrica* 55:391–407.

Engle, R., Ng, V., and Rothschild, M. (1990). Asset pricing with a factor ARCH covariance structure: Empirical estimates for treasury bills. *Journal of Econometrics* 45:213–37.

French, K., Schwert, W., and Stambaugh, R. (1987). Expected stock returns and volatility. *Journal of Financial Economics* 19:3–30.

Friedman, B., and Kuttner, K. (1992). Time-varying risk perceptions and the pricing of risky assets. *Oxford Economic Papers* 44:566–98.

Glosten, L., Jagannathan, R., and Runkle, D. (1993). On the relation between the expected value and the volatility of the nominal excess return on stocks. *Journal of Finance* 48:1779–801.

Hamao, Y., Masulis, R., and Ng, V. (1990). Correlation in price changes and volatility across international stock markets. *Review of Financial Studies* 3:281–307.

Hansen, B. (1994). Autoregressive conditional density estimation. *International Economic Review* 35:705–30.

Hentschel, L. (1995). All in the family: Nesting symmetric and asymmetric GARCH models. *Journal of Financial Economics* 39:71–104.

Higgins, M., and Bera, A. (1992). A class of nonlinear ARCH models. *International Economic Review* 33:137–58.

Kanas, A. (2000). Volatility spillovers between stock returns and exchange rate changes: International evidence. *Journal of Business, Finance and Accounting* 27:447–67.

Karolyi, G. (1995). A multivariate GARCH model of international transmissions of stock returns and volatility: The case of the United States and Canada. *Journal of Business and Economic Statistics* 13:11–25.

Kearney, C., and Patton, A. (2000). Multivariate GARCH modelling of exchange rate volatility transmission in the European monetary system. *Financial Review* 35:29–48.

Kroner, K., and Ng, V. (1998). Modeling asymmetric comovements of asset returns. *Review of Financial Studies* 11:817–44.

Laloux, L., Cizeau, P., Bouchaud, J., and Potters, M. (1999). Noise dressing of financial correlation matrices. *Physical Review Letters* 83:1467–89.

Lamoureux, C., and Lastrapes, W. (1990). Persistence in variance, structural change, and the GARCH model. *Journal of Business and Economic Statistics* 8:225–34.

Nelson, D. (1991). Conditional heteroscedasticity in asset returns: A new approach. *Econometrica* 59:347–70.

Ng, A. (2000). Volatility spillover effects from Japan and the US to the Pacific Basin. *Journal of International Money and Finance* 19:207–33.

Press, W., Teukolsky, S., Vetterling, W., and Flannery, B. (1992). *Numerical recipes in C: The art of scientific computing.* Cambridge, UK: Cambridge University Press.

Ramchand, L., and Susmel, R. (1998). Volatility and cross correlation across major stock markets. *Journal of Empirical Finance* 5:397–416.

Sentana, E. (1995). Quadratic ARCH models. *Review of Economic Studies* 62:639–61.

Silvennoinen, A., and Teräsvirta, T. (2007). Multivariate GARCH models. Working paper, University of Technology, Sydney.

Taylor, S. (1986). *Modelling financial time series.* New York: John Wiley & Sons.

Teräsvirta, T. (2006). An introduction to univariate GARCH models. Working paper, Stockholm School of Economics.

Tse, Y., and Tsui, A. (2002). A multivariate GARCH model with time-varying correlations. *Journal of Business and Economic Statistics* 20:351–62.

Tsouma, E. (2007). Stock return dynamics and stock market interdependencies. *Applied Financial Economics* 17:805–25.

Yang, S. (2005). A DCC analysis of international stock market correlations: The role of Japan on the Asian four tigers. *Applied Financial Economics Letters* 1:89–93.

Zakoian, J. (1994). Threshold heteroskedastic model. *Journal of Economic Dynamics and Control* 18:931–95.

CHAPTER 5

Detecting and Exploiting Regime Switching ARCH Dynamics in U.S. Stock and Bond Returns

Massimo Guidolin

CONTENTS

5.1 INTRODUCTION

It is now well established in the minds of investors, commentators, and academic researchers that financial markets follow up-and-down cycles that involve both mean and volatility of asset returns. Both the popular press and the empirical finance literature refer to the former phenomenon using catchy expressions, such as bull and bear markets, and to the latter phenomenon by writing about periods of financial turmoil to be contrasted to quiet times. Recently, many empirical finance researchers have also noticed that the correlations between returns on different assets often undergo massive changes, even shifting from negative to positive territory, i.e., from an average tendency of prices to move in opposite directions to a tendency to co-move.

As a result of this increasing awareness of the potential for means, variances, and covariances of asset returns to change over time, an ever-increasing and powerful array of econometric tools have been introduced that allow quantitative analysts to make inferences and predictions on the current and future means, variances, and covariances of asset returns. The first step in this direction was taken in the literature on time-varying volatility, a phenomenon commonly termed conditional heteroskedasticity (CH), a term borrowed from the statistical literature to indicate that the variances and covariances of the series of interest may change as a function of current information on the state of the economy or the financial markets. Since the seminal work by Robert Engle (see, e.g., Engle et al., 1987), we know that for most financial return series and frequencies, simple time-series models of the autoregressive moving average (ARMA) type may be used to successfully model and forecast time variation in financial volatility. In practice, this means that asset returns are much riskier at some times than others. As early as in the late 1980s, the literature on models of conditional variances has been extended to encompass multivariate applications in which ARMA models are adopted to describe and predict the dynamics of conditional covariances and hence correlations (see Bollerslev et al., 1988, for an early attempt).

During the 1990s another, different strand of the empirical finance literature developed that—initially borrowing ideas and techniques originally proposed in the macroeconomics literature by James Hamilton (see, e.g., Hamilton, 1989)—proposes that the dynamics over time of financial returns might be fruitfully modeled as mixtures (i.e., weighted combinations with weights represented by probability measures) of different but simpler conditional distribution. For instance, conditional normal densities with high mean returns and moderate variance would characterize the so-called bull markets, while conditional normal densities with low (or negative) mean returns and high variance should be used to characterize the bear states; the mixing would be governed by simple, finite-memory (Markov) unobservable state variables describing whether markets are in bull or bear conditions. Both bull and bear markets would be relatively persistent over time. After the seminal applications by Turner et al. (1989) to univariate contexts, the literature has more recently shifted toward multivariate applications (see, e.g., Ang and Bekaert, 2002; Guidolin and Timmermann, 2006) to find that Markov switching models may be helpful to understand the dynamics of markets and—so it is contended in some of these papers (see, e.g., Guidolin and Timmermann, 2007)—to time financial markets, allowing an investor to build portfolios that exploit the presence of (nonlinear) predictability patterns that Markov switching frameworks could reveal.

Our chapter proposes to bring these two strands of the literature together on an important financial application—modeling and predicting means, variances, and correlations for U.S. stock and bond returns—and proceeds to describe and estimate Markov switching (MS), vector autoregressive (VAR), autoregressive conditional heteroskedastic (ARCH) models for the bivariate conditional density of U.S. financial returns on post-WWII data. These means that we shall investigate in depth the empirical performance of dynamic time-series models that can be placed at the intersection of the two literatures we have reviewed—on multivariate ARCH models to capture volatility clustering and time-varying correlations, and on multivariate MS models to capture bull and bear market dynamics—to study a key portfolio choice problem, i.e., the strategic asset allocation between stocks and bonds for a domestic U.S. investor. In fact, our plan is to proceed to review some basic stylized facts of both data series (i.e., stock and long-term bond monthly returns) under investigation at the univariate as well as bivariate level, before gearing up to specify and estimate a variety of MS VAR ARCH models.

One small literature exists that has proposed and estimated MS ARCH models before. In fact, although generalized ARCH (GARCH) models driven by normally distributed innovations and their numerous extensions can account for substantial portions of both volatility clustering and the excess kurtosis in financial returns, GARCH-type models are usually unable to produce filtered residuals (i.e., residuals that discount time variation in volatility and covariances) that fail to exhibit clear-cut signs of nonnormality. At the same time, it has been observed that especially with reference to acute crises periods, GARCH models would display less than commendable forecasting performance (e.g., Lamoureux and Lastrapes, 1993). As a result, a number of researchers have suggested that this lack of performance of standard GARCH models may be related to the presence of structural instability in the ARCH process. Hamilton and Susmel (1994) stress that ARCH models often impute a lot of persistence to stock volatility and yet give relatively poor forecasts. One explanation is that extremely large shocks, such as the October 1987 crash, may have arisen from quite different causes and have different consequences for subsequent volatility than do small shocks. As a result, they propose and develop a *univariate* (Markov) switching ARCH model that separates out high- from low-variance periods.[*]

We depart from the existing literature in two ways. First, to our knowledge this is a first attempt at understanding and forecasting the dynamic properties of U.S. stock and long-term bond returns using models in the MS VAR ARCH class. Given that the recent interest in the literature for the strategic asset allocation decision across stocks and bonds (see, e.g., Campbell et al., 2003; Guidolin and Timmermann, 2007) has been centered around simpler, relatively unsophisticated VAR and MS VAR models, this seems to be an interesting effort. Second, we do not limit our efforts to a plain analysis of in- and out-of-sample properties of MS VAR ARCH models, but we also proceed to quantify the value of modeling and predicting regimes in ARCH dynamics by using a simple mean-variance portfolio problem that allows us to compute and report a measure of economic value. To our knowledge, both contributions are novel.

We obtain two main empirical results. First, we report strong and unequivocal evidence that regime switching in CH dynamics ought to be carefully modeled to obtain a good fit to U.S. excess stock and bond returns. We find evidence of three separate CH regimes. Regime 1 is a mildly persistent state in which bond premia are relatively high but equity premia are negative; the volatility of both excess stock and bond returns is high, while their correlation is zero; and the ARCH process is rather persistent for excess bond returns, weaker for excess stock returns, and practically absent in the covariance. Regime 2 is a persistent state that captures bull markets and periods of economic expansions: the equity premium is high, bond premia are negative, while excess bond returns become unpredictable, and excess equity returns are predictable using

---

[*] Recently, Haas et al. (2004a, 2004b) proposed finite mixtures of conditional distributions—also extended to include MS GARCH models—that appear flexible enough to include both normal and nonnormal (with thicker tails) distribution for the innovations terms. However, their focus has remained mostly of a univariate type.

lags of excess bond returns. Volatility almost completely evaporates, while the stock-bond correlation turns negative. Regime 3 can be interpreted as a normal, highly persistent state in which the equity premium is positive and statistically significant, the bond premium is positive but modest, excess stock and bond returns are hardly predictable, and ARCH effects are moderate. When we test for the presence of leverage effects in CH, the null of no leverage cannot be rejected in a bivariate setup.

Second, we find that a three-state t-Student ARCH model provides a superior forecasting performance than a number of natural competitors (a single-state bivariate t-Student ARCH, a pair of Markov switching t-Student ARCH models that imposes a constant correlation, and a simple independent, identically distributed (IID) homoskedastic model with constant means, variances, and stock-bond covariance). In a recursive, pseudo out-of-sample 1983:12–2007:11 exercise, we find that the 1-month-ahead mean, variance, and covariance predictions from the three-state model generally outperform all other candidates. In particular, the three-state ARCH model produces the minimum root-mean-square forecast error (RMSFE) for excess stock return variance. This is a consequence of the lower volatility of the MS forecast errors. In general, the constant variance benchmark tends to be the second best model. The fact that a richly parameterized MS t-Student ARCH model outperforms a simple IID homoskedastic benchmark with only five parameters is very interesting. The results for excess bond return variance are similar, if not better: the three-state model displays the lowest RMSFE.

The rest of the chapter is organized as follows. Section 5.2 introduces Markov switching ARCH models and reviews a few details on structure and estimation. Section 5.3 introduces the main features of U.S. stock and bond excess returns data, shows the presence of persistence and predictability in both variances and covariances, and provides initial, motivating evidence of structural change and instability in CH dynamics. Section 5.4 reports the main empirical results of the chapter. Section 5.5 analyzes the 1-month-ahead forecasting performance of alternative models for means, variance, and the stock-bond covariance. Section 5.6 concludes.

5.2 MARKOV SWITCHING ARCH MODELS

5.2.1 General, Multivariate Case

Let $\mathbf{r}_t$ be a $N \times 1$ vector of (excess) asset returns, in our application the returns of a broad equity index and long-term government bonds in excess of 1-month T-bills. Denote S_t as an unobserved latent random variable that can take on the values 1, 2,..., K and—for the sake of concreteness—let us suppose that S_t can be described by an irreducible, first-order Markov chain with time-homogeneous transition probabilities:

$$\Pr(S_t = j \mid S_{t-1} = i, \ S_{t-2} = l, ..., \mathbf{r}_{t-1}, \mathbf{r}_{t-2}, ..., \mathbf{r}_0) = \Pr(S_t = j \mid S_{t-1} = i) = p_{ij},$$

for $i, j, l = 1 \ 2, ..., K$. The variable S_t may be used to capture the nature of the state in which the markets are in at time t, for instance, to distinguish between bull and bear market states or quiet and volatile periods. We can write this dependence as stating that the conditional density of returns data at time t will depend on at most a finite number of lags, $q \geq 1$, of the Markov state variable S_t:

$$f(\mathbf{r}_t \mid S_t, S_{t-1}, ..., S_{t-q}, \mathbf{r}_{t-1}, \mathbf{r}_{t-2}, ..., \mathbf{r}_0), \tag{5.1}$$

where $\mathbf{r}_0$ is assumed to be fixed and known.[*] It is often convenient to collect the transition probabilities in a $K \times K$ (constant, time-invariant) transition matrix $\mathbf{P}$ in which the generic $[i \ j]$ element is the probability p_{ij}, i.e., $\mathbf{e}_i' \ \mathbf{P} \mathbf{e}_j = p_{ij}$.[†] Notice that this specification assumes that if markets were in a given state last period, the probability of switching to a different state does not depend on how long markets have been in the current state or on any other features of recent market behavior (like recent mean returns or volatility).

In the empirical literature, it is typical to replace the general specification in Equation (5.1) with simple vector autoregressive frameworks that allow for ARCH(q) effects in which—at least in principle—all the matrices collecting parameters may become a function of the Markov state S_t:[‡]

$$\mathbf{r}_t = \boldsymbol{\mu}_{S_t} + \sum_{j=1}^{p} \mathbf{B}_{j, S_t} \mathbf{r}_{t-j} + \mathbf{u}_t \quad \mathbf{u}_t \sim F(\mathbf{0}, \mathbf{H}_{t, S_t}; \mathbf{v}_{S_t}), \ E[\mathbf{u}_t] = \mathbf{0}, \ Var[\mathbf{u}_t] = \mathbf{H}_{t, S_t}$$

$$\mathbf{H}_{t, S_t} = \mathbf{A}_{0, S_t} \mathbf{A}'_{0, S_t} + \sum_{j=1}^{q_1} (\mathbf{A}_{j, S_t} \mathbf{A}'_{j, S_t}) \mathbf{u}_{t-j} \mathbf{u}'_{t-j} + \sum_{j=1}^{q_2} \mathfrak{S}_{t-j} \odot (\boldsymbol{\Upsilon}_{j, S_t} \boldsymbol{\Upsilon}'_{j, S_t}) \mathbf{u}_{t-j} \mathbf{u}'_{t-j},$$

$$\tag{5.2}$$

---

[*] The assumption of q finite in Equation (5.1) provides a rather general framework but constrains the type of conditional heteroskedastic models to be embedded within the Markov switching framework to the original Engle ARCH(q) (1982) type and prevents modeling regime shifts in Bollerslev (1986) type GARCH processes.

[†] Here $\mathbf{e}_i$ is a $K \times 1$ vector with zeros everywhere and a 1 in its i-th position; therefore, $\mathbf{e}_i' \ \mathbf{P} \mathbf{e}_j$ simply selects the element in row i and column j of the matrix $\mathbf{P}$. Correspondingly, ι_M shall be defined as a $M \times 1$ vector that collects 1s in all of its elements.

[‡] In the expression that follows, $\odot$ denotes the element-by-element (Hadamard) product. Given two conformable matrices $\mathbf{A}$ and $\mathbf{B}$, the generic element i,j of $\mathbf{A} \odot \mathbf{B}$ is $a_{ij} \times b_{ij}$. The conditional mean function is of the standard Markov switching VAR(p) case, as in Guidolin and Ono (2005).

where $\mathbf{H}_t$ is the time t conditional covariance matrix of dimension $N \times N$, $F(\cdot; \mathbf{v}_{S_t})$ denotes a generic density function parameterized by the vector $\mathbf{v}_{S_t}$, the (regime-dependent) matrices $\mathbf{A}_{0,S_t}$, $\mathbf{A}_{j,S_t}$, and Υ_{j,S_t} are matrices of rank up to to N, and the matrices $\mathfrak{I}_{t-j}$ are selector matrices with generic element $\mathbf{e}_i' \mathfrak{I}_{t-j} \mathbf{e}_l = 1$ if $u_{i,t-j} < 0, u_{l,t-j} < 0$ and 0 otherwise.

The reason why the parameter matrices all appear in the outer product format (e.g., $\mathbf{A}_{0,S_t} \mathbf{A}_{0,S_t}'$ instead of $\mathbf{A}_{0,S_t}$) is to ensure that the resulting covariance matrix $\mathbf{H}_{t,S_t}$ be positive-definite within each regime. In words, at each lag $j = 1, \ldots, q_2$, the matrices $\mathfrak{I}_{t-j}$ select elements of $\Upsilon_{j,S_t} \Upsilon_{j,S_t}' \mathbf{u}_{t-j} \mathbf{u}_{t-j}'$ that are associated with pairs of negative return shocks only. As it is well known from the literature (see Engle and Ng, 1993) that this effect ought to capture the existence of leverage (asymmetries) in asset returns, i.e., the fact that negative return shocks (or interactions of negative shocks) ought to increase variance and covariances more than positive return news does. Equation (5.2) is in fact a version of Bollerslev et al.'s (1988) multivariate VECH GARCH model, with the peculiarity that Equation (5.2) fails to include a GARCH component, while it is extended to model leverage effects, of order q_2. Of course, Equation (5.2) also generalizes the VECH ARCH(q_1, q_2) to the Markov switching case. Because p, q_1, and q_2 are all finite, the condition that the conditional density $\mathbf{r}_t$ ought at most to depend on a finite number of lags of the history of $\mathbf{r}_t$ itself is satisfied. In most applications, $F(\cdot; \mathbf{v}_{S_t})$ is either a multivariate Gaussian density (in which case $\mathbf{v}_{S_t}$ is empty, which means it can be set to 1 and this object is irrelevant in the estimation) or a multivariate Student t, in which case $\mathbf{v}_{S_t}$ collects to the degrees of freedom parameters.

5.2.2 Special Cases

Several cases of interest—which have often attracted the attention of researchers—can be derived by imposing restrictions on Equation (5.2). One application that has been considered since Hamilton and Susmel (1994) is univariate financial return series, i.e., $N = 1$. In this contingency, Equation (5.2) simplifies to a simple Markov switching AR(p) asymmetric ARCH(q_1, q_2) model with shocks drawn from a generic density $F(\cdot; v)$.

$$r_t = \mu_{S_t} + \sum_{j=1}^{p} \phi_j s_t r_{t-j} + u_t \quad u_t \sim F(0, h_{S_t}; v), \ E[u_t] = 0, Var[u_t] = hs_t$$

$$hs_t = a_0, s_t + \sum_{j=1}^{q_1} a_j, s_t u_{t-j}^2 + \sum_{j=1}^{q_2} I_{\{u_{t-j} < 0\}} \gamma_j s_t u_{t-j}^2. \tag{5.3}$$

However, Hamilton and Susmel (1994) argue that in many applications Equation (5.2) is likely to be difficult to estimate and possibly overparameterized. They propose instead a much simpler framework in which the only influence of the Markov switching state variable manifests itself through $\tilde{u}_t = \sqrt{g_{S_t}} u_t$, which now enters the conditional mean function in place of u_t, while g_{S_t} is a regime-specific scale factor that will have the role of globally scaling up and down the level of conditional variance at a given time t:

$$r_t = \mu + \sum_{j=1}^{p} \phi_j r_{t-j} + \tilde{u}_t = \mu + \sum_{j=1}^{p} \phi_j r_{t-j} + \sqrt{g_{S_t}} u_t \quad u_t \sim F(0, h_t; v)$$

$$h_t = a_0 + \sum_{j=1}^{q_1} a_j u_{t-j}^2 + \sum_{j=1}^{q_2} I_{\{u_{t-j}<0\}} \gamma_j u_{t-j}^2.$$

Usually, g_{S_t} is normalized to equal 1 in the first state, $g_1 = 1$, while $g_{S_t} \geq 1$ for $S_t = 2,\ldots,K$.

If on the contrary we set $q_1 = q_2 = 0$ in Equation (5.2), we obtain a simple Markov switching VAR(p) model with regime-dependent covariance matrix, as in Guidolin and Timmermann (2006, 2007):

$$\mathbf{r}_t = \boldsymbol{\mu}_{S_t} + \sum_{j=1}^{p} \mathbf{B}_{j,S_t} \mathbf{r}_{t-j} + \mathbf{u}_t \quad \mathbf{u}_t \sim F(\mathbf{0}, \mathbf{A}_{0,S_t} \mathbf{A}'_{0,S_t}; v).$$

When $\mathbf{A}_{0,S_t}$ is regime independent, then the model is a Markov switching, homoskedastic VAR.

One last special case of a multivariate model (in fact, bivariate) has been popularized by Hamilton and Lin (1996) and can be written as (in its simplest form)

$$\mathbf{r}_t = \boldsymbol{\mu}_{S_t} + \mathbf{B}\mathbf{r}_{t-1} + \mathbf{L}_{S_t} \boldsymbol{\varepsilon}_t \quad \boldsymbol{\varepsilon}_t \sim F(\mathbf{0}, \mathbf{I}_N; v), \tag{5.4}$$

where $\mathbf{L}_{S_t}$ is a diagonal matrix with

$$\mathbf{L}_{S_t} \mathbf{L}'_{S_t} = diag \left\{ g_{S_t} \left[a_{0i} + a_{1i} u_{i,t-1}^2 + I_{\{u_{i,t-1}<0\}} \gamma_i u_{i,t-1}^2 \right] \right\}, \tag{5.5}$$

and $g_1 = 1$, $g_{S_t} \geq 1$ for $S_t \geq 2$.

5.2.3 Estimation

All of the models discussed in Sections 5.2.1 and 5.2.2 may be represented in terms of particular parameterization for the conditional log-density of the return vector $\mathbf{r}_t$. For instance, when $f(\mathbf{r}_t \mid S_t, \mathbf{r}_{t-1}, \mathbf{r}_{t-2}, \ldots, \mathbf{r}_0)$ is multivariate Gaussian, then

$$\ln f(\mathbf{r}_t \mid S_t, \mathbf{r}_{t-1}, \mathbf{r}_{t-2}, \ldots, \mathbf{r}_{t-q}) = -\frac{1}{2} N \ln(2\pi) - \frac{1}{2} \ln |\mathbf{H}_{t,S_t}| - \frac{1}{2} \mathbf{u}'_t \mathbf{H}_{t,S_t}^{-1} \mathbf{u}_t$$

where $\mathbf{u}_t$ and $\mathbf{H}_{t,S_t}$ are defined as in Equation (5.2), and $q=\max\{q_1,q_2\}$. When the conditional density is multivariate t-Student, we have:

$$\ln f(\mathbf{r}_t\,|\,S_t,\mathbf{r}_{t-1},\mathbf{r}_{t-2},\ldots,\mathbf{r}_{t-q})=\ln\left[\frac{\Gamma\!\left(\frac{v_{S_t}+N}{2}\right)v_{S_t}^{N/2}}{(v_{S_t}\pi)^{N/2}\Gamma\!\left(\frac{v_{S_t}}{2}\right)(v_{S_t}-2)^{N/2}}\right]-\frac{1}{2}$$

$$\ln|\mathbf{H}_{t,S_t}|-\frac{1}{2}(v_{S_t}+N)$$

$$\ln\left\{1+\frac{1}{v_{S_t}-2}\mathbf{u}_t'\,\mathbf{H}_{t,S_t}^{-1}\mathbf{u}_t\right\},$$

where v_{S_t} is a regime-dependent (scalar) degree of freedom parameter, to be estimated, and $\mathbf{u}_t$ and $\mathbf{H}_{t,S_t}$ have identical definitions as in the Gaussian case. Given the conditional density and the parameters of the Markov transition matrix $\mathbf{P}$ for the overall Markov state variable S_t, it is possible to evaluate the log-likelihood function of the observed data using the methods described in Hamilton (1994).[*]

5.3 THE DATA

We use monthly data on U.S. excess stock and long-term government bond returns for the 55-year-long period 1953–2007. Excess returns are computed as the difference between total returns and 1-month Treasury bill (T-bill) yields, as common in the empirical finance literature. Stock, bond, and T-bill returns (yields) are obtained from the Chicago Research Center in Security Prices (CRSP). CRSP equity data refer to a value-weighted index that aggregates NYSE, AMEX, and (after December 1972) NADSAQ prices and distributions. CRSP long-term government bond return data are instead constructed by choosing at the end of each month a valid issue that falls closer (in terms of residual time to maturity) to the selected 10-year term.

Table 5.1 reports basic summary statistics for the data under investigation. In order for us to be able to introduce the issue of structural instability in the dynamic time-series properties of stock and bond returns,

---

[*] Numerical optimization is performed using the steepest ascent method and then switching to the BFGS algorithm in the final step of the maximization. For all models we have generated at least fifty different starting values to check whether the maximum found could only have local nature. Additional details on estimation and inference can be found in Hamilton and Susmel (1994, with explicit reference to the univariate case) and Guidolin and Ono (2005, with explicit reference to Markov switching models).

TABLE 5.1 Summary Statistics for U.S. Stock and Long-Term Government Bond Returns

	Mean	Volatility	Sharpe Ratio	Median	Skewness	Kurtosis	Jarque-Bera	LB(12)—Levels	LB(12)—Squares
Full Sample (1953–2007)									
Excess stock returns	0.565	4.215	13.405	0.908	–0.512	5.014	140.30	9.302	29.781
	(0.001)	(0.000)	(0.000)	(0.000)	(0.000)	(0.004)	(0.000)	(0.677)	(0.003)
Excess bond returns	0.114	2.127	5.360	0.020	0.273	4.554	74.621	13.366	200.53
	(0.169)	(0.000)	(0.043)	(0.351)	(0.134)	(0.048)	(0.000)	(0.343)	(0.000)
Sample (1953–1970)									
Excess stock returns	0.679	3.697	18.366	1.085	–0.413	3.190	6.458	20.666	18.140
	(0.008)	(0.000)	(0.000)	(0.001)	(0.039)	(0.746)	(0.040)	(0.055)	(0.112)
Excess bond returns	–0.075	1.754	–4.276	–0.144	0.575	6.104	98.606	16.745	68.174
	(0.529)	(0.000)	(0.679)	(0.128)	(0.039)	(0.746)	(0.000)	(0.159)	(0.000)
Sample (1971–1988)									
Excess stock returns	0.368	4.859	7.574	0.309	–0.402	5.588	66.104	9.942	6.996
	(0.267)	(0.000)	(0.073)	(0.161)	(0.034)	(0.000)	(0.000)	(0.621)	(0.858)
Excess bond returns	0.122	2.619	4.658	0.030	0.360	3.840	10.999	13.824	55.028
	(0.493)	(0.000)	(0.140)	(0.816)	(0.094)	(0.143)	(0.004)	(0.312)	(0.000)
Sample (1989–2007)									
Excess stock returns	0.645	4.026	16.021	1.093	–0.659	4.087	27.720	6.430	35.138
	(0.016)	(0.000)	(0.000)	(0.002)	(0.000)	(0.001)	(0.000)	(0.893)	(0.000)
Excess bond returns	0.286	1.913	14.950	0.325	–0.258	3.555	5.441	9.001	11.004
	(0.025)	(0.000)	(0.000)	(0.012)	(0.160)	(0.294)	(0.066)	(0.703)	(0.525)

Table 5.1 reports customary statistics not only with reference to the overall 1953–2007 sample period, but also for three subperiods of 18 years each: 1953–1970, 1971–1988, and 1989–2007. The basic properties of the series at hand are well known: equities offer on average a premium over short-term, relatively riskless T-bills of approximately 6.8% per annum, with an annualized volatility of 14.6%; the corresponding monthly Sharpe ratio is of 13% and is statistically significant. However, excess stock returns also display strong and persistent departures from normality. Their unconditional distribution is skewed to the left and has much thicker tails than a Gaussian benchmark. The left skewness is consistent with the observation that the median of excess equity returns is almost the double their mean. Overall, the Jarque-Bera test rejects the null of normality. While stock returns appear not to be serially correlated, their squares are strongly serially correlated, which is generally taken as an indication of volatility clustering. Long-term bonds pay on average a premium over short-term bonds of almost 136 basis points in annualized terms. The annualized volatility of excess bond returns is instead 7.4%, i.e., roughly half the volatility of stocks. The bond Sharpe ratio, however, is a less generous 5.4%, although this is statistically significant. Also, excess bond returns display significant departures from normality: their unconditional distribution is skewed to the right and has tails thicker than a Gaussian. Even though skewness is not excessive, the Jarque-Bera test rejects the null of normality with a p-value that is essentially zero. Once more, bond returns appear not to be serially correlated but show the typical signs of ARCH because their squares are strongly serially correlated at all lags.

The table reports basic summary statistics for U.S. stock and (long-term government) bond excess returns. Excess returns are computed as differences between stock and bond returns and 1-month T-bill yields. In the table, statistics in parentheses are the p-values associated with the null hypothesis of a zero value for the parameter or statistic under investigation. When possible, the p-values are computed for two-tailed tests of hypothesis. In the case of kurtosis, the null hypothesis is of a kurtosis that equals the Gaussian benchmark of 3. Jarque-Bera is a test of distributional normality based on deviations of skewness and kurtosis coefficients from the null of normality. LB(12) is the Ljung-Box test for zero serial correlation up to order 12 for levels and squared returns, respectively.

Table 5.1 proceeds then to split up our sample in the way described. Strikingly, all the features we have presented as typical and that have been widely documented in the literature tend to disappear in at least one of the subsamples. To start from a property of direct interest for our purposes, it is clear that while excess stock returns are characterized by strong ARCH in the more recent 1989–2007 period, this is not the case for the preceding 36 years. On the opposite, bond returns show volatility clustering in two subperiods (the initial years 1953–1988) but not in the most recent period. This is rough, but powerful evidence that proposing time-series models with variable intensity for ARCH effects may have considerable value. In a similar way, while the equity premium appears to be definitely positive and statistically significant in the 1953–1970 and 1989–2007 periods, it is positive but rather modest (4.4%) and not statistically significant in the 1971–1988 period. While in general the serial correlation structure of excess equity returns does not allow one to accurately forecast, in the 1953–1970 there are signs of a precisely estimated correlation structure. Remarks of the same type apply also to bond returns. Therefore, it appears that bear-bull-type models that imply time-varying conditional means may also have some value in forecasting and portfolio choice. However, one feature exists that reliably persists over the entire sample period: both series and the overall subsample present strong departures from normality, as shown by the Jarque-Bera tests. Interestingly, this is once more consistent with the presence of bull-bear dynamics, ARCH, as well as structural instability of ARCH features through the entire sample period.

Figure 5.1 strengthens these impressions of pervasive instability by plotting 3-year rolling window sample estimates (i.e., based on moving windows of thirty-six observations) for mean excess returns, volatilities, and the equity-bond correlation coefficient. The first panel shows pronounced swings in the equity premium estimates, which in fact reaches negative values for two prolonged periods, 1974–1976 (the first, big oil shock) and then 2001–2004 (the dot.com bubble burst and the 2001–2002 U.S. recession). Periods of euphoric (if not bubbly) stock markets are also evident, such as the mid-1950s, the mid-1980s, and especially the 1996–2000 period, with average premia in excess of 12–20% per annum for periods of 3–4 consecutive years. Interestingly, the most recent, 2006–2007, has been a period of bullish stock markets, with premia of approximately 12% per annum. Swings in bond premia are less visible but *de facto* even more persistent than equity oscillations. In practice, bond premia were nil or slightly negative on average over the long 1953–1982 period, and then become positive and relatively high (300–400 basis points per year) over the subsequent 1983–2007 sample. Of course, consistently with common empirical observations, even during the last 25 years, the yield curve has

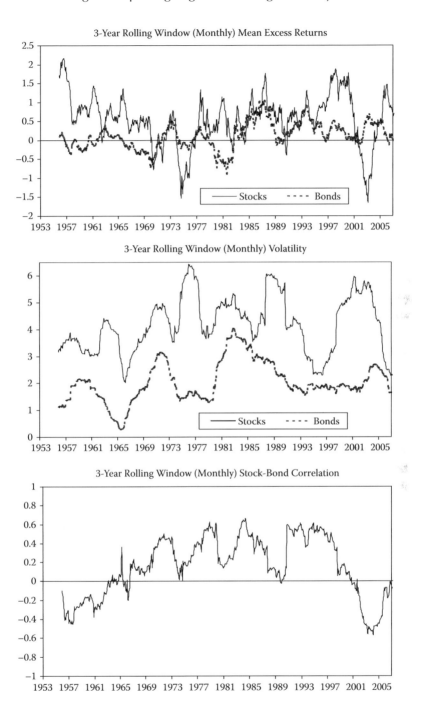

FIGURE 5.1   Three-year rolling window estimates of mean excess returns, volatilities, and correlations.

recurrently become flat or even downward sloping during recessions, such as 1990, 2001, and late 2007.

The second panel of Figure 5.1 shows rolling window volatility estimates. Consistent with Table 5.1, equity volatility tends to be double the bond volatility. However, the two series tend to approximately swing together, with an overall correlation of 0.38. Equity volatility oscillates between minima of 10–12% per year over calm periods—such as the late 1950s, the mid-1960s, 1993–1998, and the recent 2005–2007 period—and maxima in excess of 20% over the turbulent periods—such as the mid-1970s, 1987–1991, and 1999–2003. Interestingly, some of these periods also correspond to bear regimes, as shown by the first panel. Bond volatility presents instead two large historical spikes, 1970–1973 and 1982–1986, when it exceeded 7% in annualized terms; one trough in the mid-1960s, when volatility almost disappeared; and one long, protracted plateau at a moderate level of 5–6% per annum between 1992 and 2002. This plot provides additional, powerful evidence of the presence of regimes in the volatility of U.S. financial returns. The last panel of Figure 5.1 depicts instead time variation in the stock-bond correlation coefficient. There are three main regimes: in the early part of the sample (1953–1965), the correlation is slightly negative but trends up over time; in the central part of the sample (1966–1997), the correlation oscillates between zero and relatively high levels of up to 0.65; and in the final part of the sample (1998–2007), correlation falls again to negative territory, touching –0.60 around 2004, even though correlation appears to be again trending up between 2006 and 2007.

5.3.1 Preliminary Evidence of Instability in Conditional Heteroskedasticity

Figure 5.2 provides a powerful display of the presence of conditional heteroskedasticity in U.S. stock and bond monthly return data consists of plots of the squares of stock and bond returns. These show the classical signs of volatility clustering: in most periods, large squared returns tend to be followed by (many) other large squared returns, and vice versa, i.e., quiet periods tend to persist over time. At least visually, this effect seems to be more pronounced for bond returns, when volatility seems to almost completely evaporate between 1963 and 1966, and then again between 1973 and 1980. These two facts indeed match our comments on the features of the bond volatility series in Figure 5.1. This means that ARCH effects may be stronger for bond than for stock returns. The last panel of Figure 5.2 also shows the product of stock and bond returns, which is a raw measure of *covariance* for

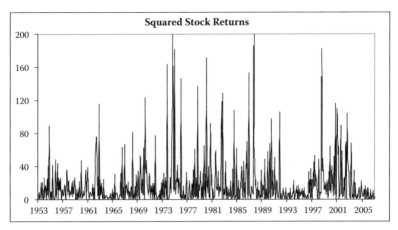

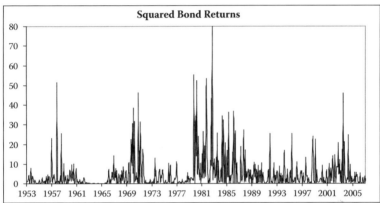

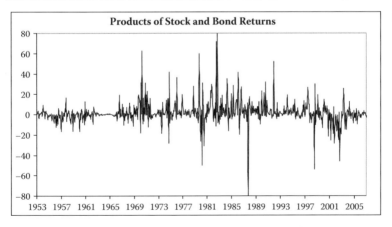

FIGURE 5.2   Plots of squared returns and of return products.

a pair of series in the same way in which squared returns are a raw measure of *variance* for each individual series. The presence of some degree of covariance clustering is as evident as in the first two panels of the figure, at least in the sense that quiet periods in which the return product gravitates around zero tend to be persistent. Consistently with Figure 5.1, the period 2001–2003 is mostly characterized by large and negative return products, leading to negative (for the period 2002–2004) correlations. This shows that ARCH-like effects are likely to extend beyond variances, to affect covariances.

One aspect of ARCH behavior the literature has been concerned with since the seminal paper by Engle and Ng (1993) is the presence of leverage (asymmetric) effects in conditional heteroskedasticity, i.e., the fact that negative returns (shocks) tend to induce to a larger, subsequent volatility reaction than equally sized positive returns (shocks). We pursue a similar empirical hypothesis and plot rolling window volatilities and average excess returns for both stocks and bonds. Even though this is not exactly the formulation popular in the empirical finance literature (see Section 5.2.1 for details), using the average quantities in Figure 5.1 will give us some ideas for the phenomenon. In the case of stocks, it is clear that there is considerable leverage: large and negative equity premia are associated with much higher volatilities than large and positive premia; in fact, the relationship seems to be approximately linear and negative (the correlation between the two rolling window sample statistics is −0.46). However, in the case of bonds there is no strong evidence of leverage: the relationship seems to describe an approximately symmetric U shape in which negative and positive premia seem to have the same effect on volatility; the correlation between risk premia and volatility is a modest 0.15. We also tried to build a similar picture with reference to return products. It plots the signed product of stock and bond premia versus rolling window stock-bond correlations. The signing of return products is performed by appending a minus sign when either stock or bond premia are negative for a given period. Here the evidence of a leverage effect is weak at best. However, it is clear that while large positive product of premia may be associated with both positive and negative correlations, large negative products of premia may only lead to zero or negative correlations. Overall, although for stock returns there is evidence of leverage, it remains to be seen whether these asymmetries derive from specific features of the CH dynamics followed by excess asset returns or whether—on the contrary—they are one result of the presence of structural instability in simpler CH functions.

As a way to introduce the issue of structural instability in CH, we compute the estimates of a simple bivariate VAR(2) VECH t-Student ARCH(3) with leverage for stock and bond returns with reference to the overall sample period and for two subsamples, 1953–1980 and 1981–2007.[*] It turns out that cross-asset and other asset effects in the CH equations all turn out to be insignificant and can be dropped from the model.[†] The full sample estimates produce results that are in general agreement with our comments to Figures 5.1 through 5.3. There is some mild, pure VAR predictability of excess returns, with pseudo R^2s in the order of 1.9–2.1%. In particular, as it is well known since Fama and French (1989), a positively sloped term structure at time $t-1$ forecasts higher subsequent excess equity returns at both time t and $t+1$, while excess bond returns are simply serially correlated. However, the corresponding coefficients are small in economic terms, and they command rather negligible R^2s if compared to those typical in the literature based on monthly returns.[‡] ARCH effects are rather strong and statistically significant in excess bond returns, and weaker for excess stock returns. However, excess stock returns offer unequivocal evidence of leverage effects. In fact—rather oddly—a time $t-1$ shock to equity return $(u_{s,t-1})$ induces effects on time t equity variance only if negative, so to activate the ARCH effect through $I_{\{u_{s,t-1}<0\}}u_{s,t-1}^2$. There is rather modest evidence of ARCH in the covariance, and while last month shocks do not seem to affect the current level of the conditional covariance, when the distance in time equals 2 or 3 months, the effect is positive and statistically significant. Finally, the estimated process implies unconditional means, volatilities, and correlations that are all rather sensible and closely match the full-sample statistics reported in Table 5.1. The estimate of the (common, for simplicity) t-Student degree of freedom parameter (8.00) illustrates that—even after accommodating for ARCH effects—some residual need for modeling tails thicker than a simple bivariate Gaussian density does remain.

---

[*] The VAR(2) VECH t-Student ARCH(3) has been selected to represent a single-state counterpart to the MS ARCH models estimated and commented on in Section 5.4.2. However, we notice that a thorough model specification search among single-state models reveals a need to use such a framework along all the information criteria.

[†] These cross-asset and other asset effects simply measure the effects of shocks to bond (stock) returns on the conditional variance of bond (stock returns), and the impact of terms obtained as the product of bond and stock shocks on the conditional variance of both bond and stock returns.

[‡] For instance, a one standard deviation increase in the excess bond return at time $t-1$ (this is a 2.13% increase) induces an increase in the excess stock return of 0.34% the following month, which is only 8% of a one standard deviation increase. This is, of course, the cause of the rather small R^2s in Table 5.2.

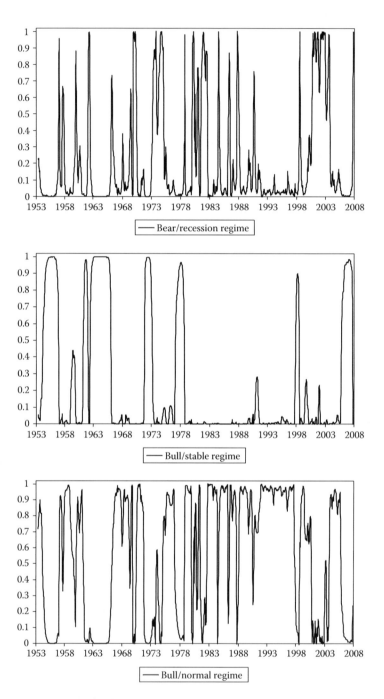

FIGURE 5.3 Smoothed state probabilities from bivariate, three-state Markov switching VAR(2) VECH t-Student ARCH(3) model for U.S. stock and bond excess returns.

It is also of interest to notice that the two lower panels of Table 5.2 display of important instabilities in parameter estimates. Although the conditional mean estimates change somewhat (but this hardly affects the resulting R^2s, which remain between 1.5 and 2.8% and seem only marginally higher for the pre-1981 sample), the most important breaks in estimated parameters concern the CH process. In practice, there is little evidence of ARCH in equity returns in the first part of the sample, apart from an oddly isolated leverage effect (i.e., it seems that only negative equity return shocks affect subsequent variance), while equity ARCH effects appear in the post-1981 sample. Excess bond returns exhibit ARCH in both subsamples, although the evidence is stronger in the early sample. ARCH in the covariance is mild in both samples, but the parameter estimates change substantially, while there is evidence of a leverage in covariance in the 1953–1980 subsample only. There seems to be sufficient evidence of structural instability in CH to motivate a formal Markov switching approach to bivariate ARCH in excess stock and bond returns.

5.4 EMPIRICAL RESULTS

In this section we try to summarize the conclusions reached after estimating hundreds of (restricted and unrestricted, as explained in Section 5.2) MS ARCH models in the attempt to isolate a model at the same time feasible and with sufficient promise in terms of forecasting performance and as a support to portfolio decisions. To gain some additional insights incremental to Table 5.2, we have first proceeded to estimate a relatively wide range of univariate MS ARCH models for excess stock and bond returns, respectively. The underlying idea is that a bivariate MS ARCH model will be justified only when the underlying series contain MS ARCH in the first instance. Moreover, the univariate MS ARCH properties may give us insights on the main properties of an adequate bivariate model. The cost of focusing on univariate series is obvious: we will be lacking a model for the CH dynamics of the covariance and therefore of correlations. In unreported results, we find a strong need for MS ARCH models. In particular, there seems to be little doubt left that U.S. excess equity return data contain strong evidence of regimes in their variance process, consistent with that reported by Hamilton and Susmel (1994). In the case of excess bond returns, the evidence leans in favor of larger, three- or even four-state ARCH models.

5.4.1 Bivariate Model Selection

Table 5.2 reports results for a bivariate specification search involving excess stock and bond returns. Besides reporting parameters that

TABLE 5.2 Model Selection: Bivariate K-State AR(2) ARCH(3) Leverage Models for U.S. Stock and Bond Excess Returns

Regimes	ARCH Order	Marginal Distribution	Leverage	Final Negative Log-Likelihood	AIC	BIC	H-Q	RCM 1	RCM 2	Pseudo-R^2 Stocks	Pseudo-R^2 Bonds	No. Parameters	Saturation Ratio
							Single-State Models						
1	1	Gaussian	No	3267.989	10.0122	10.0321	9.9853	—	—	2.365	2.051	16	82.00
1	2	Gaussian	No	3224.089	9.8875	9.9111	9.8556	—	—	2.360	1.938	19	69.05
1	3	Gaussian	No	3211.046	9.8568	9.8842	9.8199	—	—	2.329	2.142	22	59.64
1	1	t-Student	No	3230.897	9.8991	9.9190	9.8722	—	—	1.943	1.869	16	82.00
1	2	t-Student	No	3198.425	9.8092	9.8329	9.7773	—	—	2.155	1.997	19	69.05
1	3	t-Student	No	3184.847	9.7770	9.8044	9.7401	—	—	2.064	1.931	22	59.64
1	1	Gaussian	Yes	3256.608	9.9866	10.0103	9.9547	—	—	2.255	1.988	19	69.05
1	2	Gaussian	Yes	3213.302	9.8637	9.8911	9.8268	—	—	2.288	1.911	22	59.64
1	3	Gaussian	Yes	3200.862	9.8349	9.8661	9.7930	—	—	2.332	2.138	25	52.48
1	1	t-Student	Yes	3220.383	9.8762	9.8998	9.8443	—	—	1.933	1.843	19	69.05
1	2	t-Student	Yes	3188.404	9.7878	9.8152	9.7509	—	—	2.057	1.860	22	59.64
1	3	t-Student	Yes	3175.399	9.7573	9.7884	9.7154	—	—	2.115	1.934	25	52.48
							Two-State Models						
2	1	Gaussian	No	2910.534	8.9772	9.0196	8.9202	31.709	44.593	17.695	18.176	34	38.59
2	2	Gaussian	No	2904.083	8.9759	9.0257	8.9088	27.590	52.243	18.559	20.554	40	32.80
2	3	Gaussian	No	2900.633	8.9836	9.0409	8.9065	21.581	58.562	19.714	21.193	46	28.52
2	1	t-Student	No	2903.494	8.9558	8.9981	8.8987	18.612	60.554	18.222	21.323	34	38.59
2	2	t-Student	No	2896.855	8.9538	9.0036	8.8867	19.672	68.360	16.957	17.557	40	32.80
2	3	t-Student	No	2893.104	8.9607	9.0180	8.8835	17.048	73.760	13.882	20.854	46	28.52
2	1	Gaussian	Yes	2910.518	8.9955	9.0453	8.9284	17.082	64.665	15.330	22.610	40	32.80
2	2	Gaussian	Yes	2904.044	8.9940	9.0513	8.9169	16.840	58.788	17.411	18.909	46	28.52
2	3	Gaussian	Yes	2900.433	9.0013	9.0661	8.9141	13.716	55.130	15.211	21.978	52	25.23

2	1	t-Student	Yes	2903.483	8.9740	9.0238	8.9069	15.990	44.343	16.252	25.897	40	32.80
2	2	t-Student	Yes	2896.845	8.9721	9.0294	8.8949	16.588	42.946	19.152	**29.147**	46	28.52
2	3	t-Student	Yes	2893.006	8.9787	9.0434	8.8914	15.620	39.808	21.852	24.507	52	25.23
Three-State Models													
3	1	Gaussian	No	2830.976	8.7957	8.8629	8.7051	1.459	51.695	18.583	18.073	54	24.30
3	2	Gaussian	No	2474.781	7.7371	7.8156	7.6315	1.393	40.203	21.805	14.927	63	20.83
3	3	Gaussian	No	2364.358	7.4279	7.5176	7.3071	1.525	34.449	22.685	16.959	72	18.22
3	1	t-Student	No	2402.543	7.4895	7.5567	7.3989	1.390	29.761	20.873	17.437	54	24.30
3	2	t-Student	No	2312.644	7.2428	7.3213	7.1371	1.582	27.808	20.200	16.065	63	20.83
3	3	t-Student	No	2239.713	**7.0479**	**7.1376**	**6.9271**	1.729	29.461	23.111	15.006	72	18.22
3	1	Gaussian	Yes	2476.581	7.7426	7.8211	7.6369	1.593	26.719	22.857	16.290	63	20.83
3	2	Gaussian	Yes	2472.734	7.7583	7.8480	7.6376	1.703	21.810	23.017	17.212	72	18.22
3	3	Gaussian	Yes	2337.229	7.3726	7.4735	7.2368	1.382	23.981	21.189	16.428	81	16.20
3	1	t-Student	Yes	2276.728	7.1333	7.2118	7.0276	1.169	26.806	18.072	15.907	63	20.83
3	2	t-Student	Yes	2242.958	7.0578	7.1475	6.9370	1.141	24.519	21.630	15.008	72	18.22
3	3	t-Student	Yes	2237.977	7.0761	7.1770	6.9403	0.963	**20.329**	22.810	15.157	81	16.20
Four-State Models													
4	1	Gaussian	No	2708.234	8.4885	8.5832	8.3610	0.022	44.549	**28.964**	13.862	76	17.26
4	2	Gaussian	No	2441.089	7.7106	7.8202	7.5630	**0.019**	40.656	21.805	14.927	88	14.91
4	3	Gaussian	No	2375.294	7.5466	7.6712	7.3789	0.022	46.269	19.772	14.333	100	13.12
4	1	t-Student	No	2624.339	8.2327	8.3274	8.1052	0.025	46.464	19.417	13.885	76	17.26
4	2	t-Student	No	2423.363	7.6566	7.7662	7.5090	0.027	49.882	19.562	15.262	88	14.91

(Continued)

TABLE 5.2 Model Selection: Bivariate K-State AR(2) ARCH(3) Leverage Models for U.S. Stock and Bond Excess Returns (*Continued*)

Regimes	ARCH Order	Marginal Distribution	Leverage	Final Negative Log-Likelihood	AIC	BIC	H-Q	RCM 1	RCM 2	Pseudo-R² Stocks	Pseudo-R² Bonds	No. Parameters	Saturation Ratio
4	3	t-Student	No	2412.422	7.6598	7.7844	7.4921	0.022	50.136	19.361	14.087	100	13.12
4	1	Gaussian	Yes	2644.732	8.3315	8.4411	8.1839	0.023	41.221	21.805	14.927	88	14.91
4	2	Gaussian	Yes	2415.533	7.6693	7.7938	7.5016	0.021	37.940	21.805	14.927	100	13.12
4	3	Gaussian	Yes	2354.314	7.5193	7.6587	7.3314	0.023	43.214	21.805	14.927	112	11.71
4	1	t-Student	Yes	2553.425	8.0531	8.1627	7.9055	0.026	47.559	21.267	14.202	88	14.91
4	2	t-Student	Yes	2405.720	7.6394	7.7639	7.4716	0.027	41.704	23.249	14.128	100	13.12
4	3	t-Student	Yes	2378.481	7.5929	7.7324	7.4050	0.032	41.242	22.312	13.159	112	11.71

Note: The table reports summary estimation outputs for a range of bivariate models defined by the number of regimes K ($K = 1, 2, 3, 4$), the ARCH order q_1, the type of marginal density distribution for the errors (Gaussian or t-Student), and the presence of leverage effects ($q_2 = 0$ or 1). Final negative log-likelihood is -1 times the maximized model log-likelihood. The saturation ratio is the ratio between the total number of observations (for both series) available for estimation purposes and the number of parameters to be estimated. All the models are estimated setting aside the initial five observations for lagging purposes. Boldfaced values indicate which models minimize the information criteria and the regime classification measures, and which models maximize the pseudo-R² for the conditional mean function.

identify the basic model features (such as K, q_1, and q_2, and the fact that v is either finite—in which case u_t has a conditional t-Student density—or infinite—which is equivalent to the state where u_t has a conditional Gaussian density), the table shows the negative of the final, maximized log-likelihood, three information criteria now standard in the literature on nonlinear time series (see Guidolin and Ono, 2005, for a discussion), two regime classification measures, and the pseudo R^2 that gives some information on the fit of the conditional mean function.

A few remarks are in order. Reporting the negative of the final, maximized log-likelihood implies that *lower* values of such a measure are to be preferred to higher values. This is commonly done to make the logics of comparison across log-likelihood values similar to that which applies to the information criteria. In our case we report the Akaike (AIC), the Bayes-Schwartz (BIC), and the Hannan-Quinn (H-Q) criteria. It is well known that each of these is decreasing in the negative of the final log-likelihood, $-\ell(\mathbf{r}_1, \mathbf{r}_2, \ldots, \mathbf{r}_T; \boldsymbol{\theta})$, and increasing in the number of parameters to be estimated. Therefore, better-fitting models (i.e., that return a lower value for $-\ell(\mathbf{r}_1, \mathbf{r}_2, \ldots, \mathbf{r}_T; \boldsymbol{\theta})$ and more parsimonious models that imply a lower number of parameters to be estimated will yield lower values of each of three information criteria. In this sense, information criteria do trade off fit against parsimony. Numerous studies have shown that information criteria (especially BIC) may in some sense be predictors of good out-of-sample forecasting performance, as they could avoid an excessive preference on overparameterized models that may fit adequately in sample, but are unlikely to perform well out-of-sample. As a result, models are ranked in the sense that the best are the ones that offer the lowest values for most or all these information criteria.

Regime classification measures have been popularized since the early work on MS models by Hamilton (1988) and propose a rather intuitive idea: a well-specified MS model ought to be able to accurately predict in which state the system is at each point in time. Equivalently, a MS model that is always able to indicate which of the K regimes would be prevailing now, at some time origin t, is to be preferred to another model that offers imprecise indications on the nature of the current state. In simple two-regime frameworks, the early work by James Hamilton offered a rather intuitive regime classification measure:

$$RCM_1 = 100 \frac{K^2}{T} \sum_{t=1}^{T} \sum_{k=1}^{K} \Pr(S_t = k \mid \mathbf{r}_1, \mathbf{r}_2, \ldots, \mathbf{r}_T; \boldsymbol{\theta}),$$

i.e., the sample average of the products of the smoothed state probabilities. Clearly, when a MS model offers precise indications on the nature of the regime at each time t, the implication is that for at least one value of $k = 1,\ldots,K$, $\sum_{k=1}^{K} \Pr(S_t = k \,|\, \mathbf{r}_1, \mathbf{r}_2,\ldots,\mathbf{r}_T; \boldsymbol{\theta}) \simeq 0^*$ because most other smoothed probabilities are zero. Therefore, a good MS model will imply $RCM_1 \simeq 0$. Table 5.2 does report the values of RCM_1 for each estimated model provided $K > 1$. However, when applied to models such that $K > 2$, RCM_1 has one obvious disadvantage: a model can imply an enormous degree of uncertainty on the current regime, but still have $\sum_{k=1}^{K} \Pr(S_t = k \,|\, \mathbf{r}_1, \mathbf{r}_2,\ldots,\mathbf{r}_T; \boldsymbol{\theta}) \simeq 0$ for most values of t. As a result, it is rather common to witness that as K exceeds 2, almost all MS models (good and bad) will automatically imply values of RCM_1 that decline toward 0. One alternative measure that may shield against this type is:

$$RCM_2 = 100 \left\{ 1 - \frac{K^{2K}}{(K-1)^2} \frac{1}{T} \sum_{t=1}^{T} \sum_{k=1}^{K} \left[\Pr(S_t = k \,|\, \mathbf{r}_1, \mathbf{r}_2,\ldots,\mathbf{r}_T; \boldsymbol{\theta}) - \frac{1}{K} \right]^2 \right\}.$$

One can easily show that $RCM_2 \in [0,100]$.

In Table 5.2 we report statistics and summary measures useful to select a bivariate MS VAR(2) VECH ARCH(q_1) model of the type (5.2), including the single-state benchmark $K = 1$, and allowing the possibility that $q_2 = 0, 1$ (i.e., leverage) and that ν be finite (i.e., a bivariate t-Student density for the errors, with common degrees of freedom parameters). For simplicity, we do not perform any specification search involving the conditional mean function and instead set the VAR order to $p = 2$. Importantly, here the Markov switching also involves all the parameters entering the conditional mean function, i.e., the constants as well as the VAR matrix coefficients in (5.2). Clearly, there is overwhelming evidence of regimes. For instance, the (negative of the) log-likelihood function approximately drops from a level of 3,175–3,267 in the $K = 1$ case to a range of 2,893–2,910 in the case $K = 2$. This corresponds to a likelihood ratio test statistic in excess of 500, which—even with a number of restrictions ranging from 8 to 27—commands a p-value that is essentially zero. This is confirmed by all the information criteria dropping by at least 10% when going from $K = 1$ to $K = 2$. However, it is now obvious that the number of parameters to be estimated grows more than proportionally relative to K, reaching levels

---

* On the opposite, the worst possible MS model implies $\Pr(S_t = 1 | \mathbf{r}_1, \mathbf{r}_2,\ldots,\mathbf{r}_T; \theta) = \ldots = \Pr(S_t = K | \mathbf{r}1, \mathbf{r}2,\ldots,\mathbf{r}_T; \theta) = 1/K$ so that $\sum_{K=1}^{K} \Pr(S_t = 1 K | \mathbf{r}_1, \mathbf{r}_2,\ldots, \mathbf{r}_T; \theta) = 1/K^2$ and $RCM_1 = 100$. Therefore $RCM_1 \in [0,100]$ and lower values are to be preferred to higher.

easily in excess of 60 for $K = 3, 4$. In fact, despite that the log-likelihood function keeps declining as the number of regimes is increased, the information criteria all indicate that a three-state MS ARCH(3) model with no leverage and t-Student shocks achieves the best possible trade-off between fit and parsimony. The absence of leverage effects does not come as a complete surprise, because in Section 5.3.1 we had noticed already that asymmetries are weak at best in the case of excess bond returns. In the bivariate case RCM_2 is minimized by a three-state model, although of a different type (i.e., including leverage effects) when compared to the model that minimizes the information criteria (the model that excludes leverage has $RCM_2 = 29.5$). The minimal value of 20.3 reached by RCM_2 again points to the existence of difficulties in precisely classifying the current regime at all times, although this does mean that model forecasting performance or economic value have to be considered *ex ante* inadequate. However, the addition of MS effects to the conditional mean function—in short, the fact that bull and bear markets are explicitly allowed by making the process followed by the conditional mean a function of the current state—tremendously increases the pseudo R^2s achieved. The R^2 is 23% for equities and 15% for bonds under the three-state model selected by the information criteria, although higher R^2s of approximately 29% (for both assets) can be achieved using different models. The three-state MS ARCH(3) model with no leverage and t-Student errors implies the need to estimate seventy-two parameters and, as such, a saturation ratio of 18.2, which is around the boundary of what is commonly considered acceptable.

We also compute unreported estimates for a number of restricted models. The restrictions are imposed along the lines illustrated by Hamilton and Lin (1996) and lead to the bivariate Equations (5.4) and (5.5), with $g_1 = 1$, $g_{S_t} \geq 1$ for $S_t \geq 2$. Clearly, imposing these restrictions enormously reduces the number of parameters to be estimated, often by a factor as large as 4. However, the restrictions severely degrade the quality of the fit provided: while the unrestricted models can take the information criteria down to levels of 7 or even lower, when restrictions are present (and despite the fact that the number of estimated parameters collapses), the minima information criteria fall at around 8.6. All in all, we find no reason to believe that both in-sample and out-of-sample, a restricted model may provide better guidance than the unrestricted but larger models in Table 5.2.

5.4.2 Estimation Results

Table 5.3 reports parameter estimates for the best (in the information criteria metrics) model from Table 5.2, a MS VAR(2) VECH t-Student

TABLE 5.3 Estimates of a Bivariate, Three-State Markov Switching VAR(2) VECH t-Student ARCH(3) Model for U.S. Stock and Bond Excess Returns

Estimated Transition Matrix in the table, boldfaced coefficients imply an estimated coefficient p-value equal to or below 0.1.

	Regime 1 Parameter Estimates (Standard Errors in Parentheses)	Unconditional Mean [Volatility/Correlation]
Conditional mean function—stock returns	$r_{s,t} = \mathbf{-1.659} + 0.105r_{s,t-1} - \mathbf{0.199}r_{s,t-1} + \mathbf{0.576}r_{b,t-1} + \mathbf{0.012}r_{b,t-2} + u_{s,t}$ (0.660) (0.102) (0.098) (0.212) (0.204)	−1.351
Conditional mean function—bond returns	$r_{b,t} = 0.331 + 0.039r_{s,t-1} - \mathbf{0.009}r_{s,t-2} + 0.133r_{b,t-1} - \mathbf{0.476}r_{b,t-2} + u_{s,t}$ (0.286) (0.049) (0.049) (0.087) (0.113)	0.307
Conditional variance function—stock returns	$h_{s,t} = \mathbf{7.255} + \mathbf{0.527}u^2_{s,t-1} + 0.092u^2_{s,t-2} + 0.050u^2_{s,t-3}$ (0.729) (0.237) (0.088) (0.072)	21.918 [4.682]
Conditional variance function—bond returns	$h_{b,t} = \mathbf{1.694} + \mathbf{0.261}u^2_{b,t-1} + 0.007u^2_{b,t-2} + \mathbf{0.641}u^2_{b,t-3}$ (0.334) (0.140) (0.031) (0.179)	18.615 [4.315]
Conditional covariance function	$h_{sb,t} = \mathbf{1.455} + \mathbf{0.147}u_{s,t-1}u_{b,t-1} + 0.001u_{s,t-2}u_{b,t-2} - 0.001u_{s,t-3}u_{b,t-3}$ (0.412) (0.054) (0.030) (0.074)	1.706 [0.085]
t-Student degrees of freedom parameter	9.971 (1.787)	

Regime 2

Conditional mean function—stock returns	$r_{s,t} = \mathbf{1.475} - 0.040r_{s,t-1} - 0.150r_{s,t-2} + \mathbf{0.607}r_{b,t-1} + \mathbf{0.601}r_{b,t-2} + u_{s,t}$ (0.330) (0.103) (0.089) (0.298) (0.378)	1.171
Conditional mean function—bond returns	$r_{b,t} = -0.099 - 0.031r_{s,t-1} - \mathbf{0.009}r_{s,t-2} + \mathbf{0.192}r_{b,t-1} - 0.156r_{b,t-2} + u_{s,t}$ (0.087) (0.029) (0.012) (0.100) (0.115)	−0.068
Conditional variance function—stock returns	$h_{s,t} = \mathbf{8.476} + \mathbf{0.231}u^2_{s,t-1} + 0.111u^2_{s,t-2} + 0.097u^2_{s,t-3}$ (1.833) (0.118) (0.102) (0.119)	15.109 [3.887]

Conditional variance function—bond returns	$h_{b,t} = \mathbf{0.121} + \mathbf{0.203}u^2_{b,t-1} + \mathbf{0.567}u^2_{b,t-2} + \mathbf{0.172}u^2_{b,t-3}$ (0.057) (0.082)　(0.096)　　(0.066)	2.086 [1.444]
Conditional covariance function	$h_{sb,t} = \mathbf{-0.708} - 0.021u_{s,t-1}u_{b,t-1} + 0.269u_{s,t-2}u_{b,t-2} - 0.143u_{s,t-3}u_{b,t-3}$ (0.272) (0.116)　　(0.181)　　　　(0.094)	−0.791 [−0.141]
t-Student degrees of freedom parameter	**0.802** (3.557)	

Regime 3

Conditional mean function—stock returns	$r_{s,t} = \mathbf{1.394} - 0.051r_{s,t-1} + 0.015r_{s,t-2} + 0.104r_{b,t-1} + \mathbf{0.186}r_{b,t-2} + u_{s,t}$ (0.314)(0.078)　(0.058)　(0.101)　(0.092)	1.368
Conditional mean function—bond returns	$r_{b,t} = 0.222 - \mathbf{0.096}r_{s,t-1} - 0.010u_{s,t-2} + \mathbf{0.135}r_{b,t-1} + \mathbf{0.197}r_{b,t-2} + u_{s,t}$ (0.136) (0.044)　(0.033)　(0.061)　(0.094)	0.080
Conditional variance function—stock returns	$h_{s,t} = \mathbf{6.820} + \mathbf{0.200}u^2_{s,t-1} + 0.130u^2_{s,t-2} + \mathbf{0.367}u^2_{s,t-3}$ (2.018) (0.120)　(0.147)　(0.169)	22.508 [4.744]
Conditional variance function—bond returns	$h_{b,t} = \mathbf{3.161} + 0.057u^2_{b,t-1} + \mathbf{0.268}u^2_{b,t-2} + 0.085u^2_{b,t-3}$ (0.748) (0.081)　(0.147)　(0.170)	5.358 [2.315]
Conditional covariance function	$h_{sb,t} = 0.515 + \mathbf{0.091}u_{s,t-1}u_{b,t-1} + 0.207u_{s,t-2}u_{b,t-2} + \mathbf{0.177}u_{s,t-3}u_{b,t-3}$ (0.558) (0.103)　　(0.138)　　　　(0.101)	0.981 [0.089]
t-Student degrees of freedom parameter	**6.798** (2.698)	

(Continued)

TABLE 5.3 Estimates of a Bivariate, Three-State Markov Switching VAR(2) VECH t-Student ARCH(3) Model for U.S. Stock and Bond Excess Returns (*Continued*)

Estimated Transition Matrix in the table, boldfaced coefficients imply an estimated coefficient p-value equal to or below 0.1

	Regime 1	Regime 2	Regime 3	Ergodic Probs
Regime 1	**0.895**	0.031	0.074	0.143
	(0.358)	(0.042)		
Regime 2	**0.043**	**0.913**	0.044	0.331
	(0.021)	(0.177)		
Regime 3	0.001	**0.046**	0.953	0.526
	(0.010)	(0.018)		
Log-likelihood	−2,239.713			
Akaike information criterion	7.048	Saturation ratio	18.222	
Bayes-Schwartz information criterion	7.138	Regime classif. measure 1	1.729	
Hannan-Quinn information criterion	6.927	Regime classif. measure 2	29.461	
Number of parameters	72	Pseudo R^2—stocks (in %)	23.111	
		Pseudo R^2—bonds (in %)	15.006	

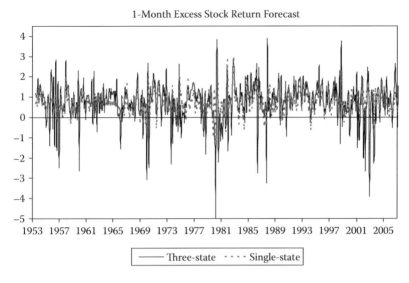

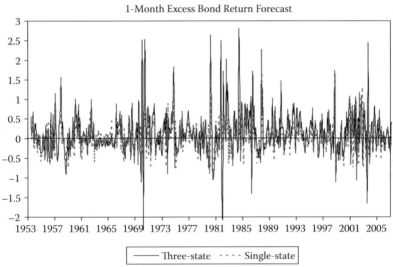

FIGURE 5.4   One-month-ahead forecasts of excess stock and bond returns from a bivariate, three-state Markov switching VAR(2) VECH t-Student ARCH(3) model.

symmetric ARCH(3) model (i.e., with no leverage). Figure 5.2 shows the smoothed probabilities of each of the three alternative regimes. Figure 5.3 shows the 1-month-ahead predictions for excess stock and bond returns, while Figure 5.4 shows 1-month-ahead predictions for equity and bond volatilities as well as the stock-bond correlation. In fact, the figures are

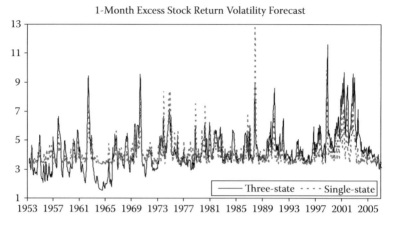

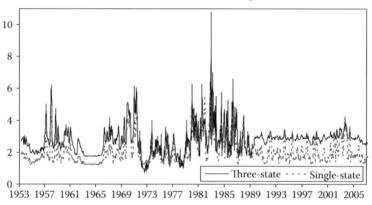

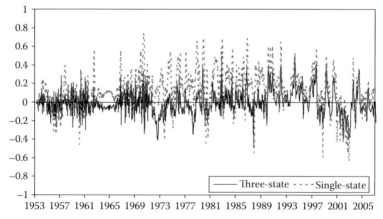

FIGURE 5.5 One-month-ahead forecasts of volatilities and correlation of excess stock and bond returns from a bivariate, three-state Markov switching VAR(2) VECH t-Student ARCH(3) model.

used to aid our effort of interpretation of the estimates presented in Table 5.3. In fact, here forecasts are used to assist our efforts to understand the properties of the models under consideration, while in Section 5.4, these forecasts are analyzed in terms of their accuracy.

Table 5.3 confirms the evidence reported in Section 5.3.2 and shows that structural instability is pervasive, affecting both parameters in the conditional mean and CH functions. Regime 1 is a mildly persistent state (the average duration is over 9 months) in which bond premia are relatively high (31 basis points per month) but equity premia are negative and rather large (–1.4% per month; in this case annualizing would be incorrect, as the regime has a duration inferior to 1 year). Equity returns are quite predictable, especially using one lag of the bond term premium and with a coefficient (positive) that is both typical in the literature and economically nonnegligible.

In unconditional terms (as implied by the regime-specific ARCH process), the volatility of both excess stock and bond returns is rather high, 4.7% and 4.3% per month, respectively, while correlation is essentially zero. The ARCH process is rather persistent for excess bond returns, and weaker for excess stock returns. Notice that this association between bear market periods and relatively high variance for both stocks and bonds also represents a type of leverage effect, able to explain the presence of pronounced asymmetries in the unconditional distribution of asset returns. Although there is some evidence of ARCH in the covariance, the effect is mild (the coefficient is 0.147 only at lag 1, and the other coefficients are practically zero). The regime-specific estimate of v (9.97) implies mild departures from normality, with slightly thicker tails in spite of having accommodated ARCH effects. Overall, this is a bear/recession regime of declining equity prices, declining interest rates (hence, of positive bond returns in excess of short-term rates), high volatility, and modest covariation (indeed) between shocks to stock and bond markets. In fact, this regime also holds all the features that have been previously identified with flight to quality phenomena in the literature: bond and stock markets seem to be affected by a different sentiment dynamics. Figure 5.2 corroborates our interpretation and shows spikes of regime 1 smoothed state probabilities in correspondence to a few major recession periods (1974–1976, 1979–1980, 1990–1991, 2001–2002, and late 2007), as well as to other spells of unrest or turmoil in the U.S. stock market (1987, 1998, 2000–2001, and again, late 2007).

Regime 2 is a persistent state (with average duration in excess of 11 months) that represents bull markets and periods of economic expansions typical of the early stages of good times. In (within-regime) unconditional terms, the equity premium is high and positive (14.1% in annualized terms), and bond premia are negative (but small, −0.82% in annualized terms) and indicate an essentially flat term structure with interest rates slowly increasing. While excess bond returns (apart from a small AR(1) component) become essentially unpredictable, excess equity returns are strongly predictable using lags of excess bond returns and with sensible coefficients (i.e., higher excess returns on long-term bonds forecast higher excess risk premia). Volatility almost completely evaporates from the financial markets, with levels of 13.5% in the equity market and 5% in the bond market (both in annualized terms), while the stock-bond correlation turns negative (albeit small, −0.14). Interestingly, the stock-bond covariance becomes completely unpredictable. Also in this regime, an estimate of v that equals 8.8 indicates only moderate departures from conditional bivariate normality. Figure 5.2 shows that this bull/stable (low volatility) regime fits market dynamics in correspondence of the early 1950s, the early 1960s and 1970s, the 1977–1978 economic rebound after the first oil shock, and the recent 2005–2006 period. Interestingly, while the probability of entering a bull/ stable period leaving the bear state is nonnegligible (0.074), once the system enters this state there is equal probability of leaving it to step back in a bear state (as happened during 2007) or to switch to the roaring third regime.

Regime 3 has interesting features on its own, but it can be best interpreted as a normal (one would like to add, textbook), highly persistent (average duration is 21 months) state in which the equity premium is positive and statistically significant (16.5% per annum), the bond premium is positive but modest (about 1% per annum), excess stock and bond returns are hardly predictable (i.e., their conditional mean function is approximately constant), and ARCH effects are moderate but conform to the general idea that the variance and covariance of financial returns are generally predictable using their own lagged values. Volatility is relatively high for both stocks (a textbook 16% per annum) and bonds (8%), and matches typical values reported in the literature as benchmarks. The unconditional correlation returns back to zero. The estimate of v also shows nonnegligible departures from bivariate normality, even after accommodating for ARCH. Figure 5.2 simply shows that the U.S. markets are roughly half of the time in this state (which is actually quite good news for long-run equity investors), with long spells of as many as 7 years during the late 1960s, most of the 1980s (but after 1983), and especially the stock market boom of 1992–1999 (with a short break in the summer of 1998, when regime 1 picks up the effects of the so-called Asian flu crises). In fact, the ergodic (long-run, steady-state) probabilities of the three regimes are 0.14, 0.33, and 0.53, respectively.

Figure 5.3 shows 1-month-ahead predictions for excess stock and bond returns obtained from the model estimated in Table 5.3. As a benchmark, we also plot predictions from the single-state VAR(2) VECH t-Student ARCH(3) model in Table 5.2, i.e., a framework that cannot infer from the data any differences between bull and bear regimes. Clearly, the two forecasts tend to somewhat co-move (their correlation is 0.61), although their variability differs considerably (the standard deviation of the single-state forecast is 0.53% versus 1.00% for the three-state forecasts). This depends on the fact that a MS model is able to capture the presence of instability in the conditional mean function, something impossible for a single-state framework. In particular, while the single-state forecast of the equity premium is positive most of the time, the three-state predictions often become and stay negative for a few consecutive months (e.g., in 2001–2002). From Figure 5.1 we know that even long periods of negative average excess equity returns have been typical of the U.S. financial history. Similar comments apply to one-step-ahead forecasts of excess bond returns, although in this case both forecasts oscillate around the zero axis, which is consistent with the basic properties of the data; in spite of their positive correlation (0.67), it remains true that three-state forecasts are much more volatile (0.54%) than single-state ones (0.37%).

Figure 5.4 depicts instead the 1-month-ahead forecasts of volatility and correlation produced by the single- and three-state models. In this case, the differences are rather important. As far as equity volatility is concerned, we immediately notice that while the single-state volatility has in practice a lower bound at around 3.4% per month, this is not the case for the three-state model. This is important, because Figure 5.1 has shown that historical periods exist (essentially, mid-1960s and then mid-1990s) in which equity volatility has fallen below the threshold of 3% per month (i.e., an annualized 10%). Otherwise, the two forecasts appear once more highly correlated (0.65), with a much higher standard deviation for the MS forecasts (1.4 versus 1%). In the case of bond volatility, the main difference lies in the heterogeneous level of the forecast series in the periods 1953–1970 and 1989–2007, when single-state forecasts are considerably lower (although they still strongly co-move, the overall correlation is 0.69), at around 2%, than three-state forecasts, which oscillate around 2.5%. Finally, the panel devoted to correlation predictions offers an unusual perspective: in this case, single-state forecasts are in fact more volatile (0.21) than three-state ones (0.14), as the latter tend to simply oscillate in a narrow range around zero with only two exceptions—the late 1980s and early 1990s, when predicted correlation becomes positive and averages almost 0.2, and 2000–2004, when the predicted correlations turn negative and gravitate around –0.2. Interestingly, both facts are consistent with Figure 5.1. The single-state forecasts gyrate much more but seem to have problems in reproducing these stylized facts. However, and consistent with the evidence in both Sections 5.3 and 5.4.1, it is once more clear that—with or without regimes—ARCH effects in correlation coefficients are rather modest for the series at hand.

5.5 FORECASTING PERFORMANCE

Ultimately, what matters of a model is not (or not mainly) its ability to produce an accurate in-sample fit, but especially its out-of-sample forecasting performance. In fact, when the models are flexible enough thanks to the presence of a high number of parameters, accuracy of fit is relatively not surprising. However, rich parameterizations are also well known to introduce large amounts of estimation uncertainty, which normally end up deteriorating the out-of-sample performance. To assess whether a three-state MS ARCH model offers any useful prediction performance, we implement the following pseudo out-of-sample recursive strategy. We obtain recursive parameter estimates over expanding samples starting with 1953:05–1983:12, 1953:05–1984:01, etc., up to 1953:05–2007:11, for the three-state bivariate MS ARCH model as well as for three sets of benchmarks:

1. A single-state VAR(2) VECH t-Student ARCH(3) model.

2. Two separate, univariate MS VAR(2) t-Student ARCH(3) models for excess stock and bond returns, respectively; the MS model for excess stock returns is a two-state one, while the MS model for excess bond returns is a four-state one, consistent with the indications of Section 5.4.

3. A constant mean and constant variance model with Gaussian IID shocks, which in practice corresponds to a random walk with drift, homoskedastic benchmark for bond and stock prices (also assuming a constant short-term interest rate).

This gives a sequence of 288 sets of parameter estimates specific to each of the models.[*] For instance, the MS ARCH (2) model generates 288 sets of regime-specific intercepts, VAR coefficients, ARCH coefficients, t-Student degrees of freedom parameters, and transition matrices. At each final date in the expanding sample, i.e., on 1983:12, 1984:01, etc., up to 2007:11, we calculate 1-month-ahead forecasts for (mean) excess stock and bond returns, their variance, and their covariance. We call $\hat{r}_{i,t}^M$, $(\hat{\sigma}_{i,t}^M)^2$, and $\hat{\rho}_t^M$ the forecast generated by model M at time t, i = stock, bond. Finally, we evaluate the accuracy of the resulting forecasts, by calculating the resulting forecast errors defined as $e_{t+1}^M \equiv r_{i,t+1} - \hat{r}_{i,t}^M$, $v_{t+1}^M \equiv u_{i,t+1}^2 - (\hat{\sigma}_{i,t}^M)^2$, and $\xi_{t+1}^M \equiv u_{stock,t+1}u_{bond,t+1} - \hat{\rho}_t^M$. In the following, we refer to forecast errors generically as ε_{t+1}^M, where ε_{t+1}^M coincides with e_{t+1}^M in the case of levels (means), with v_{t+1}^M in the case of variances, and with ξ_{t+1}^M when covariances are concerned. Of course, our chapter has devoted most of its attention to modeling MS dynamics in second moments, and much less care in producing accurate models for conditional means. However, we deem of a certain importance to also report information on the predictive accuracy of our models for the levels of excess stock and bond returns, especially in the light of the noncompletely disappointing in-sample R^2 in Table 5.3.

Table 5.4 reports summary statistics concerning the quality of the relative forecasting performance. In particular, we report three statistics illustrating predictive accuracy: the root mean-squared forecast error (RMSFE),

$$RMSFE^M \equiv \sqrt{\frac{1}{288} \sum_{t=1983:12}^{2007:11} \left(\varepsilon_{t+1}^M\right)^2},$$

the prediction bias, and the forecast error standard deviation:

$$Bias^M \equiv \sqrt{\frac{1}{288} \sum_{t=1983:12}^{2007:11} \left(\varepsilon_{t+1}^M\right)} \qquad SD^M \equiv \sqrt{\frac{1}{287} \sum_{t=1983:12}^{2007:11} \left[\varepsilon_{t+1}^M - \frac{1}{288}\sum_{t=1983:12}^{2007:11}\right]^2}.$$

---

[*] The starting date at 1983:12 corresponds to the need to have at least ten observations per parameter available for the initial estimation exercise in the case of the least parsimonious, three-state ARCH(3) model with seventy-two parameters. In fact, in correspondence of the initial 1953:05–1983:12 period, the saturation ratio is 10.22. Since our entertaining of a richly parameterized bivariate three-state MS ARCH model implies a considerable loss of data for out-of-sample evaluation, we also propose measures of forecast performance starting with the 1953:05–1974:12 period, but rely on final parameter estimates and smoothed state probabilities in the case of the three-state MS ARCH model, for a total of 396 sequential forecasts.

TABLE 5.4 Forecasting Performance Measures

Panel A: Means

	Excess Stock Returns				Excess Bond Returns			
	Three-State t-ARCH(3)	Single-State t-ARCH(3)	Univariate t-ARCH(3)	IID Homoskedastic	Three-State t-ARCH(3)	Single-State t-ARCH(3)	Univariate t-ARCH(3)	IID Homoskedastic
(Pseudo Out-of-) Sample 1983:12–2007:11 (Recursive Estimates Only)								
Root mean squared forecast error	**3.742**	3.803	3.856	3.828	1.920	1.932	1.930	**1.911**
Bias	−0.219	−0.132	−0.228	**0.083**	**0.141**	0.254	0.296	0.251
Standard deviation of forecast errors	**3.736**	3.800	3.850	3.827	1.915	1.915	1.907	**1.895**
Mean absolute error	**2.492**	2.531	2.594	2.583	**1.324**	1.337	1.340	1.324
(Pseudo Out-of-) Sample 1974:12–2007:11 (End-of-Sample Estimates for Three-State Model)								
Root mean squared forecast error	**3.867**	3.878	3.916	3.902	2.085	2.089	2.107	**2.085**
Bias	−0.118	**−0.072**	−0.164	0.130	**0.056**	0.165	0.197	0.182
Standard deviation of forecast errors	**3.865**	3.877	3.912	3.900	2.084	2.083	2.098	**2.077**
Mean absolute error	**2.619**	2.626	2.658	2.661	**1.407**	1.410	1.435	1.413

Panel B: Variances

	Excess Stock Returns				Excess Bond Returns			
	Three-State t-ARCH(3)	Single-State t-ARCH(3)	Univariate t-ARCH(3)	IID Homoskedastic	Three-State t-ARCH(3)	Single-State t-ARCH(3)	Univariate t-ARCH(3)	IID Homoskedastic
(Pseudo Out-of-) Sample 1983:12–2007:12 (Recursive Estimates Only)								
Root mean squared forecast error	**36.147**	36.967	36.602	36.217	**6.289**	6.568	6.354	7.556
Bias	−5.719	**0.187**	0.727	0.669	−0.322	−0.561	−0.138	4.582
Standard deviation of forecast errors	**35.692**	36.967	36.595	36.211	6.280	6.544	6.353	**6.008**
Mean absolute error	14.983	14.824	**14.546**	15.155	**3.638**	3.991	3.786	3.666
Log-error square	3.874	**3.854**	3.862	4.140	6.173	6.809	6.247	**6.008**
Log-error absolute	**1.239**	1.253	1.247	1.320	3.985	4.019	**3.249**	3.666
(Pseudo Out-of-) Sample 1974:12–2007:12 (End-of-Sample Estimates for Three-State Model)								
Root mean squared forecast error	**34.633**	35.411	35.045	34.977	**8.181**	8.545	8.647	9.830
Bias	−3.361	**0.690**	0.937	1.725	**−0.097**	−0.395	1.070	5.442
Standard deviation of forecast errors	**34.469**	35.405	35.032	34.934	**8.180**	8.536	8.581	8.186
Mean absolute error	15.121	15.332	15.176	15.274	**4.227**	4.577	4.296	4.354
Log-error square	2.043	**1.998**	1.999	2.044	8.292	8.718	8.411	**8.186**
Log-error absolute	1.286	1.272	**1.270**	1.310	4.626	4.575	**3.912**	4.354

(Continued)

TABLE 5.4 Forecasting Performance Measures (*Continued*)

Panel C: Covariance

	Three-State t-ARCH(3)	Single-State t-ARCH(3)	IID Homoskedastic
	(*Pseudo Out-of-) Sample 1983:12 – 2007:11 (Recursive Estimates Only*)		
Root mean squared forecast error	**10.922**	11.292	11.204
Bias	0.699	**−0.133**	−1.171
Standard deviation of forecast errors	**10.899**	11.291	11.143
Mean absolute error	5.385	5.522	**5.310**
	(*Pseudo Out-of-) Sample 1974:12–2007:11 (End-of-Sample Estimates for Three-State Model*)		
Root mean squared forecast error	**12.373**	12.699	12.524
Bias	1.765	0.531	**0.138**
Standard deviation of forecast errors	**12.246**	12.688	12.523
Mean absolute error	5.814	5.987	**5.736**

Note: Boldfaced values of the performance measures indicate that a model is superior to the competitors'.

Notice that the three statistics are not independent, as it is well known that $MSFE^M \equiv [Bias^M]^2 + [SD^M]^2$, i.e., the MSFE can be decomposed in the contribution of bias and variance of the forecast errors.

As discussed in Hamilton and Susmel (1994), the root MSFE may often be an unfair standard for data. Even though all of our estimates have produced values of the t-Student degrees of freedom parameter—which is well known to correspond to the highest existing moment under a t-Student distribution—it may be prudent to also evaluate forecasting accuracy for second moments using three alternative loss functions, the mean absolute forecast error (MAE),

$$MAE^M \equiv \sqrt{\frac{1}{288} \sum_{t=1983:12}^{2007:11} |\varepsilon_{t+1}^M|},$$

the $LogE$-square, and the $LogE$-absolute metrics,[*]

$$LogES^M \equiv \sqrt{\frac{1}{288} \sum_{t=1983:12}^{2007:11} \left[\ln\left(u_{i,t+1}^2\right) - \ln\left(\hat{\sigma}_{i,t}^M\right) \right]^2} \quad LogEA^M$$

$$\equiv \sqrt{\frac{1}{288} \sum_{t=1983:12}^{2007:11} |\ln\left(u_{i,t+1}^2\right) - \ln\left(\hat{\sigma}_{i,t}^M\right)|}.$$

Table 5.4 reports summaries of forecasting performance for means (levels), variances, covariances, and excess stock and bond returns, separately. Besides the recursive, pseudo out-of-sample 1983:12–2007:11 exercise, the table also shows results for a longer 1974:12–2007:11 recursive exercise in which—because otherwise the number of available observations would become insufficient—the three-state model is implemented using final parameter estimates and simply updating the state probabilities using smoothed state probabilities, which imparts a forward-looking bias to the forecasts. Panel A shows that a three-state VAR t-Student ARCH model produces an interesting predictive performance as far as the mean of excess stock returns is concerned. Even though RMSFEs of 3.7% per month are just below the monthly volatility of equity returns, it is well known that stock returns are very hard to accurately predict. In this sense, in line with earlier results by Guidolin and Timmermann

---

[*] Both *LogES* and *LogEA* are of course defined for variances only, because $\ln(u_{stock,t+1} \, u_{bond,t+1})$ need not be defined.

(2006) and Guidolin and Ono (2005), MS proves useful by allowing one to model bull and bear states as different statistical regimes. Interestingly, this performance remains good when a MAE criterion is employed, while the (bias² + variance) decomposition shows that the relatively low RMSFE for the three-state model mostly comes from a reduction in variance, which seems to be typical of forecasts from mixture distributions (see Guidolin and Ono, 2005, for similar remarks). Although differences are small, it seems that after the MS three-state model, the next best thing is to use a single-state VAR t-Student ARCH bivariate model, which—by comparison to the univariate model performance—confirms the importance of modeling multivariate relationships and capturing the presence of predictability in covariances. These findings fully extend to the longer 1974–2007 period, although one should be reminded that in this case the three-state model is used on the basis of full-sample estimates and state probabilities, and therefore suffers from a substantial look-ahead bias. Results are considerably more mixed for the mean of excess bond returns: in this case, a three-state model produces the lowest MAE, but not the lowest RMSFE; in a RMSFE metric, it seems that a simple, constant means, variances, and covariances benchmark may actually do better (this is similar to a standard finding that a random walk model wins in out-of-sample experiments). Also in this case, results hardly change for the longer sample period. However, it must also be stressed that in this case all the measured performances are rather close and hard to tell apart. For instance, on the 1983–2007 out-of-sample period, the random walk model produces a RMSFE of 1.91% against a RMSFE of 1.92% for the three-state model, while forecasts have typical standard deviations of 1.9%. So we read the part of panel A of Table 5.4 as mostly indicating that models are very similar in their predictive performance for mean excess bond returns.

Panel B of Table 5.4 focuses instead on the recursive, 1-month-ahead prediction of variances. In this case, besides RMSFE and MAE, we also report the two additional *LogES* and *LogEA* measures we have introduced. In both samples, the three-state ARCH model produces the minimum RMSFE for excess stock return variance. This is a consequence of the lower volatility of the MS forecast errors (but the bias tends to be rather large and negative, i.e., the model tends to overshoot the forecast of variance, on average). In the truly out-of-sample exercise, the three-state model also delivers the lowest *LogEA* measure, although differences are small. Interestingly, and despite the look-ahead bias, the three-state model does not return the lowest *LogES* and *LogEA* in the 1974–2007 exercise. In general, the constant variance benchmark tends to be the second best model. The fact that a richly parameterized MS t-Student ARCH model outperforms a simple IID homoskedastic benchmark with only five parameters is very interesting, almost surprising. The results for excess bond return variance are similar, if not better. In the truly out-of-sample recursive exercise, the three-state model displays the lowest RMSFE and MAE; in the longer sample, the performance is even superior, since the three-state model also shows minimum values for bias and standard deviation of forecast errors. Interestingly, in this case the second best is represented by a univariate t-Student ARCH model, and not by the random walk, which represents a fundamental difference between equity and bond variance dynamics. However, the homoskedastic benchmarks fare rather well in the *LogES* and *LogEA* metrics.

Finally, panel C reports performance measures related to covariances. There is some tension between RMSFE and MAE results. Under the former criterion, a three-state model turns out to be superior to the remaining benchmarks in the horse race. Once more, this is due to the fact that MS forecast errors show the minimum standard deviation among all forecast functions. However, the MAE favors the constant covariance benchmark, even though results are very close (e.g., 5.31 under constant covariance versus 5.39 under the three-state model). Also in this case, results are the same when a longer sample is employed.

5.6 CONCLUSION

This chapter has investigated the presence of Markov regimes in the conditional heteroskedastic (i.e., ARCH effects in variances and covariances) dynamics for U.S. excess stock and bond returns. We found strong evidence in favor of a three-state model in which the regimes are persistent and correspond to easily interpretable market states, as defined by conditional means, volatility, and the possibility to predict mean returns. Additionally, all attempts at simplifying the model, either by imposing tight parameter restrictions or by constraining the Markovian probabilistic structure of regimes in equity and bond markets, have led to rejections. However, persistence and predictability in the stock-bond covariance tend to be weak in at least two states out of three. We find that the three-state model outperforms a number of benchmarks in out-of-sample prediction tests concerning means, variances, and covariances.

It would be interesting to extend Guidolin and Timmermann (2007) results on optimal strategic asset allocation between stocks and bonds to the MS ARCH case. For instance, one could compute the (pseudo) out-of-sample, 1-month portfolio performance of the three-state model versus a set of sensible benchmarks under different levels of the risk aversion parameter that trades off expected portfolio returns and variance. We leave this extension for future work.

ACKNOWLEDGMENTS

Yu Man Tam provided excellent research assistance. All errors remain my own.

REFERENCES

Ang, A., and Bekaert, G. (2002). International asset allocation with regime shifts. *Review of Financial Studies* 15:1137–87.

Bollerslev, T., Engle, R. F., and Wooldridge, J. M. (1988). A capital asset pricing model with time-varying covariances. *Journal of Political Economy* 96:116–31.

Campbell, J. Y., Chan, Y. L., and Viceira, L. M. (2003). A multivariate model of strategic asset allocation. *Journal of Financial Economics* 67:41–80.

Engle, R. F. (1982). Autoregressive conditional heteroskedasticity with estimates of the variance of United Kingdom inflation. *Econometrica* 50:987–1007.

Engle, R. F., Lilien, D. M., and Robins, R. P. (1987). Estimating time varying risk premia in the term structure: The Arch-M model. *Econometrica* 55:391–407.

Engle, R. F., and Ng, V. (1993). Measuring and testing the impact of news on volatility. *Journal of Finance* 48:1749–78.

Fama, E. F., and French, K. R. (1989). Business conditions and expected returns on stocks and bonds. *Journal of Financial Economics* 25:23–49.

Guidolin, M., and Ono, S. (2005). Are the dynamic linkages between the macroeconomy and asset prices time-varying? Working paper 2005–056, Federal Reserve Bank of St. Louis, Missouri.

Guidolin, M., and Timmermann, A. (2006). An econometric model of nonlinear dynamics in the joint distribution of stock and bond returns. *Journal of Applied Econometrics* 21:1–22.

Guidolin, M., and Timmermann, A. (2007). Asset allocation under multivariate regime switching. *Journal of Economic Dynamics and Control* 31:3503–44.

Haas, M., Mittnik, S., and Paolella, M. S. (2004a). A new approach to Markov-switching GARCH models. *Journal of Financial Econometrics* 2:493–530.

Haas, M., Mittnik, S., and Paolella, M. S. (2004b). Mixed normal conditional heteroskedasticity. *Journal of Financial Econometrics* 2:211–50.

Hamilton, J. D. (1988). Rational-expectations econometric analysis of changes in regime: An investigation of the term structure of interest rates. *Journal of Economic Dynamics and Control* 12:385–423.

Hamilton, J. D. (1989a) A New Approach to the Economic Analysis of Nonstationary Time Series and the Business Cycle. *Econometrica*, 57(2): 357–384.

Hamilton, J. D. (1989b). A New Approach to the Economic Analysis of Nonstationary Time Series and the Business Cycle. *Econometrica*, 57:357–384.

Hamilton, J. D. (1994) Time Series Analysis. Princeton University Press: Princeton, NJ.

Hamilton, J. D., and Lin, G. (1996). Stock market volatility and the business cycle. *Journal of Applied Econometrics* 11:573–93.

Hamilton, J. D., and Susmel, R. (1994). Autoregressive conditional heteroskedasticity and changes in regime. *Journal of Econometrics* 64:307–33.

Lamoureux, C. G., and Lastrapes, W. (1993). Forecasting stock-return variance: Toward an understanding of stochastic implied volatilities. *Review of Financial Studies* 6:293–326.

Turner, C. M., Startz, R., and Nelson, C. R. (1989). A Markov model of heteroskedasticity, risk, and learning in the stock market. *Journal of Financial Economics* 25:3–22.

CHAPTER 6

A DCC-VARMA Model of Portfolio Risk

A Simple Approach to the Estimation of the Variance-Covariance Matrix of Large Stock Portfolios

Valerio Potì

CONTENTS

6.1 INTRODUCTION

Forecasts of asset volatilities and correlations are required inputs for the estimation of portfolio value-at-risk (VaR) (for a discussion, see Szego, 2002), for portfolio optimization and for the construction of optimal hedge ratios. Because of their clustering behavior, asset volatilities and especially the volatility of stocks and portfolios of stocks typically exhibit high persistence (Engle and Patton, 2001), especially at relatively high frequencies (such as weekly, daily, and higher-frequency data). Engle (2002), among others, argues that this is often the case of asset correlations. Volatilities and correlations, especially in stock portfolios, are therefore natural candidates to be modeled using conditional autoregressive specifications such as ARCH and GARCH models.

One of the advantages of univariate models is their ease of estimation relative to more complex multivariate specifications. Univariate exponential moving averages (EWMAs) and generalized conditional heteroskedasticity (GARCH) models are therefore routinely used in the financial industry to estimate the VaR of portfolios of stocks over a given time horizon (see, for example, Bauwens et al., 2003). The univariate model, however, must be reestimated every time portfolio weights change. This can be a serious drawback if the portfolio contains large positions in financial instruments with nonlinear payoffs, payoffs that depend on the correlation structure of asset returns, and instruments that require time-consuming numerical procedures for their pricing. This problem does not arise, or it is considerably milder, if estimates of the full variance-covariance matrix are available. The elements of the latter can be directly used to compute the portfolio variance and hence VaR for any set of asset weights.

One of the advantages of multivariate models is that they provide estimates of such a matrix. Another advantage is that suitable restrictions can be imposed to make sure that the estimated variance-covariance matrix is positive-definite, as in the well-known BEKK model put forth by Engle and Kroner (1995). In fact, the correlation matrix used in VaR estimates should be positive-definite to ensure that pair-wise correlations lie between −1 and 1, and that every subportfolio of assets under consideration has a correlation that lies between −1 and 1 with any other subportfolio. Imposing the further requirement that volatilities are nonnegative ensures that the variance-covariance matrix is positive semidefinite. This is a desirable property of any estimate of the variance-covariance matrix of asset returns, as it ensures that the variance of every variable and of every combination of the variables is always nonnegative. This rules out the often counterfactual possibility that investors can enjoy "free lunches" by forming risk-free positive-expected return arbitrage portfolios. Positive-definiteness of the variance-covariance matrix is more restrictive, and it ensures that the latter is invertible, thus making it possible to use it further in econometric (such as a weighting matrix in weighted least squares regressions) and financial (notably in asset pricing and portfolio optimization algorithms) applications.

Multivariate models, however, have the drawback of being computationally very intensive. As a consequence, a number of industry applications impose heavy restrictions on the structure of the variance-covariance matrix to curb computational requirements. For example, in JP Morgan's RiskMetrics™ procedure, each element of the conditional variance-covariance matrix is estimated, using exponential smoothing, as a univariate EWMA. Such a procedure, however, when viewed as a data generating process as opposed to a filter, is formally degenerate (Nelson, 1990). Alternatively, in the spirit of orthogonal factor models, many financial institutions adopt the simplifying assumption that most of the variation of asset returns is generated by a limited number of common factors, whereas the residual variation is attributable to purely idiosyncratic (and hence negligible) sources of variability. However, as Campbell et al. (2001) and Kearney and Potì (2008) show for portfolios of U.S. stocks and Euro-area stocks, respectively, this can be in many circumstances a somewhat heroic assumption, as the idiosyncratic portion is the main component of total volatility, and the number of stocks needed to diversify it away is large and tends to increase at times of market distress.

To reduce the computational burden while retaining the appeal of multivariate variance-covariance estimates, Engle and Sheppard (2001) and Engle (2002) proposed the dynamic conditional correlation (DCC) GARCH model. Engle and Sheppard (2001) use the DCC-GARCH model to estimate the conditional variance-covariance matrix of up to 100 assets represented by S&P sector indices and Dow Jones Industrial Average stocks and conduct specification tests using JP Morgan's RiskMetrics industry standard EWMA as a benchmark. They show that the DCC-GARCH model captures important empirical features of the conditional variances and covariances of the stock indices considered in their analysis. Morillo and Pohlman (2002) estimate the variance-covariance matrix of daily and weekly returns on the twenty-four largest international stock market indices included in the MSCI World Index using sample unconditional estimators and various conditional models. They use their variance-covariance matrix estimates in a portfolio optimization exercise, and report that the optimal portfolio based on DCC-GARCH estimates dominates the optimal portfolios based on all the other estimates.

While the DCC-GARCH model is less computationally demanding than other multivariate models, the two-step procedure recommended by Engle (2002) for its estimation can still pose substantial challenges in most financial industry applications, as the number of stocks included in financial institutions' and investors' portfolios can be very large. We thus propose a simpler way to estimate this model. To this end, we first show that its parameters can be derived from the estimated parameters of an ARMA model of the average conditional correlation process. With estimates of these parameters in hand, one can very simply recover the dynamics of the full conditional correlation and variance-covariance matrices. In the remainder of this chapter, we first provide a brief but formal outline of the DCC-GARCH model. We then recast this model in terms of the DCC-VARMA model and show how the parameters of the former can be very simply recovered by estimating the parameters of the latter. Finally, to demonstrate how the DCC-VARMA model can be used to recover the parameters of the DCC-GARCH model, we present an empirical application to a portfolio of forty-two stocks included in the Eurostoxx50 index.

6.2 THE DCC-GARCH MODEL

Engle and Sheppard (2001) and Engle (2002) formulate the DCC-GARCH model as a two-step estimator of conditional variances and correlations. The first step entails the estimation of the mean model for each asset in the portfolio under consideration, nested in a univariate GARCH model of the asset conditional variance. These univariate variance estimates are used to standardize the zero-mean return innovations for each asset. In the second step, a model of the first moments of the standardized zero-mean return innovations, nested in a scalar multivariate GARCH model of conditional second moments, is estimated. Engle and Sheppard (2001) show that this two-step procedure produces consistent maximum-likelihood parameter estimates. More formally, consider the following specification of the multivariate process of returns:

$$r_t \mid \Omega_{t-1} \sim N(0, H_t) \tag{6.1}$$

and

$$H_t \equiv D_t R_t D_t \tag{6.2}$$

Here, r_t is the $k \times 1$ vector of zero-mean return innovations conditional on Ω_{t-1}, the information set available at time $t-1$, R_t is the $k \times k$ conditional correlation matrix, and D_t is a $k \times k$ diagonal matrix. The elements on its main diagonal are the conditional standard deviations of the returns on each asset. Therefore:

$$[H_t]_{ij} = h_{ij}$$

and

$$[D_t]_{ij} = d_{t,ij} = \sqrt{h_{ij}} \quad i = j \tag{6.3}$$
$$[D_t]_{ij} = d_{t,ij} = 0 \quad i \neq j$$

Notice that, from Equations (6.1) and (6.2), $E(\frac{r_{t,i}r_{t,j}}{d_{t,i}d_{t,j}} | \Omega_{t-1}) = R_{t,ij}$. Possible simple specifications for the GARCH processes followed by D_t^2 and R_t are the following:

$$D_t^2 = \bar{D}^2(1 - A - B) + A(r_{t-1}r_{t-1}') + BD_{t-1}^2 \tag{6.4}$$

$$R_t = \bar{R}(1 - \alpha - \beta) + \alpha \varepsilon_{t-1}\varepsilon_{t-1}' + \beta R_{t-1} \tag{6.5}$$

The expressions A and B in Equation (6.4) are $k \times k$ diagonal coefficient matrices. In Equation (6.5), α and β are scalar matrices with all the elements on the main diagonal equal to a and b, respectively. $\bar{R}$ is a $k \times k$ matrix with ones on the main diagonal. It represents the long-run, baseline level to which conditional correlations mean-revert. Engle and Sheppard (2001) and Engle (2002) propose to estimate Equation (6.2) in two steps. The first step entails the estimation of univariate models[*] of the return on each asset nested in a GARCH model, described by Equation (6.4), of its conditional variance. This yields consistent, time-varying estimates of the parameters of the process followed by D_t^2. Then, Engle and Sheppard (2001) and Engle (2002) suggest to estimate the parameters of the process of R_t, conditional on the estimated D_t. This entails standardizing r_t by the estimated D_t to obtain the $k \times 1$ vector ε_t. The parameters of the process followed by R_t are found by estimating a multivariate model of ε_t nested in a multivariate scalar GARCH model, provided by Equation (6.5), of its conditional second moments.

6.3 THE DCC-VARMA MODEL

We now introduce an alternative and much simpler estimation procedure of the parameters of the DCC-GARCH model. Rewrite Equation (6.5) as

$$\varepsilon_t\varepsilon_t' = \bar{R}(1 - \alpha - \beta) + (\alpha + \beta)\varepsilon_{t-1}\varepsilon_{t-1}' - \beta(\varepsilon_{t-1}\varepsilon_{t-1}' - R_{t-1}) + (\varepsilon_t\varepsilon_t' - R_t) \tag{6.6}$$

---

[*] The presence of an intercept term ensures that the estimated residuals are zero-mean random variables.

or equivalently,

$$\varepsilon_t \varepsilon_t' = \bar{R}(1-\alpha-\beta) + (\alpha+\beta)\varepsilon_{t-1}\varepsilon_{t-1}' - \beta[\varepsilon_{t-1}\varepsilon_{t-1}' - E(\varepsilon_{t-1}\varepsilon_{t-1}')]$$

$$+[\varepsilon_t \varepsilon_t' - E(\varepsilon_t \varepsilon_t')]$$

$$= \bar{R}(1-\alpha-\beta) + (\alpha+\beta)\varepsilon_{t-1}\varepsilon_{t-1}' - \beta e_{t-1} + e_t \qquad (6.7)$$

where

$$e_t = \varepsilon_t \varepsilon_t' - E(\varepsilon_t \varepsilon_t') = \varepsilon_t \varepsilon_t' - R_t$$

Equations (6.6) and (6.7) show that we can rewrite the multivariate conditional correlation process in ARMA(1,1) form. For each pair of assets i and j, the elements e_{ijt} of the residual matrices e_t are martingale differences by construction.[*] Therefore, they are stationary and serially uncorrelated,[†] thus satisfying the requirements of standard inference procedures. This was already shown by Bollerslev, Engle, and Nelson (1994) in a univariate setting. Consider now the following transformation $P'\varepsilon_t \varepsilon_t' P$ of Equation (6.7) from $R^{k \times k}$ to R^1. As shown in the appendix,

$$P'\varepsilon_t \varepsilon_t' P = (1-\alpha-\beta)P'RP + (\alpha+\beta)P'\varepsilon_{t-1}\varepsilon_{t-1}'P - \beta u_{t-1} + u_t \qquad (6.8)$$

with

$$u_t = P'e_t P = \sum_{i=1}^{k}\sum_{j=1}^{k} p_i p_j e_{t,ij} = \sum_{i=1}^{k}\sum_{j=1}^{k} p_i p_j (\varepsilon_{t,i}\varepsilon_{t,j} - R_{t,ij}) \qquad (6.9)$$

Then

$$E(u_t) = \sum_{i=1}^{k}\sum_{j=1}^{k} p_i p_j E(e_{t,ij}) = \sum_{i=1}^{k}\sum_{j=1}^{k} p_i p_j E[\varepsilon_t \varepsilon_t' - E(\varepsilon_t \varepsilon_t')]$$

---

[*] In fact, if $\Omega_t = \{\varepsilon_{t,i}\varepsilon_{t,j}, \varepsilon_{t-1,i}\varepsilon_{t-1,j}, ..., \varepsilon_{1,i}\varepsilon_{1,j}\}$ denotes information available at time t, $E(\varepsilon_{t,i}\varepsilon_{t,j}|\Omega_{t-1}) = R_t$ by Equations (6.1), (6.2), and (6.5) $\forall$ i and $j \in [1, 2,..., k]$. Therefore, $E(e_{t,ij}|\Omega_{t-1}) = [E(\varepsilon_{t,i}\varepsilon_{t,j}) - R_t|\Omega_{t-1}] = 0$, thus satisfying the definition of martingale difference. For further details and references on relevant asymptotic results, see Hamilton (1994).

[†] The martingale difference condition is stronger than absence of serial correlation but weaker than independence, since it does not rule out the possibility that higher moments might depend on past realizations. See Hamilton (1994).

In Equation (6.8), P is a $k \times 1$ vector. If P is chosen to be a vector of fixed weights, then Equation (6.8) provides an ARMA(1,1) model for the weighted average of conditional correlations across all the assets in the sample. Therefore, using Equation (6.8), the parameters a and b of the conditional correlation process can be estimated by fitting a univariate ARMA(1,1) model to the observations on $\varepsilon_t \varepsilon_t'$ transformed using the weights vector P. By Equation (6.5), $E(\varepsilon_t \varepsilon_t') = R_t$. Therefore, we can write:[*]

$$E(ut) = \sum_{i=1}^{k} \sum_{j=1}^{k} p_i p_j E(\varepsilon_t \varepsilon_t' - R_t) = 0 \qquad (6.10)$$

Moreover,

$$Var(u_t) = E\left(u_t^2\right) = E\{P'[\varepsilon_t \varepsilon_t' - E(\varepsilon_t \varepsilon_t')]PP'\varepsilon_t \varepsilon_t' - E(\varepsilon_t \varepsilon_t')P\} \qquad (6.11)$$

By Equations (6.7) and (6.11), we can rewrite $Var(u_t)$ as follows:

$$Var(u_t) = E\left[(P'\varepsilon_t \varepsilon' P)^2 - 2P'\varepsilon_t \varepsilon' PP' R_t P + (P' R_t P)^2 \right] \qquad (6.12)$$

By Equation (6.5), R_t depends on $\varepsilon_{t-1}\varepsilon_{t-1}'$ but not on $\varepsilon_t \varepsilon_t'$. Therefore, we can rewrite Equation (6.12) as follows:

$$Var(u_t) = E\left[(P'\varepsilon_t \varepsilon_t' P)^2 \right] - (P' R_t P)^2 \qquad (6.13)$$

From Equation (6.9), the variable u_t is a weighted average of the random variables $e_{t,ij}, i = [1, \dots, k]$ and $j = [1, \dots, k]$. Under the null implied by Equation (6.7), i.e., the null that $e_t = \varepsilon_{t-1}\varepsilon_{t-1}' - E(\varepsilon_{t-1}\varepsilon_{t-1}')$, u_t is a martingale difference by construction with zero mean and variance as in Equation (6.13). Moreover, if we further assume all the $\varepsilon_{t,i} \varepsilon_{t,j}$ terms to be independently distributed across all the pairs of assets i and j in the sample (i.e., $E(\varepsilon_{t,i} \varepsilon_{t,j} \varepsilon_{t,k} \varepsilon_{t,l}) = 0$, k and $l \neq i$ and j), u_t is a sum of independently distributed random variables, and therefore, if the number of assets k is large (above thirty should suffice), it can be considered to be normally distributed (by the weak law of large numbers). The residuals in Equation (6.8) thus satisfy the assumptions about the error term typical of a standard ARMA model. As a consequence, we can use the familiar standard test statistics that have been developed by the econometric literature on ARMA models in order to test hypotheses about the parameters of Equation (6.8). Note that the variance of u_t is

---

[*] Notice also that R_t does not depend on but only on its lagged value.

the variance of the transformation of the correlation process defined by the operation of pre- and postmultiplication by P, i.e., the variance of the average correlation estimator. Also, the assumption that $\varepsilon_{t,i}\varepsilon_{tj}$ is independently distributed across the assets in the sample implies that the second moments of the asset return process are independently distributed. In other words, we exclude that there is a correlation[*] of the correlations (or of the variances). This is fully consistent with the scalar structure of the variance-covariance matrix of the DCC-GARCH model. This explains why, under the null that the DCC-GARCH(1,1) model in Equation (6.5) adequately represents the correlation in the data, $P'\varepsilon_t\varepsilon_t'P$ follows the ARMA(1,1) process described by Equation (6.8), and its error term u_t satisfies the assumptions of classical inference procedures. In other words, the assumption that the $\varepsilon_{t,i}\varepsilon_{t,j}$ are independently distributed for each pair of i and j assets in the sample means that, while we remove the assumption of independently distributed errors at the level of the mean Equation (6.1) by explicitly modeling correlations, we make this assumption at the level of the correlation matrix. Somewhat technically, in order to guarantee that u_t is normally distributed, we assume that, while $r_t \sim N(0, H_t)$, with H_t not necessarily diagonal,

$$vec(\varepsilon_t\varepsilon_t')_{k(k-1)/2} \sim \Phi(vec(R_t), diag[\varphi_{ij}]_{k(k-1)/2})$$

where $vec(.)$ is the operator that stacks the off-diagonal elements of the argument matrix in a conformable vector and Φ is some distribution (with finite variance). In the scalar standardized multivariate specification of the DCC-GARCH provided by Equation (6.5), $\phi_{ij} = 1, \forall\, i = j$.

6.4 EUROSTOXX50 INDEX STOCKS CONDITIONAL CORRELATIONS

To demonstrate how to use the DCC-VARMA model to recover the DCC-GARCH model, we estimate the parameters of the process followed by the daily conditional correlations among the returns on forty-two stocks

---

[*] Notice that, strictly speaking, there can be no correlation of correlations because the latter have not been defined as random variables, but rather as deterministic quantities that change over time, according to the dynamics specified by Equation (6.5), of the multivariate asset returns stochastic process. Instead, using a formally valid but slightly boring and less effective formulation, we should say that we exclude that standardized asset returns squares and cross-products are correlated.

TABLE 6.1   DCC-GARCH and DCC-VARMA

Model	Coefficient	Coefficient Estimate	Standard Error	T-Ratio	p-Value
DCC-VARMA	$\alpha + \beta$	.9985	.0015	659.56	.000
	$-\beta$	−.9621	.0065	−148.62	.000
	Implied DCC-GARCH Coefficients				
	α	.0364	—	—	—
	β	.9621	.0065	148.62	.000
DCC-GARCH	α	.0020	.0001	18.05	.000
	β	.9899	.0007	1,351.75	.000

Note: This table reports the estimated coefficients, standard errors, and p-values for the DCC-GARCH and DCC-VARMA models of the monthly correlations among forty-two stocks from the Eurostoxx50 index over the period February 12, 1993–November 23, 2001.

included in the Eurostoxx50 index (the leading stock market index in the Euro area) using both the DCC-GARCH in Equation (6.5) and the DCC-VARMA model in Equation (6.8). The results are reported in Table 6.1. The expression $\bar{R}$ in Equations (6.5) and (6.8) is set equal to the unconditional sample correlation matrix. Notice that in order to constraint the ARMA(1,1) process in Equation (6.8) to mean-revert to $P'RP(1-\alpha-\beta)$, we need an estimate of the sum of the α and β parameters. This sum can be recovered from the estimation of Equation (6.8) with no restriction on the constant term.[*] All the parameters of the DCC-GARCH model are identified in the estimation that uses the DCC-VARMA model, with the exception of the standard error of α. This seems to be a minor shortcoming of the DCC-VARMA procedure, more than compensated by far greater speed and simplicity. There is a slight difference between the DCC-GARCH parameter estimates obtained by first estimating the DCC-VARMA model and then solving for the parameters of the DCC-GARCH model ($\alpha = 0.0364$, $\beta = 0.9621$) and those obtained by direct estimation of the latter ($\alpha = 0.0020$, $\beta = 0.9899$). We tend to consider the former more reliable, since the direct estimation of the DCC-GARCH model relies on a numerical optimization procedure that, in turn, is heavily influenced by the value of the initial guesses and by the shape and local behavior of regions of the likelihood function.

---

[*] I also iterated this procedure. I set α and β in equal to their point estimates from the restricted estimation of Equation (6.8) and reestimated. The parameter estimates converged to the reported values after just two iterations.

6.5 FINAL REMARKS AND CONCLUSIONS

Modeling time-varying correlations among the assets included in large portfolios can be a difficult and computationally demanding task. Yet, in many portfolio and risk management applications, it is an essential one. In this chapter, to reduce the computational burden, we proposed a simple approach to the estimation of the essential parameters of the DCC-GARCH model. Further research might fruitfully attempt to apply this approach to more general versions of such model, for example DCC-GARCH specifications that allow for asymmetric reactions of the estimated volatility and correlation processes to positive and bad news, as in Cappiello, Engle and Sheppard (2006) and Kearney and Potì (2006).

APPENDIX

Consider the transformation of Equation (6.7) from R^{xk} to R^1:

$$P'\varepsilon_t\varepsilon_t'P = P'[\overline{R}(1-\alpha-\beta)+(\alpha+\beta)\varepsilon_{t-1}\varepsilon_{t-1}'-\beta(e_{t-1}) + (e_t)]P$$

$$= P'\overline{R}(1-\alpha-\beta)P+P'(\alpha+\beta)\varepsilon_{t-1}\varepsilon_{t-1}'P - P'\beta(e_{t-1})P + P'(e_t)P$$

$$= (1-\alpha-\beta)P'\overline{R}P+(\alpha+\beta)P'\varepsilon_{t-1}\varepsilon_{t-1}'P - \beta P'(e_{t-1})P + P'(e_t)P$$

$$= (1-\alpha-\beta)P'\overline{R}P+(\alpha+\beta)P'\varepsilon_{t-1}\varepsilon_{t-1}'P - \beta u_{t-1} + u_t \qquad (6.A1)$$

with

$$u_t = P'e_t P = P'[\varepsilon_t\varepsilon_t' - E(\varepsilon_t\varepsilon_t')]P = P'\left[\sum_{j=1}^k e_{t,ij}P_j\right] = \sum_{i=1}^k P_i \sum_{j=1}^k e_{t,ij}P_j \qquad (6.A2)$$

$$= \sum_{i=1}^k \sum_{j=1}^k P_i e_{t,ij}P_j = \sum_{i=1}^k \sum_{j=1}^k P_i P_j e_{t,ij} = \sum_{i=1}^k \sum_{j=1}^k P_i P_j (\varepsilon_{t,i}\varepsilon_{t,j} - R_{t,ij})$$

then

$$E(ut) = \sum_{i=1}^k \sum_{j=1}^k P_i P_j E(e_{t,ij}) = \sum_{i=1}^k \sum_{j=1}^k P_i P_j E[\varepsilon_t\varepsilon_t' - E(\varepsilon_t\varepsilon_t')] \qquad (6.A3)$$

REFERENCES

Bauwens, L., Laurent, S., and Rombouts, J. V. K. (2003). Multivariate GARCH models: A survey. CORE discussion paper, Université Catholique de Louvain, Louvain-La-Neuve, Belgium.

Bollerslev, T., Engle, R. F., and Nelson, D. B. (1994). ARCH models. In *Handbook of econometrics*, ed. R. F. Engle and D. L. McFadden. 4:2961–3038. North-Holland, Amsterdam: Elsevier Sciences.

Campbell, J. Y., Lettau, M., Malkiel, B. G., and Xu, Y. (2001). Have individual stocks become more volatile? An empirical exploration of idiosyncratic risk. *Journal of Finance* 56:1–43.

Cappiello, L., Engle, R. F., and Sheppard, K. (2006). Asymmetric Dynamics in the Correlations of Global Equity and Bond Returns, *Journal of Financial Econometrics* 4:537–572.

Engle, R. F. (2002). Dynamic conditional correlation: A simple class of multivariate GARCH models. *Journal of Business and Economic Statistics* 20:339–50.

Engle, R. F., and Kroner, K. F. (1995). Multivariate simultaneous generalised ARCH. In *Econometric theory*. 1:122–150. Cambridge, MA: Cambridge University Press.

Engle, R. F., and Patton, A. J. (2001). What good is a volatility model? *Quantitative Finance* 1:237–245.

Engle, R. F., and Sheppard, K. (2001). Theoretical and empirical properties of dynamic conditional correlation multivariate GARCH. Working paper, University of California, San Diego.

Hamilton, J. D. (1994). *Time series analysis*. Princeton, NJ: Princeton University Press.

Kearney, C., and Potì, V. (2008). Have European stocks become more volatile? An empirical investigation of idiosyncratic risk and market risk in the Euro-area. *European Financial Management* 14:419–44.

Morillo, D., and Pohlman, L. (2002). Large scale multivariate GARCH risk modelling for long-horizon international equity portfolios. Working paper, Panagora Asset Management, New York.

Nelson, D. B. (1990). Stationarity and persistence in the GARCH(1,1) model. *Econometric Theory* 6:318–34.

Potì, V., and Kearney, C. (2006). Correlation Dynamics in European Equity Markets, Research In International Business and Finance 20:305–321.

Szego, G. (2002). Measures of risk. *Journal of Banking and Finance* 26:1253–72.

CHAPTER 7

The Economic Implications of Volatility Scaling by the Square-Root-of-Time Rule

Craig Ellis and Maike Sundmacher

CONTENTS

7.1 INTRODUCTION

Traditional models of financial asset returns are based on a number of simplifying assumptions. Among these is the primary assumption that consecutive price changes (returns) follow a standard Brownian motion process, i.e., a Gaussian random walk. One critical feature of standard Brownian motions is the relationship between moments of the distribution over different time intervals. For instance, when returns follow a Gaussian random walk the temporal dimension of volatility is irrelevant, meaning that the volatility of returns measured over one time interval can be precisely estimated by linearly rescaling the volatility of returns over any other time interval using the square-root-of-time rule ($\sqrt{T}$).

One common application of volatility scaling laws in financial economics is the estimation of long-horizon volatility in models such as the capital asset pricing model (CAPM) and the Black-Scholes option pricing model, both of which are typically estimated on the basis of annual equivalent measures of asset return and volatility. Investors wishing to price positions based on a target level of volatility and preferred investment horizon also typically assume a Gaussian random walk when scaling short-horizon volatility to estimate long-horizon volatility (Celati, 2004).

When asset returns do not follow a Gaussian random walk, however, annualizing volatility by the square root of time will not correctly estimate the real level of risk associated with an investment. The misestimation of volatility has distinct implications for modeling risk/return relations. One implication for investors discussed by Holton (1992) is that investment risk is a function not only of the type of asset being considered, but also of the investor's preferred investment horizon. Mandelbrot (1971) argues, however, that statistical nonrandomness would only yield economically significant outcomes when investors have infinitely long investment horizons.

Volatility scaling laws for Gaussian processes have been previously examined by Batten and Ellis (2001), and for different financial time series using a variety of techniques by Mantegna and Stanley (1995), Mandelbrot (1997), Canning et al. (1998), and Gençay et al. (2001). Despite the widespread practice of linearly rescaling risk by the square root of time, Diebold et al. (1988) argue strongly against this on the basis that the procedure overestimates long-horizon volatility. Müller et al. (1990) alternatively show that intra-day foreign exchange volatility scales faster than the square root of time. Peters (1994) similarly finds the same result for the volatility of daily foreign exchange returns.

The objective of research in this chapter is to demonstrate the implications for scaling financial asset risk when long-term returns do not follow a Gaussian random walk. Using a selection of Australian Stock Exchange (ASX) Top 50 equities, volatility at horizons ranging from 1 day to 1 year is measured directly and for longer time horizons, by linearly rescaling short-horizon volatility. The research shows that even small deviations from pure random behavior can lead investors to significantly misestimate their real level of risk.

The remainder of the chapter is structured as follows: Section 7.2 provides an overview of linear rescaling and its applications in finance.

Section 7.3 describes the data and research methodology, and summarizes the results. Section 7.4 provides some concluding remarks.

7.2 LINEAR RESCALING AND VOLATILITY

Under the assumption that a time series is independent and identically distributed (IID), the temporal dimension of risk is irrelevant, such that the volatility of asset returns calculated over one time interval (e.g., annual) can be estimated from the volatility of returns over any other interval (e.g., monthly or weekly). According to the square-root-of-time rule the volatility of annual returns should, for example, be $\sqrt{12}$ times the volatility of monthly returns and $\sqrt{52}$ times that of weekly returns for a Gaussian random walk. Stated mathematically, linear rescaling implies that the volatility of returns X_t should scale as

$$\left[\sigma^2(\log P_t - \log P_{t-k})\right]^{0.5} = \left[k\sigma^2(\log P_t - \log P_{t-1})\right]^{0.5}$$

$$= k^{0.5}\left[\sigma^2(X_t)\right]^{0.5} \tag{7.1}$$

where P_t and P_{t-1} are the current and last price for the asset, P_{t-k} is the price k periods previous to t, and X_t is the asset return. Introducing a scale exponent h, which denotes the rate at which volatility scales over time, Equation (7.1) can be generalized to give

$$\left[\sigma^2(\log P_t - \log P_{t-k})\right]^{0.5} = (k/n)^h\left[\sigma^2(\log P_t - \log P_{t-n})\right]^{0.5} \tag{7.2}$$

where k and n are positive constants ≥ 1.

The value of the scale exponent in Equation (7.2) is $h = 0.5$ for a Gaussian time series and implies that the volatility of returns should scale precisely linearly with the square root of time. Estimating the implied volatility of annual returns from the observed volatility of monthly returns using Equation (7.2), for example, would require $k = 252$ and $n = 21$. Using observed weekly volatilities would alternatively require $k = 252$ and $n = 5$. For time series that exhibit dependence—that is, long-run autocorrelation—the exponent value is $0 \leq h \leq 1$; $h \neq 0.5$. For functions exhibiting positive long-term dependence, the value of the exponent will be $h > 0.5$. Negative long-term dependent functions are alternatively characterized by exponent values of $h < 0.5$. The volatility of returns for series characterized by $h > 0.5$ should be expected to scale faster than $\sqrt{T}$, and slower than $\sqrt{T}$ for series characterized by $h < 0.5$.

7.2.1 Example: Scaled Equity Returns

The process of linear rescaling and some basic implications for the mis-estimation of implied volatility may be illustrated by the comparison of real asset returns versus a simulated Gaussian random walk. For the purposes of this example we have selected three stocks from our portfolio of ASX equities: Alumina[*] (exhibiting long-term negative dependent returns), Macquarie Group[†] (exhibiting near Gaussian returns), and Telstra[‡] (exhibiting long-term positive dependent returns). Returns for each equity series and their volatility are calculated for intervals of $k = 1, 2,\ldots, 252$ periods (1 day, 2 days,…, 1 year). Implied volatilities are then estimated by rescaling observed n-interval volatilities ($n < k, n = k, n > k$) as per Equation (7.2) using the Gaussian exponent $h = 0.5$. A summary of observed volatilities versus implied volatilities for each equity series over selected return intervals is provided in Table 7.1. Volatility scale exponents—the value of exponent h in Equation (7.2) for which implied k-interval returns exactly equal observed k-interval returns—are also calculated. These are provided in Table 7.2. For returns series that conform to a Gaussian random walk, implied k-interval volatilities should exactly equal observed k-interval volatilities and the volatility scale exponent should be $h = 0.5$. The general findings presented in this example are characteristic of the behavior of all the equities in our sample; that is, none of the equity return series conform to a strict Gaussian random walk. The discussion of these outcomes provides the basis of our later analysis of scaling stock market volatility in Section 7.3.

Values along the diagonal (in bold) in Table 7.1 show the observed k-interval volatility of returns for each equity series. Off-diagonal values show implied k-interval volatilities from observed n-interval volatilities using Equation (7.2) given $h = 0.5$. Reading down each column in the table, the extent to which rescaled short-interval volatilities mis-estimate observed long-interval volatilities is evident. Reading across each row in the table similarly shows the extent to which rescaled long-interval volatilities mis-estimate observed short-interval volatilities. For Alumina returns exhibiting negative long-term dependence, linearly rescaled short-interval volatilities

---

[*] Alumina specializes in bauxite mining, alumina refining, and aluminum smelting.
[†] Macquarie Group is a nonoperating holding group and the parent entity of various banking and nonbanking organizations.
[‡] Telstra is a telecommunications and information products and services provider.

TABLE 7.1   Observed versus Scaled Volatility

Alumina: Exhibited Negative Dependence, $h < 0.5$					
k/n	1	5	21	63	126	252
1	**0.0185**	0.0187	0.0175	0.0166	0.0159	0.0135
5	0.0413	**0.0418**	0.0390	0.0371	0.0355	0.0302
21	0.0847	0.0856	**0.0800**	0.0761	0.0728	0.0619
63	0.1467	0.1483	0.1386	**0.1318**	0.1262	0.1072
126	0.2074	0.2097	0.1960	0.1864	**0.1784**	0.1516
252	0.2934	0.2965	0.2771	0.2636	0.2523	**0.2143**

Macquarie Group: Exhibited Near Gaussian, $h \approx 0.5$					
k/n	1	5	21	63	126	252
1	**0.0176**	0.0178	0.0167	0.0175	0.0165	0.0161
5	0.0394	**0.0398**	0.0373	0.0391	0.0369	0.0361
21	0.0808	0.0817	**0.0764**	0.0801	0.0756	0.0739
63	0.1399	0.1415	0.1323	**0.1388**	0.1309	0.1280
126	0.1978	0.2000	0.1871	0.1962	**0.1852**	0.1811
252	0.2798	0.2829	0.2645	0.2775	0.2619	**0.2561**

Telstra: Exhibited Positive Dependence, $h > 0.5$					
k/n	1	5	21	63	126	252
1	**0.0136**	0.0133	0.0142	0.0149	0.0162	0.0182
5	0.0303	**0.0298**	0.0317	0.0333	0.0363	0.0406
21	0.0621	0.0610	**0.0649**	0.0682	0.0744	0.0833
63	0.1076	0.1057	0.1124	**0.1180**	0.1288	0.1442
126	0.1522	0.1495	0.1589	0.1669	**0.1822**	0.2040
252	0.2152	0.2114	0.2247	0.2361	0.2577	**0.2885**

consistently overestimate observed long-interval volatilities, and rescaled long-interval volatilities consistently underestimate observed short-interval volatilities. For Telstra returns exhibiting positive long-term dependence, the reverse is true. In the case of Macquarie Group, with near Gaussian returns, observed long- and short-interval returns neither consistently under- nor overestimate observed short- and long-interval returns. This outcome is consistent across all combinations of return interval $k, n = \{1, 2, \ldots, 252\}$.

Rearranging Equation (7.2) to solve for the volatility scale exponent yields the value of h for which scaled n-interval volatility (i.e., implied volatility) exactly equals the observed k-interval volatility. The emerging interest in and significance of volatility scale exponents is discussed by Bouchaud (2002) in his review of stylized facts concerning financial time-series data. Volatility scale exponents for the three equities in this example are provided in Table 7.2.

TABLE 7.2 Volatility Scale Exponents

	Alumina			
k/n	**1**	**5**	**21**	**63**	**126**
5	0.5066				
21	0.4813	0.4529			
63	0.4741	0.4535	0.4543		
126	0.4688	0.4500	0.4476	0.4370	
252	0.4432	0.4172	0.3966	0.3509	0.2647

	Macquarie Group			
k/n	**1**	**5**	**21**	**63**	**126**
5	0.5069				
21	0.4816	0.4532			
63	0.4980	0.4924	0.5436		
126	0.4863	0.4761	0.4944	0.4163	
252	0.4840	0.4746	0.4869	0.4419	0.4675

	Telstra			
k/n	**1**	**5**	**21**	**63**	**126**
5	0.4887				
21	0.5141	0.5426			
63	0.5223	0.5436	0.5449		
126	0.5372	0.5614	0.5764	0.6263	
252	0.5530	0.5793	0.6006	0.6446	0.6629

Consistent with the above described outcome from Table 7.1—that short-interval volatilities rescaled by the square root of time consistently overestimate observed long-interval volatilities when returns are long-term negative dependent, and vice versa when returns are long-term positive dependent—scale exponents for the volatility of Alumina returns in Table 7.2 are less than $h = 0.5$, and are greater than $h = 0.5$ for Telstra. Characteristic of compounding in the estimation error over time, it is further noted that scale exponents in Table 7.2 diverge from $h = 0.5$ as k increases relative to n and as the values of both k and n increase. Overall these results highlight the failure of the square-root-of-time rule in general, and specifically its failure for scaling short-interval volatilities to estimate implied long-interval volatilities.

7.3   SCALING STOCK MARKET VOLATILITY

In the previous section we examined some of the statistical implications of linearly rescaling volatility by the square-root-of-time rule using individual equities by way of example. We now turn attention to some of the economic implications of volatility scaling for portfolio selection as a testable analog for scaling relations in market volatility.

7.3.1 Data and Sample

For the purposes of this analysis we employ equities listed in the S&P/ASX Top 50 Index.* The sample data comprise daily closing prices for thirty equities from the index over the 20 years to March 7, 2008. As of March 7, 2008, the total capitalization of the S&P/ASX Top 50 was USD 269,332,273.37 (AUD 289,978,761.17). The largest company in the index—BHP Billiton—had capitalization of USD 60,714,614.00 (AUD 65,368,878.12), and the smallest—Bendigo Bank—capitalization of USD 1,024,316.03 (AUD 1,102,838.10). Summary analysis of daily returns for all equities in the sample, including the moments of the distribution, indicates that all the series are highly non-Gaussian. Both the Anderson-Darling and Ryan-Joiner *p*-values reject the null hypothesis of normality for all series. This result is not surprising since leptokurtic, nonnormal distributions are common in financial time series (Pagan, 1996). Unit root tests are also conducted for each series. The results of these tests show that all the difference series are stationary. Test results for the log levels of each series were not able to reject the unit root null hypothesis.† While the random walk model assumes the presence of a single unit root, such that the time series being observed can be decomposed into a set of stationary increments, the test results for the log levels data do not prove that the returns series were random. Overall, these results suggest that scaling short-interval volatility by the square root of time should yield inaccurate estimates of long-interval volatility.

For the purposes of our analysis of stock market volatility scaling laws, we consider two portfolios of equities: one low capitalization and one high capitalization portfolio. The ranking of equities by capitalization in this study follows from Hawawini (1983), who presents evidence that systematic volatilities for smaller than average capitalization equities are expected to increase as the time horizon over which they are measured (daily, weekly, etc.) increases. Systematic volatilities for equities with higher than average capitalization are alternatively expected to decline as the time horizon

---

* Representing approximately 75% of total equity market capitalization, the S&P/ASX 50 comprises the fifty largest shares—by capitalization—on the Australian Stock Exchange (ASX). The S&P/ASX 50 is also included as part of the S&P Global 1200 Index.
† Full test results and summary statistics for individual equities are available on request.

TABLE 7.3 Observed k-Interval Volatilities

k =		1	5	21	63	126	252
Low Capitalization Equities							
Mean		0.0178	0.0382	0.0736	0.1243	0.1790	0.2636
Standard error of mean		0.0013	0.0028	0.0054	0.0088	0.0129	0.0197
Maximum		0.0291	0.0627	0.1193	0.1932	0.2814	0.4216
Minimum		0.0130	0.0269	0.0493	0.0867	0.1258	0.1736
Mean scale exponent	0.4963						
Standard error of mean	0.0100						
High Capitalization Equities							
Mean		0.0161	0.0361	0.0705	0.1182	0.1674	0.2360
Standard error of mean		0.0007	0.0017	0.0035	0.0062	0.0105	0.0213
Maximum		0.0202	0.0459	0.0958	0.1627	0.2473	0.3977
Minimum		0.0117	0.0267	0.0520	0.0872	0.1096	0.1413
Mean scale exponent	0.4754						
Standard error of mean	0.0126						

increases. The low capitalization portfolio herein comprises the bottom 50% of S&P/ASX Top 50 Index equities studied, and the high capitalization, the top 50%. The total capitalization of the low capitalization portfolio is USD 25,118,339.31 (AUD 27,043,862.31), and USD 179,495,798.82 (AUD 193,255,597.35) for the high capitalization portfolio.[*] Summary statistics for the volatility of high and low capitalization equities over intervals of 1 day ($k = 1$) to 1 year ($k = 252$) are provided in Table 7.3.

While observed k-interval volatilities for high capitalization equities are consistently larger than for low capitalization equities at all intervals, the difference is not statistically significant at the 0.10 level. That volatility scale exponents using Equation (7.3) for low versus high capitalization equities are not statistically different, and that they are not significantly different from the Gaussian null $h = 0.5$, suggests that capitalization in itself bears no consistent relation to scaled volatility.

7.3.2 Some Economic Implications of Scaling Market Volatility

Under the assumption that the underlying returns series conforms to a Gaussian random walk, the relation described by Equation (7.2) gives rise to the concept of the temporal irrelevance of volatility, following which the length of the investor's investment horizon is irrelevant, such that for a given level of volatility all investors are rewarded with an identical risk

---

[*] Capitalization as of March 7, 2008.

premium whether they are investing over the short term or the long term. What are some of the implications for investors, however, if volatility does not scale according to the Gaussian null?

Consider the case of a long-term versus a short-term investment in a single (homogeneous) asset where the risk premium is determined using a CAPM-type model by the asset's total risk and the market risk premium. Assuming that the returns conform to a Gaussian random walk, the temporal dimension of volatility is irrelevant and the annualized risk premium is the same at both investment horizons, short term and long term. In the case, however, where asset returns are negative long term dependent with $h < 0.5$, investments held in the long term are relatively less risky and investments held in the short term are relatively more risky than predicted by the square-root-of-time rule. Alternatively, where asset returns are positive long term dependent with $h > 0.5$, investments held in the short term are relatively less risky and investments held in the long term are relatively more risky than predicted by the square-root-of-time rule. Since the relative volatility of the investment depends on the investor's horizon, how do we determine which investment (short-term or long-term) should offer the highest return?

Within the context of the CAPM, Figure 7.1 uses the Sharpe ratio (the slope of the capital market line) to demonstrate the relative level of mispricing to investors from rescaling short-interval volatility for equal weighted portfolios of low capitalization and high capitalization equities. Herein observed k-interval volatilities are annualized using Equation (7.2) given $h = 0.5$—the square root of time—and the difference in Sharpe ratios using the observed annual volatility and k-interval annualized volatilities is recorded.

Consistent with the earlier suggestion that series exhibiting long-term negative dependence appear relatively less risky in the long run, results for both portfolios show that linearly rescaled short-interval (1-day, 1-week, 1-month) volatility consistently understates the observed annual volatility, and hence the Sharpe ratio. Furthermore, the degree of mispricing increases with increasing values of the market risk premium, $R_p - R_f$. In strong contrast, however, to the prior finding of no significant difference in mean volatility scale exponents of low versus high capitalization equities, results in the figure indicate highly significant differences in relative error for low versus high capitalization portfolios.

As is well known for portfolio volatility in general, this latter result suggests that portfolio volatility scale exponents likewise depend on the

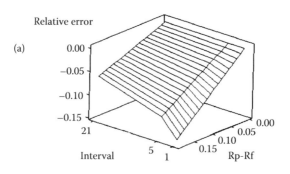

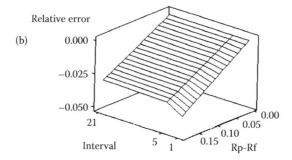

FIGURE 7.1   Sharpe ratios for equal weighted portfolios. (a) Low capitalization. (b) High capitalization.

covariance of equity pairs, or more specifically on how equity covariances scale over different interval lengths, i.e., the covariance scale exponent(s). That is, the portfolio volatility scale exponent (0.4805 and 0.4897 for equal weighted low and high capitalization portfolios, respectively) is not simply the weighted sum of the equity volatility scale exponents.

Incorporating equity covariance, we now turn to consider the implications of volatility scaling for estimating efficient portfolios using Markowitz portfolio theory (Markowitz, 1952). Using observed k-interval returns $k = 1, 2,…, 252$, we estimate the minimum variance set (MVS) and identify the minimum variance portfolio (MVP). The MVS and MVP are then reestimated using annualized k-interval returns, volatilities, and covariances. Summary findings for return intervals of $k = 1, 5, 21,$ and 252 periods are provided in the following tables and figures. Tables 7.4 and 7.5 present the mean, volatility, and volatility scale exponent (measured over all intervals) for the MVP comprising low capitalization equities and high capitalization equities, respectively. Figures 7.2 and 7.3 meanwhile show the relative positions of the MVSs calculated on the basis of annualized

TABLE 7.4   Mean Return and Volatility of Low Capitalization Minimum Variance Portfolios

$k =$		1	5	21	252
		Observed			
Mean return		0.0002	0.0010	0.0044	0.0658
Volatility		0.0074	0.0151	0.0289	0.0612
		Annualized			
Mean return		0.0514	0.0519	0.0533	0.0658
Volatility		0.1179	0.1075	0.1002	0.0612
Scale exponent	0.3945				

mean returns, volatility, and covariances for portfolios of low and high capitalization equities, respectively.

Relative to minimum variance portfolios estimated from observed k-interval returns, annualized MVPs in Table 7.4 exhibit significantly overstated levels of both mean return and volatility. The coefficient of variation for annualized MVPs can be shown, however, to be approximately equal to the annualized coefficient of variation of MVPs estimated from observed k-interval returns, suggesting that while rescaling by the square root of time does significantly overstate the real risk of the portfolio, it does not necessarily mis-state the level of return received per unit of volatility borne by the investor.* That the trade-off between return and volatility remains proportionate, not only at the MVP but for all points along the MVS, can be inferred from comparison of the slopes of the annualized MVSs in Figure 7.2.

Given the mean volatility scale exponent for low capitalization equities in Table 7.3, linearly rescaled k-interval volatilities ($k < 252$) are expected to produce close estimates of the observed annual volatility of low capitalization minimum variance portfolios. That linearly rescaled k-interval volatilities are significantly different from the observed annual volatility for all lengths of k demonstrates the additional contribution to mispricing of rescaled short-interval covariances when used in conjunction with rescaled short-interval volatility for the purposes of estimating long-term portfolio volatility.

The volatility of annualized high capitalization MVPs in Table 7.5 consistently overestimates observed annual volatility by a higher amount relative to the estimation error for low capitalizations MVPs in Table 7.4. This

---

* Using values from the table, $\frac{\mu_{k,annualized}}{\sigma_{k,annualized}} \approx \frac{\mu_k}{\sigma_k}\sqrt{\frac{252}{k}}$ indicating both mean returns and volatility are equally overstated.

TABLE 7.5　Mean Return and Volatility of High Capitalization Minimum Variance Portfolios

k =	1	5	21	252
		Observed		
Mean return	0.0003	0.0017	0.0090	0.1540
Volatility	0.0082	0.0179	0.0358	0.0654
		Annualized		
Mean return	0.0779	0.0843	0.1075	0.1540
Volatility	0.1305	0.1271	0.1241	0.0654
Scale exponent 0.3755				

result is contrary to that for equal weighted portfolios in Figure 7.1, wherein high capitalization portfolios demonstrated a lower relative forecast error than low capitalization portfolios. The result is consistent, though, with the lower relative volatility scale exponent for high capitalization MVPs

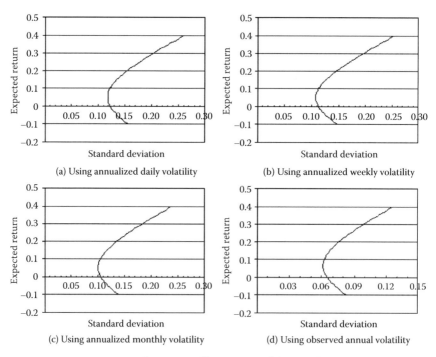

FIGURE 7.2　Low capitalization efficient portfolios. (a) Using annualized daily volatility. (b) Using annualized weekly volatility. (c) Using annualized monthly volatility. (d) Using observed annual volatility.

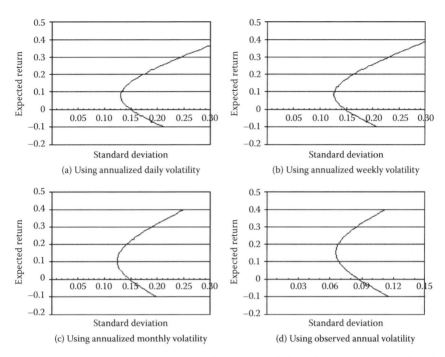

FIGURE 7.3 High capitalization efficient portfolios. (a) Using annualized daily volatility. (b) Using annualized weekly volatility. (c) Using annualized monthly volatility. (d) Using annualized annual volatility.

(0.3755 in Table 7.5 versus 0.3945 in Table 7.4), suggesting that different equity weights may intensify the aforementioned impact of rescaled volatilities and covariances on estimates of long-term portfolio volatility.

As is the case for low capitalization MVPs, the coefficient of variation for annualized high capitalization MVPs in Table 7.5 is not significantly different from the annualized coefficient of variation of MVPs estimated from observed k-interval returns. Irrespective then of differences in capitalization, equity weightings, covariances, and equity scale exponents, the level of portfolio return received per unit of volatility borne by the investor is largely invariant to errors due to the inappropriate scaling of short-horizon volatility.

7.4 CONCLUSION

Traditional financial modeling predicts that all investors are mean-variance optimizers and, in sharing a common economic view of the world, value investments identically irrespective of their individual preferred investment horizons. When asset returns conform to a Gaussian random walk the volatility of an asset over a long return interval can be precisely calculated by scaling the observed volatility from shorter return intervals using the square-root-of-time rule. However, since the traditional measure of volatility used in portfolio selection models (i.e., standard deviation) transpires from the Gaussian distribution, it fails to account for dependence in returns and the actual volatility of long-horizon returns may be misspecified.

This chapter examines some of the implications of statistical long-term dependence for scaling volatility at different investment horizons. Having formally defined the concept of linear rescaling and shown by way of example the implications of long-term dependence for scaling equity volatilities, we examine the scaling properties of portfolios comprising equities included in the S&P/ASX Top 50 Index. The general results for all the equity series indicated that the equity returns series did not follow a Gaussian random walk. That rescaled volatilities tended to overestimate the true level of risk is consistent with the equity series scaling at less than the square root of time. While contrary to prior research by Müller et al. (1990) and Peters (1994), our findings are consistent with those of Diebold et al. (1988) and suggest the presence of structured short-term dependence (e.g., GARCH) as well as possible long-term dependence. With respect to the implications for investors, the use of rescaled volatility estimates in various models of financial risk and return implies that any rankings derived from these models would depend on time horizon used to annualize the observed short-horizon volatility.

REFERENCES

Batten, J., and Ellis, C. (2001). Scaling properties of Gaussian processes. *Economics Letters* 72:291–96.

Bouchaud, J. (2002). An introduction to statistical finance. *Physica A* 313:238–51.

Canning, D., Amaral, L. A. N., Lee, Y., Meyer, M., and Stanley, H. E. (1998). Scaling the volatility of GDP growth rates. *Economics Letters* 60:335–41.

Celati, L. (2004). *The dark side of risk management: How people frame decisions in financial markets.* Harlow, UK: Financial Times Prentice Hall.

Diebold, F., Hickman, A., Inoue, A., and Schuermann, T. (1998). Scale models. *Risk* 11:104–7.

Gençay, R., Selçuk, F., and Whitcher, B. (2001). Scaling properties of foreign exchange volatility. *Physica A* 289:249–66.

Hawawini, G. (1983). Why beta shifts as the return interval changes. *Financial Analysts Journal* 39:73–77.

Holton, G. A. (1992). Time: The second dimension of risk. *Financial Analysts Journal* 48:38–45.

Mandelbrot, B. B. (1971). When can price be arbitraged efficiently? A limit to the validity of the random walk and martingale models. *Review of Economics and Statistics* 53:225–36.

Mandelbrot, B. B. (1997). *Fractals and scaling in finance: Discontinuity, concentration, risk.* Berlin: Springer.

Mantegna, R. N., and Stanley, H. E. (1995). Scaling behaviour in the dynamics of an economic index. *Nature* 376:46–49.

Markowitz, H. M. (1952). Portfolio selection. *Journal of Finance* 7:77–91.

Müller, U. A., Dacorogna, M. M., Olsen, R. B., Pictet, O. V., Schwarz, M., and Morgenegg, C. (1990). Statistical study of foreign exchange rates, empirical evidence of a price change scaling law, and intraday analysis. *Journal of Banking and Finance* 14:1189–208.

Pagan, A. (1996). The econometrics of financial markets. *Journal of Empirical Finance* 3:15–102.

Peters, E. E. (1994). *Fractal market analysis.* Hoboken, NJ: John Wiley & Sons.

CHAPTER 8

Jumps and Microstructure Noise in Stock Price Volatility

Rituparna Sen

CONTENTS

8.1 INTRODUCTION

Accurate specification of volatility is of crucial importance in several financial and economic decisions, such as portfolio allocation, risk management using measures like value at risk, and pricing and hedging of derivative securities. In the past, squared returns have been a frequently used proxy for volatility. However, as pointed out in Andersen and Bollerslev (1998), squared returns are a very noisy estimate for volatility. Another candidate is implied volatility, which is obtained by inverting option prices. But it is model dependent and incorporates some price of risk, since it actually measures expected future volatility.

Hence, there is a need for accurate model free measures of volatility. Note that the volatility of a price process is fairly model free. "Any log-price process subject to a no-arbitrage condition and weak auxiliary assumptions will constitute a semi-martingale that may be decomposed into a locally predictable mean component and a martingale with finite second moment" (Andersen et al., 2005). The predictable quadratic variation of the martingale is the volatility.

A new proxy for volatility, termed realized volatility, has been introduced concurrently in Andersen et al. (2003b) and Barndorff-Nielsen and Shephard (2004). If prices have continuous paths and are not contaminated by microstructure noise, then realized volatility is a consistent estimator of daily integrated volatility. Andersen et al. (2004) show that simple reduced-form time-series models for realized volatility using high-frequency data outperform the commonly used GARCH and related stochastic volatility models in forecasting future volatility.

It is believed, though, that log price processes may display jumps, due to, for example, macroeconomic and financial announcement effects. Recent studies have highlighted the significance of allowing different treatments of the jump and continuous sample path components, in estimating parametric stochastic volatility models (e.g., Andersen et al., 2002; Chernov et al., 2003; Eraker et al., 2003; Ait-Sahalia, 2004), in nonparametric realized volatility modeling (e.g., Andersen et al., 2003a; Barndorff-Nielsen and Shephard, 2004, 2006; Huang and Tauchen, 2005), and in empirical option pricing (e.g., Bates, 1991). More specifically, in the stochastic volatility and realized volatility literatures, the jump component is observed to be significantly less predictable than the continuous sample path component, evidently demonstrating separate roles for these in a forecasting context. Barndorff-Nielsen and Shephard (2004) have recently introduced a new realized measure, called bipower variation, which is consistent for integrated volatility when the underlying price process exhibits occasional jumps.

The problem arises when we observe data at high frequency and microstructure noise becomes important. So, in most cases, even though data are available tick by tick, current practice is to use a moderate number of intraday returns, e.g., 30 or 5 minutes in computing realized volatility. This has two problems. First, we are throwing away a lot of the data. Second, sampling at log horizons may limit the value of the asymptotic approximations derived under the assumption of an infinite number of intraday returns. Zhang et al. (2005) have suggested a new realized measure that is consistent for integrated volatility when the prices are contaminated by microstructure noise. However, how this measure performs in the presence of jumps has not been studied.

The contribution of this chapter is twofold. For the first time we provide a method to deal with both microstructure noise and jumps in the same framework. We demonstrate, using simulated and real data, that trying to predict jumps by ignoring the noise can lead to unrealistic results. On the other hand, it is not clear how to incorporate jumps into other estimation methods that do take care of noise. Second, our method enables us to separate the smooth noise and jump components of volatility, as well as the part of drift that contributes to volatility for finite frequency. This separation can lead to better understanding and prediction of the components separately.

The rest of the chapter is organized as follows. In Section 8.2 we present a brief review of functional data analysis and outline how we apply this technique to data on stock price processes. In Section 8.3 we describe the method for detecting jumps using the functional data analysis (FDA) technique and study the performance of this method with simulated and real data. In Section 8.4 we use this method to separate and study the components of volatility. Finally, in Section 8.5 we present our conclusions.

8.2 FDA OF VOLATILITY PROCESS

Muller et al. (2007) introduce the functional volatility process as a tool for modeling volatility trajectories. Consider the volatility trajectory of each day to be a realization from the distribution of functions resulting from a smooth functional volatility process in combination with a multiplicative white noise. The proposed nonparametric approach requires no assumptions from the functional volatility process beyond smoothness and integrability. An important tool for the analysis of trajectories of volatility within the framework of functional data analysis (FDA) is functional principal component analysis (Castro et al., 1986; Rice and Silverman, 1991). Functional volatility processes can be characterized by their mean function and the eigenfunctions of the autocovariance operator. This is a consequence of the Karhunen-Loève representation of the functional volatility process. Individual trajectories of volatility can then be represented by their functional principal component scores. One can then use the functional principal component scores for subsequent statistical analysis. The rest of this section describes this procedure. For a more detailed exposition of this material and asymptotic results, refer to Muller et al. (2007).

8.2.1 The Model

We consider the following underlying model with random drift and volatility functions for the stock price process $X(t, \omega)$:

$$d \log X(t,\omega) = \bar{\mu}(t,\omega)dt + \bar{\sigma}(t,\omega)dW(t,\omega), \quad t \in [0,T] \tag{8.1}$$

Here $\bar{\mu}(t,\omega), \bar{\sigma}(t,\omega)$ and $W(\omega,t)$ are independent stochastic processes, none of them necessarily stationary, where both the drift $\bar{\mu}(.)$ and the volatility $\bar{\sigma}(.)$ are assumed to have smooth (twice differentiable) sample paths. This is a generalization of the Black-Scholes model. Suppose the price process is observed at times $t_1, \ldots, t_j$, which are at regular intervals Δ apart for n days. Let $X_i(t_j)$ denote the price for the i-th day at time t_j. Define the scaled log returns Z_{ij} as

$$Z_{ij} = \frac{1}{\sqrt{\Delta}} \log \left(\frac{X_i(t_j + \Delta)}{X_i(t_j)} \right), \quad i = 1, \ldots, n, j = 1, \ldots, \left[\frac{T}{\Delta} \right] \tag{8.2}$$

We allow for the presence of multiplicative errors in the transaction recordings. Specifically, transaction recordings are assumed to be contaminated by independent nonnegative errors $e_{ij} > 0$ with the properties

$$E\left[\log\left(e_{ij}^2 \right) \right] = 0, \quad E\left[\log\left(e_{ij}^2 \right) \right]^2 < \infty \tag{8.3}$$

in such a way that the contaminated observations are

$$Z_{ij} = \frac{1}{\sqrt{\Delta}} \log \left(\frac{X_i(t_j + \Delta)}{X_i(t_j)} \right) e_{ij}, \quad i = 1, \ldots, n, j = 1, \ldots, \left[\frac{T}{\Delta} \right] \tag{8.4}$$

In addition, on a small fraction of days there are big jumps in the price process. That is, for certain days,

$$Z_{ij} = \frac{1}{\sqrt{\Delta}} \log \left(\frac{X_i(t_j + \Delta)}{X_i(t_j)} \right) e_{ij} J_{ij}, \quad i \in \Im, j = 1, \ldots, \left[\frac{T}{\Delta} \right] \tag{8.5}$$

where $\Im$ is the set of days on which jumps happen. Our aim is to identify these days.

8.2.2 Separating the Drift

The first step is to decompose noisy functional data into a smooth random process $\tilde{\mu}$ and additive noise R:

$$Z_{ij} = \tilde{\mu}(t_{ij})\sqrt{\Delta + R_{ij}}, \quad i=1,\ldots,n, \quad j=1,\ldots,\left[\frac{T}{\Delta}\right] \tag{8.6}$$

$R_{ij}, R_{i'k}$ are independent for all $i \neq i'$, $E(R_{ij})=0$, $var(R_{ij})=\sigma_R^2 < 1$. Note that the noise R_{ij} within the same subject or item i may be correlated. We estimate individual drifts $\tilde{\mu}_i(t)\sqrt{\Delta}$, by smoothing scatter plots $\{(t_j, Z_{ij}), j=1,\ldots,[T/\Delta]\}$, for each fixed $1 \leq i \leq n$. For the initial smoothing step we used a cross-validation bandwidth choice. Denoting the smoothed trajectories obtained from this smoothing step by $\hat{z}_i(t)$, which are substituted for $\tilde{\mu}_i(t)\sqrt{\Delta}$, one then forms $Z'_{ij} = Z_{ij} - \hat{z}_i(t_j)$. This is a finite sample correction and is necessary only because the price process is not observed at all points of time but on a discrete grid. The contribution of the drift term to the realized volatility goes away in the limit as the grid size Δ grows smaller.

8.2.3 Modeling the Noise Component

We now work with the estimated noise Z'_{ij} obtained from Section 8.2.2. We model the noise component as

$$Y_{ij} = \log\left(\{Z'_{ij}\}^2\right) - q_0 = V(t_{ij}) + W_{ij} \tag{8.7}$$

where V is the functional variance process, which is smooth; i.e., it has a smooth mean function μ_V and a smooth covariance structure

$$G_V(s,t) = cov(V(s), V(t)), \quad s,t \in [0,T] \tag{8.8}$$

The W_{ij} are white noise:

$$E(W_{ij})=0, \quad var(W_{ij})=\sigma_W^2, W_{ij} \perp W_{ik} \text{ for } j \neq k, W \perp V, W \perp S \tag{8.9}$$

Note that the adjustment by the constant $q_0 = -1.27$ has the consequence that $E(W_\Delta(t))=0$ for all t, while $Cov(U_\Delta(s), W_\Delta(t))=0$ for $|t-s| > \Delta$ (independent increments property). The smooth functional volatility process V does not depend on Δ. This decomposition implies

$$E(Y_{ij}) = E(V(t_{ij})) = \mu_V(t_{ij}) \tag{8.10}$$

$$\mathrm{cov}(Y_{ij},Y_{ik})=\mathrm{cov}(V_i(t_{ij}),V_i(t_{ik}))=G_V(t_{ij},t_{ik}),j\neq k \qquad (8.11)$$

for the functional variance process V. The autocovariance operator associated with the symmetric kernel G_V has smooth eigenfunctions ψ_k with nonnegative eigenvalues ρ_k and implies that we have representations

$$G_V(s,t)=\sum_k \rho_k \psi_k(s)\psi_k(t), s,t \in T \qquad (8.12)$$

$$V(t)=\mu_V(t)+\sum_k \zeta_k \psi_k(t) \qquad (8.13)$$

with functional principal component (FPC) scores $\zeta_k, k\geq 1$ with $\mathrm{E}(\zeta_k)=0$, $\mathrm{var}(\zeta_k)=\rho_k$, $\zeta_k = \int_0^T (V(t)-\mu_V(t))\psi_k(t)dt, \zeta_k$ uncorrelated, $\Sigma \rho_k < \infty$.

8.2.4 Estimation of Model Components

Apply functional principal component analysis (principal analysis of random trajectories (PART) algorithm) to the sample of transformed residuals Y_{ij}:

- Estimate mean function μ_V (smoothing of cross-sectional averages).

- Estimate smooth covariance surface by smoothing of empirical covariances (omitting the diagonal).

- Obtain eigenvalues/eigenfunctions, choosing number of components M by cross-validation.

- From diagonal of covariance surface, obtain $\mathrm{var}(W_{ij})=\sigma_W^2$.

- Obtain individual FPC scores ζ_{ij} by integration.

8.3 JUMP DETECTION

8.3.1 The Method

For each day i we calculate

$$\Xi_i = \sum_{j=1}^{[T/\Delta]}\left(\exp\{Y_{ij}\}-\left(1+\frac{\sigma_W^2}{2}\right)\exp\{V_{ij}\}\right) \qquad (8.14)$$

Conditioning on the V process and using the independence of V and W processes and the delta method we obtain

$$E(\Xi_i) = 0$$

$$\text{Var}(\Xi_i) = \exp(V_i)^T G_V \exp(V_i) + \frac{\sigma_W^4}{4} \text{Var}(RV) \qquad (8.15)$$

$\text{Var}(RV)$ can be estimated using $\frac{TP}{2[T/\Delta]}$, where TP is the tripower variation defined in Barndorff-Nielsen and Shephard (2006) as follows:

$$TP_i = M\mu_{4/3}^{-3}\left(\frac{M}{M-2}\sum_{j=3}^{M}|r_{i,j-2}|^{4/3}|r_{i,j-1}|^{4/3}|r_{i,j}|^{4/3}\right) \qquad (8.16)$$

$\mu_a = E(|Z|^a)$ is a normalizing constant where Z is the standard normal random variable, $M = [\frac{T}{\Delta}]$ is the number of intervals in a day, and $r_{i,j} = \sqrt{\Delta Z_{i,j}}$ is the log return for the j-th interval of the i-th day. As Δ goes to zero, under the null hypothesis of no jumps, the asymptotic distribution of

$$\frac{E_i}{\sqrt{\exp(V_i)^T G_V \exp(V_i) + \frac{\sigma_W^4 TP}{8[T/\Delta]}}} \qquad (8.17)$$

is standard normal. The proof goes in the lines of Barndorff-Nielsen and Shephard (2002, 2003). We compute this quantity for each day. Those days for which this quantity exceeds a preset quantile of the standard normal distribution are detected to have jumps.

8.3.2 Empirical Application

The first data set consists of 5- and 1-minute data on the S&P500 index from November 11, 1997 to March 3, 2006 (see Figure 8.1). We have eliminated days when trading was thin or the market was open for a shortened session. Huang and Tauchen (2005) study the same instruments over a period from 1997 to 2002. They use 5-minute data after applying an adjustment that consists of the following: Regress 5-minute absolute returns on the time dummies. Then keep the predicted absolute returns and call them $\hat{a}$. Divide the original 5-minute returns by the corresponding $\hat{a}$. Scale the adjusted data to have variance 1.

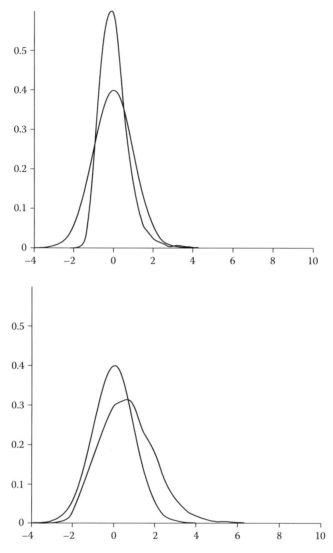

FIGURE 8.1 Density estimate of superimposed with standard normal density for S&P500. The top panel uses FDA. The bottom panel uses bipower variation. The figures on the top are for 5-minute data, and those on the bottom are for 1-minute data.

It is unclear how to incorporate this adjustment into the general setting of Barndorff-Nielsen and Shephard (2004). In Figure 8.2 we present the distribution of the test statistics for both methods along with the standard normal density functions for sampling frequencies at 1 and 5 minutes. It is clear

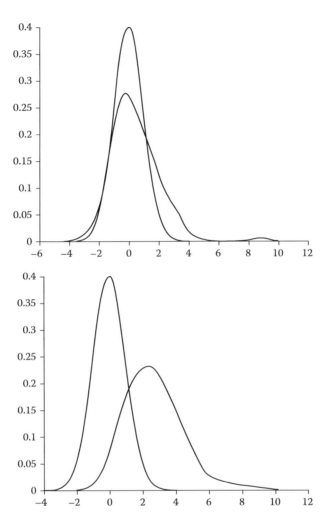

FIGURE 8.2   Density estimate of superimposed with standard normal density for Japanese yen–U.S. dollar exchange rate. The top panel uses FDA. The bottom panel uses bipower variation. The figures on the top are for 5-minute data, and those on the bottom are for 1-minute data.

from these plots that our method is more consistent with the result that the statistics has a standard normal distribution on most days, and there are a few high values on days when there are jumps. The sampling distribution of the statistics obtained by the Huang-Tauchen method using bipower variation has a large positive bias. This becomes more pronounced when the sampling frequency is higher. For our method, the sampling frequency does

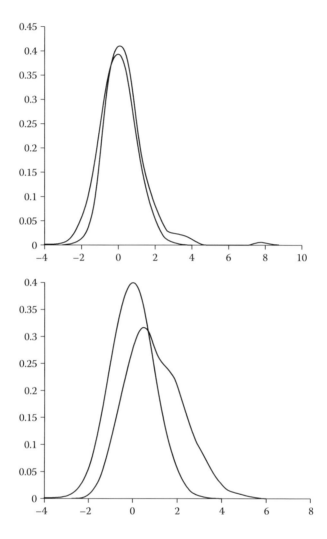

FIGURE 8.3 Density estimate of superimposed with standard normal density for Euro–U.S. dollar exchange rate. The top panel uses FDA. The bottom panel uses bipower variation. The figures on the top are for 5-minute data, and those on the bottom are for 1-minute data.

not affect the distribution too much, which should be the case. We carry out the same exercise with two other data sets: the Japanese yen–to–U.S. dollar exchange rate and the euro–to–U.S. dollar exchange rate.

The results are displayed in Figure 8.3. It is very similar to Figure 8.2 and strengthens our point that we cannot ignore microstructure noise while detecting jumps. The Barndorff-Nielsen and Shephard (2004) methodology does not account for microstructure noise. Realized variance measures the combination of volatility, microstructure noise, and jumps, while bipower variance measures volatility. Hence, the difference, which is the test statistic used in Huang and Tauchen (2005), is not a measure of jump, but of jump and microstructure noise.

8.4 COMPONENTS OF REALIZED VOLATILITY

Realized volatility can be split up into four components: the part due to drift, smooth time-varying volatility, microstructure noise, and jump. Our approach enables us to separate these four components. Drift is the component we are not worried about, because it should go away in the limit and not affect the volatility of interest in other spheres, like option pricing. The smooth time-varying volatility part is predictable from previous observations. The microstructure noise is not predictable, but has a more or less fixed level. The jump component can be modeled separately as in Tauchen and Zhou (2005). Methods that ignore noise essentially put noise and jump components together (e.g., Huang and Tauchen, 2005; Fan and Wang, 2007). However, these have entirely different dynamics. Both have low predictability. While noise is at a fixed level every day, jumps are rare and large and might be accompanied by arbitrage opportunities if detected early.

8.4.1 Methodology

Realized volatility is defined as the sum of squared returns. Hence, in the notation of Section 8.2.1, realized volatility equals $\Delta\Sigma Z_{ij}^2$. The part of this that is attributable to drift is

$$\text{Vol}_{drift} = \Delta\sum \tilde{z}_{ij}^2$$

which is an estimate for $\Delta^2\Sigma\bar{\mu}_i^2(t_j)$, and hence converges to zero as Δ goes to zero. The remaining part is

$$\Delta\sum Z_{ij}'^2 = \Delta\,\exp(q_0)\sum \exp Y_{ij} = \Delta\,\exp(q_0)\sum \exp(V_{ij})\times\exp(W_{ij}) \quad (8.18)$$

Since the V and W processes are independent, this approximately equals the product of

$$\text{Vol}_{smooth} = \Delta\sum \exp(V_{ij})$$

TABLE 8.1   Statistics of the Four Components for S&P500 1-Minute Data

Statistics	Drift	Volatility	Noise	Signed Root of Nonzero Jump
Mean	2.5369	1.1224	1.3190	0.6408
Median	1.4711	0.7003	1.3171	5.0815
Standard	3.4629	1.2234	0.1988	1.8752
Min	0.1745	0.0789	0.8055	−50.2395
Max	48.7390	12.0745	2.3703	20.9168
Skewness	5.5209	3.1719	0.1599	−0.4981
Kurtosis	51.3136	17.5904	1.1064	2.8732

and

$$\text{Vol}_{noise} = \exp(q_0) \sum \exp(W_{ij})/M$$

The former is the part due to the smooth underlying volatility process, and the latter is due to the microstructure noise and jumps. We use the procedure outlined in Section 8.3.1 to detect the days that have jumps. The contribution of jumps to realized volatility for such days is estimated by subtracting the average level of microstructure noise from Vol_{noise} and equals

$$\text{Vol}_{jump} = \exp(q_0) \left(\sum \exp(W_{ij})/M - \left(1 + \frac{\sigma_W^2}{2}\right) \right)$$

8.4.2 Size of the Components in Empirical Data

We present the statistics of the four components computed for the three real data sets under consideration in Tables 8.1 to 8.3. We observe that the noise is much higher (high mean) than the smooth part of volatility, but

TABLE 8.2   Statistics of the Four Components for JPYA0 1-Minute Data

Statistics	Drift	Volatility	Noise	Signed Root of Nonzero Jump
Mean	0.0122	0.6423	1.9972	1.1777
Median	0.0062	0.5652	1.9972	6.8208
Standard	0.0159	0.3632	0.5849	15.1604
Min	0.0001	0.1013	1.0185	−28.3315
Max	0.1143	2.3059	6.6730	28.8391
Skewness	2.8278	1.2346	1.0465	−0.0277
Kurtosis	13.4732	4.9742	7.6253	0.4697

TABLE 8.3   Statistics of the Four Components for EURA0 1-Minute Data

Statistics	Drift	Volatility	Noise	Signed Root of Nonzero Jump
Mean	0.0179	0.6161	1.7776	0.9931
Median	0.0070	0.5643	1.7777	7.1392
Standard	0.0329	0.3229	0.3752	12.2518
Min	0.0004	0.0719	1.0122	−26.9250
Max	0.3726	1.8916	3.3409	23.2150
Skewness	6.1082	0.9309	0.3047	−0.1023
Kurtosis	58.5171	3.8252	1.5363	0.5176

is less variable (low standard deviation). On the whole, the two exchange rates have pretty similar behavior, while the index is very different from either of them. The relative contribution of drift to the total volatility is higher for the index. There are fewer jumps, and the size of jumps is smaller for the index. The relative level of noise is much lower for the index than for the exchange rates. As expected, for all the data sets, the level of noise has very little variability, whereas that of jumps has very high variability.

8.5   CONCLUSION

We present a procedure, based on principal component analysis of functional data, to detect days on which a price process displays jumps. This procedure takes into account microstructure noise and performs better than existing methods that ignore the noise. Andersen et al. (2002) estimate jumps happen three to four times a year, which is consistent with our findings. Since the microstructure noise level goes to infinity, the other methods will ultimately recognize all days as having jumps. In practice, when information arrives, there is not one single big jump, but a series of small jumps. These have a cumulative effect that is higher than the microstructure noise level, though they might not be very big individually. Our method can capture this kind of behavior, while other methods, like Fan and Wang (2007), cannot. Our procedure provides a tool for separating the different components that contribute to the total volatility. Thus, we can study these components separately and make better predictions.

REFERENCES

Ait-Sahalia, Y. (2004). Disentangling diffusion from jumps. *Journal of Financial Economics* 74:487–528.

Andersen, T. G., L. Benzoni, and J. Lund. (2002). An empirical investigation of continuous-time equity return models. *Journal of Finance* 57:1239–84.

Andersen, T. G., and T. Bollerslev. (1998). Answering the skeptics: Yes, standard volatility models do provide accurate forecasts. *International Economic Review* 39:885–905.

Andersen, T. G., T. Bollerslev, P. F. Christoffersen, and F. X. Diebold. (2005). Volatility forecasting. In *Handbook of economic forecasting*, ed. G. Elliott, C. W. J. Granger, and A. Timmermann. Amsterdam: North Holland 778–878.

Andersen, T. G., T. Bollerslev, and F. X. Diebold. (2003a). Some like it smooth, and some like it rough: Untangling continuous and jump components in measuring, modeling, and forecasting asset return volatility. Working paper, Duke University, Durham, NC.

Andersen, T. G., T. Bollerslev, F. X. Diebold, and P. Labys. (2003b). Modeling and forecasting realized volatility. *Econometrica* 71:579–625.

Andersen, T. G., T. Bollerslev, and N. Meddahi. (2004). Analytical evaluation of volatility forecasts. *International Economic Review* 45:1079–110.

Barndorff-Nielsen, O. E., and N. Shephard. (2002). Econometric analysis of realized volatility and its use in estimating stochastic volatility models. *Journal of Royal Statistical Society* 64B:253–80.

Barndorff-Nielsen, O. E., and N. Shephard. (2003). Realized power variation and stochastic volatility. *Bernoulli* 9:243–65.

Barndorff-Nielsen, O. E., and N. Shephard. (2004). Power and bipower variation with stochastic volatility and jumps [with discussion]. *Journal of Financial Econometrics* 2:1–48.

Barndorff-Nielsen, O. E., and N. Shephard. (2006). Econometrics of testing for jumps in financial economics using bipower variation. *Journal of Financial Econometrics* 4:1–30.

Bates, D. S. (1991). The crash of '87: Was it expected? The evidence from options markets. *Journal of Finance* 46:1009–44.

Castro, P. E., W. H. Lawton, and E. A. Sylvestre. (1986). Principal modes of variation for processes with continuous sample curves. *Technometrics* 28:329–37.

Chernov, M., A. R. Gallant, E. Ghysels, and G. Tauchen. (2003). Alternative models of stock price dynamics. *Journal of Econometrics* 116:225–57.

Eraker, B., M. Johannes, and N. Polson. (2003). The impact of jumps in volatility and returns. *Journal of Finance* 58:1269–300.

Fan, J., and Y. Wang. (2007). Multi-scale jump and volatility analysis for high-frequency financial data. *Journal of American Statistical Association* 102:1349–62.

Huang, X., and G. Tauchen. (2005). The relative contribution of jumps to total price variation. *Journal of Financial Econometrics* 3:456–99.

Muller, H. G., R. Sen, and U. Stadtmuller. (2007). Functional data analysis for volatility process. Working paper, University of California, Davis.

Rice, J. A., and B. W. Silverman. (1991). Estimating the mean and covariance structure non-parametrically when the data are curves. *Journal of Royal Statistical Society Series* 53B:233–43.

Tauchen, G., and H. Zhou. (2005). Identifying realized jumps on financial markets. Working paper, Federal Reserve Board, Washington, DC.

Zhang, L., P. A. Mykland, and Y. Ait-Sahalia. (2005). A tale of two time scales: Determining integrated volatility with noisy high-frequency data. *Journal of the American Statistical Association* 100:1394–411.

II

Portfolio Management and Hedge Fund Volatility

CHAPTER 9

Mean-Variance versus Mean-VaR and Mean-Utility Spanning

Laurent Bodson and Georges Hübner

CONTENTS

9.1 INTRODUCTION

Since the seminal mean-variance (*VAR*) framework of Markowitz (1952) appeared, a large part of the financial literature has investigated new approaches to capture the non-Gaussian distribution of financial asset returns. Indeed, Samuelson (1970) has given prominence to the deviation from the normal distribution of different classes of financial assets. Frequently, the optimal allocations deduced from the Markowitz framework differ largely from the optimal allocations obtained using a risk measure that integrates higher-moment estimates. Nowadays, practitioners do not integrate systematically the impact of extreme risks on the optimal allocations of their portfolios. Generally, the implications of the extreme risks treatment are not well followed by portfolio managers, and the recent financial crises are good proofs of this observation.

In order to adjust the Markowitz framework, some authors have developed specific risk metrics that take into consideration the higher moments of the return distributions. One interesting measure, proposed by Favre and Galeano (2002), is the modified value-at-risk (*MVaR*) that corrects the quantile estimate used in the formulation of the Gaussian VaR.

One of the drawbacks of the *VAR* and the *MVaR* is that they do not integrate the investor's preferences and perception of risk. It is generally approved that the *MVaR* permits to consider different risk aversion of the investor through the confidence interval chosen (alpha), but this risk measure assumes that all investors have the same vision of risk, or more precisely, the same compromise between higher statistical moments.

In this chapter, we propose to analyze the empirical effects of the risk measure choice on the efficient frontiers (in a risk-return framework) and the impacts of tradable hedge fund strategies on these specific efficient sets of portfolios. We focus our analysis on three major risk measures: the *VAR* (Markowitz, 1952) and two risk measures based on the higher-order moments, the *MVaR* (Favre and Galeano, 2002) and a utility-based risk (*UBR*) measure (Bell, 1988). We compare the efficient frontiers deduced for each risk metric (with and without hedge funds) over two distinct periods, a bear market period (from January 2000 to December 2002, a total of 36 months) and a bull market period (from January 2003 to December 2005, also a total of 36 months).

The chapter proceeds as follows. In Section 9.2, we describe the risk measures implemented to construct the different efficient frontiers. We introduce in Section 9.3 the data set and develop our methodology. Section 9.4 presents the empirical results, and Section 9.5 concludes.

9.2 RISK MEASURES

9.2.1 Variance

The risk measure chosen by Markowitz is the *VAR* of the portfolio p return (VAR_p). The major disadvantage of this risk measure is its indifference between downside risk and upside potential. In other words, this metric penalizes upside potential, which is implicitly recognized as not relevant to investor behavior.

9.2.2 Modified Value-at-Risk

Contrarily to the *VAR*, the *MVaR* takes only into consideration the downside risk of the portfolio. The *MVaR* of a portfolio p is defined as the portfolio value-at-risk where the quantile used to calculate the value-at-risk of the portfolio has been adjusted (using the Cornish-Fisher expansion (1937)) to the higher moments of the portfolio return distribution. In fact, the Gaussian value-at-risk assumes that the returns are normally distributed, and therefore, the skewness is null and the standardized kurtosis is equal to 3. However, the return distributions of several financial assets generally exhibit large deviations from these Gaussian values of the skewness and the kurtosis. For this reason, Favre and Galeano (2002) propose to adjust the quantile used in the traditional Gaussian value-at-risk.

In this chapter, we only focus on the risk part of the *MVaR*. More precisely, we do not integrate the expected return in the formulation of the portfolio *MVaR*. Indeed, we want to separate clearly the first moment of the return distribution from the three other moments (the variance, the skewness, and the kurtosis) of the portfolio return distribution. The expression of the *MVaR* of portfolio p at the confidence level α (over the investment horizon defined by the frequency of the returns) is defined by

$$MVaR_{p\alpha} = \left(z_\alpha - \frac{1}{6}\left(z_\alpha^2 - 1\right)S_p^* + \frac{1}{24}\left(z_\alpha^3 - 3z_\alpha\right)K_p^* - \frac{1}{36}\left(2z_\alpha^3 - 5z_\alpha\right)S_p^{*2} \right)V_p^{0.5}$$

$$(9.1)$$

where z_α is the normal quantile value such as $P(X \geq z_\alpha) = \alpha$ (knowing that X follows a standard normal distribution (i.e., a mean equal to 0 and a standard deviation of 1)), S_p^* is the standardized skewness (i.e., the third centered moment divided by the third power of the portfolio standard deviation), K_p^* is the standardized excess kurtosis (i.e., the fourth centered moment divided by the fourth power of the standard deviation, this quotient minus 3), and V_p is the variance of the portfolio return distribution.

9.2.3 Utility-Based Risk

The *UBR* measure that we use in our empirical comparison is the risk measure inferred from Bell's linear and exponential (linex) utility functions (1988, 1995). After some mathematical manipulations and integrating the

Taylor series expansion, we obtain from Bell's utility functions the following UBR for a portfolio p:

$$UBR_{pC} = \frac{1}{2}V_p - \frac{C}{6}S_p + \frac{C^2}{24}K_p \qquad (9.2)$$

where C is the global risk perception of the investor,[*] V_p is the portfolio return variance, S_p is the skewness of the portfolio return distribution, and K_p is the kurtosis of the portfolio return distribution. Note that we use for this UBR measure the basic centered moments and not the standardized ones.

The intuition underlying this risk metric is quite simple. Investors with high Cs put more emphasis on the possibility of bad outcomes than investors with low Cs.

9.3   DATA AND METHODOLOGY

We propose to compare the different efficient frontiers built from a set of industry portfolios and from the same set of portfolios augmented by global hedge fund indices. For each risk measure, the optimal allocation is computed minimizing the risk measure of the portfolio (VAR_p, $MVaR_{p\alpha}$, and UBR_{pC}) for a given average return. We consider in our analysis that the weights of the portfolio components add up to 1 and that they are not negative (i.e., short positions are not allowed). We compute the efficient frontiers for each risk measure and for the two sets of securities (with and without hedge funds) over two subperiods.

We consider the monthly returns of each security from January 2000 to December 2005. In order to study the sensitivity of the optimal allocations to different market conditions, we distinguish the bearish and the bullish subperiods of our sample. We divide our period of analysis in the following way: a first bearish subperiod covering the 3 years from January 2000 to December 2002 and a second bullish subperiod including the 3 subsequent years from January 2003 to December 2005. Indeed, if we look, for instance, at the S&P 500 evolution (in Figure 9.1) from January 2000 to December 2005, we observe that this period exhibits two main market trends: a first bear market period (from January 2000 to December 2002) and a second bull market period (from January 2003 to December 2005).

To represent the equity universe of securities available over the two subperiods, we take the forty-eight industry portfolios proposed

---

[*] The range of this parameter depends on the asset classes analyzed and the risk aversion of the investor.

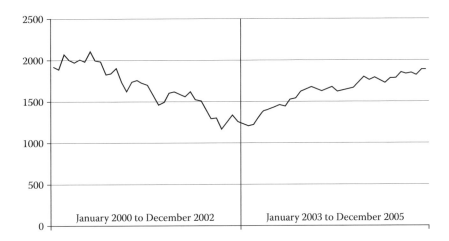

FIGURE 9.1   Historical value (in US$) of the S&P 500 Composite (total return) Index from January 2000 to December 2005 and presentation of the two subperiods of analysis.

on Kenneth R. French's website.* The hedge funds benchmarks are represented by the thirty-seven HFRX global indices.†

9.4   EMPIRICAL RESULTS

First, we optimize the portfolio allocation minimizing the *VAR* of the portfolio return for a given average return with the two constraints on the optimal weights exposed supra (sum up to 1 and not negative). Figure 9.2 shows the results of this optimization for the two sets of assets (the industry portfolios with and without the hedge funds indices) and for the two subperiods. Obviously, the upper efficient frontier is always the efficient frontier integrating the hedge fund indices because the optimization has a larger set of potential assets.

We observe in Figure 9.2 that in bear market conditions, the integration of hedge funds does not permit the investor to get a better portfolio diversification. Effectively, the difference between the two efficient frontiers is almost nonexistent. The major reason is that in bear market conditions, the asset returns exhibit higher correlation. Therefore, the diversification effect is more limited. In bull market conditions, investors have the opportunity to diversify their portfolios using hedge funds to take advantage of better allocations.

---

* http://mba.tuck.dartmouth.edu/pages/faculty/ken.french/data_library.html.
† www.hedgefundresearch.com.

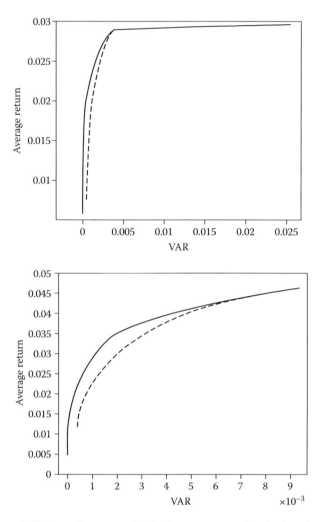

FIGURE 9.2 Efficient frontiers (including or not the hedge fund class) based on the portfolio variance for the bear market and bull market subperiods.

Second, we compute the efficient frontiers minimizing the *MVaR* of the portfolio ($MVaR_{p\alpha}$). We fix α at its traditional value of 1%. Figure 9.3 exhibits the efficient frontiers for the two sets of assets (with and without hedge funds) for the two subperiods such that the optimal allocation obtained minimizes the portfolio *MVaR* and respects the two constraints on the optimal weights exposed supra (sum up to 1 and not negative).

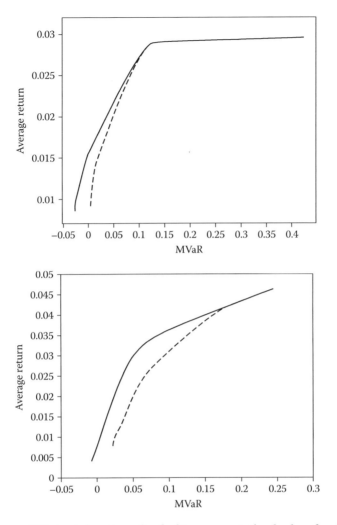

FIGURE 9.3 Efficient frontiers (including or not the hedge fund class) based on the portfolio modified value-at-risk (1%) for the bear market and bull market subperiods.

Figure 9.3 shows that the efficient portfolios obtained using the *MVaR* as risk measure instead of the *VAR* are completely different. The possibility to benefit from the diversification opportunity is more pronounced using the *MVaR*. The *MVaR* succeeds in capturing the upside potential of hedge funds, differentiating clearly the efficient frontier without hedge funds indices from the efficient frontier with hedge funds indices. The results in Figure 9.3 emphasize the fact that the *MVaR* adjusts the risk measure to take notably into consideration the "good" extreme events.

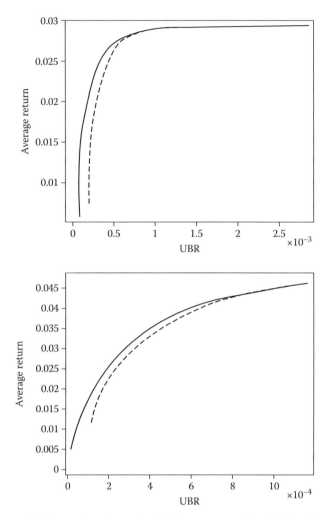

FIGURE 9.4 Efficient frontiers (including or not the hedge fund class) based on the portfolio utility-based risk ($C = 0.5$) for the bear market and bull market subperiods.

Third, we build the efficient frontiers for each set of assets and for each subperiod minimizing the portfolio *UBR*. For this last risk measure, we propose to take three different values for the parameter C ($C = 0.5$, $C = 0.75$, and $C = 1$) to take into consideration three different risk perception profiles of the investor. In Figures 9.4 to 9.6, we plot for each value of C the efficient frontiers of the two sets of data for the two subperiods

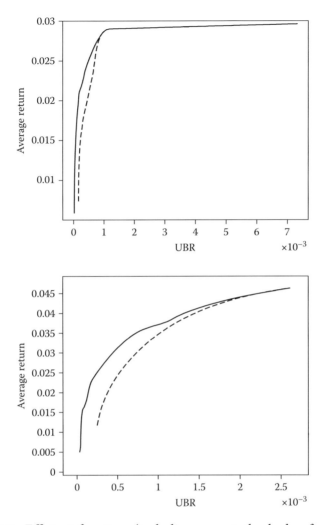

FIGURE 9.5 Efficient frontiers (including or not the hedge fund class) based on the portfolio utility-based risk ($C = 0.75$) for the bear market and bull market subperiods.

minimizing the *UBR* of the portfolio and meeting the constraints on the optimal weights (sum up to 1 and not negative).

The results reported in Figures 9.4 to 9.6 demonstrate the obvious necessity to integrate the risk perception of the investor in the portfolio optimization. Indeed, according to the risk perception profile of the investor, the efficient frontiers vary largely from the *VAR* or *MVaR* optimizations but also between *UBR* optimizations.

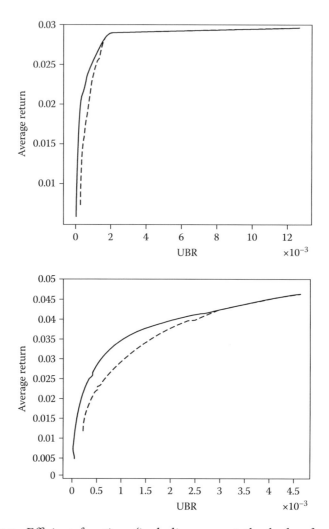

FIGURE 9.6 Efficient frontiers (including or not the hedge fund class) based on the portfolio utility-based risk ($C = 1$) for the bear market and bull market subperiods.

The efficient frontiers computed minimizing the *UBR* metric are more reliable because they integrate the investor profile. For different values of the parameter *C*, we observe that the movement of the efficient frontier is really particular, and we also note a different treatment of the hedge fund opportunity. We note in this new framework (mean − *UBR*) that the separation theorem does not hold because the optimal combination of risky assets cannot be determined without knowledge of the investor risk perception.

TABLE 9.1   Efficient Portfolio Allocation for Each Efficient Frontier
(with Hedge Funds)

	VAR	$MVaR_{1\%}$	UBR		
C	—	—	**0.5**	0.75	**1**
		Bear (constant required return)			
Equity	31.88%	63.87%	65.51%	43.16%	32.79%
HF	42.25%	15.20%	23.42%	35.78%	40.57%
VAR	0.05%	0.10%	0.14%	0.07%	0.05%
$MVaR_{1\%}$	4.75%	3.69%	8.01%	4.98%	4.80%
		Bull (constant required return)			
Equity	13.88%	11.89%	72.20%	43.96%	23.67%
HF	86.12%	88.11%	27.79%	55.96%	76.33%
VAR	0.11%	0.11%	0.22%	0.16%	0.12%
$MVaR_{1\%}$	5.36%	5.00%	11.22%	8.21%	5.88%

The shape of the efficient frontier is not exactly a robust test because we can have an identical shape with absolutely different optimal allocations. For this reason, we propose to compare the optimal allocations of a specific efficient portfolio. Based on our results, we propose to consider, for each case, the efficient portfolio that earns a constant rate of return in bear market conditions and the efficient portfolio that earns a higher constant rate of return in bull market conditions.

In Table 9.1, we compare the efficient portfolio compositions in each case, i.e., for each efficient frontier integrating the hedge fund indices presented in this chapter. We have only reported the results of the efficient frontiers with hedge funds to decompose the allocations in terms of equity and hedge funds parts. Indeed, we are forced to distinguish only two categories because we cannot report each equity weight (forty-eight industry portfolios) and each hedge fund weight (thirty-seven HFRX global indices).

We also find that the $MVaR$ tend to produce lower hedge fund allocations in bear market conditions and slightly higher hedge fund allocations in bull market conditions than the mean-variance approach.

According to the results reported in Table 9.1, the optimal allocation of the VAR is close to the optimal allocation of the UBR_1 (high C), and the optimal allocation of the $MVaR_{1\%}$ is close to the optimal allocation of $UBR_{0.5}$ (low C). These observations show the importance of the investor risk perception in the portfolio optimization. The UBR measure covers all the range of optimal allocations (in our case from the $MVaR_{1\%}$ to the VAR optimal allocations) simply defining the risk perception of the investor.

The results confirm the insights obtained from the previous graphs. In particular, the efficient allocations corresponding to the very same target—a mean return equal to a constant—feature dramatically different allocations. Furthermore, in bearish market conditions, one can see that a mean-variance investor or, quite similarly, an investor with a high C in the linex function will use more extensively the money market instrument (ca. 26–27% of the total allocation) than an investor who cares more about extreme risks. Under bullish market conditions, the risk-free rate logically vanishes from all efficient allocations.

Of course, the risk measure corresponding to each criterion is minimized when the optimization is performed with that same criterion. For instance, VAR optimized portfolios produce the lowest variance of all allocations. But the striking element to consider is the existence of a continuum of equity/hedge fund combinations that provide the same expected returns under the UBR. This finding is important, because one has to consider that the linex utility function represents much more adequately the investors' preferences than the MVaR, which is essentially an arbitrary risk measure, or the variance, which rests on unrealistic assumptions about investors' preferences. In other words, any allocation between UBR_1 and $UBR_{0.5}$ likely represents the optimal portfolio for a type of representative investor, under the very same market conditions. There is no such thing as a market portfolio, but a wide variety—and a large array—of market portfolios.

9.5 CONCLUSION

This chapter has examined the difference between three risk measures: the variance of the portfolio return (VAR), the modified value-at-risk ($MVaR$) of the portfolio return, and a utility-based risk measure (UBR) deduced from Bell's utility function.

The higher-moments-based risk measures are coherent with the risk-averse preferences of the investors and offer new tools for portfolio and risk managers. The adjustment of the risk introduced by the higher moments must be dependent of the investor risk perception, and the optimal allocations must finally reflect the investor's preferences toward the risk-return trade-off.

The piece of evidence presented in our analysis suggests that the utility-based risk measure yields higher customized optimal allocation and integrates a fundamental element in the optimization algorithm, the risk perception of the investor. In other terms, a portfolio optimization cannot ignore the perception of the investor of the different facets of risk.

We consider our analysis as a rather illustrative comparison of three risk measures. It is not meant to present a robust statistical side but shows the impact and importance of the investor's preferences toward risk and return in portfolio and risk management.

Future research should probably focus on the difference in statistical properties of these three measures and especially of the utility-based measure. Another approach would be the study of the stability and persistence of these risk measures.

REFERENCES

Bell, D. E. (1988). One-switch utility functions and a measure of risk. *Management Science* 34:1416–24.

Bell, D. E. (1995). A contextual uncertainty condition for behavior under risk. *Management Science* 41:1145–50.

Cornish, E., and Fisher, R. (1937). Moments and cumulants in the specification of distributions. *Review of the International Statistical Institute* 5:307–20.

Favre, L., and Galeano, J.-A. (2002). Mean-modified value-at-risk optimization with hedge funds. *Journal of Alternative Investment* 5:21–25.

Markowitz, H. M. (1952). Portfolio selection. *Journal of Finance* 7:77–91.

Samuelson, P. (1970). The fundamental approximation theorem of portfolio analysis in terms of means, variances and higher moments. *Review of Economic Studies* 37:537–42.

CHAPTER 10

Cyclicality in Stock Market Volatility and Optimal Portfolio Allocation

Jason C. Hsu and Feifei Li[*]

CONTENTS

10.1 CYCLICALITY IN MARKET VOLATILITY

In standard finance applications, asset class volatilities are usually assumed to be constant over time for simplicity. For example, Markowitz's mean-variance optimization requires that asset class volatilities are known and constant over the holding horizon. While this simplifying assumption reduces the complexity of the models and their calculations, it could also lead to suboptimal portfolio and risk management solutions. If equity market volatility is time varying and is negatively correlated with equity market returns, ignoring

---

[*] The authors acknowledge Micah Allred, Vitali Kalesnik, and Lillian Wu for their assistance in completing this chapter.

this countercyclicality could lead to excess allocation to stocks when forward-looking risk for stocks is high. Furthermore, if equity market volatility is positively correlated with the volatilities of other asset classes, ignoring this correlation would again lead to excess allocation to risky assets.

In Table 10.1, we show the U.S equity market volatility in an average bull market versus an average bear market. We use a classic bull/bear market definition, where a bull market is defined as a period of general price appreciation, during which the cumulative market return exceeds 20%. A bear market, by contrast, is a period of price decline, during which the cumulative market negative return exceeds –20%. For simplicity, the market is classified to be in either a bull or bear market phase. Additionally, we show the volatility of other mainstream asset classes over the same equity market cycles. Furthermore, to illustrate the robustness of the finding, we also show, in Table 10.2, the volatilities of these asset classes in different phases of the business cycle (expansion versus recession). We employ the National Bureau of Economic Research (NBER) definitions for expansions and recessions, which uses GDP growth/decline and other macroeconomic factors to classify business cycles.

Notice that equity market volatility is significantly higher in bear markets and recessions. The increase in volatility in down/contracting markets can be attributed to a variety of reasons. Down/contracting markets may be triggered by instability in the macroeconomy. Under this assumption, down/contracting markets are likely to be times where shocks to the productive factors in the economy are more severe and more frequent

TABLE 10.1   Asset Class Volatilities over Equity Bull/Bear Market Cycles

Asset Class Volatility (ann.)	Bull	Bear
U.S. equities (S&P 500)	13.33%	17.13%
International equities (MSCI EAFE)	15.54%	16.36%
Bond (Lehman Agg)	5.57%	6.92%
Commodities (DJ AIG)	11.76%	13.83%
Real estate (FTSE NAREIT)	13.01%	15.60%
Asset Class Return (ann.)	**Bull**	**Bear**
U.S. equities (S&P 500)	21.09%	−19.09%
International equities (MSCI EAFE)	20.45%	−15.96%
Bond (Lehman Agg)	8.26%	11.29%
Commodities (DJ AIG)	5.14%	−0.69%
Real estate (FTSE NAREIT)	16.57%	2.47%

TABLE 10.2   Asset Class Volatilities over NBER Expansion/Recession Cycles

Asset Class Volatility (ann.)	Expansion	Recession
U.S. equities (S&P 500)	14.06%	19.00%
International equities (MSCI EAFE)	15.08%	23.24%
Bond (Lehman Agg)	4.84%	10.96%
Commodities (DJ AIG)	11.84%	13.83%
Real estate (FTSE NAREIT)	12.80%	18.84%
Asset Class Return (ann.)	**Expansion**	**Recession**
U.S. equities (S&P 500)	14.04%	11.59%
International equities (MSCI EAFE)	15.85%	−2.16%
Bond (Lehman Agg)	7.55%	20.07%
Commodities (DJ AIG)	5.82%	−9.46%
Real estate (FTSE NAREIT)	13.41%	22.25%

Note: A recession is a significant decline in economic activity spread across the economy, last-ing more than a few months, normally visible in real GDP, real income, employment, industrial production, and wholesale-retail sales. A recession begins just after the econ-omy reaches a peak of activity and ends as the economy reaches its trough. Between trough and peak, the economy is in an expansion. Expansion is the normal state of the economy; most recessions are brief, and they have been rare in recent decades. The National Bureau's Business Cycle Dating Committee places particular emphasis on two monthly measures of activity across the entire economy: (1) personal income less transfer payments, in real terms, and (2) employment. In addition, the committee refers to two indicators with coverage primarily of manufacturing and goods: (3) industrial production and (4) the volume of sales of the manufacturing and whole-sale-retail sectors adjusted for price changes. The committee also looks at monthly estimates of real GDP such as those prepared by Macroeconomic Advisers (see http://www.macroadvisers.com). Although these indicators are the most important mea-sures considered by the NBER in developing its business cycle chronology, there is no fixed rule about which other measures contribute information to the process.

than usual. In down/contracting markets, leveraged investments are likely to face margin calls, which increase liquidity-driven asset sale; these liquidating transactions tend to induce additional price volatility. Lastly, market-making agents and noise traders who engage in market liquid-ity provision, and who trade against informed flows, are likely to become more risk averse in down/contracting markets. In these markets, where market participants have experienced wealth decline, their ability to bear risks declines as a result (their local risk aversion increases).

From Figure 10.1, we observe that asset class volatilities appear to co-move over time, suggesting that common macro factors may drive volatil-ities for various risky assets. Specifically, we observe from Table 10.1 that the volatilities of other risky asset classes seem to also increase noticeably during equity bear markets. This increase in volatility suggests that the

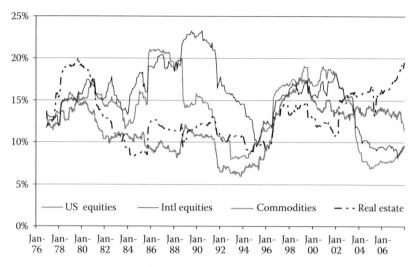

FIGURE 10.1   Asset class rolling 36-month volatilities.

increased shocks to equity valuation often spill over to other markets, and that liquidity-driven selling and the reduction in liquidity provision in the capital market are often systemic across various asset classes. Not surprisingly, equity bear markets and recessions can often have significant overlaps and have similar influences on asset return characteristics.

In this chapter, we argue that the countercyclical nature of equity market volatility (high volatility in down markets), combined with positive correlations between asset class volatilities, has a significant impact on optimal portfolio allocation. We first present a simple model of time-varying asset class volatilities. We then illustrate how to calibrate the model and integrate the method with the classic mean-variance approach. We compare our proposed optimal portfolio solution to the standard static portfolio solution where the time-varying volatility is ignored and argue that a dynamic mean-variance approach is superior to the standard approach.

10.2   LITERATURE REVIEW ON MARKET VOLATILITY

Before we introduce our model on cyclical equity market volatility, we explore the literature on market volatility and examine the drivers for the level and variation for market variance. Using a simple present value model, Shiller (1981) finds that the level of stock market volatility is too high relative to the variation in the underlying micro and macro fundamentals.

Specifically, he finds that the changes in real dividends and real interest rates cannot explain the level of market volatility. Studies that examine the variation in market volatility also conclude that standard macro factors and corporate characteristics cannot explain the time-varying nature of equity volatility. Specifically, Officer (1973), Black (1976), and Christie (1982) find that financial leverage only weakly explains the variation in market volatility. Schwert (1989) finds that standard macroeconomic variables, such as inflation, money growth, and industrial production, also do not sufficiently explain the variation in the market volatility. Therefore, nonfundamentally based volatility drivers likely exist and may have better explanatory powers.

Behavioral finance literature points to information herding (cascading), noise trading, and liquidity-driven transactions as potential reasons for the higher level of market volatility, relative to the volatility in the underlying information flow. Theoretical work by Banerjee (1992) and Bikhchandani et al. (1992) suggests that information cascade can lead to price overshooting, which would inject additional volatility, in excess of the contribution from the existing volatility drivers. Campbell and Kyle (1993) and DeLong et al. (1990) study the effect of noninformed trading (uninformed speculation by noise trader or portfolio trading driven by liquidity shocks to the investor). They suggest that these uninformed trading activities create a new source of shocks to prices. This additionally creates excess equity market volatility.

The return predictability literature and the value premium literature offer rational pricing models as well as behavioral explanations for time-varying market volatility. Ferson and Harvey (1991) find that expected stock market return and volatility vary over time in a predictable way. Lettau and Ludvigson (2001), Chordia and Shivakumar (2002), and Zhang (2005) offer models that relate variation in aggregate risk aversion to decline in aggregate wealth. Intuitively, a period of negative returns driven by shocks to fundamentals will lead to aggregate wealth destruction; this can increase the aggregate risk aversion, which further decreases prices today and increases forward-looking return and increases volatility contemporaneously.

Equilibrium models of cyclical volatility are often difficult to apply; in addition, they often do not match well to data or offer insufficient degrees of freedom for empirical calibration. For this reason, statistical models are often relied upon for modeling stochastic volatility; these statistical models can be used with great flexibility for asset pricing or asset allocation exercises. Various statistical volatility models have been developed specifically to capture and measure time-varying volatilities. Engle (1982) and Bollerslev (1986) provide the basic framework for such modeling with the ARCH/GARCH process (autoregressive conditional heteroskedasticity/generalized autoregressive conditional heteroskedasticity). The technique has been applied widely to the estimation of the time-varying equity market volatility. Recent researches have proposed new techniques that could improve forecasting power through the usage of high-frequency tick-by-tick data. Anderson et al. (2001, 2003, 2005) use 5-minute realized volatility with a vector autoregressive model of log standard deviation, which eliminates much of the serial dependence in the volatilities and appears to outperform the traditional ARCH/GARCH specifications. Ghysels et al. (2006) also use higher-frequency data but propose a regression model using a beta weighting function to estimate and forecast volatility. Their model appears to be easier to parameterize and provides better forecasts against traditional ARCH/GARCH models. Vasilellis and Meade (1996) show that the implied stock volatility from option prices is an efficient forecast for future volatility. Poon and Granger (2003, 2005) show that option-implied volatility provides the best forecast for future volatility; they used option-implied volatility data from the last 20 years and compare against volatility models such as time-weighted volatility, rolling volatility, ARCH/GARCH, and other stochastic volatility models.

So why should we care about time-varying market volatility? If we do not properly characterize the time-varying nature of volatility and covariance for the various capital markets we invest in, our asset pricing model would be flawed, our portfolio allocation would be suboptimal, and our *ex ante* risk assessment would be incorrect. Bentz (2003) and Bollerslev et al. (1988) show that using a time-varying covariance estimate (beta estimate) can improve the application of the capital asset pricing model for forecasting returns. Horasanh and Fidan (2007) show that applying GARCH estimates for volatility can improve portfolio allocation efficiency. Blake and Timmermann (2002) find evidence that some pension funds seem to vary asset allocation to take advantage of time-varying asset class volatilities and risk premia. Myers (1991) finds that using GARCH models can improve the effectiveness of hedging fixed-income exposure relative to traditional regression approach with constant variance. Baillie and Myers (1991) extend the study into the commodities market and find that GARCH-based hedging provides a substantial improvement in risk reduction effectiveness.

10.3   A SIMPLE MODEL OF TIME-VARYING VOLATILITY

We introduce in this section a simple model that captures the counter-cyclicality nature of asset class volatilities. This approach is more intuitive and more tractable than other models of time-varying volatilities and leads to greater intuition and ease of calibration. The world is assumed to follow a two-state, two-stage Markov chain. The world can either be in a bull market state (U for upmarkets) or in a bear market state (D for downmarkets) at time t. For example, if we are currently in a bull market, for the next period, the economy can either transition into a bear market with the transition probability $P_{U \to D}$ or remain in the current bull state with probability $P_{U \to U} = 1 - P_{U \to D}$. If we transition to the bear market state at time $t + 1$, then for $t + 2$, we could transition to the bull market state with probability $P_{D \to U}$ or remain in the bear market state with probability $1 - P_{D \to U}$. Figure 10.2 illustrates graphically this Markov process.

Following the empirical results shown in Tables 10.1 and 10.2, the bull market state (U) is characterized by lower volatilities and higher returns for the asset classes, while the bear market state (D) is characterized by high volatilities and lower returns. We let Σ_U denote the vector of bull market volatilities $\{\sigma_1^U, \sigma_2^U, \ldots, \sigma_k^U\}$ and Σ_D denote the vector of bear market volatilities $\{\sigma_1^D, \sigma_2^D, \ldots, \sigma_k^D\}$; note that we assume an investment opportunity set with k assets. Similarly, μ_U and μ_D denote the vector of bull and bear market mean returns $\{\mu_1^U, \mu_2^U, \ldots, \mu_k^U\}$ and $\{\mu_1^D, \mu_2^D, \ldots, \mu_k^D\}$.

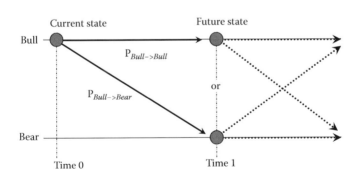

$P_{Bull->Bull}$ is the probability of starting in a bull market state and remaining in the bull market state next period.

$P_{Bull->Bear}$ is the probability of starting in a bull market state and transitioning to the bear market state next period.

FIGURE 10.2   A Markov two-state (bull/bear market) transition model.

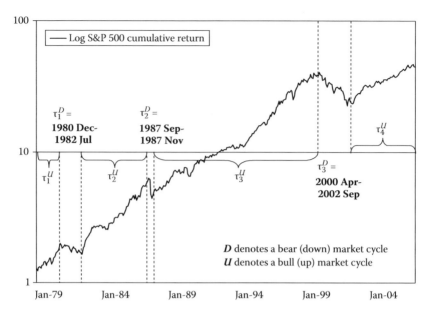

FIGURE 10.3 Identifying bear market periods (January 1979–December 2007).

10.3.1 Model Parameter Calibration

We now illustrate how to calibrate this Markov model to data. First, we classify our time period into equity bull and bear market periods (using the common definitions of bull and bear markets presented earlier). For the data time span T, we decompose T into nonoverlapping bull/bear time segments as illustrated in Figure 10.3. We denote the bull market time segments as $\{T_1^U, T_2^U, \ldots, T_m^U\}$ and the bear market time segments as $\{T_1^D, T_2^D, \ldots, T_n^D\}$, where $T = \sum_{i=1}^m T_i^U + \sum_{i=1}^n T_i^D$. The average duration for an equity bull market is empirically estimated by $\tau_U = \frac{1}{m} \sum_{i=1}^m T_i^U$, and the average duration for a bear market is $\tau_D = \frac{1}{n} \sum_{i=1}^n T_i^D$. Using S&P 500 return data from January 1976 through June 2008, we have encountered four bear market cycles, each averaging about 17 months, whereas the four bull market cycles average about 81 months each.[*]

To compute the Markov transition probabilities $P_{U \to D}$ and $P_{D \to U}$, we make use of the derived relationships, where $P_{U \to U} = 1 - \frac{1}{\tau_U}$ with

---

[*] Certainly, the more data that are used in the estimation, the more reliable and robust the estimation. Because there have not been many bull/bear market cycles, the estimation error will always be a concern when applying this calibration exercise.

$P_{U \to D} = 1 - P_{U \to U}$ and $P_{D \to D} = 1 - \frac{1}{\tau_D}$ with $P_{D \to U} = 1 - P_{D \to D}$ (see Meyn and Tweedie (1993) for a complete theoretical treatment on Markov models). Again, using data from 1976 through June 2007, conditioning on starting in a bull market, the probability for transitioning to a bear market by next year is $P_{U \to D} = 15\%$, and the probability for remaining in a bull market next year is $P_{U \to U} = 85\%$. Similarly, conditioning on starting in a bear market, the probability for remaining in a bear market next year is $P_{D \to D} = 27\%$, and the probability for transitioning to a bull market next year is $P_{D \to U} = 73\%$.

For each asset class, the time series of returns $r = \{r_1, r_2, ..., r_T\}$ is divided into bull market returns $r_U = \{r_1^U, r_2^U, ...\}$ and bear market returns $r_D = \{r_1^D, r_2^D, ...\}$. The return volatility and expected return corresponding to the bull and bear market cycles are then estimated by the subsample volatility and average return. Using S&P 500 data from 1976 through 2007, the bull market volatility is 13% while the bear market volatility is 17%. The bull market average return is 21% versus −19% for the bear market average.

10.4 OPTIMAL PORTFOLIO ALLOCATION

With the economy characterized and calibrated as a two-state Markov chain, we are now ready to examine the optimal portfolio exercise. Like the classic Markowitz portfolio analysis, we are seeking a set of portfolio weights that maximize the portfolio expected return given a volatility constraint. The portfolio optimization requires that we supply the expected returns for all of the assets in the investment opportunity set and the covariance matrix governing returns. In the context of our two-state Markov model, first, we must determine the current state of the economy before we can compute these asset return moments. This can be a difficult exercise, as we need to identify whether we are currently in a bull or bear market state; there may be no clear evidence suggesting a bull or bear market condition. In the next section, we discuss how to refine the model to overcome this uncertainty in our knowledge regarding the current state of the economy. We continue with the basic model for the time being.

Next, we need to use the calibrated model parameters from the previous section to compute the moments required for mean-variance optimization. Again, recall that we have k assets. In our simple model, we have two possible future states with conditional probability $P_{S \to U}$ of transitioning to a bull market from the current state S and $P_{S \to D}$ of transitioning to a bear market. The expected return vector and covariance matrix depends upon the future regime. Let μ_U and μ_D each be a $1 \times k$ vector of expected returns, and Ω_U and Ω_D be the covariance matrix for the bull and bear states, respectively. The vector of expected asset returns given that we are in state S is $\mu(S) = P_{S\to U}\mu_U + P_{S\to D}\mu_D$.

The derivation of the covariance term is a bit more complex. We are interested in computing $\Omega(S) = E[(r-\mu)'(r-\mu)|S]$. From the law of iterated expectations:

$$E[(r-\mu)'(r-\mu)|S]$$
$$= P_{S\to U}E[(r-\mu)'(r-\mu)|U] + P_{S\to D}E[(r-\mu)'(r-\mu)|D]$$
$$= P_{S\to U}E[(r_u-\mu)'(r_u-\mu)|U] + P_{S\to D}E[(r_D-\mu)'(r_D-\mu)|D]$$

To simplify the above expression, we note

$$E[(r_U-\mu)'(r_U-\mu)|S]$$
$$= E[(r_U-\mu_U+P_{S\to D}(\mu_U-\mu_D))'(r_U-\mu_U+P_{S\to D}(\mu_U-\mu_D))]$$
$$= \Omega_U + P_{S\to D}^2(\mu_U-\mu_D)'(\mu_U-\mu_D)$$

The covariance matrix then becomes:

$$\Omega(S) = P_{S\to U}\left[\Omega_U + P_{S\to D}^2(\mu_U-\mu_D)'(\mu_U-\mu_D)\right]$$
$$+ P_{S\to D}\left[\Omega_D + P_{S\to U}^2(\mu_D-\mu_U)'(\mu_D-\mu_U)\right]$$

The mean-variance optimal portfolio is then determined by the standard Markowitz optimal portfolio solution taking $\mu(S)$ and $\Omega(S)$ as inputs. Since the expected returns and volatilities are assumed to be time varying, the portfolio optimization exercise needs to be revisited frequently as the current state of the market changes. The resulting mean-variance optimal portfolio is then state dependent rather than static (as in the traditional solution). In particular, when the economy transitions from a bull market phase with low volatility to a bear market phase with high volatility, the optimal portfolio will also change and will shift to reduce risk in the bear market state.

10.5 SIMPLE MODEL EXTENSION

We noted previously that it may be difficult to determine exactly the current state of the economy. Generally, one does not know with a high degree of certainty whether one is in a bull market or bear market state (until after the market has fully run its course, which would eliminate the information advantage of this approach).

The lack of perfect knowledge about the current state means that we need to adjust for this uncertainty in our calculation. Hsu and Kalesnik (2008) show the benefits of properly adjusting for model uncertainty in portfolio construction and risk management. Suppose that there is a probability P_U that we are in a bull market environment, and $P_D = 1 - P_U$ that we are in a bear market environment. These probabilities will likely depend on a set of macroeconomic observables; as the macro variables change over time, the probabilities will also shift. The computation of the asset class return moments becomes more involved now; first, we need to repeat the exercise described in the last section for the bull and bear market states independently. Then we formulate a model for characterizing P_U and P_D. The uncertainty-adjusted moments for the mean-variance optimization are then computed as $\mu = P_U \mu(U) + P_D \mu(D)$ and $\Omega = P_U \Omega(U) + P_D \Omega(D)$. Finally, the mean-variance optimal portfolio is determined by the standard Markowitz optimal portfolio solution.

Since the probabilities P_U and P_D change in response to the changes in the macroeconomy, the optimal portfolio also changes with observed changes in the macro variables. As we observe signs that suggest greater likelihood that we have entered a bear market, P_D will increase and the optimal portfolio will take on a lower risk posture given the potentially higher volatility and lower forward returns.

10.6 CONCLUSION

Equity market volatility is time varying, as is the equity risk premium. Additionally, other risky asset volatilities appear to also be time varying and positively correlated with equity market volatility. Specifically, we find that volatilities for various risky asset classes tend to be low in equity bull markets and high in equity bear markets. Capturing this time-varying characteristic of joint asset class volatilities is important in order to properly execute mean-variance portfolio optimization.

We introduce in this chapter a simple and intuitive model of time-varying volatility and risk premia using the Markov state switching modeling technique. In our simple model, the state of economy switches between bull and bear markets. Asset classes have distinct volatility and risk premium characteristics in the two states of the market. By properly formulating the conditional moments, the traditional mean-variance optimization becomes a conditional optimization, and the traditional static optimal portfolio solution becomes a dynamic one. This results in a more efficient asset allocation, which takes advantage of the time-varying nature of market risk characteristics.

Applying this simple modeling technique improves portfolio characteristics over time. In the traditional constant volatility and risk premium model, optimal portfolio allocation remains constant over time. The state switching modeling approach has significant advantages when market volatilities and risk premia are time varying. Specifically, when we are in a state of bull equity market, where volatility has been low, properly assessing the probability for transitioning into a bear equity market, where the volatility would be substantially higher, would lead to a risk reduction portfolio. Reciprocally, in a bear market state, this approach would suggest greater risk taking. Relative to classic constant volatility models and static portfolio solutions, the time-varying approach with its associated dynamic optimal portfolio solution leads to better long-term portfolio efficiency and therefore a higher portfolio Sharpe ratio.

REFERENCES

Anderson, T. G., et al. (2001). The distribution of realized stock return volatility. *Journal of Financial Economics* 61:43–76.

Anderson, T. G., et al. (2003). Modeling and forecasting realized volatility. *Econometrica* 71:579–625.

Anderson, T. G., et al. (2005). Practical volatility and correlation modeling for financial risk management. Working paper 11069, NBER.

Baillie, R., and Myers, R. J. (1991). Bivariate GARCH estimation of the optimal commodities futures hedge. *Journal of Applied Econometrics* 6:109–24.

Banerjee, A. V. (1992). A simple model of herd behavior. *Quarterly Journal of Economics* 107:797–818.

Bentz, Y. (2003). Quantitative Equity Investment Management with Time-Varying Factor Sensitivities, *Applied Quantitative Methods for Trading and Investment*: 213–237.

Bikhchandani, S., Hirshleifer, D., and Welch, I. (1992). A theory of fads, fashion, custom, and cultural change as informational cascades. *Journal of Political Economy* 100:992–1026.

Black, F. (1976). Studies of stock price volatility changes. *Proceedings of the 1976 Meetings of the Business and Economics Statistics Section*. American Statistical Association: 177–81.

Blake, D., and Timmermann A. (2002). Performance Benchmarks for institutional Investors: Measuring, Monitoring and Modifying Investment Behavior. *Performance Measurement in Finance: Firms, Funds and Managers*, Butterworth Heinemann, Oxford: 108–141.

Bollerslev, T. (1986). Generalized autoregressive conditional heteroskedasticity, *Journal of Econometrics* 31:307–327.

Bollerslev, T., Engle, R. F., and Wooldridge, J. M. (1988). A capital asset pricing model with time-varying covariances. *Journal of Political Economy* 96:116–31.

Campbell, J., and Kyle, A. (1993). Smart money, noise trading, and stock price behavior. *Review of Economic Studies* 60:1–34.

Chordia, T., and Shivakumar, L. (2002). Momentum, business cycle, and time-varying expected returns. *Journal of Finance* 57:985–1020.

Christie, A. A. (1982). The stochastic behavior of common stock variances, value, leverage and interest rate effects. *Journal of Financial Economics* 10:407–32.

De Long, B. J., Shleifer, A., Summers, L. H., and Waldmann, R. J. (1990). Noise trader risk in financial markets. *Journal of Political Economy* 98:703–38.

Engle, R. F. (1982). Autoregressive conditional heteroscedaticity with estimates of the variance of United Kingdom inflation, *Econometrica* 50:987–1008.

Ferson, W. E., and Harvey, C. R. (1991). The variation of economic risk premiums. *Journal of Political Economy* 99:385–415.

Ghysels, E., Santa-Clara, P., and Valkanov, R. (2006). Predicting volatility: Getting the most out of return data sampled at different frequencies. *Journal of Econometrics* 131:59–95.

Horasanh, M., and Fidan, N. (2007). Portfolio selection by using time varying covariance matrices. *Journal of Economic and Social Research* 9:1–22.

Hsu, J., and Kalesnik, V. (Forthcoming). Risk-managing the uncertainty in VaR model parameters. In *The VaR implementation handbook*, ed. G. N. Gregoriou. New York: McGraw-Hill.

Lettau, M., and Ludvigson, S. (2001). Resurrecting the (C)CAPM: A cross sectional test when risk premia are time-varying. *Journal of Political Economy* 109:1238–87.

Meyn, S. P., and Tweedie, R. L. (1993). *Markov chains and stochastic stability*. London: Springer-Verlag.

Myers, R. J. (1991). Estimating time-varying optimal hedge ratios in futures markets. *Journal of Futures Markets* 11:39–53.

Officer, R. R. (1973). The variability of the market factor of New York Stock Exchange. *Journal of Business* 46:434–53.

Poon, S., and Granger, C. (2003). Forecasting volatility in financial markets: A review. *Journal of Economic Literature* 41:478–539.

Poon, S., and Granger, C. (2005). Practical issues in forecasting volatility. *Financial Analysts Journal* 61:45–56.

Schwert, W. G. (1989). Why does stock market volatility change over time? *Journal of Finance* 44:1115–53.

Shiller, R. J. (1981). Do stock prices move too much to be justified by subsequent changes in dividends? *American Economic Review* 71:421–36.

Vasilellis, G., and Meade, N. (1996). Forecasting volatility for portfolio selection. *Journal of Business Finance and Accounting* 23:125–43.

Zhang, L. (2005). The value premium. *Journal of Finance* 60:67–103.

CHAPTER 11

Robust Portfolio Selection with Endogenous Expected Returns and Asset Allocation Timing Strategies

Wolfgang Breuer, Marc Gürtler, and Olaf Stotz

CONTENTS

11.1 INTRODUCTION

The application of portfolio optimization according to Markowitz (1952, 1959) traditionally proceeds as follows. In a first step, input parameters of the optimization algorithm (expected returns and covariance matrix) are estimated from a time series of historical returns (i.e., sample means and covariances). The second step then applies these estimates in the mean-variance optimization in order to obtain optimal portfolio holdings. It is well known that using sample estimates of the mean and covariance to obtain optimal portfolio allocations over a large number of assets is problematic (Michaud, 1989; Green and Hollifield, 1992; Britten-Jones, 1999). The resulting portfolios usually contain extreme long and short positions, are poorly diversified, and produce a poor out-of-sample performance. Michaud (1989) argues that the mean-variance optimization has a tendency to maximize the effects of errors in input parameters. Moreover, small changes in input parameters (in particular in expected returns) can lead to large changes in optimal portfolio weights (e.g., Jobson and Korkie, 1980; Chopra and Ziemba, 1993). Within the mean-variance framework, several methods have been proposed to reduce the sensitivity of optimal portfolio weights with respect to variations of input parameters. The first line of research modifies the estimation procedure for input parameters (e.g., Jorion, 1986; Black and Litterman, 1992). The second line adjusts the selection procedure for optimal security weights of the mean-variance optimization method, for example, by imposing financially meaningful constraints (Frost and Savarino, 1988).

This chapter contributes to the first line. We argue that the cause for the potential contradiction between the theoretical recommendation of quite sensitive portfolio weights and the practical finding of their comparatively high stability does not lie in properties of the Markowitz portfolio theory in itself. In contrast, this contradiction can be resolved, when return expectations are treated as endogenous in mean-variance optimization. Endogenous returns are then modeled on the basis of an inverted dividend discount model. Mean-variance optimization can then account for changes in return expectations that are either common knowledge (endogenous expectations) or insider knowledge (exogenous expectations). This should have a different impact on both the investor's optimal portfolio and the market portfolio. In addition to these two extreme cases, there might be situations termed as mixed cases, with the investor's expectations being only partially reflected by corresponding changes in dividend expectations of the capital market as a whole.

Our approach leads to several conclusions. First, we show that real-life sensitivity of portfolio weights can be explained by investors acting according to the Markowitz portfolio theory, with their changes in return expectations being reflected by altered market expectations to a good deal. Thus, investors' expectations seem to reflect market's expectations to some extent. Second, volatility of investors' portfolio weights is still greater than seems to be reasonable. The higher volatility is caused by investors' dividend expectations that deviate from market expectations (but need not be of superior quality). The reason for the higher volatility of portfolio weights is, however, not the application of the classical Markowitz portfolio optimization per se. We show this by introducing an indicator variable *ik* in the dividend discount model, with *ik* > 1 (*ik* < 1) characterizing situations in which an investor is more (less) confident than the market as a whole regarding future dividends. Third, we are able to utilize this indicator variable to define easy-to-follow portfolio selection timing strategies like momentum or contrarian behavior for asset allocation problems and apply them to the capital markets of Germany, Japan, the UK, and the United States.

The rest of our chapter is organized as follows. Section 11.2 reviews the common Markowitz optimization approach with exogenously given expected returns, derives the approach with endogenous returns, and presents the mixed case. Section 11.3 introduces the quantitative indicator variable *ik*, which may be used as a starting point for the adjustment of expectations. Section 11.4 presents numerical and empirical evidence that there is no contradiction between portfolio weight sensitivity implied by the simple Markowitz approach and real-life observable investor behavior. Moreover, the efficiency of different timing strategies in asset allocation problems based on variations of *ik* over time is examined for different national capital markets. Section 11.5 concludes the chapter.

11.2 PORTFOLIO SELECTION WITH EXOGENOUS RETURNS AND ENDOGENOUS RETURNS

11.2.1 The Exogenous Case

We consider an investor with preference function $\Phi(\mu,\sigma^2)=\mu-0.5\cdot k\cdot\sigma^2$ defined in expected portfolio return μ and corresponding return variance σ^2. She invests her wealth for one period and can borrow or lend money at a certain interest rate r_f. Additionally, she can invest in N risky assets with expected one-period return μ_i and risk σ_i $(i = 1,\ldots, N)$. Then, optimal portfolio weights are given by (e.g., Campbell and Viceira, 2002)

$$x_i = \frac{1}{k} \cdot \sum_{j=1}^{N} \sigma_{ij}^{-1} \cdot (\mu_j - r_f), \ i = 1,\ldots, N, \tag{11.1}$$

with σ_{ij}^{-1} being an element (i, j) of the inverse of the variance-covariance matrix of returns. $x_0 = 1 - \sum_{j=1}^{N} x_j$ denotes the relative holding of the risk-free asset $j = 0$. The preference parameter k enters the optimal portfolio only via the scalar term $1/k$. In Equation (11.1), expected returns are given exogenously, and it is implicitly assumed that the supply of all assets is infinitely elastic. A higher demand for asset i would thus not change expected one-period return μ_i.

To become more specific we relate expected returns to prices by the assumption that market participants form their return expectations according to the cash flow (or dividend) discount model (e.g., Gordon, 1962):

$$\hat{\mu}_j = \frac{E(CF_j)}{P_j} + g_j, \tag{11.2}$$

with $\hat{\mu}_j$ as the estimation for expected one-period return of risky security j $(j = 1,\ldots, N)$, $E(CF_j)$ as the expected cash flow for the end of the period, g_j as the constant expected cash flow growth rate, and P_j as the price of security j. One weakness of the dividend discount model (Equation (11.2)) is the assumption of a constant growth rate for dividends that could be relaxed by using the dynamic dividend discount model of Campbell and Shiller (1988).

We are particularly interested in the consequences of changing cash flow expectations. To this end, let cash flow expectations for security j ceteris paribus change from $E(CF_j)$ to $c_j^{(I)} \cdot E(CF_j)$, with $c_j^{(I)} \neq 1$. Inserting Equation (11.2) in Equation (11.1) and replacing $E(CF_j)$ with $c_j^{(I)} \cdot E(CF_j)$ yields

$$x_i = \frac{1}{k} \cdot \sum_{j=1}^{N} \sigma_{ij}^{-1} \cdot \left[\left(\frac{c_j^{(I)} \cdot E(CF_j)}{P_j} + g_j \right) - r_f \right], \ i = 1,\ldots, N. \tag{11.3}$$

In order to reduce complexity we will mainly focus on the classical asset allocation problem with security $i = 1$ denoting stocks and security $i = 2$ standing for risky bonds.

Moreover, without loss of generality, we restrict ourselves to changes in cash flow expectations with respect to asset 1. As, for given prices, such a change in cash flow expectation immediately carries over to a change in return expectations μ_1, we look at the following derivatives for comparative static analysis:

$$\frac{\partial x_1}{\partial \mu_1} = \frac{1}{k} \cdot \sigma_{11}^{-1} = \frac{1}{k} \cdot \frac{\sigma_2^2}{\sigma_1^2 \cdot \sigma_2^2 - \sigma_{12}^2} > 0,$$

$$\frac{\partial x_2}{\partial \mu_1} = \frac{1}{k} \cdot \sigma_{21}^{-1} = -\frac{1}{k} \cdot \frac{\sigma_{12}}{\sigma_1^2 \cdot \sigma_2^2 - \sigma_{12}^2} < 0,$$

(11.4)

with the latter inequality being valid in the (plausible) case of a positive return covariance σ_{12} between asset classes 1 and 2, i.e., positively correlated stock and bond returns. Not very surprisingly, better return expectations regarding stocks will *ceteris paribus* lead to a greater stock investment and reduced purchases of risky bonds.

11.2.2 The Endogenous Case

While the above approach can be followed by an individual investor, this usually does not hold true for a representative investor representing total capital market behavior, as market equilibrium conditions and countervailing price-return effects due to investors' demand have to be considered. Therefore, expected returns must be made endogenous.

Total initial wealth W is identical to total market capitalization of risky securities and the investment in the riskless asset. Capital market equilibrium requires $x_j^{(M)} = x_j$, where $x_j^{(M)}$ denotes the market weight of asset $j = 0,\ldots, N$. For each risky security $j = 1,\ldots, N$, the total number of shares issued shall be one, while—for the sake of simplicity—we assume a net supply of the riskless asset of zero. Moreover, in equilibrium we thus must have $W := W(P_1,\ldots,P_N) = \Sigma_{m=1}^{N} P_m$, since investors' initial wealth is now determined by the current market value of all (risky) assets. Therefore, the equilibrium price of a share in security j is

$$P_j = x_j^{(M)} \cdot \sum_{m=1}^{N} P_m$$

(11.5)

Moreover, we are once more especially interested in consequences of changing cash flow expectations $c_j^{(I)} \neq 1$. With this as well as Equations (11.2) and (11.5), Equation (11.1) becomes

$$
x_i^{(M)} = \frac{1}{k} \cdot \sum_{j=1}^{N} \sigma_{ij}^{-1} \cdot \left[\left(\frac{c_j^{(I)} \cdot E(CF_j)}{x_j^{(M)} \cdot \sum_{m=1}^{N} P_m} + g_j \right) - r_f \right], \quad i = 1, \dots, N. \quad (11.6)
$$

In Equation (11.6), expected returns are determined endogenously, as a change in optimal portfolio weights affects expected returns through Equations (11.2) and (11.5) via P_j.

Again, we focus on the simple two-asset case and allow only for alterations in cash flow expectations regarding security 1, i.e., stocks. We now look at changes of optimal portfolio holdings caused by modifications in cash flow expectations. Thereby, these alterations in cash flow expectation are to be assumed in such a way so as to imply a new (equilibrium) value of μ_1 that is identical to a certain reference situation in the exogenous case with the same change in expected stock return. To be more precise, we thus look at two values for $c_1^{(I)}$ in the exogenous case and the endogenous case that lead to the same new expected stock return μ_1. For such a situation changes in portfolio holdings in the endogenous case are determined by the signs and scale of the following derivatives:

$$
\frac{\partial x_1^{(M)}}{\partial \mu_1} = \frac{1}{k} \cdot \left(\sigma_{11}^{-1} + \underbrace{\sigma_{12}^{-1}}_{<0} \cdot \underbrace{\frac{\partial \mu_2}{\partial \mu_1}}_{>0} \right) < \frac{1}{k} \cdot \sigma_{11}^{-1},
$$

$$
\frac{\partial x_2^{(M)}}{\partial \mu_1} = \frac{1}{k} \cdot \left(\underbrace{\sigma_{21}^{-1}}_{>0} + \underbrace{\sigma_{22}^{-1}}_{} \cdot \underbrace{\frac{\partial \mu_2}{\partial \mu_1}}_{>0} \right) > \frac{1}{k} \cdot \sigma_{21}^{-1}.
$$

(11.7)

In the endogenous case, we will typically have $\partial \mu_2 / \partial \mu_1 > 0$, as the price for risky bonds will *ceteris paribus* decrease if stock and bond returns are sufficiently positively correlated so that both security classes interact as substitutes. In fact, with a net supply of zero for the riskless asset we can immediately conclude that the partial derivatives $\partial x_1^{(M)}/\partial\mu_1$ and $\partial x_2^{(M)}/\partial\mu_1$ must sum up to 1, as we always have $x_1^{(M)} + x_2^{(M)} = 1$. From this and Equation (11.7) we can calculate $\partial\mu_2/\partial\mu_1$ as

$$\frac{\partial x_1^{(M)}}{\partial\mu_1} + \frac{\partial x_2^{(M)}}{\partial\mu_1} \overset{!}{=} 0$$

$$\Leftrightarrow \sigma_{11}^{-1} + \sigma_{12}^{-1} \cdot \frac{\partial\mu_2}{\partial\mu_1} + \sigma_{21}^{-1} + \sigma_{22}^{-1} \cdot \frac{\partial\mu_2}{\partial\mu_1} = 0 \qquad (11.8)$$

$$\Leftrightarrow \frac{\partial\mu_2}{\partial\mu_1} = -\frac{\sigma_{11}^{-1} + \sigma_{21}^{-1}}{\sigma_{12}^{-1} + \sigma_{22}^{-1}} = -\frac{\sigma_2^2 - \sigma_{12}}{\sigma_1^2 - \sigma_{12}} = -\frac{\sigma_2 \cdot (\sigma_2 - \rho_{12} \cdot \sigma_1)}{\sigma_1 \cdot (\sigma_1 - \rho_{12} \cdot \sigma_2)},$$

with ρ_{12} as the correlation coefficient between stock and bond returns. Without loss of generality we may (reasonably) assume $\sigma_1 > \sigma_2$ so that the denominator of the last fraction in Equation (11.8) immediately becomes positive. Then, for sufficiently high positive correlation ρ_{12} ($>\sigma_2/\sigma_1$) the corresponding numerator in Equation (11.8) gets negative, thus implying $\partial\mu_2/\partial\mu_1 > 0$.

The finding $\partial\mu_2/\partial\mu_1 > 0$ for sufficiently high positive correlation between stock and bond returns indicates that, in such a situation, for constant prices increasing stock return expectations will result in an excess demand for stocks and an excess supply of risky bonds that lead to a falling price P_2 and a rising expected bond return μ_2. As a consequence, a certain change in expected security return μ_1 will lead to less sensitive reactions of optimal portfolio weights $x_1 = x_1^{(M)}$ and $x_2 = x_2^{(M)}$. The economic intuition behind this finding is that an increasing expected return μ_2 due to a falling price P_2 weakens the incentive to portfolio revision in the wake of rising values μ_1. This effect should be distinguished from the straightforward countervailing consequences of an increasing value P_1 for a certain modification $c_1^{(I)}$ of expected cash flow on stocks in the endogenous case compared to the exogenous case with the same value $c_1^{(I)}$. Certainly, a *ceteris paribus* variation of stock cash flow expectations leads to smaller amounts of portfolio revisions in the endogenous case than in the exogenous one, as changes in expected stock returns now will differ in both situations. However, the derivatives according to Equation (11.7) imply the more interesting and practically relevant result that even for a given change in expected stock return, portfolio revisions in the exogenous case will be more pronounced than in the endogenous one. We will illustrate this finding via a numerical example in Section 11.2.3.

11.2.3 The Mixed Case

The endogenous case can be interpreted as a situation where the investors' cash flow expectations as represented by $c_1^{(I)}$ are identical to market cash flow expectations $c_1^{(M)}$: $c_1^{(I)} = c_1^{(M)}$. On the contrary, the exogenous case can be described as a situation with $c_1^{(M)} = 1$ and $c_1^{(I)} \neq 1$. Certainly, one can imagine mixed cases with $1 \neq c_1^{(I)} \neq c_1^{(M)} \neq 1$.

In fact, one may think of investors' expectations being the result of the aggregation of changes in autonomously expected dividends, $c_1^{(D)}$, as well as changes in market expectations, $c_1^{(M)}$. Let us assume the simple case of $c_1^{(I)}$ being just a weighted arithmetic mean of $c_1^{(D)}$ and $c_1^{(M)}$:

$$c_1^{(I)} = \lambda \cdot c_1^{(D)} + (1 - \lambda) \cdot c_1^{(M)}. \tag{11.9}$$

Apparently, $\lambda = 0$ describes the endogenous case, while $\lambda = 1$ in connection with $c_1^{(M)} = 1$ implies the exogenous case. However, based on Equation (11.9) it is possible to analyze mixed cases with only a fraction of the investor's (original) expectation based on individual assessments, and the other fraction being based on market expectations.

In such a mixed case one has to proceed by two steps. First, Equations (11.5) and (11.6) (for $N = 2$) have to be used in order to determine new equilibrium holdings of stocks and bonds as well as their equilibrium prices. Thereby, $c_1^{(I)}$ in Equation (11.6) must be replaced with $c_1^{(M)} \neq 1$ and $c_2^{(I)}$ is set to 1. Second, for these new equilibrium prices P_1 and P_2, Equation (11.3) of the exogenous case has to be applied. If we especially look at a situation with $c_1^{(I)} > c_1^{(M)} > 1$, we will get greater individual stock holdings than in the purely endogenous case with the same value for $c_1^{(M)}$ but $c_1^{(I)} = c_1^{(M)}$. The reason is that in the former case one can divide changes in dividend expectations into two terms $c_1^{(M)}$ and $\Delta c_1^{(I)}$. Portfolio adjustment induced by the first term in connection with the identical change in market expectations is determined by Equation (11.6), but additional individual portfolio changes as a consequence of $\Delta c_1^{(I)}$ follow according to Equation (11.4), thus leading to *ceteris paribus* greater stock holdings (and smaller bond holdings) than in the purely endogenous case.

11.3 PORTFOLIO OPTIMIZATION AND INSIDER KNOWLEDGE

Obviously, the quotient

$$ik_1 := c_1^{(I)}/c_1^{(M)} = \lambda \cdot \frac{c_1^{(D)}}{c_1^{(M)}} + (1-\lambda) \tag{11.10}$$

can be interpreted as some measure of (subjectively felt) insider knowledge on the investor's side. While $\lambda = 0 \Leftrightarrow ik_1 = 1$ stands for no insider knowledge at all, the relative importance of insider knowledge increases as λ approaches 1, and ik_1 thus $c_1^{(D)}/c_1^{(M)}$. To be more precise: if we observe situations with ik_1 smaller than 1, an investor overestimates reductions in expected cash flows and underestimates increases in expected cash flows in comparison to the market assessments, and vice versa. In short, the greater the deviation of ik_1 from 1, the greater the relevance of the investor's presumed insider knowledge compared to market expectations.

In fact, Equation (11.10) may be used as a starting point for a new kind of a portfolio management approach. Thereby, for the sake of generality, we define ik_2 for bonds in the same ways as ik_1 for stocks. Consider now a certain incoming information that may induce market participants to modify their cash flow expectations. The amount of this change in dividend expectation can be calculated from resulting price changes. To this end, for given new prices and thus market portfolio structure $(P_1, P_2, x_1^{(M)}, x_2^{(M)})$, it is only necessary to solve Equation (11.6) with respect to variables $c_j^{(M)}$ (that replace variables $c_j^{(I)}$).

In the special two-asset case we get

$$c_j^{(M)} \cdot E(CF_j) = \left(k \cdot \frac{x_j^{(M)} \cdot \sigma_{hh}^{-1} - x_h^{(M)} \cdot \sigma_{12}^{-1}}{\sigma_{11}^{-1} \cdot \sigma_{22}^{-1} - \left(\sigma_{12}^{-1}\right)^2} - (g_j - r_f) \right) \cdot P_j \quad (h, j = 1, 2, h \neq j). \tag{11.11}$$

Taking the definitions of ik_1 and ik_2 into account and Equation (11.11) for stocks and bonds, Equation (11.3) becomes

$$x_i = \sum_{j=1}^{2} \sigma_{ij}^{-1} \cdot f_j, \quad i = 1, 2, \tag{11.12}$$

with parameter f_j being defined in the following way:

$$f_j := \frac{1}{k} \cdot \left[\left(\frac{ik_j \cdot c_j^{(M)} \cdot E(CF_j)}{P_j} + g_j \right) - r_f \right]$$

$$\underset{(11.11)}{=} \frac{1}{k} \cdot \left[\left(\frac{ik_j \cdot \left(k \cdot \dfrac{x_j^{(M)} \cdot \sigma_{hh}^{-1} - x_h^{(M)} \cdot \sigma_{12}^{-1}}{\sigma_{11}^{-1} \cdot \sigma_{22}^{-1} - \left(\sigma_{12}^{-1} \right)^2} - (g_j - r_f) \right) \cdot P_j}{P_j} + g_j \right) - r_f \right] \quad (11.13)$$

$$= ik_j \cdot \frac{x_j^{(M)} \cdot \sigma_{hh}^{-1} - x_h^{(M)} \cdot \sigma_{12}^{-1}}{\sigma_{11}^{-1} \cdot \sigma_{22}^{-1} - \left(\sigma_{12}^{-1} \right)^2}.$$

Obviously, for given parameter values ik_1 and ik_2, optimal portfolio weights x_1 and x_2 for stocks and bonds can be determined independently of risk aversion parameter k and expected cash flows for stocks and dividends. We just need estimators for return variances and the covariance and for growth rates of stock and bond cash flows in order to apply the portfolio selection rule according to Equation (11.12) by choosing investor-specific values ik_1 and ik_2. In what follows, we consequently set ik_2 equal to 1, implicitly assuming that it is not possible to outperform the capital market with respect to bond cash flow estimates, as there is too little volatility in the market (corresponding to our prior statement of being mainly interested in changing dividend expectations). The last assessment will be verified in the empirical part of our chapter, below.

However, we are able to construct strategies for time-varying determinations of ik_1. In particular, as is well known from the empirical literature on asset pricing, capital markets may underreact in the short run, thus giving investors opportunities to follow a so-called momentum strategy (see, for example, Jegadeesh and Titman, 1993). The conceptual framework outlined in this paper now offers the option for a new kind of such a strategy. By the help of Equation (11.11), it is possible to compute whether market cash flow estimations rise or fall from one period to another. To establish a momentum strategy, an investor should apply

$ik_1 > 1$ in Equations (11.12) and (11.13) for rising market expectations and $ik_1 < 1$ for falling ones. Furthermore, by exchanging $ik_1 > 1$ with $ik_1 < 1$ and vice versa, it is also possible to define a contrarian strategy that could also be advantageous for longer portfolio holding periods (see DeBondt and Thaler (1985) for the winner-loser effect and Lakonishok et al. (1994) for the glamour effect as possible theoretical backgrounds for contrarian strategies). Additionally, simple strategies with fixed values $ik_1 \neq 1$ over all periods under consideration could be examined. In what follows we want to give numerical and empirical illustrations of all our results of Section 11.3. Thereby, we will also return to the analysis of the performance of momentum and contrarian strategies as defined in this subsection.

11.4 NUMERICAL AND EMPIRICAL ANALYSIS

11.4.1 Numerical Analysis

We start by analyzing the European capital market situation at the beginning of November 2004 as our base scenario. The annual risk-free rate, approximated by the average 1-year yield of government bonds in the Euro area, was $r_f = 2.3\%$. We consider two asset classes ($N = 2$): $i = 1$ denotes stocks and $i = 2$ denotes bonds in general. According to Datastream, the total market capitalization of traded stocks in Europe amounted up to €4,841 billion, while the total market capitalization of traded bonds was €2,990 billion. As pointed out earlier, for simplicity, we assume that the (net) investment in the risk-free asset is zero. Therefore, total initial wealth W in our base scenario equals €7,831 billion. Expected cash flows at the end of the period (November 2005) are assumed to be €152 billion for stocks and €115 billion for bonds (approximated from realized cash flows). Expected growth rates for dividends on stocks are set to 6% (approximated from the historical growth rate over the last 20 years). Expected growth rates for cash flows on bonds are assumed to be 0%. According to Equation (11.2), we estimate $\hat{\mu}_1 = 9.14\%$ for stocks and $\hat{\mu}_2 = 3.87\%$ for bonds. Risks of stocks and bonds are estimated by the implied volatility of option prices: $\sigma_1 = 15\%$ (approximately the implied volatility of DJ Euro Stoxx 50 index options) and $\sigma_2 = 4\%$ (approximately the implied volatility of Bund future options). Preference parameter k and covariance σ_{12} are implicitly calculated according to Equation (11.6) by assuming that the current capital market situation is in equilibrium. This yields a correlation between stock and bond returns of 79.38% (greater then $\sigma_2/\sigma_1 = 26.67\%$) and a preference parameter k of 4.3497.

We now compare optimal portfolio weights given by Equations (11.1) and (11.6), respectively, by changing expected returns of stocks. We assume that investors form return expectations according to Equation (11.2). Thereby, the expected cash flow of stocks is varied by setting $E(CF_1) = c \cdot €152$ billion, with $c \in C = \{0.5, 0.8, 0.9, 1.1, 1.2, 1.5, 2\}$. For each $c \in C$ optimal portfolio weights are calculated for the exogenous case (Equation (11.1)) and the endogenous case (Equation (11.6)). Panel A of Table 11.1 displays the results of the exogenous case. As expected, portfolio weights take extreme values and react in a very sensitive way to changes in expected returns. For example, if expected cash flows are estimated 100% higher than in the base scenario (resulting in an expected return of 12.28%), the optimal weights in stocks and bonds equal 148.53% and −219.96%, respectively. For the endogenous case, portfolio weights react by far less sensitively to changes in input parameter c (Panel B). If expected cash flows are estimated 100% higher than in the base scenario, this leads to an optimal portfolio of 72.65% in stocks and 27.35% in bonds.

Certainly, one might deem the findings of Table 11.1 not to be too surprising, because it is straightforward to see that for given changes in *dividend* expectations the countervailing price effects of the endogenous case will partially undo changes in expected *returns*, and thus lead to more moderate portfolio adjustments than in the exogenous case. However, portfolio adjustments are actually more moderate in the endogenous case than in the exogenous one even if we examine the same change in expected stock return, as has been shown in general in Section 11.3. In fact, our numerical example might highlight this finding as well.

For example, in the exogenous case an expected stock return of 10.71% yields $c = 1.5$. In the endogenous scenario, the same expected stock return of 10.71% corresponds to $c = 3.9765$. Optimal shares of stocks and bonds amount to 82.16% and 17.84% (endogenous case) compared to 105.17% and −90.88% in the exogenous case. Again, the endogenous case is accompanied with quite modest portfolio adjustments, while in the exogenous case extreme weights are observed. The cause for this discrepancy in spite of an identical expected stock return lies in the alteration of expected bond returns. In the endogenous scenario, expected bond returns rise from 3.87% to 4.13% as a consequence of falling bond prices

TABLE 11.1 Optimal Mean-Variance Portfolio Weights and Corresponding Expected Security Returns (Percent): x_1 = Allocation in Stocks, x_2 = Allocation in Bonds

c	0.5	0.8	0.9	(Base scenario, c = 1)	1.1	1.2	1.5	2.0
Panel A: Exogenous case								
Expected stock return	7.57	8.51	8.83	9.14	9.45	9.77	10.71	12.28
Portfolio structure	$x_1 = 18.44$	44.46	53.13	61.80	70.48	79.15	105.17	148.53
	$x_2 = 167.28$	89.83	64.01	38.20	12.38	−13.44	−90.88	−219.96
Panel B: Endogenous case								
Expected stock return	8.33	8.87	9.01	9.14	9.26	9.36	9.63	9.98
Portfolio structure	$x_1^{(M)} = 51.26$	58.32	60.16	61.80	63.32	64.70	68.22	72.65
	$x_2^{(M)} = 48.74$	41.68	39.84	38.20	36.68	35.30	31.78	27.35

in equilibrium, while in the exogenous case expected bond returns are not changed.

11.4.2 Empirical Analysis

11.4.2.1 The Efficiency of German Special Funds' Portfolio Selection

Equation (11.7) implies that changes in portfolio weights should be more volatile for a portfolio manager with insider knowledge about future cash flows, and thus future returns and the standard deviation of portfolio weights should be larger for a specific portfolio manager (or a group of specific managers) than for the market, i.e., $\sigma(x_i^{(P)}) > \sigma(x_i^{(M)})$, where $\sigma(x_i^{(P)})$ is the standard deviation of weights of portfolio manager P and $\sigma(x_i^{(M)})$ is the standard deviation of weights of market M. In addition to our numerical example we want to examine this implication for the European capital market empirically. We analyze portfolio weights for a specific group of German investors, in particular institutional investors that invest in special investment funds (Spezialfonds). We cover portfolio weights of special funds over the period 1973–2005, looking at yearly data (as of January 1). Data are provided by Deutsche Bundesbank. Portfolio weights for the market are approximated by Datastream Total Market Indexes, for both equities and bonds. Figure 11.1 displays the weights over the sample period. It can be seen that bond and stock weights of the market and that of the special funds change in the same direction, although the average weight in stocks (bonds) is higher (lower) for special funds

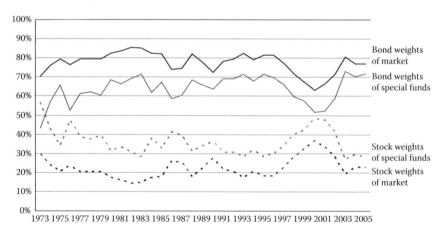

FIGURE 11.1   Development of the market portfolio and portfolios held by managers of special funds over time.

than for the market.[*] In fact, the average weight of stocks is 36.17% for special funds and only 22.70% for the market.

The time-series standard deviation suggests that special funds change their weights in a more volatile manner than the market. For the special funds, the standard deviation amounts to $\sigma(x_{bonds}^{(P)}) = \sigma(x_{stocks}^{(P)}) = 7.11\,\%$, for the market, and the standard deviation equals $\sigma(x_{bonds}^{(M)}) = \sigma(x_{stocks}^{(M)}) = 5.40\%$. Nevertheless, these standard deviations are rather similar so that we might conjecture once again the higher relevance of the endogenous case than the exogenous one.

However, only if special funds indeed had superior forecasts of future returns (future cash flows) compared to the market would the higher volatility of portfolio weights according to Equation (11.7) be justified. To investigate this implication, we calculate the Sharpe ratio for returns produced by special funds and the market. Over the sample period, the average special fund has achieved an average yearly excess return of 2.30%; the average yearly excess return for the market is 2.00%. Thus, special funds achieve a higher expected excess return than the market. However, the higher excess return cannot be explained by a superior knowledge of special fund managers, but rather with a higher risk of the portfolio composition of special funds. In particular, the standard deviation of the excess return of a special fund is 10.99%, and thus substantially higher than that for the market, which is 7.87%. Hence, on average, special funds take on more equity risk than the market does. As a result, the Sharpe ratio for the market is almost five percentage points higher than the Sharpe ratio for the special fund (i.e., $SR^{(M)} = 25.38\% > SR^{(P)} = 20.96\%$[†]). Although special funds adjust their portfolio weights with higher volatility than the market, they produce an inferior risk/return trade-off. As a

---

[*] Investment in the risk-free asset has to be neglected, as data on investments in the money market (proxy for the risk-free investment) exist only since 1993. However, the investment in the money market is small (the portfolio weight never exceeds 0.5%), and therefore has a negligible impact on the standard deviation of stock and bond weights. The analysis is thus restricted to the sole comparison of risky subportfolios. Therefore, weights in bonds and stocks always add up to one. However, resulting Sharpe ratios computed below are independent of the actual amount of riskless lending and borrowing.

[†] The z-statistic of Jobson and Korkie (1981), which has been corrected by Memmel (2003), has a value of 0.74, thus indicating that the difference in the Sharpe ratio is not significant even at a 10% significance level. However, it has already been noted by Jobson and Korkie (1981) that their test has only a very small power. According to them, for an underlying number of sixty portfolio optimizations, a difference of 0.1 between the two Sharpe ratios will lead to a rejection of the null hypothesis only in 10% of all cases.

consequence, fund managers should be very careful when deciding to choose a parameter value $ik_1 > 1$, as their knowledge need not really be superior to that of the market. In any case, as highlighted by our theoretical considerations as well as our numerical and empirical examples so far, the reason for the excess volatility is not a potential shortcoming of the Markowitz portfolio theory, but investors seem to be too confident regarding the quality of their own forecasts. However, if used wisely, the variable ik_1 can be applied to more sophisticated asset allocation strategies, as the following subsection will examine in more detail.

11.4.2.2 Portfolio Optimization for Different Values of Parameter ik_1

Even if average values of ik_1 near to 1 seem to be favorable according to our analysis so far, there might be possibilities to exploit market inefficiencies by a more sophisticated approach for the determination of optimal risky portfolios. To this end, we monthly apply Equations (11.12) and (11.13) over the time period from January 1, 1973 to January 1, 2006, for different strategies with respect to the (possibly time-dependent) choice of ik_1 for stocks. At each point in time we use the last thirty-six historical monthly excess return realizations to estimate return variances and covariances. Moreover, we hold annual growth rates of stock cash flows and bond cash flows constant at 6% and 0%, respectively. Current prices P_j as well as market portfolio weights are directly observable. Finally, we utilize 1-month interest rates from the German money market as proxies of the riskless interest rate. For any pair of parameter values ik_1 and ik_2 it is then possible to determine an investor's optimal structure of risky securities. Moreover, we are able to compute the standard deviation of relative changes in $c_1^{(M)}$ and $c_2^{(M)}$ over time. We arrive at a standard deviation of 180.034% for stock parameter $c_1^{(M)}$ and only of 6.138% for the bond parameter $c_2^{(M)}$. This confirms our implicit assumption of setting $ik_2 = 1$ for all portfolio selection problems under consideration.

However, with respect to ik_1 it pays to take a closer look at settings that imply deviations from market expectations. Table 11.2 gives an overview of our empirical results. Thus, besides strategies with a fixed value for ik_1 over time, we also allow for strategies that distinguish between settings for ik_1 after a positive change in market cash flow expectation ($ik_1^{(up)}$) and those after a corresponding negative change ($ik_1^{(down)}$). This enables us to examine the performance of several kinds of momentum and contrarian strategies. In line with our rolling-window approach, Sharpe ratios for all

TABLE 11.2 Monthly Performance (Sharpe Ratio) and Standard Deviation ($\sigma(x_1)$) of Monthly Stock Holdings in Germany for Different Portfolio Selection Strategies and the Time Period from January 1, 1973 to January 1, 2006

Strategy	**(1)**	**(2)**	**(3)**	**(4)**	**(5)**
$ik_1^{(down)}$	0.1	0.5	1	1.5	10
$ik_1^{(up)}$	0.1	0.5	1	1.5	10
Sharpe ratio	0.099730	0.109401	0.090151	0.071347	0.027115
$\sigma(x_1)$	0.086834	0.064789	0.051228	0.062810	0.684987
Strategy	**(6)**	**(7)**	**(8)**	**(9)**	**(10)**
$ik_1^{(down)}$	1.6702	0.5	1.5	0.5	1
$ik_1^{(up)}$	1.6702	1.5	0.5	1	0.5
$avg(ik_1)$	1.6702	0.9685	1.0315	0.7343	0.7657
Sharpe ratio	0.066596	0.121149	0.041166	0.122044[a]	0.072374
$\sigma(x_1)$	0.071098	0.179324	0.169891	0.104995	0.094959

[a] Significantly different from the Sharpe ratio of strategy (3) on the 10% level.

strategies are estimated out of sample on the basis of the actually resulting portfolio returns at future times $t + 1$ for fixed portfolios at times t. This means that we get 396 out-of-sample monthly return realizations from January 1, 1973 to January 1, 2006, for each strategy under consideration. For ik_1 not being constant, Table 11.2 additionally gives the average value of ik_1 ($avg(ik_1)$) over the whole time period under consideration. In total, Table 11.2 explicitly presents ten different asset allocation strategies and their resulting Sharpe ratios based on monthly return data as well as standard deviations $\sigma(x_1)$ of monthly stock holdings.

As a first result for fixed values $ik_1 \neq 1$, moderate deviations from $ik_1 = 1$ (i.e., from Equation (11.3)) between 0.5 and 1.5 lead only to fluctuations of stock holdings over time that are of similar magnitude as those of the whole market. This finding is supported by the fact that a setting of $ik_1 = 1.6702$ over the whole time period under consideration exactly reproduces the volatility of stock holdings of special funds as presented in the preceding subsection. Moreover, the resulting annual Sharpe ratio of this strategy (computed in the same way as that of the special funds in the preceding section) is 19.4609%, and thus almost identical to the empirically observable one. This gives additional evidence that portfolio managers' reactions to changing market expectations are too strong in the case of bullish market expectations, but too weak in the case of a bear market just expressing some kind of overconfidence. Nevertheless, portfolio managers' reactions are far from the extreme consequences as implied by the exogenous case of Section 11.3.1, which are approximated in Table 11.2 by setting $ik_1 = 10$.

Other constant values $ik_1 > 1$ lead to poorer performance outcomes than that of the market as well according to Table 11.2, while for moderate settings $ik_1 < 1$ an improvement in the Sharpe ratio can be obtained (with the best performance of a Sharpe ratio of 10.9923% for a setting $ik_1 = 0.373$).

The last finding hints at the possibility that market reactions themselves are too strong in the case of rising expectations or too weak in the case of decreasing cash flow expectations. This supposition is verified by the performance of momentum strategies characterized by $ik_1^{(up)} > 1$ and $ik_1^{(down)} < 1$, which is far better than those of contrarian strategies (that is, strategies with $ik_1^{(up)} < 1$ and $ik_1^{(down)} > 1$). Nevertheless, a semi-momentum strategy with $ik_1^{(up)} = 1$ and $ik_1^{(down)} = 0.5$ turns out to be even more successful. In fact, a maximum Sharpe ratio of 13.7314% could be achieved for the rather extreme timing strategy $ik_1^{(up)} = 1$ and $ik_1^{(down)} = 0$. Such semi-momentum strategies are based on the assumption that market reactions in phases of rising cash flow expectations are nearly adequate, while falling market expectations tend to underestimate real decreases in cash flow expectations. The special setting $ik_1^{(up)} = 1$ and $ik_1^{(down)} = 0$ implicitly recommends to refrain from any stock holding at all when market cash flow expectations are declining, and to reproduce the market portfolio in situations with rising market expectations. In contrast, certainly, it is not too surprising that a semi-contrarian strategy leads to rather poor performance.

Moreover, from all values in Table 11.2, only the Sharpe ratio of the semi-momentum strategy (Equation (11.9)) is significantly different from the Sharpe ratio of the market portfolio on a 10% basis. The concept developed in this section can thus even be applied to practical problems of portfolio selection. Certainly, as suggested by Table 11.2, there may be timing strategies that are able to outperform the market, but this finding is only helpful if successful timing strategies do not vary much across time and space. To elaborate on these aspects somewhat further, we reexamine Equations (11.7) to (11.10) of Table 11.2 for the time period from February 1, 1985 to January 1, 2006. Moreover, for this time period, we are able to account additionally for Japan, the UK, and the United States. As a proxy for the riskless interest rate we use the respective 1-month money market rate for each country as provided by Datastream (due to data availability before 1993, the Japanese riskless interest rate is approximated by the Japanese monetary policy rate). Portfolio weights for the market (equities and bonds) are approximated by Datastream Total Market Indexes.

TABLE 11.3   Monthly Performance (Sharpe Ratio) and Standard Deviation ($\sigma(x_1)$) of Monthly Stock Holdings in Germany, Japan, the UK, and the United States for Different Portfolio Selection Strategies and the Time Period from February 1, 1985 to January 1, 2006

Strategy	Best	Market	Momentum	Contrarian	Semi-momentum	Semi-contrarian
			Germany			
$ik_1^{(down)}$	0	1	0.5	1.5	0.5	1
$ik_1^{(up)}$	1.3674	1	1.5	0.5	1	0.5
$avg(ik_1)$	0.67285	1	0.99206	1.00794	0.74603	0.75397
Sharpe ratio	0.197791	0.135280	0.184589	0.058852	0.176691[a]	0.093194
$\sigma(x_1)$	0.221317	0.051094	0.167599	0.158318	0.095312	0.088255
			Japan			
$ik_1^{(down)}$	0.3418	1	0.5	1.5	0.5	1
$ik_1^{(up)}$	0.3618	1	1.5	0.5	1	0.5
$avg(ik_1)$	0.3515	1	0.9841	1.0159	0.7421	0.75397
Sharpe ratio	0.112316	0.064547	0.053782	0.055358	0.076603	0.075461
$\sigma(x_1)$	0.071716	0.120202	0.432921	0.403499	0.095312	0.205508
			UK			
$ik_1^{(down)}$	1.0986	1	0.5	1.5	0.5	1
$ik_1^{(up)}$	0.4299	1	1.5	0.5	1	0.5
$avg(ik_1)$	0.7444	1	1.0296	0.9704	0.7648	0.7352
Sharpe ratio	0.125657	0.102374	0.049007	0.123265	0.074609	0.124945
$\sigma(x_1)$	0.439007	0.073818	0.607292	0.622367	0.31071	0.49636
			United States			
$ik_1^{(down)}$	1.0642	1	0.5	1.5	0.5	1
$ik_1^{(up)}$	0.7120	1	1.5	0.5	1	0.5
$avg(ik_1)$	0.8672	1	1.0595	0.9405	0.7798	0.7202
Sharpe ratio	0.168519	0.163614	0.124871	0.160098	0.140373	0.165679
$\sigma(x_1)$	0.190741	0.113234	0.444754	0.452217	0.237370	0.245028

[a] Significantly different from the Sharpe ratio of the market strategy on the 10% level.

As can be seen in Table 11.3, the semi-momentum strategy performs quite well in Germany even for this restricted time period, suggesting a rather high stability of our findings for Germany. However, in Japan, it would have been best to choose both parameters $ik_1^{(up)}$ and $ik_1^{(down)}$ considerably smaller than 1, which means to take a much more pessimistic view than the market with respect to future returns even in bullish periods.

In fact, it would have been even better to choose the higher value $ik_1^{(up)}$ only if both c_1 and c_2 had been increased. For such a strategy (with $ik_1^{(up)} = 0.8347$ and $ik_1^{(down)} = 0.3118$) it would be possible to reach a quite impressive Sharpe ratio of 13.2277%, which beats the market performance on a significance level of 5%. Nevertheless, different national capital markets seem to behave quite differently, so that a certain timing strategy may work in one market but fail in another.

This conjecture is verified by the results in Table 11.3 for the British and American stock markets. In fact, based on the timing strategies considered in this section, it is not possible to outperform the UK or the U.S. capital market. We may interpret this as an indirect evidence of higher British and American stock market efficiency in comparison to that of Germany and Japan. Clearly, this finding might reflect the relative importance and size of the respective markets.

Summarizing, our approach helps to avoid portfolio selection strategies that imply overconfident behavior. Moreover, this approach may be utilized to employ certain timing strategies, like a semi-momentum one with the most extreme recommendations of abandoning all stock holdings in situations with falling market expectations and holding the market portfolio when aggregate expectations are improving. This may be regarded as a further indirect indicator for the adequacy of the Markowitz portfolio selection approach even without additional *ad hoc* restrictions on portfolio weights in order to reduce portfolio sensitivity. Our approach may also be helpful in assessing the degree of inefficiency of different national stock markets.

11.5 CONCLUSION

It is often argued that applications of the Markowitz portfolio theory imply a sensitivity of portfolio weights with respect to changes in return expectations, which contradicts empirical evidence of actual investors' behavior. We show that this contradiction results from the consideration of mean-variance optimization problems with exogenously given expected returns. This exogenous case can be interpreted as a situation with new information being private. Therefore, portfolio reactions are significantly more pronounced, as the investor tries to exploit her informational advantage. Given high-quality private information, the resulting extreme portfolio weights might actually be justified.

The contradiction between theoretical recommendation and empirical finding vanishes if mean-variance optimization with endogenously given expected returns is assumed. In such a situation, all new information on return expectations is common knowledge. This setting leads to countervailing price-return effects when investors optimize their portfolio structures. As a consequence, reactions of optimal portfolios are far less sensitive than in the case of mean-variance optimization with exogenously given expected returns. This endogenous case seems to correspond fairly well with general real-life findings of robust portfolio structures.

To elaborate somewhat further on this issue, we introduced an indicator variable ik with $ik > 1$ ($ik < 1$) implying an investor to be more (less) confident with respect to future expected security cash flows than the market as a whole. We find that professional portfolio managers act in fact as if they set $ik > 1$. Moreover, by varying ik depending on current changes in market expectations we identified semi-momentum strategies with $ik = 1$ in periods with rising market expectations and $ik = 0$ in periods with falling market expectations, leading to the highest attainable Sharpe ratio on the German capital market. This hints at the possibility that in Germany investors are overconfident in periods when cash flow expectations are falling. Resulting volatility in stock portfolio weights for this semi-momentum strategy nevertheless remains far from the values in the case of the exogenous situation, thus verifying once again that the Markowitz approach in itself does not lead necessarily to too sensitive portfolio weights. A similar (and even more pronounced) conclusion with respect to market overconfidence may be drawn from the analysis of the Japanese capital market. For the UK and the United States, however, no profitable timing strategies might be developed by the approach suggested in this chapter. This last finding can be interpreted as evidence for higher stock market efficiency in the UK and the United States than in Germany and Japan.

REFERENCES

Black, F., and Litterman, R. (1992). Global portfolio optimization. *Financial Analysts Journal* 48:28–43.

Britten-Jones, M. (1999). The sampling error in estimates of mean-variance efficient portfolio weights. *Journal of Finance* 54:655–71.

Campbell, J. Y., and Shiller, R. (1988). The dividend-price ratio and expectations of future dividends and discount factors. *Review of Financial Studies* 1:195–227.

Campbell, J. Y., and Viceira, L. M. (2002). *Strategic asset allocation.* New York: Oxford University Press.

Chopra, V. K., and Ziemba, W. T. (1993). The effects of errors in means, variances, and covariances on optimal portfolio choice. *Journal of Portfolio Management* 19:6–11.

DeBondt, W. M. F., and Thaler, R. H. (1985). Does the stock market overreact? *Journal of Finance* 40:793–808.

Frost, P., and Savarino, J. (1988). For better performance: Constrain portfolio weights. *Journal of Portfolio Management* 15:29–34.

Gordon, M. (1962). *The investment, financing, and valuation of the corporation.* Homewood, IL: Irwin.

Green, R. C., and Hollifield, B. (1992). When will mean-variance efficient portfolios be well diversified? *Journal of Finance* 47:1785–809.

Jegadeesh, N., and Titman, S. (1993). Returns to buying winners and selling losers: Implications for stock market efficiency. *Journal of Finance* 48:65–91.

Jobson, J. D., and Korkie, B. (1980). Estimation for Markowitz efficient portfolios. *Journal of the American Statistical Association* 75:544–54.

Jobson, J. D., and Korkie, B. (1981). Performance hypothesis testing with Sharpe and Treynor measures. *Journal of Finance* 36:889–908.

Jorion, P. (1986). Bayes-Stein estimation for portfolio analysis. *Journal of Financial and Quantitative Analysis* 21:279–92.

Lakonishok, J., Shleifer, A., and Vishny, R. (1994). Contrarian investment, extrapolation and risk. *Journal of Finance* 49:1541–78.

Markowitz, H. (1952). Portfolio selection. *Journal of Finance* 7:77–91.

Markowitz, H. (1959). *Portfolio selection: Efficient diversification of investments.* New York: Wiley & Sons.

Memmel, C. (2003). Performance hypothesis testing with the Sharpe ratio. *Finance Letters* 1:21–23.

Michaud, R. (1989). The Markowitz optimization enigma: Is "optimized" optimal? *Financial Analysts Journal* 45:31–42.

CHAPTER 12

Alternative to the Mean-Variance Asset Allocation Analysis

A Scenario Methodology for Portfolio Selection

Michael Schyns, Georges Hübner, and Yves Crama

CONTENTS

12.1   INTRODUCTION

Value-at-Risk (VaR) and its variants, like conditional value-at-risk (CVaR), are very popular concepts for measuring the risk associated with a portfolio of securities. Extensive research has been performed on this subject, with much of the emphasis being placed on the CVaR model (see, e.g., Krokhmal et al., 2002). The CVaR is attractive for two main reasons: it displays nice financial properties—it is a coherent measure of risk as defined by Artzner et al. (1999)—and it is easier to compute than the VaR, which uses a quantile of the portfolio returns distribution. However, since the adoption of the Comprehensive Basel II Accord by the Bank for International Settlements (BIS, 2006), the role of the VaR has become more central for regulatory purposes. Financial institutions of BIS member countries have to comply strictly with VaR requirements regarding credit risk, operational risk, and market risk management. Thus, limitations in VaR exposures typically represent binding constraints for active portfolio management.

The extant literature on portfolio management with VaR requirements has mostly focused on *estimating* the VaR associated with a *predefined* portfolio. However, this approach does not explicitly address the central problem of asset managers, whose objective is to account directly for VaR in the *selection* of optimal complex portfolios.[*]

The main purpose of this chapter is to develop a model for the selection of an optimal portfolio of stocks and options subject to value-at-risk

---

[*] Indeed, risk estimation methods based on analytical parametric estimation, such as RiskMetrics, on semiparametric approximation like the Cornish-Fisher formula, or on the extreme value theory (EVT), can be quite easily applied in linearly constrained optimization. In this context, there is not much innovation to be brought besides a precise estimation of the parameters entering the VaR. Yet, such conceptually simple methods cannot effectively handle strategies involving options. The behavior of these derivative instruments dramatically changes over time, depending on the evolution of the underlying asset, and requires a dynamic treatment of return distributions.

constraints, to demonstrate the computational feasibility of the approach, and to show the performances of different strategies.

In general terms, we are interested in the following situation. A fund manager considers investing a total budget B into a portfolio of stocks or options for a given horizon. The problem she faces is to select the quantities to be invested in each asset so as to optimize the expected value of the portfolio at the end of the horizon, while satisfying predetermined value-at-risk constraints at the end of each subperiod.

Although this optimization problem is rather easily stated, modeling it in a rigorous and meaningful way turns out to be quite challenging. Assets under consideration in this chapter are stocks and options. We restrict here our attention to a one-period model. Even with only one period, the problem remains complex since the VaR cannot be computed a priori without knowing the probability distribution of the portfolio returns, because the distribution of the portfolio returns cannot be specified before selecting the optimal quantities of its components (and before specifying their respective probability distributions), and also because the optimal quantities must be set so as to satisfy the VaR requirement for the whole portfolio. This problem is therefore far more complex than computing the VaR associated with a *predefined* portfolio. The complexity is reinforced by the fact that we need to model simultaneously the future prices of several stocks and their derivatives. The first part of the problem has been addressed by Schyns et al. (2008), and a special emphasis is set on the second part in this chapter.[*]

The inherent complexity of the model and of the optimization process justify our use of a methodology based on a *multinomial tree of scenarios*. More precisely, the formulation of the model relies on a collection of scenarios that provide a representative sample of values for the returns of a market index.

We then consider factor models to define the stock prices for each scenario. Three specifications are considered: the simple regression of the stock return on a proxy for market return, the capital asset pricing model, and the Fama and French model. Since the values of the options are indirectly determined by the scenarios, this approach allows capturing completely the random nature of the problem.

---

[*] Note that adding more underlying assets would not really challenge the methodology proposed hereafter, although it may drastically increase the size of the resulting optimization problem. Dealing with only one underlying asset simplifies and clarifies the construction of the model.

In order to model the distribution of the market returns, we do not restrict ourselves to standard parametric statistical distributions (like the normal or the Student distribution), but we propose to sample from more general distributions of returns. This approach has the advantage of providing great flexibility in the construction of the model, allowing in particular integration of different types of distributions and various realistic constraints. We illustrate this with the computation of a distribution of returns implied by the prices of the options available on the market at the time of the investment.

All in all, our work appears to be unique in its simultaneous consideration of multiperiod scenarios, of several stocks, of options, and of a broad range of realistic financial constraints, including the VaR measure. In the literature, these four features have been mostly developed in isolation, and so cannot provide reasonable insights into realistic portfolio selection problems. In particular, we pay special attention to the calibration of the parameters defining the tree of scenarios, an issue that turns out to be quite tricky, but extremely relevant when dealing with options and several stocks. Emphasis is put on the behavior of the model when the portfolio consists only in stocks such as to measure more precisely the impact of the factor models. The extension to options is briefly presented and tested. Detailed results for options written only on one underlying asset, the market index, are available in Schyns et al. (2008). This chapter also presents extensions to a multiperiod model.

Our first goal is to show that this approach is valid and tractable in practice. But thanks to our integrative framework, we can also test it on real data and show the returns an investor could have reached.

The chapter is organized as follows. Section 12.2 presents the framework of the portfolio management problem with an emphasis on its main financial features and on trees of scenarios. Section 12.3 gives a complete description of the optimization model to be solved, which is formulated here as a mixed-integer linear programming problem, and Section 12.4 discusses the difficulties that arise when instantiating the set of scenarios. Section 12.5 presents our case study. Finally, Section 12.6 draws the main conclusions of our work and presents some perspectives for future research.

12.2 FRAMEWORK

Before presenting a mathematical formulation of the optimization model to be solved, we first discuss its various components.

12.2.1 Risk Measure

The value-at-risk (VaR) of a portfolio at level P is defined as the maximal loss of the portfolio value with probability P, over a specific horizon:

$$\text{Prob}[loss \leq \text{VaR}] = P$$

or under slightly more general assumptions concerning the distribution of losses,

$$\text{VaR} = \min\{V \mid \text{Prob}[loss \leq V] \geq P\}$$

This risk measure, initially proposed by Edgeworth (1888), became popular when introduced by JP Morgan in RiskMetrics™ (1996). It has subsequently been proposed in the 1996 Amendment to the Basel I Capital Accord, included in the Comprehensive Basel II Accord, and is fully applicable nowadays.

In addition to value-at-risk requirements, risk management systems often impose stop-loss procedures in order to limit the extent of the losses incurred on an individual position. This is equivalent to setting a guaranteed amount at the end of the investment horizon.[*] The same requirement is met for insured portfolios with the option-based portfolio insurance (OBPI) technique or the constant proportion portfolio insurance (CPPI) technique proposed by Black and Jones (1987), as the portfolio becomes entirely invested in risk-free securities when the loss incurred in the risky part reaches a given level (see Bertrand and Prigent, 2005, for a discussion).

12.2.2 Market

Many professionally managed equity portfolios involve optional securities. They allow traders or fund managers to shape the future payoff of their portfolios, for instance, to ensure a floor in the terminal payoff, which is easily achieved thanks to the intrinsic properties of options. We assume that the manager's main goal is to maximize the expected value of the portfolio at the end of the investment horizon. To track real market

---

[*] Note that the guarantee constraint is a special case of the value-at-risk constraint where the probability P is set to 100%. We could actually impose several VaR constraints in our model. The methodology would be similar for each of them, even if some simplifications could take place (indeed, several VaR constraints would imply overlapping lower bounds). Alternatively, the manager could impose a limit on the conditional value-at-risk (a.k.a. expected shortfall) beyond the VaR level. This requirement limits the extent of the expected losses when a disaster occurs. We discuss the implications of this requirement in Section 12.5.

conditions as closely as possible, we take into account the bid-ask spreads and transaction costs.

Our model is specially intended for large investments, and this allows us to formulate some simplifying hypotheses. In particular, if the initial budget is large, then the fixed commission cost is small with respect to the total invested amount, and it can be neglected in the model. Also, since the amount invested in each option can be assumed to be large, we can consider only one proportional tax rate, namely, the rate that applies to the largest trading amounts.

12.2.3 Multiperiod Trees of Scenarios

Trees of scenarios constitute a generic, relatively simple approach to represent future states of the world in stochastic optimization problems (see, e.g., Birge and Louveaux (1999) or Prekopa (1995) for a broad introduction to stochastic programming). In finance, such trees have been used in numerous computational models, both in applied and in theoretical frameworks, as in Dembo (1991), Dybvig (1988a, 1988b), Gülpinar et al. (2004), Larsen et al. (2002), Mulvey (1994), Muzzioli and Torricelli (2005), Rockafellar and Uryasev (2000), Rubinstein (1994, 1998), etc. Each node of a scenario tree represents a possible state of the world at a particular date; i.e., each node is explicitly associated with the value of the underlying asset (but not with the portfolio value, since the composition of the portfolio is unknown at the outset). The intrinsic quality of a tree of scenarios depends on the process used to instantiate each node and on the number of nodes.

Grinold (1999) ponders some of the relative advantages and drawbacks of scenario-based approaches versus mean-variance approaches when dealing with portfolio optimization problems. In Grinold's view, scenario-based models are mostly useful—and even indispensable—or portfolio management problems involving options or assets with alternative distributions. They deal with the entire distribution of outcomes and thereby allow for a broad variety of objectives. However, they also have to respond to several major challenges. In particular, setting up a tree of scenarios requires the solution of several complex numerical problems, including the specification of the entire distribution of all assets.

In this chapter, we consider trees where each node (or scenario) corresponds to a possible value of the market index at the end of a subperiod. The scenarios are viewed as equiprobable; i.e., they constitute a *representative sample of market values*, which can be obtained by sampling from the probability distribution function of the market values (see Section 12.4.1). It would be straightforward to extend the model to the case where the scenarios are not equiprobable. These values can be used, in turn, to price the stocks and options written on them. In order to price the stocks, we rely on classical factor models: OLS approach, capital asset pricing model, and Fama and French model.

All options considered here have a one-period maturity, i.e., a maturity coinciding with the horizon of investment.[*] For the first period, the initial characteristics of traded options are directly observable on the market. However, some adjustments could be required to avoid numerical difficulties due to arbitrage opportunities.

12.3 MODEL

Informally, the manager's problem is to select a portfolio with maximum expected value at the end of the horizon, under the following constraints:

- Budget: The initial cost of settlement does not exceed the available budget.

- Guarantee: The value of the portfolio at the end of each period cannot be less than a predefined fraction of the initial budget, under no circumstances.

- Value-at-risk: With a predefined probability P, the final value of the portfolio cannot be less than a predefined fraction of the initial budget.

In this section, we turn to a mathematical formulation of this portfolio optimization problem.

12.3.1 Notations

We consider here a one-period tree. We denote by $i = 1,\ldots, nbS$ the terminal nodes of the tree, while node 0 corresponds to the initial state.

At each node $i, j = 1, \ldots, N$ stocks are available and each stock j serves as the underlying of a number of $nbOpt_j$ options.

---

[*] Extensions to other maturities could be included in the model, but they complicate its formulation as well as the interpretation of the results.

For $i = 0, 1,\ldots, nbS$, $j = 1,\ldots, N$, and $k_j = 1,\ldots,\ nbOpt_j$ the parameters of the model are:

B: the available budget

M_i: the price of the market index at node i

S_j^i: the price of stock j at node i

K_k^j: the strike price of option k written on stock j

$popt_k^j$: the initial ask price of option k written on stock j

T: the taxation rate

r_f: the risk-free rate

Some of the main decision variables are:

Q_j: the (positive) quantity of stock j

$Qopt_k^j$: the (positive) quantity of option j written on stock j

z: the amount invested in the risk-free asset at node i

12.3.2 Portfolio Value

The portfolio value can be easily determined at each node of the tree if we know the quantities invested in each security. Indeed, the portfolio value at a node is essentially the sum of the securities values weighted by the invested quantities. By construction of the tree, the value of each security at each node is known (see Section 12.4 for details). This general scheme just has to be slightly adapted to integrate the transactions costs, i.e., a given percentage of the option values, and the bid, ask, or maturity prices.

Mathematically, at maturity, the value of an option is given by

$$vopt(S,K) = \begin{cases} S-K, & \text{for a call option with } S > K \\ K-S, & \text{for a put option with } K > S \\ 0, & \text{otherwise} \end{cases} \qquad (12.1)$$

where S is the underlying asset price and K is the option strike price.

The final value of the portfolio composed at node i is therefore given by

$$vport_i = z(1+r_f) + (1-T)\sum_{j=1}^{N}\left(Q_j S_{i,j} + \sum_{k=1}^{nbOpt_j} Qopt_{j,k} vopt(S_{i,j}, K_{j,k})\right) \quad (12.2)$$

Similarly, the cost incurred in order to initially compose the portfolio is given by

$$vportinit = z + (1+T)\sum_{j=1}^{N}\left(Q_j S_{0,j} + \sum_{k=1}^{nbOpt_j} Qopt_{j,k} popt_{j,k}\right) \quad (12.3)$$

The expected portfolio value is obtained as the sum of the portfolio value at each leaf, multiplied by the probability of the associated scenario.

12.3.3 Risk and Investment Limits

The VaR constraint can be redefined as a minimum payoff, denoted λB, to be reached with probability P (where λ is the percentage of the initial budget to preserve, and $P = 95\%$ or $P = 99\%$ is the usual value). A tree of scenarios is very convenient to model this constraint, if we consider that the set of leaves represents all the possible outcomes. The VaR constraint is then satisfied if and only if, for each second-period subtree, the value of the portfolio is greater than or equal to λB in at least $(P \times nbS)$ scenarios, where nbS is the number of scenarios in each subtree. Of course, this requires that nbS must be large enough to faithfully represent all possible outcomes.

The model obtained when we impose a minimum portfolio value at each leaf, i.e., when $P = 100\%$, is easy to solve: indeed, it is a continuous linear programming problem. Our main challenge arises instead when P is smaller than 1, i.e., when the threshold λB must be achieved at a subset of the leaves only. This VaR constraint leads to a formulation involving binary variables that express that, for a given leaf, either the threshold is reached or not. The number of leaves where the threshold is reached must be larger than or equal to $(P \times nbS)$. This becomes a complex mixed-integer programming (MIP) problem.

Since the guarantee constraint is the special case of the VaR constraint where P is equal to 1, we can use the same formulation in both cases and simply impose the minimum guarantee level at each leaf of the tree. Note that the guarantee will only be satisfied in the future if a scenario happens exactly like it was defined. Note that if the tree does not represent faithfully the possible outcomes and no scenario finally corresponds to the reality, there is no guarantee that the portfolio value will be larger than the required threshold. Another approach, based on Dert and Oldenkamp's paper (2000) and detailed in Schyns et al. (2008), could be adapted to reinforce the constraint. This could, however, increase drastically the size of the problem without providing a total guarantee.

VaR Model

$$max\ ER = \frac{1}{nbS} \sum_{i=1}^{nbS} vport_i$$

subject to

(*budget:*)	$vportinit \le B$	
(*guarantee:*)	$vport_i \ge \theta B$	$i = 1..nbS$
(*VaR:*)	$vport_i \ge (1 - \alpha_i)\lambda B + \alpha_i \theta B$	$i = 1..nbS$
(*VaR confidence level:*)	$\sum_{j=1}^{nbS} \alpha_j \le nbS(1 - P)$	$i = 1..nbS$

The core of the VaR optimization model can now be expressed as a VaR model, where *ER* denotes the expected value of the portfolio, θ is the percentage of the budget required at the end of the period, and λ is the percentage of the budget required with probability *P*. Each variable belongs to IR^+ except for the variables $\alpha_{i,j}$, which are binary variables with the interpretation that $\alpha_i = 0$ if the value of the portfolio exceeds the VaR threshold λB at node *i*, and $\alpha_i = 1$ otherwise. Note that when $\alpha_i = 1$, the corresponding VaR constraint is implied by the guarantee constraints, as discussed above.

12.3.4 Extensions

The model VaR, based on a scenario tree, is a very general framework. It can easily be completed to integrate other financial realities. Short sales could be allowed by relaxing to *IR* the quantity variables and by integrating bid option prices. A preliminary tuning process described in Schyns et al. (2008) is, however, required to detect and avoid arbitrage opportunities. Q and $Qopt$ could be restricted to N since real contracts deal with entire quantities. More advanced transaction cost schemes could also be incorporated, e.g., with minimal transaction costs. It is also easy to model minimal and maximal bounds on quantities of stocks or groups of stocks corresponding to different sectors. Similarly, we can impose a minimal number of different stocks in the portfolio to ensure diversification and at the same time set a maximal cardinality to avoid a too large fragmentation in small stock amounts. Since the factorial model presented below assumes diversification, this last constraint will be considered in the case study:

$$C_j \leq Q_j \qquad j=1...N$$

$$C_j \frac{B}{S_{0,j}} \geq Q_j \qquad j=1...N$$

$$\sum_{j=1}^{N} C_j \geq MinS \qquad\qquad (12.4)$$

$$\sum_{j=1}^{N} C_j \leq MaxS$$

where C_j are binary variables with the interpretation that $C_j = 1$ if the stock j belongs to the optimal portfolio and $C_j = 0$ otherwise. $\frac{B}{S_{0,j}}$ represents the largest quantity of stock j that could be obtained with the budget B and is therefore the implied upper bound on the stock j quantities.

Together with the previous constraints, we can impose a lower bound on the quantities in the case when the stock is purchased. Without this protection, some negligible quantities of stocks could be artificially added in the portfolio to satisfy the cardinality constraint. Moreover, it is coherent with practice since it is not attractive to buy tiny quantities of a stock due to minimal transaction costs applied. This gives

$$Q_j \geq C_j Q \min_j \quad j=1,...,N$$

where $Q\min_j$ is the minimal quantity of stock j that must be considered in the case of a purchase.

The other propositions, while intrinsically interesting, do not fundamentally affect our approach and will not be discussed here.

12.4 INSTANTIATION OF THE TREE OF SCENARIOS

In this section, we discuss a specific difficulty that arises when the VaR model described in Section 12.3 is to be instantiated: how to construct a representative tree of scenarios with a limited and computationally manageable number of scenarios. This question must be answered before we can solve any instance of the VaR model.

12.4.1 Distribution of Asset Prices

Before generating the tree of scenarios, we first need to model the probability distribution of the values of the underlying assets. This important question is not exclusively linked to the VaR problem under consideration, but constitutes a broad topic in itself. The most usual way is to consider a normal distribution of the returns. The two first moments are extracted from past data, directly or using smoothing and predictive schemes. A first improvement is to consider generalized Student distributions to also take into account the skewness and the kurtosis. This is again a parametric approach where the four moments can be computed from historical data. In Schyns et al. (2005), we describe how to precisely construct these distributions. We also present a third attractive approach: the construction of an implied probability density function (pdf) derived from observed option prices.

The implied distribution approach is based on papers of Breeden and Litzenberger (1978) and Shimko (1993). This has the advantage of relying only on current pricing information, rather than requiring long, outdated time series, and to preserve consistency with the observed market prices. Practically, it has been applied to a market index for which several options were available. All information is extracted from options available on the market at the time of the investment and with a maturity corresponding to the horizon of the tree.

When a pdf has been computed, we use stratified sampling in order to obtain a sample of equiprobable states. Stratified sampling preserves information relative to the distributions in the parameterization of the tree. Thanks to this modeling choice, complex continuous problems can be faithfully represented by a relatively small number of nodes, which makes them computationally tractable.[*]

---

[*] See Schyns et al. (2005) for a complete description of the procedure for a one-period tree.

12.4.2 Factor Models

The previous section is dedicated to the construction of a specific tree of scenarios for one asset. However, in the VaR model we work simultaneously with N assets and the corresponding options. We are not interested in N trees but in a unique tree for which the prices of the N stocks are defined at each node. This is a far more complex problem.

The direct approach would be to consider every combination of the returns of the different stocks, i.e., to adjust at each node of the tree of a first stock, the tree for a second stock, and so on with all the other stocks. There are two main drawbacks with this approach. First, by construction, each scenario tree is equiprobable. The combination of such trees gives also an equiprobable tree. However, stocks prices are clearly not independent. We should then recompute the probabilities of each final nodes according to joint probabilities of the stock distribution and it is not obvious. The second problem arises when a real-size case is considered. If 100 scenarios are required to model faithfully the distribution of one stock, then the recombined tree for 200 stocks consists of 100^{200} nodes, which is computationally hard to manage.

Jamshidian and Zhu (1997) propose a multifactorial approach to try to overcome these two difficulties. They model the correlation between the assets, perform a principal component analysis (PCA), and use the principal eigenvectors to explain each stock return. They illustrate the approach for the computation of the yield curve using up to five principal factors. The depth of the tree is therefore reduced to a maximum of five levels. Moreover, since the eigenvectors are orthogonal to each other, each factor is independent. This greatly simplifies the computation of the node probabilities.

Jamshidian and Zhu then extend their methodology to a portfolio of a few assets. Note, however, that this approach is based on factors difficult to interpret, on a historical covariance matrix and based on normality assumptions. Moreover, the size of the final tree, even though it is far smaller than the one obtained with the basic approach, could remain large when lots of stocks are under consideration.

For the reasons mentioned above, we have decided in this chapter to focus on a one-factor approach. The goal is to construct a small tree such that the optimization problems keep reasonable dimensions, with a factor, are easily interpretable, and are instantiable with some flexibility. Of course, we also have to check that a one-factor model is accurate enough to provide a sufficiently good representation of each stock return. Three variations of a linear one-factor model are considered in the following.

The factor selected is simply the market index, consistent with the capital asset pricing model, whose simplest empirical counterpart is the market model. In its *ex post* form, the model is written as

$$r_i - r_f = \beta_i(r_m - r_f) \qquad (12.5)$$

Current option prices can also be picked up in order to construct an instantaneous implied nonparametric distribution.

Consistent with the equilibrium framework underlying the CAPM, it is assumed that the unsystematic risk of the stock will fade away through a proper diversification of the portfolio. This assumption is important and should be taken into account in the main portfolio model we want to tackle in order to ensure a good representation of the scenario tree.

A first simple variant of the approach is to consider a more general regression scheme. In the *ex post* CAPM with excess returns, such as in Equation (12.5), there is only one parameter since the intercept is supposed to be equal to r_f. We could relax this financial assumption and compute a second parameter, α_i, for the linear model:

$$r_i = \beta_i r_m + \alpha_i \qquad (12.6)$$

Finally, we can wonder if this one-factor approach is not too basic. Ample empirical evidence suggests that the explanatory power of the CAPM could be greatly improved, especially when considering individual stocks. The R-squared fit of the regression is usually low with respect to statistical standards. We can still argue that the final goal is to construct a well-diversified portfolio, and that the lack of perfect representation at the stock level will be corrected at the portfolio level. We can, however, try to adjust the model and slightly alleviate this issue. To achieve this goal, we consider here the basic extension of the CAPM to size and value effects proposed by Fama and French (1992). They add two factors in the CAPM to provide a better estimation for small caps (*SMB*) and for stocks with a high book-value-to-price ratio (*HML*):

$$r_i - r_f = \beta_i^{FF}(r_m - r_f) + \chi_i^{FF} SMB + \delta_i^{FF} HML + \alpha_i^{FF} \qquad (12.7)$$

Note that β_i^{FF} is not equal to β_i since the two other factors also contribute to the estimation, but these two betas are usually close. The proposition now is not to construct a multivariate three-dimensional probability density function where each dimension would correspond to one factor. Indeed, we would again face difficulties similar to the ones mentioned at the beginning of this section. We would have first to find the adequate probability distribution to use, e.g., a multivariate normal distribution or a nonparametric kernel pdf. We can then only rely on historical data to instantiate it and not anymore on implied instantaneous information. Finally, we must be able to sample faithfully this distribution into a reduced set of nodes. The probability of each node must be worked out. This is usually not trivial.

Our proposal is to use the two added factors for what they were initially basically designed for, as corrections of the CAPM estimates. We therefore proceed as follows. The four coefficients of Equation (12.7) are obtained for each stock by a regression on historical values. We still only build the scenario tree to represent the market index. We attach to each node of this tree the conditional expected values of the *SMB* and *HML* factors at the horizon of investment. It could be simply the last observed values, a simple regression with respect to the past, or a more advanced econometric model. Since the three factors are not independent, it is clear that any of these choices would represent a rough approximation. Yet, we contend that the error on (only) the adjustment terms of the Fama and French model certainly does not offset the benefit of these two terms with respect to the CAPM approach. It is particularly true when the expected values of *SMB* and *HML* are carefully built. The empirical part of this chapter will compare the three approaches to check this assumption.

12.5 CASE STUDY

12.5.1 Experimental Settings

The theoretical model presented above is tested on a set of real-world option data. In order to check the stability and the quality of the approach, the same experiment was reproduced each month from January 1996 up to April 2007. Market and stock returns are already collected in the Thomson Datastream database, from January 1990 up to April 2008, i.e., 220 periods. The T-bill returns are also collected each month for the same period as a proxy of the risk-free rate.

As a proxy of the market, we use the S&P 500 index. Its behavior is modeled by a normal distribution and a skewed t distribution. For each experiment, the parameters of the distributions are based on the S&P 500 returns for the 6 preceding years. They are computed until April 2007 to leave at least 1 year of observed returns to check the performance of the investments. It has proved to be impracticable to collect enough option prices to construct the implied densities for the 136 resulting experiments we consider. Afterwards, the continuous pdfs are sampled in 80 discrete scenarios.

When the S&P 500 index is used as a proxy of the market, it is natural to consider the S&P 500 stocks as the set of representative securities. We have kept the 341 stocks for which there were no missing data for the whole period. For each of them and for the 136 months, the coefficients for the three regression models are computed.

The goal is to find the optimal portfolio of stocks at each period and then to look 1 and 2 years after what would have been the real returns. We assume a 1-year horizon of investment, an initial budget of US$500,000, and costs of transaction of 0.17%. For each of the 136 optimization problems, we require that the final portfolio value is at least at 99% of the initial budget (guarantee) and also above the risk-free return of the period with a probability of 95%. Note that the T-bill can belong to the optimal portfolio, and this is generally the case in a large part when VaR constraints are under consideration. When a stock is incorporated in the portfolio, it must be an investment of at least 1% of the initial budget (taking into account the T-bill). In the first set of results, no constraints are applied on the number of stocks in the portfolio.

12.5.2 Experimental Results

The 1-year optimal investments for each of the 136 periods are computed for the three regression strategies. A first important result is about the composition of the portfolios. A large portion of the budget is invested in T-bills and the remaining in only a few stocks. This is not unusual for this kind of risk constraints. The T-bills ensure the guarantee level and the stocks allow us to reach the VaR level at least in 95% of cases. It implies also that with this kind of constraint, diversification is not naturally achieved. If we also compare the portfolios obtained when considering a normal distribution with respect to the one achieved with the skewed t density, there are few differences. Even while skewness and kurtosis are observed at each period, the two figures are not significantly different to imply highly different final portfolios most of the time. Therefore, in the following discussions we only present results for the most general pdf, i.e., the skewed t one.

We also measure the performance of the optimal portfolios. First, we check the expected ones, and then the observed ones. Remember that the returns during these 18 years are far from being stationary. Risk-free rates have significantly decreased over time, and some events, like the dot com crisis, 9/11, and the subprime mortgage meltdown, had a large impact on the market at some specific periods. Therefore, we used a relative performance indicator defined as the percentage of portfolio excess return with respect to the risk-free investment of each period:

$$Eperf = \frac{(ER/B-1)-rf}{rf}$$

where ER is the expected portfolio value obtained after 1 year with one of the three regression strategies.

Results for the skewed t distribution are represented by the box plots of Figure 12.1.

Each box plot represents the distribution of the excess returns. The bold horizontal line corresponds to the median value. The bottom and the top of the box indicate, respectively, the first and third quartiles. The lower (upper) whisker goes from the lower (upper) quartile up to

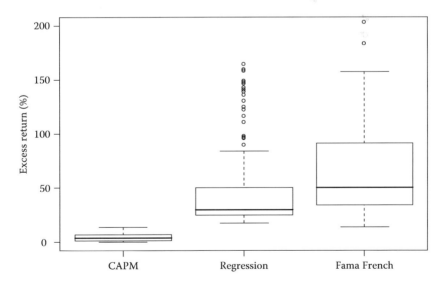

FIGURE 12.1   Distribution of the expected returns for the 136 periods.

the smallest (largest) observation within a range of 1.5 times the inter-quartile range. Excess returns that are outside of this range are represented by circles.

Not surprisingly, for the three strategies, positive excess returns are expected. It is in fact required by the VaR constraint. More interesting is the fact that the Fama and French approach leads usually to higher expected returns than the simple regression scheme, which itself already outperforms the CAPM strategy. The higher degree of sophistication of the Fama and French approach therefore appears to be valuable. We observe that while the guarantee and VaR constraints are quite conservative, the investor can expect most of the time at least 1.5 times the risk-free rate. This seems economically very significant with so strict constraints, and without any particular managerial skill for the stock picking or market timing ability.

12.5.3 Observed Results

The previous results seem to outline the superiority of the Fama and French approach and the good performances of the methodology. However, we have not yet proven that the model itself is a good approximation of reality. Instead of computing an expected mean return over a whole set of theoretical scenarios, we can wonder what happens in reality when one specific scenario materializes. Therefore, we now measure the performance of the portfolio based on its observed value 1 year after the investment. Since 1 year can be considered a rather short horizon, we also look at what happens when we keep the same portfolio during an additional year:

$$Perf_1 = \frac{(P_1/B - 1) - rf}{rf}$$

$$Perf_2 = \frac{\sqrt{P_2/B - 1} - 1 - rf}{rf}$$

where P_1 and P_2 are the observed historical portfolio values 1 and 2 years after the initial investment, respectively.

The results (see Figure 12.2) are less appealing but look more realistic. Most of the time, a return larger than the risk-free rate is observed, while at the same time the possible loss remains very limited (guarantee constraint). The CAPM strategy is still outperformed by the other two regression approaches, but the domination of the Fama and French strategy is not obvious anymore.

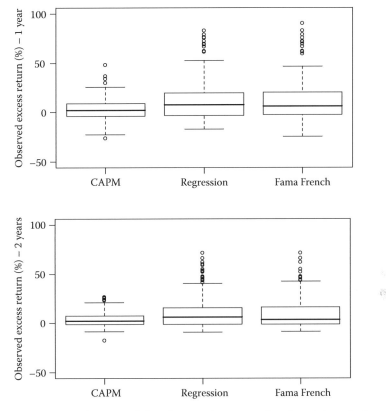

FIGURE 12.2   Distribution of the observed returns for the 136 periods.

The complete distribution of absolute returns for the 2-year horizon is depicted in Figure 12.3. On each of the 124 dates of investment, the return we would have obtained by investing in the optimal portfolios is plotted. It is interesting to note that something important seems to have happened around period 50. It corresponds to the beginning of the year 2000 and the Internet bubble crisis. The CAPM strategy worked only before this period and not very well afterwards. The Fama and French model, and especially the regression model, worked fine around this period. The regression approach was more stable. During the year 2000, no strategy gives good results. This period is in fact too special to be correctly modeled by a classical parametric probability density function based on historical parameters. The tree of scenarios is not representative. A better alternative would probably be to use an implied distribution based on prevailing option prices during this period.

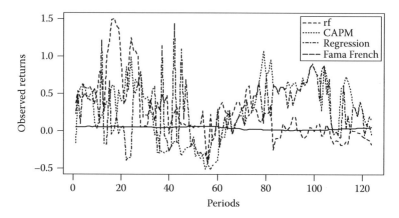

FIGURE 12.3   Evolution of the absolute observed returns for 2 years.

12.5.4 Variations

As mentioned before, the optimal portfolio is not diversified. This property is, however, an assumption of the regression models. Therefore, we restart the same experiments but with an additional constraint on the number of stocks in the final portfolio. The optimal portfolios must contain at least ten stocks.

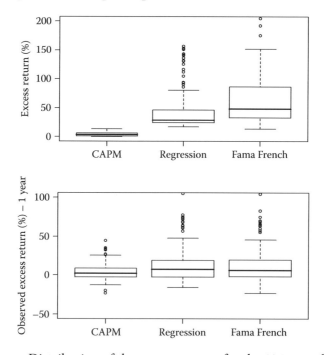

FIGURE 12.4   Distribution of the excess returns for the 136 periods.

Since these problems are more constrained, the expected portfolio values should be lower. The question is to check if this extra care will lead to a better approximation of the portfolio expected values, and therefore to less risky investments. Figure 12.4, where the expected and observed excess returns are depicted, suggests that this variation has a very small impact on the results.

12.6 CONCLUSION

This chapter has developed a scenario tree method for a rather complex problem, namely, the computation of an optimal portfolio involving stocks and options, subject to value-at-risk (VaR) management constraints. We have shown that formulating this problem as a mathematical optimization problem requires some care, especially for the determination of scenarios and of associated option prices.

Our case study on the S&P 500 constituents tests the simultaneous account for strong VaR and guarantee constraints, on the one side, and the instantiation of the scenario tree with different asset pricing approaches on the other side. The results unambiguously suggest that the quality of the asset pricing approach provides a substantial—and economically significant—improvement of the performance of the optimal portfolio strategy. This intersection between asset pricing and constrained optimal asset allocation opens up the way for a set of practical applications of advances in asset pricing models into institutional portfolio management applications, for which VaR and guarantee constraints are common features.

ACKNOWLEDGMENTS

Financial support from Deloitte Luxembourg is gratefully acknowledged.

REFERENCES

Artzner, P., Delbaen, F., Eber, J.-M., and Heath, D. (1999). Coherent measures of risk. *Mathematical Finance* 9:203–28.

Bank for International Settlements (BIS). (2006). International convergence of capital measurement and capital standards—A revised framework. Comprehensive version. Available at http://www.bis.org/publ/bcbs107.htm.

Bertrand, P., and Prigent, J.-L. (2005). Portfolio insurance strategies: OBPI versus CPPI. *Finance* 26:5–32.

Birge, J. R., and Louveaux, F. (1999). *Introduction to stochastic programming.* Springer Series in Operations Research. New York: Springer-Verlag.

Black, F., and Jones R. (1987). Simplifying portfolio insurance. *Journal of Portfolio Management* 14:48–51.

Breeden, D. T., and Litzenberger, R. H. (1978). Prices of state-contingent claims implicit in option prices. *Journal of Business* 51:621–51.

Dembo, R. S. (1991). Scenario optimization. *Annals of Operations Research* 30:63–80.

Dert, C., and Oldenkamp, B. (2000). Optimal guaranteed return portfolios and the casino effect. *Operations Research* 48:768–75.

Dybvig, P. H. (1988a). Inefficient dynamic portfolio strategies or how to throw away a million dollars in the stock market. *Review of Financial Studies* 1:67–88.

Dybvig, P. H. (1988b). Distributional analysis of portfolio choice. *Journal of Business* 61:369–93.

Edgeworth, F. Y. (1888). The mathematical theory of banking. *Journal of the Royal Statistical Society* 51:113–27.

Fama, E., and French, K. (1992). The cross-section of expected stock returns. *The Journal of Finance* 47:427–465.

Grinold, R. C. (1999). Mean-variance and scenario-based approaches to portfolio selection. *Journal of Portfolio Management* 25:10–22.

Gülpinar, N., Rustem, B., and Settergren, R. (2004). Simulation and optimization approaches to scenario tree generation. *Journal of Economic Dynamics and Control* 28:1291–315.

Jamshidian, F., and Zhu, Y. (1997). Scenario simulation: theory and methodology. *Finance and Stochastics* 1:43–67.

Krokhmal, P., Palmquist, J., and Uryasev, S. (2002). Portfolio optimization with conditional value-at-risk objective and constraints. *Journal of Risk* 4:43–68.

Larsen, N., Mausser, H., and Uryasev, S. (2002). Algorithms for optimization of value-at-risk. In *Financial engineering, e-commerce and supply chain*, ed. P. Pardalos and V. K. Tsitsiringos, 129–57. Applied Optimization Series. Boston: Kluwer Academic Publishers.

Mulvey, J. M. (1994). Financial planning via multi-stage stochastic programs. In *Mathematical programming: State of the art 1994*, ed. J. R. Birge and K. G. Murty, 151–71. Ann Arbor, MI: University of Michigan.

Muzzioli, S., and Torricelli, C. (2005). The pricing of options on an interval binomial tree. An application to the DAX-index option market. *European Journal of Operational Research* 163:192–200.

Morgan Guaranty Trust Company. (1996). *RiskMetrics*. 4th ed. New York: Morgan Guaranty Trust Company. Available at http://www.riskmetrics.com/research.

Prekopa, A. (1995). *Stochastic programming*. Malden, MA: Kluwer Academic Publishers.

Rockafellar, R. T., and Uryasev, S. (2000). Optimization of conditional value-at-risk. *Journal of Risk* 2:21–41.

Rubinstein, M. (1994). Implied binomial trees. *Journal of Finance* 69:771–818.

Rubinstein, M. (1998). Edgeworth binomial trees. *Journal of Derivatives* 5:20–27.

Schyns, M., Crama, Y., and Hübner, G. (2005). Grafting information in scenario trees: Application to option prices. Working paper, HEC Management School, University of Liège. Revised November 2005. Available at http://ssrn.com/abstract=418520.

Schyns, M., Crama, Y., and Hübner, G. (2008). Optimization of a portfolio of options under Value-at-Risk constraints: A scenario approach. Working Paper, HEC Management School, University of Liège. April 2008. Available at http://www.hec.ulg.ac.be/FR/recherche/activites/working-papers.php.

Shimko, D. (1993). Bounds of probability. *Risk* 6:33–37.

CHAPTER 13

The Black and Litterman Framework with Higher Moments

The Case of Hedge Funds

Giampaolo Gabbi, Andrea Limone, and Roberto Renò[*]

CONTENTS

13.1 INTRODUCTION

The Black-Litterman asset allocation model gained a wide consensus in several financial applications after its publication in 1990. Black and Litterman developed a model that "provides the flexibility to combine the market equilibrium with additional market views of the investor. (···) In the

---

[*] This chapter is the product of the cooperation of the three authors. However, Section 13.1 is mostly due to R. Renò, Sections 13.2 and 13.3 to A. Limone, and Section 13.4 to G. Gabbi.

Black-Litterman model, the user inputs any number of views or statements about the expected returns of arbitrary portfolios, and the model combines the views with equilibrium, producing both the set of expected returns of assets as well as the optimal portfolio weights" (He and Litterman, 1999).

A substantial difference with the traditional mean-variance approach is that the user inputs a complete set of expected returns (views), and the portfolio optimizer generates the optimal portfolio weights. The Black-Litterman model was developed to provide a systematic resolution to the necessity to consider specific investor's insights; in particular, the optimal portfolio weights are moved in the direction of assets favored by the investor.

Another difficulty arises when considering funds of hedge funds. Both the Markowitz approach and the original Black and Litterman approach have been conceived in a mean-variance world, in which risk premium is generated by exposition to variance only. This is a serious limitation for hedge funds, which are typically characterized by dynamic management strategies that sometimes employ a multitude of complex products, which makes them difficult to capture with the linear models that are classically used in finance (Martellini and Ziemann, 2005).

It has generally been recognized that financial asset returns are nonnormal. Strong empirical evidence suggests that returns are driven by asymmetric and fat-tailed distributions (Jondeau and Rockinger, 2004). In particular, hedge fund returns display peculiarities that are not commonly associated with traditional investment vehicles. Specifically, hedge funds seem more inclined to produce return distributions with significantly nonnormal skewness (downside risk) and kurtosis (fat tails).

In our study we consider the problem of portfolio allocation in which the underlying investment instruments are hedge funds. We provide evidence that the extending to four moments is a substantial improvement to the classic Black and Litterman model (Jondeau and Rockinger, 2004; Martellini et al., 2005).

In the Black and Litterman model, equilibrium expected returns are adjusted to reflect the investor's view about the potential performance (absolute or relative) of one or more hedge funds. The adjustment should reflect the confidence that the manager has in his views (Idzorek, 2004).

13.2 THE BLACK AND LITTERMAN FRAMEWORK AND THE EXTENSION TO FOUR MOMENTS

In this section we introduce the Black-Litterman formula and provide a brief description of its components. In what follows, the number of views is given by K and the number of assets is given by N. The Black and

Litterman formula is

$$E(R_p) = \left(\tau\Sigma^{-1} + P'\Omega^{-1}P\right)^{-1} \cdot \left(\Pi\tau\Sigma^{-1} + P'\Omega^{-1}Q\right) \tag{13.1}$$

where $E(R_p)$ is the combined equilibrium return vector ($N \times 1$ column vector); Σ is the historical covariance matrix of excess returns ($N \times N$ matrix); τ is a scalar that expresses the relative weight of the historical covariance matrix with respect to the covariance of the views, also known as the shrinkage parameter; P is a matrix that identifies the assets involved in the views ($K \times N$ matrix); Ω is the diagonal covariance matrix of the expressed views representing the uncertainty in each view ($K \times K$ matrix); Π is the implied equilibrium return vector ($N \times 1$ column vector); and Q is the view vector ($K \times 1$ column vector).

In the Black and Litterman model, the implied equilibrium return vector Π is determined using reverse engineering. In the classical model, reverse engineering is performed using the mean-variance equation

$$\Pi = \lambda\Sigma w_{mkt} \tag{13.2}$$

where Π is the implied excess equilibrium return vector ($N \times 1$ column vector); λ is the risk aversion coefficient; Σ is the historical covariance matrix of excess returns ($N \times N$ matrix); and w_{mkt} is the market capitalization weight ($N \times 1$ column vector) of hedge funds.

The determination of Π using the above formula implies a two-moments CAPM equilibrium in which the risk premium is determined by variance only. Unfortunately, hedge funds seem more inclined to produce return distributions with significantly nonnormal skewness and kurtosis, which are likely to appear in the risk premium.

In order to preserve the Black-Litterman framework and adapt it to the case of allocation in hedge funds, it is advisable to consider higher moments in the equilibrium return vector equation. In other words, we explicitly involve skewness and kurtosis as additional risk measures.[*] In this respect, we follow the approaches of Hwang and Satchell (1999), Jondeau and Rockinger (2004), and Martellini et al. (2005).

---

[*] In what follows, we assume that investors prefer lower variance and kurtosis, and higher skewness.

We assume that the investor utility function is given by

$$u(w) = -e^{-\lambda w} \tag{13.3}$$

The pricing equation we use is then

$$\Pi = \alpha_1 \beta(2) + \alpha_2 \beta(3) + \alpha_3 \beta(4) \tag{13.4}$$

where

$$\beta(2) = \frac{\Sigma \omega}{\mu_2(R_p)} \quad \beta(3) = \frac{\Omega_\omega}{\mu_3(R_p)} \quad \beta(4) = \frac{\Psi_\omega}{\mu_4(R_p)} \tag{13.5}$$

Ω_ω is the vector of co-skewness for the weighting vector; Ψ_ω is the vector of co-kurtosis for the weighting vector; μ_2, μ_3, and μ_4 are the portfolio second, third, and fourth moments, respectively:

$$\mu_2 = \omega' \Sigma \omega \quad \mu_3 = \omega' \Omega_\omega \quad \mu_4 = \omega' \Psi_\omega \tag{13.6}$$

and the sensitivities α are given by

$$\alpha_1 = \frac{\lambda \mu_2(R_p)}{A} \quad \alpha_2 = -\frac{\lambda^2 \mu_3(R_p)}{2A} \quad \alpha_3 = \frac{\lambda^3 \mu_4(R_p)}{6A} \tag{13.7}$$

where

$$A = 1 + \frac{\lambda^2 \mu_2(R_p)}{2} - \frac{\lambda^3 \mu_3(R_p)}{6} + \frac{\lambda^4 \mu_4(R_p)}{24} \tag{13.8}$$

The factors β have the same interpretation of the β of the CAPM, but in this case they measure the exposure to systematic skewness and systematic kurtosis, respectively, and as we shall see, they are natural measures of systematic risk, or exposure, of an asset to market variance, skewness, and kurtosis. The variables α_i can be interpreted as the risk premia associated with covariance, co-skewness, and co-kurtosis, respectively. Hence, we obtain a four-moment Black and Litterman model (Hwang and Satchell, 1999; Martellini et al., 2005; Jondeau and Rockinger, 2005).

The pricing formula, Equation (13.4), can be used in two directions. Given the equilibrium portfolio weights (typically, the average market allocation), it can be used to obtain the implied equilibrium average returns. These returns can be used in the Black and Litterman formula,

Equation (13.4), together with specific views, to obtain the adjusted equilibrium returns $E(R_p)$. Finally, Equation (13.4) can be inverted to get the portfolio weights that generate the returns $E(R_p)$. This inversion can be accomplished by solving numerically the system of N equations and N unknowns represented by Equation (13.5). These portfolio weights are the target allocation, which is a modification of the market equilibrium returns compatible with specific views, and which takes properly into account the skewness and the kurtosis of the portfolio.

13.3 APPLICATION

In our application, we consider the problem of allocating wealth among nine hedge fund indices, specifically the CSFB/Tremont Hedge Fund indices. The nine indices are constructed by using the different fund strategies, and they are convertible arbitrage, equity market neutral, event driven, fixed income arbitrage, global macro, equity long/short, managed futures, emerging markets, and dedicated short bias.

The data set is composed by the time-series monthly expected excess return[*] of the hedge funds, from December 1993 to January 2007, for a total of 158 observations. We set the risk aversion parameter λ equal to 3, and the shrinkage τ equal to 0.02, and first use the time series of return to compute the sample variance, skewness, and kurtosis matrices.

Table 13.1 reports summary statistics on the moments of the nine indices. All numbers are expressed on a monthly basis. The most performing indices have been the global macro and the equity long/short, and not surprisingly, the least performing has been the dedicated short bias, which is anticorrelated with the others. Regarding all indices, the values of skewness and excess kurtosis document a substantial departure from normality, which might be a concern for a risk-averse manager.

For our application, we need starting equilibrium weights. To attain realistic equilibrium returns, we need an allocation that is representative of the market preferences. To obtain these market weights, we regress the global HFI index on the nine indices, and we use as weights the (normalized) absolute values of the regression coefficients on each index. Obtained market weights are reported in the first column of Table 13.2. Almost 80% of the market portfolio is allocated in the global macro and equity long/short indices.

Now we can determine the implied excess return Π needed to obtain the equilibrium returns that represent a useful neutral starting point for

---

[*] Excess returns are computed by difference with the 3-month U.S. Treasury bill rates.

TABLE 13.1   Mean, Skewness, and Kurtosis of Monthly Expected Excess Return of the Hedge Funds

	Mean	Variance	Skewness	Excess Kurtosis
HFI convertible arbitrage	0.48%	0.018%	−1.38	3.42
HFI equity market neutral	0.54%	0.007%	0.34	0.45
HFI event driven	0.69%	0.026%	−3.46	25.16
HFI fixed income	0.27%	0.011%	−3.12	17.23
HFI global macro	0.85%	0.096%	0.03	3.17
HFI equity long/short	0.74%	0.084%	0.21	4.07
HFI managed futures	0.33%	0.116%	0.02	0.41
HFI emerging markets	0.58%	0.212%	−0.70	4.94
HFI dedicated short bias	−0.34%	0.239%	0.84	2.16

the Black and Litterman model. Using the market weights and the moment matrices, we get an estimate of the market equilibrium returns using Equation (13.4) when considering the two-moments Black and Litterman framework, and Equation (13.1) when considering the four-moments Black and Litterman framework. The market weights and the obtained equilibrium returns in the two cases are displayed in Table 13.2. The negative value of the dedicated short bias equilibrium return depends on the very little allocation of the market weight (only 0.11%). In both the two-moments and the four-moments equilibrium, this little allocation can be explained only by a very bad expected performance of the dedicated short bias index. This is also consistent with the positive performance expected by the equity long/short index, which has a well-documented long bias, and thus is negatively correlated with the dedicated short bias index.

TABLE 13.2   Market Weights, Two and Four Moments Implied Equilibrium Return Vectors

	Market Weight w_{mkt}	Implied Equilibrium Return Vector Π	Implied Equilibrium Return Vector—Four Moments Π_4
HFI convertible arbitrage	6.32%	0.0345%	0.0355%
HFI equity market neutral	5.01%	0.0188%	0.0183%
HFI event driven	10.44%	0.0677%	0.0700%
HFI fixed income	1.69%	0.0291%	0.0300%
HFI global macro	39.76%	0.1672%	0.1666%
HFI equity long/short	30.95%	0.1457%	0.1456%
HFI managed futures	1.54%	0.0344%	0.0310%
HFI emerging markets	4.18%	0.1901%	0.1929%
HFI dedicated short bias	0.11%	−0.1520%	−0.1550%

As previously discussed, the Black-Litterman approach combines equilibrium returns with an explicit set of views. Expected returns can be interpreted as a Bayesian weighted average of the equilibrium returns and investors' views. We will then focus on the importance of the confidence in the view, by introducing a bold and a mild view, and the ability of the four-moments extension to penalize allocation in indices with high kurtosis and negative skewness.

For simplicity, we introduce only one view at time, and we assume that the manager has a single specific view on HFI event driven (the fund with higher kurtosis and lower skewness) or, alternatively, on HFI equity market neutral (a fund close to a normal distribution). In both cases the view is of 1% monthly excess return, thus higher than the equilibrium mean. In case of the bold view, the manager is supposed to input a variance of 0.005% on the view; in the mild view, the manager is supposed to input a variance of 0.05% (corresponding standard deviations are nearly 0.7% and 2.2%).

In our example we have $K = 1$ and $N = 9$; the view vector (Q) becomes a scalar value (0.01) as well as the covariance matrix (Ω) that represents the uncertainty of the view (0.005% and 0.05%, respectively). If the view is expressed on HFI event driven, P is the row vector (1×9): $P = [001000000]$; if the view is expressed on the HFI equity market neutral, $P = [010000000]$.

Corresponding asset allocation for the two-moments and four-moments equilibrium are reported in Table 13.3, when the view is expressed on HFI event driven, and Table 13.4 when the view is expressed on HFI

TABLE 13.3   View on the HFI Event Driven

	Skewness	Excess Kurtosis	w_{mkt}	Two Moments Bold View	Two Moments Mild View	Four Moments Bold View	Four Moments Mild View
HFI convertible arbitrage	−1.38	3.42	6.32%	2.97%	5.63%	3.53%	5.73%
HFI equity market neutral	0.34	0.45	5.01%	2.35%	4.46%	4.87%	4.84%
HFI event driven	−3.46	25.16	10.44%	57.92%	20.25%	47.79%	18.66%
HFI fixed income	−3.12	17.23	1.69%	0.80%	1.51%	3.39%	1.81%
HFI global macro	0.03	3.17	39.76%	18.68%	35.40%	19.68%	35.63%
HFI equity long/short	0.21	4.07	30.95%	14.54%	27.56%	16.79%	27.94%
HFI managed futures	0.02	0.41	1.54%	0.73%	1.37%	1.92%	1.53%
HFI emerging markets	−0.70	4.94	4.18%	1.97%	3.73%	2.08%	3.76%
HFI dedicated short bias	0.84	2.16	0.11%	0.05%	0.10%	−0.06%	0.09%

TABLE 13.4　View on the Equity Market Neutral

				Two Moments		Four Moments	
	Skewness	Excess Kurtosis	w_{mkt}	Bold View	Mild View	Bold View	Mild View
HFI convertible arbitrage	−1.38	3.42	6.32%	2.78%	5.59%	2.48%	5.53%
HFI equity market neutral	0.34	0.45	5.01%	58.22%	15.97%	58.41%	16.07%
HFI event driven	−3.46	25.16	10.44%	4.59%	9.23%	3.61%	9.03%
HFI fixed income	−3.12	17.23	1.69%	0.74%	1.50%	1.32%	1.62%
HFI global macro	0.03	3.17	39.76%	17.49%	35.17%	17.68%	35.16%
HFI equity long/short	0.21	4.07	30.95%	13.61%	27.38%	13.72%	27.38%
HFI managed futures	0.02	0.41	1.54%	0.68%	1.37%	0.77%	1.38%
HFI emerging markets	−0.70	4.94	4.18%	1.84%	3.70%	1.81%	3.68%
HFI dedicated short bias	0.84	2.16	0.11%	0.05%	0.10%	0.19%	0.13%

equity market neutral. The results confirm our intuitions and highlight the importance of including higher moments in the pricing equation. The view on HFI event driven is placed on a fund with high kurtosis and large negative skewness. If we use the classic two-moments Black and Litterman approach, the allocation in that index would be nearly 60% for the bold view and 20% for the mild view. If we include the third and fourth moments, these numbers decrease to nearly 48% and 18%, respectively: it is clear that the allocation in this fund has been penalized by its higher moments.

On the contrary, if the manager has a view on HFI equity market neutral, a fund with return distribution close to a normal one, the allocation in the case of two moments and four moments is almost unchanged, for both the bold and the mild view. There is instead a marginal increase, which is due to the positive skewness of this fund, which is favored by the risk properties of an investor with utility $u(w) = -e^{-\lambda w}$.

13.4　HOW TO IMPLEMENT BLACK-LITTERMAN FINANCIAL FORECASTS

In order to put into practice the Black-Litterman model the analyst should generate a vector of forecasts expressed in two different ways: (1) in absolute terms for a single asset class and (2) in relative terms, with an asset class compared to another one. In case of hedge fund strategies, most of them require the same ability to forecast changes in economies, typically depending upon shifts in economic and monetary policies. This implies the knowledge of fundamental and logic relations among the real and the financial system.

First, analysts look at the interventions of the central banks, which should be coherent with their purposes in terms of inflation, growth, or unemployment. Second, to anticipate other players' choices, a large use of leading indicators has been recently done. In the following sections we find out central banks' strategies and styles, then variables devoted to efficiently forecast in financial markets and put into action the Black-Litterman model.

13.4.1 Central Banks' Strategies and Styles

Monetary policy authorities are oriented to pursue the economic system welfare.

> In the long run, the central bank cannot influence economic growth by changing the money supply. Related to this is the assertion that inflation is ultimately a monetary phenomenon. Indeed, prolonged periods of high inflation are typically associated with high monetary growth. While other factors (such as variations in aggregate demand, technological changes or commodity price shocks) can influence price developments over shorter horizons, over time their effects can be offset by some degree of adjustment of the money stock. In this sense, the longer-term trends of prices or inflation can be controlled by central banks. (European Central Bank, 2004, 42)

Within the Eurosystem, this purpose is codified in Article 2 of the Maastricht Treaty: "to promote economic and social progress and a high level of employment and to achieve balanced and sustainable development, in particular through the creation of an area without internal frontiers, through the strengthening of economic and social cohesion and through the establishment of economic and monetary union."

Thus, the three final purposes quoted by Article 2 are growth, price stability, and employment. Internal equilibrium should be added as an external variable, such as balance of payments or the exchange rate stability. The decisions of authorities are not always characterized by disclosure, due to potential mechanism conflicts, elections, and financial crises. This ambiguity is amplified by the incompatibility among final targets, in particular between growth and unemployment on one side, and inflation on the other. This relation (Phillips, 1958) has been proved in many cases (United Kingdom during 1861–1957; United States for the 1960s and

1970s, while after 1984 it is weakened). In the United States, the Federal Reserve changes its policy priority depending on the economic situation, passing from a growth- to a price stability–oriented strategy (Atkeson and Ohanian, 2001).

According to Rumler (2005), the Phillips curve is on the basis of the decision to make the price stability the priority of the monetary policy, explicated in Article 105 of the Maastricht Treaty: "The primary objective of the ESCB shall be to maintain price stability. Without prejudice to the objective of price stability, the ESCB shall support the general economic policies in the Community with a view to contributing to the achievement of the objectives of the Community as laid down in Article 2."

The European Central Bank justifies its priority as follows:

> First, price stability makes it easier for people to recognise changes in relative prices, since such changes are not obscured by fluctuations in the overall price level. As a result, firms and consumers do not misinterpret general price level changes as being relative price changes and can make better informed consumption and investment decisions…. Second, if creditors can be sure that prices will remain stable in the future, they will not demand an "inflation risk premium" to compensate them for the risks associated with holding nominal assets over the longer term…. Third, the credible maintenance of price stability also makes it less likely that individuals and firms will divert resources from productive uses in order to hedge against inflation…. Fourth, tax and welfare systems can create perverse incentives which distort economic behaviour…. Fifth, inflation acts as a tax on holdings of cash. This reduces household demand for cash and consequently generates higher transaction costs. Sixth, maintaining price stability prevents the considerable and arbitrary redistribution of wealth and income that arises in inflationary as well as deflationary environments, where price trends change in unpredictable ways. (European Central Bank, 2004, 42–43)

Once a central bank will define the inflation target, the spread between the target and the actual trend suggests the authority behavior: restricting when the actual value is higher than the target, expansive in the opposite

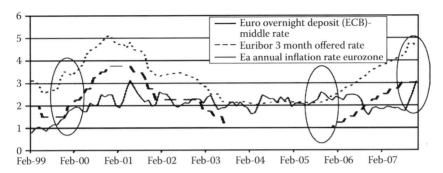

FIGURE 13.1   CPI and interest rates in the Euro area (1999–2007). (From ECB.)

case. In order to forecast the scenario, the analyst should anticipate the variable trend.

In case of the Euro area, the

> strategy is based on a quantitative definition of price stability, namely that an annual increase in the Harmonised Index of Consumer Prices (HICP) of below 2% can be considered as being compatible with this primary objective of monetary policy.... The phrase "below 2%" clearly delineates the upper bound for the rate of measured inflation in the HICP which is consistent with price stability. At the same time, the use of the word "*increase*" in the definition clearly signals that deflation, i.e. prolonged declines in the level of the HICP index, would not be deemed consistent with price stability. (European Central Bank, 1999, 9 and 46)

This target has never been corrected.

Correlation between interest and inflation rates in the Euro area can be seen in Figure 13.1. The three circles in the figure show inflationary periods. The reader must consider that all the inflation values higher than 2% will generate an expectation of higher interest rates.

Albeit the HICP dynamics depends upon various factors, the ECB makes its forecasts modeling interest rates, exchange rates, oil price, and the public deficit national policies. Table 13.5 shows forecasts and actual values for GDP and inflation.

The same process must be implemented with different countries when the target is the growth or some other variable. The forecasting ability, especially for the U.S. GDP, declined during these last years. According to

TABLE 13.5 Forecasts and Actual Values for GDP and Inflation (2003–2006)

	GDP			Inflation		
	2003	**2004**	**2005**	**2003**	**2004**	**2005**
Dec-03	0.2–0.6	1.1–2.1	1.9–2.9	2.0–2.2	1.3–2.3	1.0–2.2
Jun-04	0.5	1.4–2.0	1.7–2.7	2.1	1.9–2.3	1.1–2.3
Sep-04	0.5	1.6–2.2	1.8–2.8	2.1	2.1–2.3	1.3–2.3
	2004	**2005**	**2006**	**2004**	**2005**	**2006**
Dec-04	1.6–2.0	1.4–2.4	1.7–2.7	2.1–2.3	1.5–2.5	1.0–2.2
Mar-05	1.8	1.2–2.0	1.6–2.6	2.1	1.6–2.2	1.0–2.2
Jun-05	1.8	1.1–1.7	1.5–2.5	2.1	1.8–2.2	0.9–2.1
Sep-05	1.8	1.0–1.6	1.3–2.3	2.1	2.1–2.3	1.4–2.4
	2005	**2006**	**2007**	**2005**	**2006**	**2007**
Dec-05	1.2–1.6	1.4–2.4	1.4–2.4	2.1–2.3	1.6–2.6	1.4–2.6
Mar-06	1.4	1.7–2.5	1.5–2.5	2.2	1.9–2.5	1.6–2.8
Jun-06	1.4	1.8–2.5	1.3–2.3	2.2	2.1–2.5	1.6–2.8

D'Agostino et al., (2006), the relative MSFE* from 1985 to 1999 increased for two relevant forecasters: the Federal Reserve and the Survey of Professional Forecasters of the Philadelphia FRB. The random walk forecasts show a higher magnitude of error.

13.4.2 Leading Indicators

The fundamental analysis rarely is able to generate a forecast when it is useful in order to put into practice the Black-Litterman model, since most of the variables are lagged. This is directly stated in the first strategic document written in January 1999 by the European Central Bank:

> Although the monetary data contain information vital to informed monetary policymaking, on their own they will not constitute a complete summary of all the information about the economy required to set an appropriate monetary policy for the

---

* The relative MSFE is:

$$\text{Relative MSFE} = \frac{\sum_{t=T_1}^{T_2-h}\left(P_{t+h}^h - P_{i,t+h|t}^h\right)^2}{\sum_{t=T_1}^{T_2-h}\left(P_{t+h}^h - P_{0,t+h|t}^h\right)^2}$$

where $P_{i,t+h|t}^h$ is the out-of-sample forecast and $P_{0,t+h|t}^h$ is the benchmark forecast over the same period, and T_1 and $T_2 - h$ are the first and last date, respectively, of the out-of-sample period.

TABLE 13.6   Indicators Useful to Identify Target Variables
(Not Comprehensive)

Target Variables	Proxy Variables
Economic growth	Industrial production
	Private and public consumption
	Real investments
	Supply change
Employment	New working places
	Hourly earnings
	Working cost per production unit
	Unemployment subsidy
Inflation	GDP deflator
	Wholesale inflation
Foreign equilibrium	Commercial trade balance/GDP
	International capital flows

maintenance of price stability. Therefore, in parallel with the analysis of monetary growth in relation to the reference value, a broadly based assessment of the outlook for price developments and the risks to price stability in the euro area will play a major role in the Eurosystem's strategy. This assessment will be made using a wide range of economic indicators. This wide range of indicators will include many variables that have leading indicator properties for future price developments. (European Central Bank, 1999, 49).

First, analysts should check a set of data that, even though not leading, can define target variables more precisely (Jones and Ferris, 1993), as shown in Table 13.6.

Some other variables, known as leading indicators, help the market analyst to estimate the phenomenon dynamics (Table 13.7).

The ability to anticipate the target variables depends on many factors (Lahiri and Moore, 1991), such as the ability to elaborate and connect data to each other. This way, it is possible to recognize the information useful to the portfolio manager so she takes the right financial position.

There are three leading indicator categories (Niemira and Klein, 1994):

1. Pressure indicators (Pericoli and Sbracia, 2003)

2. Diffusion indicators (Burns, 1969)

3. Synthetic indicators

TABLE 13.7   Indicators Useful to Forecast Target Variables (Not Comprehensive)

Target Variables	Leading Indicators
Economic growth	Manufacturers' new orders for consumer goods and materials
	The vendor performance component of the ISM index
	The average level of weekly initial claims for unemployment insurance
	Building permits
	Index of consumer expectations
	Manufacturers' new orders for nondefense capital goods
Employment	The average manufacturing workweek; demographic dynamics
Inflation[*]	Consumer price by sector and region
	Commodities' prices
	Money change (M2 and M3)[†]

[*] Among the leading inflation indices we mind the Commodity Research Bureau Index, the Journal of Commerce Index, the Center for International Business Cycle Research Index, and the Paine Webber Index. See (Garner, 1995).
[†] See Becketti and Morris (1992) and Hetzel (1992).

They can be useful for the Black-Litterman model and portfolio management. The underlying hypothesis is that the economic and financial cycles are multifactorial events. Basis indicators should be selected as follows:

1. Through causal relations, statistically robust and economically logical

2. With high-frequency data

3. Choice of data with time series deep enough

4. Eliminating autocorrelation phenomenon

The most used and simple method to build up the synthetic indicator (IS) is

$$\Delta IS = \sum_{i=1}^{n} w_i \cdot c_i \cdot \Delta IB_i$$

where w is the weight of the basic indicator (IB); c is the correlation factor needed to standardize the different measure units; and n is the total number of basic indicators.

In the United States, from 1960 to 2005, six recessions have been experienced (Table 13.8).

Table 13.9 shows the same forecasting for eight countries.

In case of the Black-Litterman model, leading indicators are particularly useful when they help to forecast financial market data, especially

TABLE 13.8 Forecasting Capacity of the Composite Index
of Leading Indicators for the United States (1960–2005, in months)

Recession Period	Leading Months before the Beginning	Leading Months before the End
1960	11	3
1970	8	7
1974	9	2
1980	15	3
1981–1982	3	8
1990–1991	6	2

for equity indices. Since in some cases (e.g., United States) the equity index is considered a leading indicator, there is a risk of loop in the analysis.

For the United States, United Kingdom, Italy, and Europe, during the period January 1998–June 2006 we compare the consumer expectations and the manufacturing change in order to evaluate the forecasting capability for the equity index.

In the U.S. analysis (Figure 13.2) there is a significant coincidence among the changing points of equity and consumer indices, while the manufacturing one was able to anticipate both the 2000 and the 2002 crises, 18 months before.

In Italy, the equity and consumer indices showed a negative correlation (−0.27505), whereas the manufacturing investments led 10 months before the beginning of the crisis and 12 months before the upturn (Figure 13.3).

TABLE 13.9 Forecasting Capacity of the Composite Index of Leading
Indicators for Different Countries (1960–2005, in months)

Country	Number of		Months before		
	Min	Max	Min	Max	Mean
United States	9	9	6	11	8
Canada	2	2	14	12	13
Germany	4	4	10	10	10
France	4	4	2	9	6
United Kingdom	3	3	13	20	17
Italy	3	2	11	12	11
Switzerland	4	4	15	13	14
Japan	2	3	12	10	11

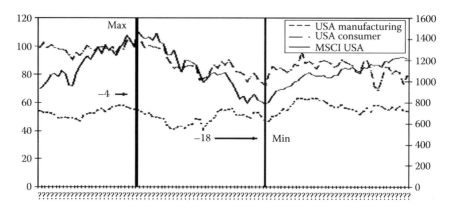

FIGURE 13.2   Equity index (MSCI USA), consumer expectation and manufacturing index (1998–2006).

In the United Kingdom, the equity market was anticipated by the consumer index of 1 and 5 months, respectively (Figure 13.4). The investment index is coincident in the 2000 crisis and anticipates (18 months) the 2003 crisis.

Finally, the European index seems not to be anticipated in the case of the 2000 downturn. In fact, the consumer index changed its direction 2 months later, and the manufacturing investments 5 months

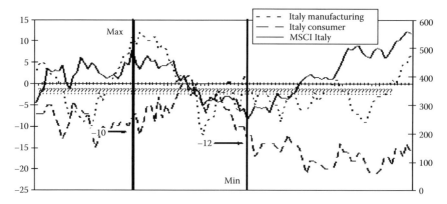

FIGURE 13.3   Equity index (MSCI Italy), consumer expectation and manufacturing index (1998–2006).

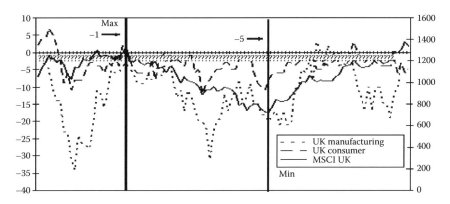

FIGURE 13.4   Equity index (MSCI UK), consumer expectation and manufacturing index (1998–2006).

later. Better was the 2003 performance, led 1 month before using the consumer variable and 16 months before by means of the manufacturing index.

Our empirical analysis makes evident that some real indicators lead financial markets indices too. Analysts must supervise their performance during the different changing points in order to make the forecasting model more effective.

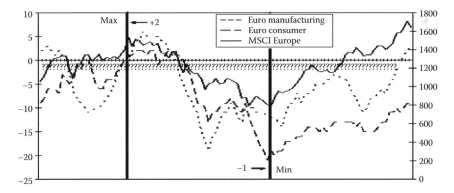

FIGURE 13.5   Equity index (MSCI Europe) consumer expectation and manufacturing index (1998–2006).

13.5 CONCLUSION

In this chapter, we extend the Black and Litterman model to four moments, explicitly involving skewness and kurtosis as additional risk measures, and we show how this can generate significant benefits in the context of hedge fund investing. The results confirm our intuitions and highlight the importance of including higher moments in the pricing equation. Extending to four moments, the allocation on a high-kurtosis fund decreases from 60% to 48%; for a normally distributed fund it remains stable. Our results can potentially be used for other nonnormally distributed assets, as commodities single stocks.

REFERENCES

Atkeson, A., and Ohanian, L. E. (2001). Are Phillips curves useful for forecasting inflation? *FRB Minneapolis Quarterly Review* 25:2–11. Available at http://minneapolisfed.org.

Becketti, S., and Morris, C. (1992). Does money still forecast economic activity? *Federal Reserve of Kansas City Economic Review* 4:65–77.

Black, F., and Litterman, R. (1990). *Asset allocation: Combining investor views with market equilibrium.* Fixed income research. New York: Goldman Sachs & Co.

Burns, A. F. (1969). *The business cycle in a changing world.* New York: Columbia University Press.

D'Agostino, A., Giannone, D., and Surico, P. (2006). *(Un)predictability and macroeconomic stability.* Working Paper Series ECB 605, Frankfurt.

European Central Bank. (January 1999). Monthly bulletin. Frankfurt.

European Central Bank. (2004). *The monetary policy of the ECB.* Frankfurt.

Garner, C. A. (1995). How useful are leading indicators of inflation? *Economic Review Federal Reserve Bank of Kansas City* 2:5–18.

He, G., and Litterman, R. (1999). *The intuition behind Black-Litterman model portfolios.* Investment management research. New York: Goldman Sachs & Co.

Hetzel, R. L. (1992). How useful is M2 today? *Economic Review, Federal Reserve Bank of Richmond,* September/October, 12–25.

Hwang, S., and Satchell, S. (1999). Modelling emerging risk premia using higher moments. *International Journal of Finance and Economics* 4:271–96.

Idzorek, T. (2004). A step-by-step guide to the Black-Litterman model. Working paper, Zephyr Associates, Zephyr Cove, NV.

Jondeau, E., and Rockinger, M. (2004). Optimal portfolio allocation under higher moments. EFMA 2004 Basel Meetings paper. Available at SSRN, January: http://ssrn.com/abstract=498322.

Jones, M., and Ferris, K. (1993). *Market movers.* London: McGraw-Hill.

Lahiri, K., and Moore, G. H. (1991). *Leading economic indicators. New approaches and forecasting records.* Cambridge, MA: Cambridge University Press.

Martellini, L., Vaissié, M., and Ziemann, V. (2005). Investing in hedge funds: Adding value through active style allocation decisions. Edhec Risk and Asset Management Research Centre. Available at http://www.edhec-risk.com.

Martellini, L., and Ziemann, V. (2005). The benefits of hedge funds in asset liability management. Edhec Risk and Asset Management Research Centre. Available at http://www.edhec-risk.com.

Niemira, M. P., and Klein, P. A. (1994). *Forecasting financial and economic cycles*. Hoboken, NJ: John Wiley & Sons.

Pericoli, M., and Sbracia, M. (2003). A primer on financial contagion. *Journal of Economic Surveys* 17:571–608.

Phillips, A. W. (1958). The relationship between unemployment and the rate of change of money wages in the United Kingdom 1861–1957. *Economica* 5:283–99.

Rumler, F. (2005). *Estimates of the open economy new Keynesian Phillips curve for Euro area countries*. ECB Working Paper Series 496, Frankfurt.

CHAPTER 14

Dampening Hedge Fund Volatility through Funds of Hedge Funds

Jodie Gunzberg and Audrey Wang

CONTENTS

14.1 INTRODUCTION

In this chapter, the authors focus the discussion on annualized volatility and dispersion of volatility over different time periods, strategies, and number of funds in a fund of hedge funds. First, the data universes will be described followed by the methodology used to arrive at the conclusions. The authors evaluated the simulations measuring the median annualized volatility and dispersion of volatility over 3, 5, 7, and 10 years for each strategy's individual hedge funds as well as for funds of funds within those strategies. Next, a similar exercise was performed to examine the volatility reduction by combining strategies.

There are a few key takeaways to point out regarding the analysis of the simulation results.

When evaluating hedge funds or funds of hedge funds within a strategy, the minimum track record will have little influence on the volatility of the average fund. However, there is evidence of a higher-volatility period included in the 10-year annualized volatility numbers that drops off as time decreases. So, one must be careful to consider the track record length when thinking about the possible dispersion in volatility. One may expand the analysis of return history by representing the distribution as a combination of peaceful times and eventful, more volatile times with greater correlation among strategies that occur during crises (Till and Gunzberg, 2005).

Although volatility of funds of hedge funds varies widely within the directional space, there is a general pattern among all the strategies of decreasing volatility as the number of funds increases. The most dramatic reduction occurs when moving from one fund to five funds. Further, there is support for combining up to fifteen hedge funds within a strategy, but minimal benefit is obtained by further additions. However, one must consider return when making an investment decision, and according to a study by Patel (2008), forty hedge fund managers are sufficient to consistently beat a benchmark of T-bills + 2.5% over 5 years.

When evaluating funds of hedge funds equally weighted across strategies, the results show that levels of annualized volatility depend much more on strategy weighting than on number of funds. Heavier weightings in directional strategies yield higher volatility and dispersion in volatility.

Lastly, there is a qualitative review of considerations that fund of hedge fund managers should take into account when selecting hedge funds. The authors conclude with a review of the portfolio construction methodology and statistics of closely followed funds of hedge funds at the Marco Consulting Group.

14.2 DATA UNIVERSE

The initial data universe consisted of all hedge funds in the Hedge Fund Research (HFR) database of approximately 5,000 funds as of March 31, 2008. The hedge funds were then separated into three broad strategy universes: directional, event driven, and relative value. The directional universe consisted of hedge funds in the HFR categories equity hedge, equity manager, equity nonhedge, and short selling. The event-driven universe consisted of hedge funds in the HFR categories event driven, distressed

TABLE 14.1   Universe Composition
(with $100 Million or More in AUM)

Universe by Strategy	Number of Funds
10 years: total	**153**
Directional	68
Event driven	29
Relative value	56
7 years: total	**305**
Directional	142
Event driven	51
Relative value	112
5 years: total	**488**
Directional	223
Event driven	85
Relative value	180
3 years: total	**687**
Directional	332
Event driven	108
Relative value	247

securities, and merger arbitrage. The relative value universe consisted of hedge funds in the HFR categories relative value arbitrage, convertible arbitrage, fixed income, and equity market neutral.

Each strategy universe was split into four sets by a screen for at least a full 10-, 7-, 5-, and 3-year track record, and then further narrowed to funds with at least $100 million in assets under management (AUM). The screen for funds with at least $100 million in assets narrowed the universe considerably. Table 14.1 shows the numbers of hedge funds in each universe. As the track record length requirement increased, naturally the number of funds diminished. Also, the universe for directional funds was largest, followed by event driven, then relative value.

14.3   SIMULATION METHODOLOGY

First, within each of the strategy universes, directional, event driven, and relative value, and in each of the minimum track record periods, 10, 7, 5, and 3 years, the annualized returns and annualized standard deviations were measured for each hedge fund. For the sets limited by a minimum 10-year track record, the 10-, 7-, 5-, and 3-year annualized returns and annualized standard deviations were measured; for the sets limited by a minimum 7-year track record, the 7-, 5-, and 3-year annualized returns and annualized standard deviations were measured; for the sets limited by a minimum 5-year track record, the 5- and 3-year annualized returns and annualized standard deviations were measured; and for the sets limited by a minimum 3-year track record, only the 3-year annualized returns and annualized standard deviations were measured. Next, the returns and standard deviations were divided into 95th, 90th, 75th, 50th, 25th, 10th, and 5th percentiles, and the dispersion between each percentile and the 5th percentile was calculated and evaluated.

Next, for each of the strategy universes in each of the minimum track record periods, the authors simulated portfolios of fund of funds using different numbers of underlying funds. Ten thousand sets of funds of 5, 10, 15, 20, and 25 equally weighted hedge funds were created (for a total of 50,000 per strategy per minimum track record). Going forward these universes will be referenced to as funds of 5, 10, 15, 20, or 25 hedge funds. First, the strategy was selected, followed by the minimum track record length, and then the number of hedge funds to be included in the fund of hedge funds. The annualized rates of return as well as the annualized standard deviations were calculated for each of the funds of hedge funds within the strategy universe. Then, the returns and standard deviations (within each strategy and time period and number of underlying funds) were divided into 95th, 90th, 75th, 50th, 25th, 10th, and 5th percentiles, and the dispersion between each percentile and the 5th percentile was calculated and evaluated.

The number of combinations available was determined by the choose function, n choose k:

$$C(n, k) = n!/((k!)^{*}(n - k)!)$$

where n = the number of funds available and k = the number of funds chosen for the fund of hedge funds.

While some actual funds of hedge funds have more than twenty-five underlying managers, the authors chose 25 as the maximum number of underlying funds because there were only twenty-nine hedge funds in the event-driven strategy with at least 10-year track records and $100 million of AUM.

All of the above data were then used to determine the reduction in volatility and the reduction in dispersion of volatility as the number of funds to be combined within a strategy universe and minimum track record increased. The second step was creating multistrategy funds of funds by equally weighting underlying funds from the directional, event-driven, and relative value categories, in order to measure how volatility and the dispersion of volatility behaved in a multistrategy context.

To create multistrategy funds of funds, all of the hedge funds were grouped by strategy and by minimum track record length. Within each track record length (minimum 10, 7, 5, and 3 years), 10,000 funds of 15, 30, 45, 60, and 75 hedge funds each (for a total of 50,000) were randomly sampled from $C(n, k)$ samples, where the number of hedge funds chosen from each strategy was equal. Equally weighting across strategies neutralizes the significantly larger number of directional hedge funds in the universe. So, for example, each of the 10,000 funds of 15 hedge funds constructed with a minimum 10-year track record consisted of 5 directional, 5 event-driven, and 5 relative value funds, each with a minimum 10-year track record. Again, annualized returns and annualized standard deviations were calculated for each multistrategy fund of hedge funds and divided into 95th, 90th, 75th, 50th, 25th, 10th, and 5th percentiles. The dispersion between each percentile and the 5th percentile was also calculated and evaluated. Then, the statistics from the multistrategy funds of fifteen hedge funds with a minimum 10-year track record were compared with the single-strategy funds of five hedge funds with minimum 10-year track records.

14.4 SIMULATION RESULTS

First, the difference in annualized 3-year volatility at the 50th percentile among hedge funds with at least 10-, 7-, 5-, or 3-year minimum track records was examined. The conclusion is that the difference is minimal, with the greatest being about 50 basis points between funds with 10-year and 3-year track records in the relative value strategy, as shown in Table 14.2. This is important because in the analysis later in this chapter,

TABLE 14.2   Annualized 3-Year Volatility at 50th Percentile

No. of Funds	Directional				Event Driven				Relative Value			
	Minimum Years Track Record				Minimum Years Track Record				Minimum Years Track Record			
	3	5	7	10	3	5	7	10	3	5	7	10
1	10.0%	10.0%	10.1%	10.6%	6.9%	6.9%	6.8%	6.7%	5.3%	5.4%	5.2%	4.9%
5	8.0%	8.1%	8.3%	7.9%	5.6%	5.4%	5.2%	5.2%	3.7%	3.9%	3.8%	3.2%
10	7.7%	7.8%	7.9%	7.3%	5.1%	5.1%	5.0%	4.8%	3.3%	3.5%	3.4%	2.7%
15	7.6%	7.7%	7.7%	7.1%	5.0%	4.9%	4.9%	4.7%	3.1%	3.3%	3.2%	2.6%
20	7.5%	7.6%	7.7%	7.0%	4.9%	4.9%	4.9%	4.6%	3.0%	3.2%	3.2%	2.5%
25	7.5%	7.5%	7.6%	7.0%	4.8%	4.8%	4.8%	4.6%	2.9%	3.1%	3.1%	2.4%

TABLE 14.3   Annualized 3-Year Volatility Dispersion between the 5th and 95th Percentiles

	Directional				Event Driven				Relative Value			
	Minimum Years Track Record				Minimum Years Track Record				Minimum Years Track Record			
No. of Funds	3	5	7	10	3	5	7	10	3	5	7	10
1	16.1%	16.3%	15.8%	10.7%	14.1%	11.6%	8.0%	8.3%	16.3%	18.8%	20.1%	10.0%
5	7.5%	7.3%	7.1%	6.1%	4.5%	4.2%	4.4%	2.8%	4.6%	4.9%	5.3%	3.0%
10	5.1%	4.8%	4.9%	4.5%	2.8%	2.6%	2.6%	1.7%	3.1%	3.3%	3.7%	2.0%
15	4.1%	3.9%	3.9%	3.5%	2.1%	2.0%	1.9%	1.2%	2.4%	2.6%	2.9%	1.6%
20	3.5%	3.3%	3.3%	3.0%	1.7%	1.6%	1.5%	0.8%	2.1%	2.2%	2.5%	1.3%
25	3.1%	2.9%	2.9%	2.5%	1.5%	1.4%	1.3%	0.5%	1.8%	2.0%	2.1%	1.1%

where levels of volatility are measured as the number of funds increase, only the minimum 10-year track record universes will be used to display the results for 10-, 7-, 5-, and 3-year annualized volatility.

Next, the dispersion in annualized 3-year volatility between the 5th and 95th percentiles among hedge funds with at least 10-, 7-, 5-, or 3-year minimum track records was examined. While the level of volatility differs very little depending on the minimum track record, the dispersion in volatility varies pretty significantly, where the dispersion is greater for funds with shorter minimum track records, as illustrated in Table 14.3. However, the pattern of decreasing dispersion is similar for each of the minimum track records, so again the results for only the minimum 10-year track record will be displayed.

The key takeaway is that when evaluating hedge funds or funds of hedge funds within a strategy, the minimum track record will have little influence on how volatile the average fund is. Nonetheless, one must be careful to consider the track record length when thinking about the possible dispersion in volatility.

In Figures 14.1 to 14.3 annualized volatility levels at the 50th percentile over 3-, 5-, 7-, and 10-year periods ending on March 31, 2008 will be examined for each of the strategies. While there is a general pattern among all of the strategies of decreasing volatility as the number of funds increase, the most dramatic reduction occurs when moving from one fund to five funds. Also, there is evidence of a higher-volatility period included in the 10-year annualized volatility numbers that drops off as time is decreased. In the following evaluation, there is support for combining up to fifteen hedge funds, but minimal benefit through reduction in volatility is obtained by further additions.

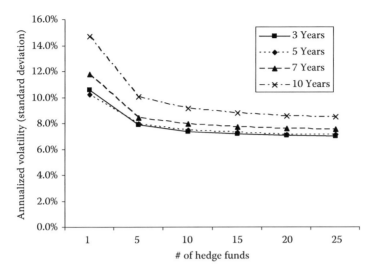

FIGURE 14.1 Directional strategy, funds of hedge funds with 10-year minimum track records, period ending March 31, 2008, 50th percentile volatility.

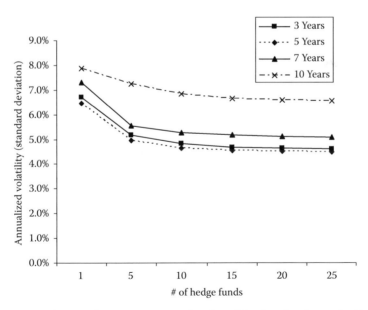

FIGURE 14.2 Event-driven strategy, funds of hedge funds with 10-year minimum track records, period ending March 31, 2008, 50th percentile volatility.

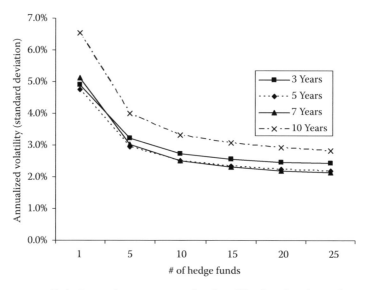

FIGURE 14.3   Relative value strategy, funds of hedge funds with 10-year minimum track records, period ending March 31, 2008, 50th percentile volatility.

In the directional strategy depicted in Figure 14.1, when moving from one to five funds, the volatility is reduced by 2.7%, 2.2%, 3.3%, and 4.7% annualized over 3, 5, 7, and 10 years, respectively. Diversifying to ten hedge funds reduces annualized volatility by about another 50 basis points over 3, 5, and 7 years, and 90 basis points annualized over 10 years. Beyond that the annualized volatility reduction is minimal, where adding another ten hedge funds for a total of twenty funds only reduces annualized volatility by 30, 30, 40, and 60 basis points annualized over 3, 5, 7, and 10 years, respectively. Finally, adding another five hedge funds for a total of twenty-five funds does not reduce the volatility for any periods except for by 10 basis points annualized over 10 years.

Next, the results of the event-driven strategy showed a similar pattern of annualized volatility reduction, where the most reduction occurred when moving from one hedge fund to five hedge funds. However, the annualized volatility savings over 10 years had the least reduction, whereas it had the most reduction in the directional strategy, as illustrated by the relatively flat slope of the 10 years line in Figure 14.2. This may be due to the payoff distributions that are discontinuous and skewed within the strategy that the volatility alone cannot measure. A value-at-risk (VAR) measure that summarizes a forward-looking distribution of portfolio profits and losses based on current positions by estimating the probability of success for each deal, the payoffs from success and failure, and the joint correlations across deals would be more sufficient (Jorion, 2008).

Increasing the number of hedge funds from one to five reduced annualized volatility over 3, 5, 7, and 10 years by 1.5%, 1.5%, 1.8%, and 0.6%, respectively. Adding five more hedge funds saved another 30 to 40 basis points of annualized volatility; however, there were almost no incremental volatility reductions by having more than ten funds.

As consistent with the directional and event-driven strategies, the relative value strategy shows the most reduction in annualized volatility by increasing the number of hedge funds from one to five. However, on a relative basis, the annualized volatility from increasing the number of hedge funds from one to five reduces the level by about 40%, which is much greater than for directional and event driven, where the reduction is about 25%. On an absolute basis, the annualized volatility over 3, 5, 7, and 10 years for the relative value strategy was reduced by 1.7%, 1.8%, 2.1%, and 2.5%, respectively. Further increasing the number of hedge funds to

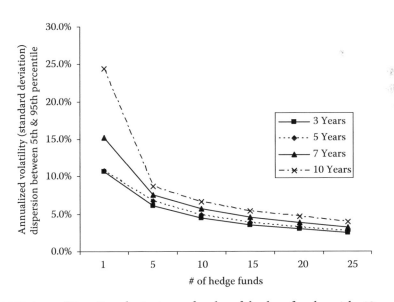

FIGURE 14.4 Directional strategy, funds of hedge funds with 10-year minimum track records, period ending March 31, 2008, 5th–95th percentile volatility dispersion (Marco Consulting Group, 2008).

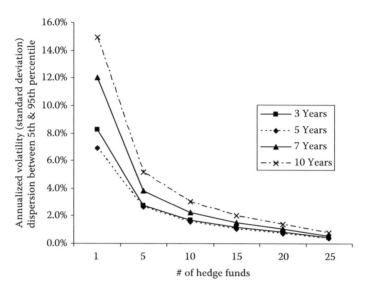

FIGURE 14.5 Event-driven strategy, funds of hedge funds with 10-year minimum track records, period ending March 31, 2008, 5th–95th percentile volatility dispersion (Marco Consulting Group, 2008).

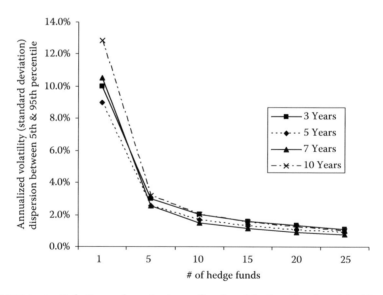

FIGURE 14.6 Relative value strategy, funds of hedge funds with 10-year minimum track records, period ending March 31, 2008, 5th–95th percentile volatility dispersion (Marco Consulting Group, 2008).

ten funds reduced the annualized volatility over 3, 5, 7, and 10 years by 50, 40, 50, and 70 basis points, respectively. Adding more hedge funds to get totals of 15 and 20 funds reduced the annualized volatility another 20 and 10 basis points, but there was no volatility-dampening benefit by adding 5 more hedge funds for a total of 25.

Next, the dispersion of annualized volatility between the 5th and 95th percentiles over 3-, 5-, 7-, and 10-year periods ending on March 31, 2008, will be discussed. As one would expect, the dispersion of volatility is far greater for single hedge funds than for funds of hedge funds. The widest dispersion among hedge funds is within the directional strategy, which might also be expected given it is the strategy with the highest volatility. In fact, even with twenty-five funds there may still be between a 2.5% and 4% discrepancy in annualized volatility from the 5th to 95th percentiles over 3, 5, 7, and 10 years. However, it is a different story for the dispersion of annualized volatility in the event-driven and relative value strategies, where beyond ten funds there is very little dispersion of 2% or less (with the exception of the 3% annualized 10-year dispersion in event driven). The conclusion is that funds of hedge funds volatility can vary widely within the directional space and not as much in the other categories.

In the next part of the analysis, the combination of strategies to create a fund of hedge funds equally weighted by underlying fund and strategy will be examined. As illustrated in Figure 14.7, there is almost no difference in annualized volatility at the 50th percentile in equally weighted funds of hedge funds with minimum 10-year track records, no matter how many funds are included. There is only marginally higher volatility of 50 basis points or less for a fund of fifteen hedge funds.

Another consistent theme when analyzing the levels of annualized volatility between funds of hedge funds within a strategy and multistrategy funds of hedge funds was that the annualized volatility of multistrategy funds of hedge funds, labeled "ALL" in Figure 14.8, always fell in between directional and relative value, somewhere near event driven. This was true regardless of how many funds were used in the strategy and minimum track record universes. The conclusion is that levels of annualized volatility are affected more by strategy weighting than by number of funds.

Contrary to the level of annualized volatility, shown in Figure 14.7, Figure 14.9 shows that the dispersion in volatility based on the number of funds is reduced by more than half when moving between fifteen and seventy-five funds. However, with the exception of the funds of fifteen funds, the dispersion of annualized volatility is never much greater than 2%.

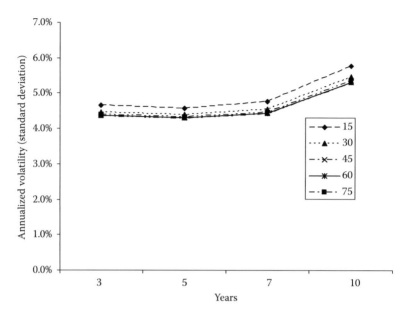

FIGURE 14.7   Equally weighted funds of hedge funds, 10-year minimum track records, period ending March 31, 2008, 50th percentile volatility.

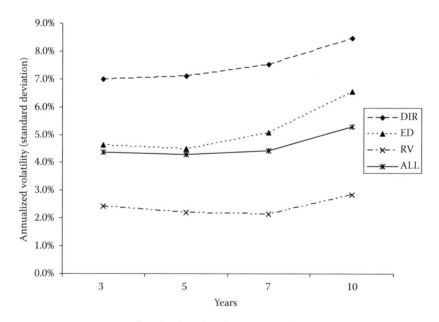

FIGURE 14.8   Seventy-five hedge funds—twenty-five directional, twenty-five event-driven, twenty-five relative value—10-year minimum track records, period ending March 31, 2008, 50th percentile volatility.

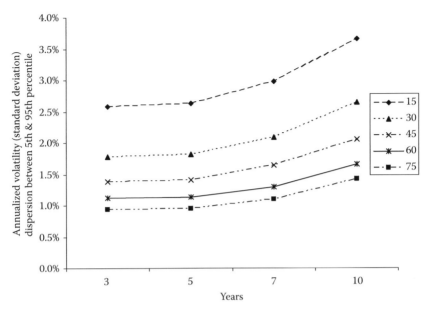

FIGURE 14.9 Equally weighted funds of hedge funds, 10-year minimum track records, period ending March 31, 2008, 5th–95th percentile volatility dispersion.

The greatest dispersion in annualized volatility occurs over a 10-year period, which makes sense given the higher volatility exhibited over that period by all of the strategies.

Figures 14.10 and 14.11 show the dispersion in annualized volatility of the funds of hedge funds combined across strategies (multistrategy) versus the single-strategy funds of funds. Although these charts include the set of equally weighted funds of seventy-five hedge funds, the results are similar despite the number of funds. The multistrategy funds of hedge funds with minimum track records of 10 years have annualized volatility dispersion just above the event-driven and relative value strategies. The point of interest is that in the 10-year minimum track record universe, directional volatility dispersion is significantly greater than the other strategies, so even though two-thirds of the "ALL" funds of funds is comprised of event driven and relative value, the combined dispersion is still wider than the dispersion of each strategy.

As opposed to the 10-year minimum track record universe, combining funds with a minimum track record of 7 years adds the benefit of reduced

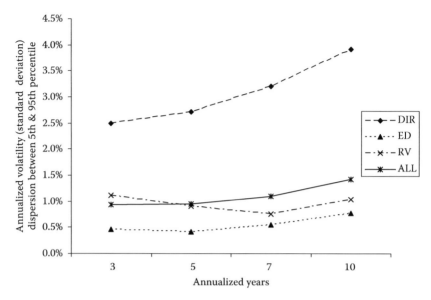

FIGURE 14.10   Equally weighted funds of seventy-five hedge funds, 10-year minimum track records, period ending March 31, 2008, 5th–95th percentile volatility dispersion.

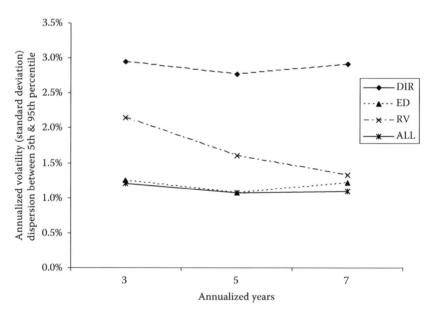

FIGURE 14.11   Equally weighted funds of seventy-five hedge funds, 7-year minimum track records, period ending March 31, 2008, 5th–95th percentile volatility dispersion.

annualized volatility dispersion. This volatility reduction is expected given that the lower dispersion over 10 years is higher for the directional strategy. Also, the difference in volatility dispersion between strategies is much less within the 7-year minimum track record universe, so the combination creates funds of "ALL" hedge funds with lower dispersion than any of the individual strategies. The lower dispersion may be rationalized by recent rises in correlations among hedge fund returns that are explained by declines in overall volatility associated with average covariance (Adrian, 2007).

14.5 FUND OF HEDGE FUNDS PORTFOLIO CONSTRUCTION

Investors in funds of hedge funds must evaluate each fund of funds manager and its investment process, including strategy and manager selection, portfolio construction, risk management, and monitoring. Fund of funds managers typically first form a top-down view of the global economy and markets, focusing on major secular trends and their macroeconomic outlook. These managers also attempt to discover market inefficiencies, such as a supply/demand imbalance in commodities, and tilt the portfolio toward the strategies expected to outperform. A top-down view is often a starting point in a fund of fund's investment process, but it often makes up only a small portion of the manager's process.

The bottom-up analysis, or manager selection, is where the typical fund of funds manager spends most of her time. The initial universe of investable hedge funds is sourced from a variety of venues, including the firm's network of industry contacts, current underlying managers, prime brokers, and public databases. Then the manager will separate the hedge funds into broad strategy categories, such as directional, event driven, and relative value. Depending on the risk and return goals, the underlying managers are analyzed based on qualitative and quantitative measures to form a universe of targeted managers. In most cases, the 20–30% of the funds with the extreme returns account for 80% of the total return variation within each style, whereas the 70–80% of funds in the bulk of the distribution account for only around 20% of the variation in total returns in each style (Mackey, 2006). The final bottom-up step, due diligence, is at the core of the process, where the firm will evaluate in detail each underlying hedge fund manager's strategy and investment process, risk management, back office capabilities, and documentation.

To construct funds of hedge funds managers must pay close attention to exposures such as sector and regional allocations, and also risk measures beyond volatility, such as drawdowns, beta, skewness, and kurtosis. In fact, failure to account for skewness and kurtosis can result in suboptimal performance, as shown by the domination over the mean-variance optimal portfolio of a portfolio optimization that maximizes the probability of a benchmark return and minimizes the expected shortfall (Popova et al., 2007). There are also many other optimization techniques managers use along with liquidity considerations to create the best portfolio. Once portfolio construction is completed, monitoring the underlying managers is an ongoing and continual process that consists of revisiting and reviewing each manager's status.

Investing via a fund of hedge funds gives investors access to a diversified allocation across broad strategies as well as substrategies and underlying managers. This is important because the cross-sectional variation and the range of individual hedge fund returns are far greater than they are for traditional asset classes. Thus, investors in hedge funds take on a substantial risk of selecting a dismally performing fund or, worse, a failing one (Malkiel and Saha, 2005). Multistrategy fund of funds portfolios are often preferred as the initial allocation for an institution investing in hedge funds for the first time. Strategy-specific funds of funds can be another way to diversify an investor's allocation via a core-satellite approach. Investors in a fund of funds will want to evaluate the manager's ability to implement the aforementioned investment process and portfolio construction. Investors should evaluate the manager's ability to select the appropriate strategy allocations/tilts and best-in-class managers, as well as their ability to perform high-quality risk management, portfolio construction, and monitoring of the underlying managers. Investors should seek portfolios that are diversified by strategy and number of funds, with low to moderate volatility profiles.

Based on the Marco Consulting Group's universe of approved and prospective (closely monitored) multistrategy funds of funds, standard deviation on a 5-year basis (for those funds that have 5-year track records ending March 31, 2008) ranges from 2.25% to 5.8%. On a 3-year basis, volatility ranges from 2.5% to 7.5%, and on a 2-year basis from 2.9% to 8.4%. The portfolios with volatility above 5% on a 3- and 5-year basis tend to have larger allocations to directional and certain event-driven strategies. On a 2-year basis, volatility has been skewed by the heightened volatility that the overall markets have experienced since mid-2007. Three-year Sharpe ratios range from –0.2 to 1.8. These portfolios range in diversification, from twenty to seventy-five underlying funds.

14.6 CONCLUSION

In this chapter, the authors have attempted to convey the dampening of hedge fund volatility through investing in a fund of hedge funds by evaluating multiple simulated portfolios of unbiased and randomly selected hedge funds. The authors measured the median annualized volatility and dispersion of volatility over 10-, 7-, 5-, and 3-year periods for individual hedge funds as well as for fund of funds within each broad strategy category of directional, event driven, and relative value.

Although the minimum track records of hedge funds must be considered when analyzing volatility, the authors found that the difference in volatility and dispersion in volatility among funds with 10-, 7-, 5-, and 3-year minimum track records was minimal, and thus displayed only the minimum 10-year track record results in this chapter. Contrary to the belief that some investors may have about younger hedge funds being more volatile, the minimum track record of a hedge fund will have little influence on how volatile the average fund is.

The rule of diversification applies to funds of hedge funds in that, generally, adding more hedge funds to a portfolio decreases the volatility of that portfolio, which is most noticeable when comparing an investor's portfolio of one fund to five funds. However, the authors found that the added benefit of diversification to the dampening of volatility is minimal as one combines more than five funds, and the effect is even less so as one combines more than fifteen funds. The large reduction in volatility from one to five funds is especially notable in relative value, where the volatility level was reduced by about 40%, which was much greater than for directional and event driven, where the reduction was only about 25%.

By comparing the dispersion between the strategy-specific funds of hedge funds and the multistrategy funds of hedge funds, the authors also conclude that the volatility of a portfolio will largely be determined by strategy weightings rather than the degree of diversification among funds. The higher the weighting in directional strategies, typically the higher the volatility of the overall fund of hedge funds will be. Of course, most actual funds of hedge funds that exist in the industry are actively managed and not portfolios of randomly selected funds. Therefore, if an investor's objective is to select a conservative, low-volatility fund of funds, many factors other than diversification must be considered, including the manager's ability to make top-down and bottom-up as well as quantitative and qualitative decisions.

ACKNOWLEDGMENTS

We thank Allan Sievert, Julie Austin, and Imran Zahid of the Marco Consulting Group for their help in preparing this chapter.

REFERENCES

Adrian, T. (2007). Measuring risk in the hedge fund sector. *Hedge Fund Journal* 29:32–36.

Jorion, P. (2008). Risk management for event-driven funds. *Financial Analysts Journal* 64:61–73.

Mackey, S. (2006). Estimating risk premiums of individual hedge funds. *Journal of Alternative Investments* 8:1–9.

Malkiel, B., and Saha, A. (2005). Hedge funds: Risk and return. *Financial Analysts Journal* 61:80–88.

Patel, K. (2008). How many fund managers does a fund-of-funds need? *Pensions: An International Journal* 13:61–69.

Popova, I., Morton, D., Popova, E., and Yau, J. (2007). Optimizing benchmark-based portfolios with hedge funds. *Journal of Alternative Investments* 10:35–55.

Till, H., and Gunzberg, J. (2005). Survey of recent hedge fund articles. *Journal of Wealth Management*, Winter, 81–98.

CHAPTER 15

Information Transmission across Stock and Bond Markets

International Evidence

Charlie X. Cai, Robert Faff, David
Hillier, and Suntharee Lhaopadchan

CONTENTS

15.1 INTRODUCTION

Considerable evidence exists that information flow, as proxied by stock return correlations, transcends national boundaries. The works of Robichek et al. (1972), Ripley (1973), and Panton et al. (1976) have all pointed to significant informational relationships between equity markets. In contrast, Granger and Morgenstern (1970), Agmon (1972), Hilliard (1979), and others have presented evidence of poor information flow across countries.

With respect to international volatility transmission, Eun and Shim (1989), Hamao et al. (1990), Theodossiou and Lee (1993), Lin et al. (1994), Koutmos and Booth (1995), and Koutmos (1996) have all reported strong spillovers in volatility across global equity markets. Prior research has examined the transmission of volatility based on local, regional, and global spillovers and then expanded the scope of their study to other markets. For example, Koch and Koch (1991) suggest that regional interdependencies have grown over time. Bakaert and Harvey (1997) distinguish between global and local shocks in emerging stock markets, while Ng (2000) identified Japan (the United States) as a regional (global) contributor to world equity market volatility.

The main focus of existing research has been to examine information transmission between markets of the same asset class. However, very little work has investigated information flows, in particular volatility spillovers, between markets of different asset classes. To illustrate why this would be of interest, consider the relationship between debt and equity securities. Equities require accurate information on the cost of capital and interest rates. Likewise, when the stock market produces any abnormal return, investors may rebalance their portfolio to achieve such a return. If there is transmission of information between stock and bond markets, investors should pay attention to their interaction, and it is of interest to understand this dynamic interdependence. Understanding the transmission of volatility between complementary assets such as stocks and bonds allows portfolio investors to diversify their portfolio more effectively. The benefits of diversification for bonds are often overstated, especially within the mean-variance approach (Fang, 2005). Given the possibility of stock and bond information flows, it is incumbent upon investors to pay attention to the pricing and volatility relationships between both markets.

The purpose of this chapter is to investigate the dynamic interdependence of domestic stock and bond markets. Six research questions are considered. First, are there volatility spillovers across domestic stock and bond markets? As discussed above, scant evidence of this cross-market spillover is available. Second, what is the direction of the transmission? Third, do the currency markets affect volatility spillovers? We do this by investigating the sensitivity of the empirical results to the use of alternative currency denominations. When a common currency (the U.S. dollar) is used, it is likely that some of the co-movement observed among returns in different markets is caused by changes in the fundamentals driving the exchange rates of the dollar.[*] Fourth, do the volatility spillover coefficients vary significantly across time? The nature of any linkage over time is one factor bearing on asset allocation decisions. Fifth, are spillovers more prevalent in certain geographical regions? Monetary policy cooperation between national governments in the form of economic agreements may produce interesting dynamics in information spillovers. Lastly, what are the determinants of volatility spillovers? Knowing the explanatory factors of information flow can assist investors in the construction of financial portfolios.

Our study presents an empirical framework for analyzing return and volatility spillovers across equity and bond markets of twelve countries over a 15-year period between January 1990 and December 2004. The empirical analysis employs generalized autoregressive conditional heteroskedastic (GARCH) models.

We make four main empirical contributions to the literature. First, we examine the interaction of volatility spillovers across domestic stock and bond markets. Second, the analysis spans all the major geographical market regions of the world. The advantage of this is that macroeconomic factors contributing to information transmission can be cogently examined. Third, the role of currency fluctuations is investigated to ascertain the effect of foreign exchange on information transmission across markets. Fourth, we assess whether macroeconomic factors influence the strength of information transmission across asset classes.

The results confirm the view that information transmission is important in international equity and bond markets. Volatility spillovers are bidirectional for both equity and debt markets and the results are not sensitive to the currency base. However, in general, we detect that equity markets tend to export volatility to the debt markets (once the covariance between equity and debt market returns is incorporated into the analysis). From an intertemporal perspective, volatility spillover coefficients do not vary over time. While information flow from bonds to stocks does not differ in each region, the flow from stocks to bonds does present some regional differences. Finally, from panel data analysis, the information linkages between the two asset markets are higher in countries with low economic

---

[*] Hamao et al. (1990), Koch and Koch (1992), Lau and Diltz (1994), and Lee et al. (2004) also analyzed the stock market interdependencies in both USD and local currency.

risk (high GDP per capita) and higher financial market integration (high percentage of foreign debt per GDP).

The outline of the chapter is as follows: In Section 15.2 we describe our model of information transmission across asset classes and data. The empirical findings are discussed in Section 15.3, and Section 15.4 concludes.

15.2 MODEL AND DATA

15.2.1 Bivariate GARCH Model

The multivariate model jointly describes the volatility of several time series, and the general model for a k-dimensional process $\varepsilon_t = (\varepsilon_{1t}, \ldots, \varepsilon_{kt})'$ is expressed as $\varepsilon_t = z_t H_t^{1/2}$, where z_t is a k-dimensional i.i.d. process with mean zero and covariance matrix equal to the identity matrix.

To complete the n-dimensional multivariate model, the parameterization for n different mean estimates, n conditional variance equations, and the $\frac{n^2-n}{2}$ conditional covariance matrix needs to be specified. In our case, we focus on the diagonal BEKK model since it guarantees a positive-definite covariance matrix.[*]

$$R_{s,t} = \alpha_{s,0} + \alpha_{s,1} R_{s,t-1} + \alpha_{b,1} R_{b,t-1} + \varepsilon_{s,t} \tag{15.1}$$

$$R_{b,t} = \alpha_{b,0} + \alpha_{b,1} R_{b,t-1} + \alpha_{s,1} R_{s,t-1} + \varepsilon_{b,t} \tag{15.2}$$

$$\sigma_{s,t}^2 = a_s + b_s \varepsilon_{s,t-1}^2 + c_s \sigma_{s,t-1}^2 + d_s \varepsilon_{s,t-1}^2 \tag{15.3}$$

$$\sigma_{b,t}^2 = a_b + b_b \varepsilon_{b,t-1}^2 + c_b \sigma_{b,t-1}^2 + d_s \varepsilon_{s,t-1}^2 \tag{15.4}$$

$$\sigma_{s,b,t} = \rho_{s,b} \sigma_s \sigma_b \tag{15.5}$$

where an s (b) subscript represent stocks (bonds).

15.2.2 Data

The data used in this study are the daily local closing figures of the aggregate stock and DataStream bond indices for twelve countries in four geographical regions. These are North America (Canada and the United States), Western Europe (Austria, France, Ireland, Italy, Spain, Sweden, Portugal,

---

[*] To test the asymmetric effect of shocks, most studies have used a bivariate or multivariate EGARCH model. However, the bivariate EGARCH model with both markets failed to converge, and as a result, we focused on the GJR-GARCH model.

and the United Kingdom), Southeast Asia (Japan), and Africa (Israel). The countries are chosen on the basis of bond and equity data availability.

The stock indices are the TSE60 (Canada), S&P500 Composite (United States), ATE (Austria), CAC40 (France), ISEQ Overall (Ireland), MIB General (Italy), Madrid SE (Spain), OMX Stockholm (Sweden), Lisbon PSI General (Portugal), FTSE-All Share Index (UK), Nikkei 225 (Japan), and Israel General (Israel). All bond indices are generated by DataStream. The sample period ranges from January 2, 1990 to December 31, 2004, for most data, except for the DataStream bond indices of Portugal, of which the availability starts from December 31, 1992.

Table 15.1 presents several statistics for daily domestic returns on the stock indices (Panel A) and bond indices (Panel B), both denominated in the domestic currency. They include the annualized mean, annualized standard deviation, skewness, and kurtosis. For stock indices, the annualized means for all markets, except Japan, are positive and range between 1.07 (Austria) and 15.00 (Ireland). The annualized standard deviations of returns vary between 13.67 (Portugal) and 37.55 (Austria). The skewness and kurtosis indicate that daily stock returns are not normally distributed. Most stock returns present negative skewness. In local currency, only Sweden and Japan show positive skewness. The distribution of returns for all stock markets is leptokurtotic relative to the normal distribution.

For bond indices (Panel B), the annualized means for all markets are positive and range between 0.72 (United States) and 9.10 (Israel). The standard deviations of returns vary between 2.49 (Austria) and 5.14 (UK). Only one country (Sweden) in local currency shows positive skewness, while three countries from Western Europe (Austria, France, and Portugal) and the countries in Asia/Pacific in USD have positive skewness. Like stock indices, all bond indices have a distribution of returns that is leptokurtotic relative to the normal distribution.

15.3 EMPIRICAL ANALYSIS

15.3.1 Joint Estimation of Return and Volatility Spillovers

We first estimate the bivariate VAR-GARCH model for each debt and equity market within a country. The coefficients and test statistics are presented in Table 15.2 (local currency) and Table 15.3 (USD). The principal idea behind this model is to consider the conditional covariance between the stock and bond market, while simultaneously estimating the GARCH model. Also, the recent past return of the other market is included as an independent variable in the conditional mean equation.

TABLE 15.1   Preliminary Statistics Based on Local Currency

Exchanges	No. of Observations	Annualized Mean (%)	Annualized Standard Deviation (%)	Skewness	Kurtosis
		Panel A: Stock Index Returns			
		I. North America			
Canada	3,914	6.08	15.32	−0.62	11.19
United States	3,914	7.88	16.06	−0.10	6.89
		II. Western Europe			
Austria	3,914	1.07	37.55	−16.66	680.06
France	3,914	4.13	21.26	−0.09	5.83
Ireland	3,914	8.03	15.33	−0.33	7.86
Italy	3,914	4.88	19.30	−0.44	6.26
Spain	3,914	7.50	18.89	−0.25	6.61
Sweden	3,914	8.10	23.34	0.20	6.69
Portugal	3,914	5.45	13.67	−0.38	12.94
United Kingdom	3,914	4.43	14.64	−0.14	6.59
		III. Asia/Pacific			
Japan	3,914	−7.80	23.36	0.20	6.35
		IV. Africa/Middle East			
Israel	3,914	15.00	20.96	−0.55	9.49
		Panel B: Bond Index Returns			
		I. North America			
Canada	3,914	1.02	4.99	−0.38	6.30
United States	3,914	0.72	4.43	−0.34	4.89
		II. Western Europe			
Austria	3,914	1.12	2.49	−0.61	10.33
France	3,914	1.38	3.48	−0.22	5.82
Ireland	3,914	2.24	4.20	−0.12	10.83
Italy	3,914	1.96	3.48	−0.54	10.45
Spain	3,914	1.93	3.04	−0.35	7.77
Sweden	3,914	1.50	4.84	0.70	35.75
Portugal	3,131	1.56	2.98	−0.54	12.45
United Kingdom	3,914	1.29	5.14	−0.03	6.59
		III. Asia/Pacific			
Japan	3,914	1.00	3.06	−0.46	6.93
		IV. Africa/Middle East			
Israel	3,914	9.10	3.44	−0.65	33.33

Note: This table reports descriptive statistics for our data set. Panels A and B report the statistics for stock indices and bond indices, respectively. The sample period is January 2, 1990 to December 31, 2004 (except for Portugal's bond index: December 31, 1992 to December 31, 2004).

TABLE 15.2 Applying a Bivariate GARCH Model to Stock and Bond Markets—Local Currency Results

	as	bs	cs	db	ab	bb	cb	ds
I. North America								
Canada	1.01E-03 *** (10.819)	0.201 *** (28.883)	0.978 (710.577)	0.072 ** (2.002)	-1.09E-04 *** (-105.955)	2.256 (173.985)	-0.353 (-360.011)	0.066 *** (114.929)
United States	-6.32E-05 (-0.009)	0.019 (0.094)	1.002 *** (63,780.890)	1.159 *** (24.407)	-1.86E-05 *** (-12.404)	0.066 (427.771)	0.963 *** (5,942.467)	-0.005 *** (-121.503)
II. Western Europe								
Austria	-0.034 *** (-14.069)	0.073 (1.426)	-0.204 (-0.649)	0.720 (0.278)	-1.20E-04 *** (-31.139)	0.255 (50.987)	0.966 *** (905.119)	0.002 *** (3.602)
France	-0.001 *** (-11.954)	0.217 *** (27.618)	0.972 (450.171)	-0.007 (-0.007)	-3.98E-05 *** (-3.887)	0.217 (31.140)	0.975 (725.749)	-0.009 *** (-9.208)
Ireland	0.001 *** (17.503)	0.250 *** (29.549)	0.961 (362.055)	0.003 (0.003)	3.56E-05 *** (3.622)	0.229 (39.605)	0.974 (776.944)	0.019 *** (10.521)
Italy	-0.001 *** (-12.064)	-0.216 *** (-32.868)	-0.972 (-501.155)	0.001 (0.000)	8.49E-05 *** (6.636)	0.261 (28.078)	0.967 *** (442.849)	0.009 *** (11.693)
Spain	-0.001 *** (-19.632)	-0.167 *** (-28.766)	0.980 (725.581)	-0.104 (-0.809)	-3.02E-04 *** (-18.619)	0.508 (44.303)	0.851 (173.970)	0.031 *** (21.743)
Sweden	4.34E-04 *** (7.413)	0.295 *** (36.720)	-0.957 (-414.773)	-0.281 *** (-8.657)	0.001 *** (21.982)	0.296 (27.430)	-0.902 (-116.392)	2.55E-04 *** (0.001)
Portugal	-1.20E-04 *** (-2.578)	0.330 *** (40.681)	0.946 (418.413)	0.444 *** (14.037)	-2.41E-04 *** (-15.989)	0.244 (36.098)	0.962 (399.402)	0.006 *** (6.716)
United Kingdom	0.001 *** (12.337)	0.314 *** (24.530)	-0.937 (-180.615)	0.003 (0.002)	4.97E-04 *** (13.451)	0.192 (25.903)	0.968 *** (364.811)	-0.019 *** (-4.701)

(*Continued*)

TABLE 15.2 Applying a Bivariate GARCH Model to Stock and Bond Markets—Local Currency Results (*Continued*)

	as	bs	cs	db	ab	bb	cb	ds
				III. Asia/Pacific				
Japan	-9.22E-05	0.262 ***	0.965 ***	0.667 ***	1.75E-04 ***	0.250 ***	0.965 ***	1.07E-05 ***
	(-1.143)	(32.831)	(509.722)	(17.826)	(15.811)	(34.136)	(572.226)	(3.64E-6)
				IV. Africa/Middle East				
Israel	0.003 ***	0.282 ***	-0.927 ***	-0.804 ***	2.65E-04 ***	0.297 ***	0.949 ***	0.010 ***
	(23.883)	(33.036)	(-251.181)	(-17.463)	(58.203)	(53.775)	(673.803)	(8.085)

Note: This table presents summary statistics of volatility parameter estimates for the sample period January 2, 1990 to December 31, 2004 (except for Portugal: December 31, 1992 to December 31, 2004), where prices are expressed in local currency terms. The model estimated is given by

$$R_{s,t} = \alpha_{s,0} + \sum_{i=1}^{5} \alpha_{s,i} R_{s,t-i} + \sum_{i=1}^{5} \alpha_{b,i} R_{b,t-i} + \varepsilon_{s,t}$$

$$R_{b,t} = \alpha_{b,0} + \sum_{i=1}^{5} \alpha_{b,i} R_{b,t-i} + \sum_{i=1}^{5} \alpha_{s,i} R_{s,t-i} + \varepsilon$$

$$\sigma_{s,t}^2 = a_s + b_s \varepsilon_{s,t-1}^2 + c_s \sigma_{s,t-1}^2 + d_s \varepsilon_{b,t-1}^2$$

$$\sigma_{b,t}^2 = a_b + b_b \varepsilon_{b,t-1}^2 + c_b \sigma_{b,t-1}^2 + d_s \varepsilon_{s,t-1}^2$$

$$\sigma_{s,b,t} = \rho_{s,b} \sigma_s \sigma_b$$

where $a > 0$, $\Sigma b + \Sigma c < 1$, $\varepsilon_{jt} = z_t h_{jt} \sim$ i.i.d. (0,1), and σ_t^2 is the variance of ε_{jt} conditional on the past information Ω_{t-1}. Maximum likelihood estimation is employed. The t-statistics are shown in parentheses. ***, **, and * indicate statistical significance at the 1%, 5%, and 10% levels.

TABLE 15.3 Applying a Bivariate GARCH Model to Stock and Bond Markets—USD Results

	as		bs		cs		db		ab		bb		cb		ds	
I. North America																
Canada	2.72E-04	***	0.223	***	0.975	***	0.086	***	4.89E-04	***	0.178	***	0.979	***	0.014	***
	(5.703)		(40.680)		(830.778)		(7.363)		(8.472)		(23.651)		(517.998)		(2.897)	
United States	4.76E-04	***	0.222	***	0.974	***	-0.060	***	5.84E-04	***	0.183	***	0.978	***	0.035	***
	(7.428)		(29.241)		(565.189)		(-5.096)		(9.186)		(23.192)		(499.346)		(5.511)	
II. Western Europe																
Austria	0.032	***	0.074		-0.083		0.046		1.04E-03	***	0.190	***	0.970	***	-0.006	**
	(54.175)		(1.301)		(-0.483)		(0.053)		(10.880)		(17.419)		(269.601)		(-2.544)	
France	0.001	***	-0.235	***	-0.963	***	-0.189	***	0.001	***	-0.156	***	-0.984	***	0.023	***
	(7.987)		(-29.422)		(-368.347)		(-13.902)		(11.950)		(-24.164)		(-837.144)		(5.219)	
Ireland	0.001	***	0.236	***	0.963	***	-0.120	***	5.46E-04	***	0.167	***	0.982	***	0.038	***
	(11.533)		(30.136)		(406.087)		(-11.421)		(13.732)		(24.431)		(764.494)		(7.325)	
Italy	0.002	***	0.247	***	0.963	***	0.073	***	4.64E-04	***	0.195	***	0.978	***	-0.031	***
	(10.744)		(38.046)		(473.424)		(2.658)		(11.372)		(34.733)		(849.898)		(-7.898)	
Spain	-0.001	***	0.261	***	0.962	***	-0.005	***	0.000	***	0.130	***	-0.979	***	0.021	***
	(-9.282)		(27.448)		(341.860)		(-0.001)		(-9.269)		(13.944)		(-297.275)		(8.245)	
Sweden	8.46E-04	***	-0.253	***	0.963	***	0.165	***	0.001	***	-0.220	***	0.961	***	-4.78E-02	***
	(12.539)		(-36.311)		(518.201)		(14.732)		(11.523)		(-24.837)		(274.778)		(-9.389)	
Portugal	n.a.		n.a.		n.a.		n.a.		n.a.		n.a.		n.a.		n.a.	
United Kingdom	-0.001	***	-0.254	***	0.964	***	0.151	***	-8.71E-04	***	-0.191	***	0.975	***	0.024	***
	(-8.590)		(-33.175)		(421.925)		(12.441)		(-10.685)		(-24.953)		(463.536)		(4.138)	

(Continued)

TABLE 15.3 Applying a Bivariate GARCH Model to Stock and Bond Markets—USD Results (*Continued*)

	as	bs	cs	db	ab	bb	cb	ds
III. Asia/Pacific								
Japan	2.51E-03 ***	0.273 ***	0.948	0.169 ***	5.29E-04 ***	0.164 ***	0.981 ***	0.035 ***
	(13.940)	(27.103)	(248.307)	(6.976)	(12.463)	(28.976)	(992.733)	(17.510)
IV. Africa/Middle East								
Israel	0.002 ***	0.259 ***	0.948	0.170 ***	2.56E-04 ***	0.271 ***	0.959 ***	0.042 ***
	(21.246)	(33.326)	(316.511)	(6.703)	(15.047)	(45.662)	(918.658)	(23.395)

Note: This table presents summary statistics of volatility parameter estimates for the sample period January 2, 1990 to December 31, 2004 (except for Portugal: December 31, 1992 to December 31, 2004), where prices are expressed in USD terms. The model estimated is given by

$$R_{s,t} = \alpha_{s,0} + \sum_{i=1}^{5} \alpha_{s,i} R_{s,t-i} + \sum_{i=1}^{5} \alpha_{b,i} R_{b,t-i} + \varepsilon_{s,t}$$

$$R_{b,t} = \alpha_{b,0} + \sum_{i=1}^{5} \alpha_{b,i} R_{b,t-i} + \sum_{i=1}^{5} \alpha_{s,i} R_{s,t-i} + \varepsilon_{b,t}$$

$$\sigma_{s,t}^2 = a_s + b_s \varepsilon_{s,t-i}^2 + c_s \sigma_{s,t-1}^2 + d_b \varepsilon_{b,t-1}^2$$

$$\sigma_{b,t}^2 = a_b + b_b \varepsilon_{b,t-1}^2 + c_b \sigma_{b,t-1}^2 + d_s \varepsilon_{s,t-1}^2$$

$$\sigma_{s,b,t} = \rho_{s,b} \sigma_s \sigma_b$$

where $a > 0$, $\Sigma b + \Sigma c < 1$, $\varepsilon_{jt} = z_j h_{jt}$, ~ i.i.d. (0,1), and σ_t^2 is the variance of ε_{jt} conditional on the past information Ω_{t-1}. Maximum likelihood estimation is employed. The t-statistics are shown in parentheses. ***, **, and * indicate statistical significance at the 1%, 5%, and 10% levels, respectively.

The results are striking. Almost all return and volatility spillover coefficients are statistically significant. Volatility spillovers from stock to bond markets are found in almost every country (exceptions are Sweden and Japan) in local currency and all countries in USD. The reciprocal transmission of information from bond to stock markets is not so strong, with only six countries experiencing transmission in this direction in local currency and nine countries in USD.

Overall, the results indicate that almost every stock market exports its volatility to the bond market. In sum, when the interaction between stock and bond market is taken into account, recent return and volatility innovations in both markets have a significant impact on its current volatility.

15.3.2   Are Volatility Spillovers Time Varying?

Volatility spillovers may change over time. To test for time-varying characteristics of volatility spillover, the bivariate GARCH model is run for each month and each year, and the coefficients of volatility spillover are analyzed using analysis of variance (ANOVA). The hypothesis is that the means of volatility spillover coefficients in each month/year are equal.

The results are reported in Panel A of Table 15.4. From the table it can be seen that none of the tests show significant differences—either across months or across years. In other words, the volatility spillover of each month and each year does not vary significantly over time.

15.3.3   Are There Regional Effects in Volatility Spillovers?

The degree of stock and bond market development in each region is different. As a result, the level of information transmission may differ also. In other words, the volatility spillover may differ according to the location of market. The z-test of ANOVA will be used to test whether the means of volatility spillover coefficients differ between different regions. To perform this test, the volatility spillover coefficients are obtained from the bivariate GARCH models, which are estimated from a sample of every month. The hypothesis is that all of the means of the volatility spillover coefficients in each region are equal. Generally, we find that from bonds to stocks, there is no difference in means among the four regions. Likewise, in general, volatility spillovers from stocks to bonds do not differ across regions.

Panel B of Table 15.4 reports tests of pairwise differences in volatility spillover coefficients between regions. When the pairwise analysis of two regions is applied, the means of volatility spillover coefficients from bond to stock markets are not significantly different. However, the pairwise

TABLE 15.4   Tests for the Difference of Mean of Volatility Spillovers

Panel A: Month/Year Results

	Month		Year	
	From Bonds to Stocks	**From Stocks to Bonds**	**From Bonds to Stocks**	**From Stocks to Bonds**
Austria	0.979	1.252	1.469	0.481
Canada	0.998	1.150	1.001	1.415
France	1.025	1.212	1.017	0.266
Ireland	0.953	0.782	0.896	0.868
Israel	0.998	0.325	1.069	0.263
Italy	0.962	0.725	0.982	0.983
Japan	1.012	1.061	0.975	0.162
Portugal	0.997	0.832	0.719	0.667
Spain	1.046	0.996	0.957	0.935
Sweden	1.486	0.888	1.184	0.795
United Kingdom	1.017	1.391	0.992	1.183
United States	0.983	0.878	0.983	1.733

Panel B: Region Results

	North America	**Western Europe**	**Asia/Pacific**	**Among Four Regions**
		From Bonds to Stocks		
Western Europe	1.105			
Asia/Pacific	1.114	0.130		
Africa/Middle East	1.094	−0.305	−0.284	
Among four regions				1.948
		From Stocks to Bonds		
Western Europe	−0.631			
Asia/Pacific	0.988	2.298 **		
Africa/Middle East	−0.026	0.970	−2.465 ***	
Among four regions				0.400

Note: This table presents summary statistics of the difference in the mean of volatility spill-over coefficient estimates of different months and years (Panel A) and of different regions (Panel B), based on ANOVA analysis. The sample period is January 2, 1990 to December 31, 2004 (except for Portugal: December 31, 1992 to December 31, 2004). The volatility spillover parameters are estimated from bivariate GARCH within each month:

$$H_0: \mu_1 = \mu_2 = \cdots \mu_i, H_a:$$

Not all population means of volatility spillover coefficients in each month/year/region are equal.

*, **, *** indicate the rejection of the null hypothesis at the 10, 5, and 1% levels of significance, respectively.

analysis of volatility spillover mean coefficients from stock to bond markets shows a difference in two cases: Western Europe versus Asia/Pacific, and Asia/Pacific versus Africa/Middle East. Specifically, while the mean spillover coefficients from stock to bonds of any region cannot be rejected to differ from those of North America, the means of both Western Europe and Africa/Middle East are significantly higher than that of Asia/Pacific.

In sum, the information transmission from bond to stock markets is likely the same in each pair of regions, whereas the transmission from stock to bond markets is statistically similar to that of North America. The mean of spillover coefficients in the Asia/Pacific group from stock to bond markets is the lowest. In the Asia/Pacific group, Japan is the largest in terms of market capitalization and the most active.

15.3.4   What Are the Determinants of Volatility Spillovers?

It is well known that individual asset prices are influenced by a wide variety of unanticipated events and that some events have a more pervasive effect on asset prices than do others (Chen et al., 1986). Importantly, asset prices are commonly believed to react sensitively to the arrival of economic news. The theory of efficient markets and rational expected intertemporal asset pricing suggests that asset prices should depend on their exposures to the state variables describing the economy (Merton, 1974; Cox et al., 1985; Ross, 1976).

Accordingly, in this section the volatility spillover coefficients from the bivariate GARCH model will be related to a set of economic and financial variables that may influence the spillover effect. In prior literature (for example, Chen et al., 1986), several economic variables are found to be significant in explaining expected stock returns when the market is highly volatile. In general, stock markets seem less affected by macroeconomic news than bond markets (McQueen and Roley, 1993). Nevertheless, should the volatility spillover from bond to stock market be significant, it is worth investigating the relevant economic variables influencing the information linkage between those markets.

A number of economic variables are considered. First is the ratio of equity market capitalization to GDP (MCAPG). The MCAPG ratio is often used as proxy for equity market development (Baele, 2005). Theory does not provide either a unique concept or a common measure of stock market development to guide empirical research (Demirguc-Kunt and Levine, 1995). The characteristics of stock market development may be related to size, activity, or integration. In this study, the size of stock market is used because of the availability of data comparison with earlier work. The assumption is that the size of the stock market is positively correlated with the ability to mobilize capital and diversify risk. Information is likely to flow more actively in developed financial markets, which are, on average, more liquid, diversified, and better integrated with each other (Bekaert and Harvey, 1995; Ng, 2000). Specifically, in larger stock markets, a recently informed investor will find it easier to trade at quoted prices (Grossman and Stiglitz, 1980; Kyle, 1984; Holmstrom and Tirole, 1994). Therefore, equity market development is expected to be positively related to volatility spillover.

Second is the ratio of current account to GDP (CAG). The CAG ratio is normally viewed as a proxy for economic integration (Rivera and Romer, 1990). Economic integration means increasing not only trade, but also the flow of ideas between two different economies. Countries with heavier bilateral trade with a region also tend to have higher return correlations with that region (Chen and Zhang, 1997) and are generally better integrated with world capital markets overall (Bakaert and Harvey, 1995). The more economies are linked, the more they will be exposed to common shocks, and the more companies' cash flows will be correlated. In other words, the movement of capital due to market integration may alter the information flow between asset classes in the capital market. CAG is thus expected to be positively related to the strength of volatility spillovers.

The remainder of the variables are employed by rating agencies to set country risk ratings. Country risk reflects the ability and willingness of a country to service its foreign financial obligations. All essential features of country risk are a function of a number of interrelated and dynamic structural factors (Carment, 2001), and they may be prompted by country-specific and regional economic, financial, political, and composite factors (Hoti, 2005).

The factors include GDP per head of population, GDPC (a measure of the country's productivity); the percentage of budget balance per GDP, BUDG; the percentage of foreign debt per GDP, FDG (gross foreign debt in a given year as a percentage of the gross domestic product); and international liquidity, LIQ (total official reserves for a given year divided by the average monthly merchandise import cost). The GDP per head of population and budget balance per GDP reflect the economic risk components, whereas foreign debt per GDP and net liquidity proxy for financial risk components.

In summary, the panel data equation that we estimate is given by

$$VSpill_{i,t} = a + b_{i,t}GDPC + c_{i,t}MCAPG + d_{i,t}CAG + e_{i,t}BUDG$$
$$+ f_{i,t}FDG + g_{i,t}LIQ + \varepsilon_{i,t} \tag{15.6}$$

The hypothesis is that there is a positive relationship between tentative explanatory variable and the strength of volatility spillovers between bond and stock markets. The results are reported in Table 15.5.

TABLE 15.5   A Test of the Determinants of Volatility Spillovers

	From Bonds to Stocks		**From Stocks to Bonds**	
GDPC	0.365	**	0.271	*
	(2.41)		(1.8)	
MCAPG	0.002		0.001	
	(1.21)		(0.98)	
CAG	0.034		0.020	
	(0.92)		(0.67)	
BUDG	0.038	*	0.027	
	(2.05)		(1.46)	
FDG	0.009	***	0.010	**
	(3.27)		(2.7)	
LIQ	−0.009		−0.010	
	(−0.97)		(−0.94)	
Constant	−0.004	**	−3.020	*
	(−2.54)		(−1.9)	
F-statistic	2.30	*	2.50	*

Note: This table reports estimation results for a model of the potential determinants of volatility spillover between stock and bond markets:

$$VSpill_{i,t} = a + b_{i,t}GDPC + c_{i,t}MCAPG + d_{i,t}CAG + e_{i,t}BUDG + f_{i,t}FDG + g_{i,t}LIQ + \varepsilon_{i,t}$$

The volatility spillover parameters (VSpill) are estimated from bivariate GARCH within each year. The independent variables are GDP per capita (GDPC), equity market capitalization per GDP (MCAPG), the percentage of current account per GDP (CAG), the percentage of budget balance per GDP (BUDG), the percentage of foreign debt per GDP (FDG), and international liquidity (LIQ) (the official reserves of the individual countries including the official gold reserves calculated at current free market prices but excluding the use of IMF credits and the foreign liabilities of the monetary authorities). The source of these data is the international country risk guide. The sample period is January 2, 1990 to December 31, 2004 (except for Portugal: December 31, 1992 to December 31, 2004). The t-statistics are shown in parentheses. ***, **, and * indicate the statistical significance at the 1%, 5%, and 10% levels, respectively.

Interestingly, the surrogates of equity market development (equity market capitalization to GDP) and the indicator of market integration (current account to GDP) are not statistically significant in both directions. The percentage change in GDP per capita and the percentage of foreign debt per GDP are common factors for the information transmission between stock and bond market. Additionally, volatility spillovers from bonds to stocks are also determined by the percentage of budget balance per GDP. All the significant signs of coefficients are positive, as expected.

15.4   CONCLUSION

In this study, we have tested for volatility spillovers between debt and equity markets within twelve countries: Canada, the United States, Austria, France, Ireland, Italy, Spain, Sweden, Portugal, the United Kingdom, Japan, and Israel. The test covers the period from January 2, 1990 to December 31, 2004.

The dynamic transmission of volatility between debt and equity markets exists in most countries in our sample. Evidence on the direction of transmission suggests that information flows more readily from stock to bond markets. Spillovers between equities and debt take place throughout the world, and market location is not particularly important. Instead, country risk and development may be more important in predicting the extent of volatility spillovers between asset classes. In conclusion, investors should consider volatility between stock and bond markets in addition to common factors in returns when forming financial portfolios.

ACKNOWLEDGMENTS

We thank Rob Hudson, Iain Clacher, Kevin Keasey, and Andrew Marshall for their valuable comments. All errors are our own.

REFERENCES

Agmon, T. (1972). The relations among equity markets in the United States, United Kingdom, Germany, and Japan. *Journal of Finance* 27:839–55.

Bekaert, G., and Harvey, C. R. (1995). Time-varying world market integration. *Journal of Finance* 50:403–44.

Bekaert, G., and Harvey, C. R. (1997). Emerging equity market volatility. *Journal of Financial Economics* 43:29–77.

Chen, N. F., Roll, R., and Ross, S. A. (1986). Economic forces and the stock market. *Journal of Business* 59:383–403.

Chen N. and Zhang F. (1997). Correlations, trades and stock returns of the Pacific-Basin markets. *Pacific-Basin Finance Journal* 5:559–577.

Cox, J., Ingersoll, J., and Ross, S. (1985). An intertemporal general equilibrium model of asset prices. *Econometrica* 53:363–84.

Demirguc-Kunt, A., and Levine, R. (1995). Stock market development and financial intermediaries. Working paper, World Bank, Washington.

Eun, C. S., and Shim, S. (1989). International transmission of stock market movements. *Journal of Financial and Quantitative Analysis* 24:241–56.

Fang, M. (2005). Bond return, spread change, and the momentum effect in corporate bond and stock markets. Working paper, Yale School of Management, New Haven, CT.

Granger, C., and Morgenstern, O. (1970). *Predictability of stock market prices.* Lexington, MA: Heath and Co.

Grossman, S., and Stiglitz, J. (1980). On the impossibility of informationally efficient markets. *American Economic Review* 70:393–408.

Hamao, Y., Masulis, R., and Ng, V. (1990). Correlations in price changes and volatility across international stock markets. *Review of Financial Studies* 3:281–308.

Hilliard, J. E. (1979). The relationship between equity indices on world exchanges. *Journal of Finance* 34:103–14.

Holmstrom, B., and Tirole, J. (1993). Market liquidity and performance monitoring. *Journal of Political Economy* 101:678–709.

Hoti, S. (2005). Modelling country spillover effects in country risk ratings. *Emerging Market Reviews* 6:324–45.

Koch, P. D., and Koch, T. W. (1991). Evolution in dynamic linkages across daily national stock indices. *Journal of Multinational Financial Management* 10:231–51.

Koutmos, G. (1996). Modeling the dynamic interdependence of major European stock markets. *Journal of Business Finance and Accounting* 23:975–88.

Koutmos, G., and Booth, G. G. (1995). Asymmetric volatility transmission in international stock markets. *Journal of International Money and Finance* 14:747–62.

Kyle, A. S. (1984). Market structure, information, future markets, and price formation in international agricultural trade. In *Advanced readings, price formation, market structure, and price instability,* ed. G. S. Gary, A. Schmitz, and A. H. Sarris. Boulder, CO and London: Westview Press 45–64.

Lau, S. T., and Diltz, J. D. (1994). Stock return and the transfer of information between the New York and Tokyo stock exchanges. *Journal of International Money and Finance* 13:211–22.

Lee, B., Rui, O. M., and Wang, S. S. (2004). Information transmission between NASDAQ and Asian second board markets. *Journal of Banking and Finance* 28:1637–70.

Lin, W. L., Engle, R. F., and Ito, T. (1994). Do bulls and bears move across borders? International transmission of stock returns and volatility. *Review of Financial Studies* 7:507–38.

McQueen, G., and Roley, V. (1993). Stock prices, news, and business conditions. *Review of Financial Studies* 6:683–707.

Merton, R. C. (1974). On the pricing of corporate debt: The risk structure of interest rates. *Journal of Finance* 29:449–70.

Ng, A. (2000). Volatility spillover effects from Japan and the US to the Pacific-Basin. *Journal of International Money and Finance* 19:207–33.

Panton, D., Lessig, V. P., and Joy, O. M. (1976). Comovement of international equity markets: A taxonomic approach. *Journal of Financial and Quantitative Analysis* 3:415–32.

Ripley, D. (1973). Systematic elements in the linkage of national stock market indices. *Review of Economics and Statistics* 3:356–61.

Rivera, L. A., and Romer, P. M. (1991). Economic integration and endogenous growth. *Quarterly Journal of Economics* 106:531–55.

Robichek, A. A., Cohn, R. A., and Pringle, J. J. (1972). Returns on alternative investment media and implications for portfolio construction. *Journal of Business* 3:427–43.

Ross, S. A. (1976). The arbitrage theory of capital asset pricing. *Journal of Economic Theory* 13:341–60.

Theodossiou, P., and Lee, U. (1993). Mean and volatility spillovers across major national stock markets: Further empirical evidence. *Journal of Financial Research* 16:337–50.

III

Developed Country Volatility

CHAPTER 16

Predictability of Risk Measures in International Stock Markets

Turan G. Bali and K. Ozgur Demirtas

CONTENTS

16.1 INTRODUCTION

Predictability of risk during normal and highly volatile periods of the stock market has important implications in both asset pricing and risk management. Although there is some work regarding the persistency of variance estimates in the U.S. stock market (see, e.g., Bali et al., 2007), there is lack of evidence in international stock markets. Furthermore, modern risk management requires a solid understanding of higher-order moments and the tails of empirical return distributions. Therefore, persistency of tail risk measures such as value-at-risk (VaR) needs to be examined.

This chapter investigates the predictability of variance and VaR in international stock markets. We use daily stock index returns for G7 countries (the United States, the United Kingdom, Germany, Japan, Canada, France, and Italy) and generate the realized variance and VaR estimates. We then compute the proportion of the 1-month-ahead variance and VaR measures that can be explained by the lagged variance and VaR obtained from the past 1 to 6 months of daily data to determine the predictability of these risk parameters.

We find that for all G7 countries considered in the chapter, persistency in variance is significantly higher than the persistency in VaR. Variance of the stock market indices for Germany and Italy has the highest persistence, whereas the persistence is low for the United States and Canada. However, different than the case of variance, the strongest predictability of VaR is obtained for Japan. We conclude that although the second moment of the stock return distribution is highly predictable for Germany and Italy, the left tail of the return distribution is more persistent for Japan.

Specifically, the U.S. variance estimates computed using the past 1 to 6 months of daily data explain between 4.5% and 6% of the monthly realized variance. This ratio is much higher for Germany and Italy: between 16.8% and 25.1% of the monthly variance can be explained by the independent variables for Germany, and the corresponding figures are similar for Italy. We conclude that autocorrelations of the second-order moment of return distributions show some similarity according to their proximity in geographic locations. These autocorrelation patterns provide striking resemblance for the United States and Canada (low), and they are also similar within European countries (high). Overall, the result provides support for the integration and significant linkages among stock market volatility of industrialized countries.

Although variance is the most commonly used risk parameter, nonnormality of stock returns all over the world and the nonquadratic utility preferences of market participants make the higher-order moments of the return distributions an essential part of the asset allocation and risk management decisions. Hence, we repeat our analysis for the value-at-risk and find that the predictability of VaR in general is lower, which gives us the expected result that the extreme events are harder to predict. Moreover, Japan stands out as the country with the strongest predictability of tail risk.

Finally, although not reported in the chapter, findings of predictability in variance and VaR are similar when we control for the variables associated with the business cycles (such as term and default premium, aggregate dividend yield, and stochastically detrended riskless rate).

The chapter is organized as follows. Section 16.2 describes the methodology used to obtain the volatility and VaR estimates. Section 16.3 provides the summary statistics of the data. Section 16.4 presents the empirical results, and Section 16.5 concludes the chapter.

16.2 METHODOLOGY

We follow French et al. (1987), Schwert (1989), Goyal and Santa-Clara (2003), and Bali et al. (2005) when we compute the monthly variance of a market index:

$$\sigma_{i,t}^2 = \sum_{s=1}^{S_t} r_{i,s}^2 + 2 \sum_{s=1}^{S_t} r_{i,s} \cdot r_{i,s-1} \tag{16.1}$$

Specifically, we use within-month daily return data to compute the monthly variance of the index returns for country i, denoted by $\sigma_{i,t}^2$. S_t is the number of trading days in month t, and $r_{i,s}$ is the portfolio's return on day s. The second term on the right-hand side adjusts for the autocorrelation in daily returns using the approach of French et al. (1987). Note that the realized variance measure given in Equation (16.1) is not, strictly speaking, a variance measure since daily returns are not demeaned before taking the expectation. However, as pointed out by French et al. (1987) and Goyal and Santa-Clara (2003), the impact of subtracting the means is trivial for short holding periods.

We use nonparametric VaR as a measure of tail risk. VaR determines "how much the value of a portfolio could decline over a given period of time with a given probability as a result of changes in market rates. For example, if the given period of time is one day and the given probability is 1%, the VaR measure would be an estimate of the decline in the portfolio value that could occur with a 1% probability over the next trading day" (Hendricks, 1996).

We compute nonparametric VaR as the lowest daily return observed in a certain period; hence we use different probability levels. For example, each month is assumed to have 21 trading days, which implies about 5% VaR (5% = 1/21) when daily returns over the past 1 month are used to measure downside risk, about 2.5% VaR (2.5% = 1/42) when daily returns over the past 2 months are used to measure downside risk, and about 1.5% VaR (1.5% = 1/63) when daily returns over the past 3 months are used to measure downside risk, etc.

After we obtain the realized variance and VaR measures as just described, we run the following regressions:

$$\sigma_{i,t}^2 = \alpha + \beta\sigma_{i,t-n}^2 + \zeta$$
$$VaR_{i,t} = \alpha' + \beta'VaR_{i,t-n} + \zeta' \tag{16.2}$$

where n changes from 1 to 6, and β and β' measure the persistence in variance and VaR estimates, respectively.

16.3   DATA

We obtain daily stock index return data for G7 countries: the United States, the United Kingdom, Germany, Japan, Canada, France, and Italy. For comparison we use the same sample period for all countries: from January 1973 through February 2005.

Table 16.1 reports the descriptive statistics for daily and monthly stock returns. As explained above, daily returns are directly used to compute the risk parameters. Although monthly returns are not directly used in the chapter, we report the corresponding statistics to observe the nonnormality in different frequencies. We report the mean, median, 25th percentile, 75th percentile, standard deviation, skewness, and kurtosis statistics for the daily and monthly returns.

Both daily and monthly returns for all countries show excess kurtosis, and although monthly kurtosis levels are more subtle (due to time diversification), they are still above their normal values. Furthermore, return distributions are generally skewed to the left. These results once more show that empirical return distributions are far from normal. This finding, combined with the existence of loss-averse investors, indicates the importance of tail risk in asset pricing. Next, we discuss the empirical findings.

16.4   EMPIRICAL RESULTS

We first examine the significance of persistence in realized variance estimates by regressing the 1-month-ahead variance on the lagged variance estimates for all G7 countries. Table 16.2 reports the parameter estimates from these regressions. In each panel, n denotes the number of months used to compute the realized variance estimates. Each month is assumed

TABLE 16.1   Descriptive Statistics

| | **Panel A: Daily Returns** | | | | | | |
	Mean	**Median**	**25%**	**75%**	**Standard Deviation**	**Skewness**	**Kurtosis**
United States	0.332	0.114	−4.567	5.263	9.874	−0.802	22.641
UK	0.375	0.296	−5.199	5.999	10.037	−0.080	10.818
Germany	0.256	0.069	−4.429	5.429	9.925	−0.505	10.283
Japan	0.204	0.000	−4.329	4.789	10.215	−0.145	14.213
Canada	0.320	0.275	−3.520	4.475	8.186	−0.592	16.121
France	0.410	0.053	−5.347	6.532	11.415	−0.233	7.567
Italy	0.468	0.000	−5.183	7.204	13.284	−0.145	7.535
	Panel B: Monthly Returns						
	Mean	**Median**	**25%**	**75%**	**Standard Deviation**	**Skewness**	**Kurtosis**
United States	7.180	9.288	−18.939	38.203	45.125	−0.346	4.787
UK	8.713	11.207	−18.657	38.506	59.690	1.280	17.690
Germany	5.860	8.106	−22.921	38.025	51.857	−0.503	5.121
Japan	4.593	4.467	−24.439	34.664	50.944	−0.013	4.312
Canada	7.289	8.654	−18.595	35.204	45.625	−0.544	5.560
France	9.396	12.152	−31.820	48.207	61.374	−0.101	4.026
Italy	11.863	3.980	−35.613	54.048	72.975	−0.596	4.649

Note: This table reports the summary statistics for daily and monthly stock index returns for all G7 countries considered in the chapter. We report the mean, median, 25th percentile, 75th percentile, standard deviation, skewness, and kurtosis statistics for the daily and monthly returns. The countries considered are the United States, the United Kingdom, Germany, Japan, Canada, France, and Italy. Mean, median, 25th percentile, 75th percentile, and standard deviation are multiplied by a thousand before presentation.

to have 21 days. For each of the seven countries, we report the coefficient estimates and the Newey-West (1987) adjusted t-statistics. To correct for the autocorrelation and heteroskedasticity in standard errors, we use the number of overlapping periods plus 1 as the optimal lag in Newey-West estimation (further robustness checks indicate that our conclusions do not change when different lags are used in Newey-West estimation). For each estimation we also report the R-square and the number of observations.

For the United States, a small proportion of the monthly variance can be explained by the lagged variance estimates. Specifically, the R-squares range from 4.5% to 6.01%. The Newey-West adjusted t-statistics are high, and except for the 1-month horizon, the coefficient estimates are

TABLE 16.2   Predictability of Variance

	United States					
n	1	2	3	4	5	6
Coefficient	0.212	0.155	0.120	0.089	0.074	0.069
t-statistic	(2.07)	(2.69)	(3.25)	(3.07)	(3.11)	(3.38)
R-square	4.5	5.84	6.01	4.82	4.57	4.96
No. of observations	385	384	383	382	381	380

	UK					
n	1	2	3	4	5	6
Coefficient	0.472	0.272	0.189	0.143	0.117	0.097
t-statistic	(5.70)	(5.97)	(5.55)	(4.14)	(3.55)	(3.15)
R-square	22.3	23.47	22.02	19.45	18.23	16.73
No. of observations	385	384	383	382	381	380

	Germany					
n	1	2	3	4	5	6
Coefficient	0.502	0.261	0.180	0.141	0.116	0.102
t-statistic	(8.01)	(6.03)	(5.84)	(6.58)	(7.13)	(7.28)
R-square	25.14	20.93	18.57	17.51	16.79	17.26
No. of observations	385	384	383	382	381	380

	Japan					
n	1	2	3	4	5	6
Coefficient	0.405	0.224	0.168	0.136	0.118	0.099
t-statistic	(5.49)	(6.36)	(7.55)	(7.45)	(6.84)	(5.94)
R-square	16.41	14.75	15.28	15.57	16.59	15.60
No. of observations	385	384	383	382	381	380

	Canada					
n	1	2	3	4	5	6
Coefficient	0.274	0.190	0.141	0.107	0.093	0.085
t-statistic	(4.62)	(4.09)	(3.95)	(3.60)	(3.38)	(3.40)
R-square	7.52	8.77	8.36	7.14	7.35	7.78
No. of observations	385	384	383	382	381	380

	France					
n	1	2	3	4	5	6
Coefficient	0.387	0.228	0.148	0.112	0.089	0.075
t-statistic	(7.15)	(9.43)	(7.44)	(7.24)	(6.16)	(5.85)
R-square	14.99	14.67	11.38	9.79	8.38	7.68
No. of observations	385	384	383	382	381	380

TABLE 16.2   Predictability of Variance (*Continued*)

	Italy					
n	1	2	3	4	5	6
Coefficient	**0.516**	**0.261**	**0.170**	**0.121**	**0.097**	**0.083**
t-statistic	(3.61)	(4.35)	(4.41)	(3.92)	(3.19)	(2.77)
R-square	**26.58**	**20.28**	**15.98**	**12.31**	**10.67**	**10.01**
No. of observations	385	384	383	382	381	380

Note: This table reports the parameter estimates from the regressions of 1-month-ahead realized variance on the lagged variance estimates. As described in Section 16.2, realized variance in a certain period is computed by using the daily returns in that period. *n* denotes the number of months used to compute the variance estimates. Each month is assumed to have 21 days. For each of the G7 countries, we report the coefficient estimates and the Newey-West (1987) adjusted t-statistics. To correct for the autocorrelation and heteroskedasticity in standard errors, we use the number of overlapping periods plus 1 as the optimal lag in Newey-West estimation. For each estimation, we also report the *R*-squares and the number of observations.

significantly different from zero with very small *p*-values. Hence, for the United States, we conclude that the realized variance shows statistically significant persistence; however, the economic significance of this persistence is rather low. A similar pattern is observed for Canada. This finding, combined with our later discussions, points to a geographical tie.

In the case of Germany and Italy, predictability of variance is significant both statistically and economically. For example, in the case of Italy, the R-squares range from 10% (for the 6-month horizon) to 26.58% (for the 1-month horizon). This means that past months' variance explains 26% of the variance of the current month's variance. Thus, we conclude that there is a significant persistence in variance across all G7 countries considered in the chapter. However, this persistence is stronger mainly in countries that are not located in the North American continent.

As discussed earlier, (1) the existence of loss-averse investors who feel a greater pain from losses than the utility they obtain from gains of the same magnitude and (2) nonnormal asset distributions make the prediction of tails risk an essential part of asset allocation and risk management. Therefore, we examine the predictability of VaR for G7 countries as well.

As shown in Equation (16.2), Table 16.3 presents the parameter estimates from the regressions of 1-month-ahead VaR on the lagged tail risk measures. Similar to Table 16.2, in each panel, *n* denotes the number of months used to compute the VaR estimates.

TABLE 16.3   Predictability of Value at Risk

	United States					
n	1	2	3	4	5	6
Coefficient	0.224	0.175	0.193	0.149	0.127	0.130
t-stat	(2.07)	(2.69)	(3.25)	(3.07)	(3.11)	(3.38)
R-square	4.79	5.04	6.78	5.96	5.19	5.74
No. of observations	385	384	383	382	381	380
	UK					
n	1	2	3	4	5	6
Coefficient	0.383	0.341	0.287	0.254	0.223	0.199
t-stat	(5.79)	(5.26)	(3.70)	(3.39)	(3.21)	(2.73)
R-square	8.12	8.32	7.63	7.21	6.67	6.24
No. of observations	385	384	383	382	381	380
	Germany					
n	1	2	3	4	5	6
Coefficient	0.290	0.233	0.192	0.182	0.150	0.136
t-stat	(3.21)	(2.74)	(2.59)	(2.54)	(2.17)	(2.06)
R-square	8.40	7.56	6.19	6.41	4.88	4.46
No. of observations	385	384	383	382	381	380
	Japan					
n	1	2	3	4	5	6
Coefficient	0.393	0.312	0.247	0.196	0.183	0.163
t-stat	(5.04)	(5.02)	(4.24)	(3.36)	(2.84)	(2.42)
R-Square	15.41	13.24	10.35	7.62	7.52	6.47
No. of observations	385	384	383	382	381	380
	Canada					
n	1	2	3	4	5	6
Coefficient	0.273	0.223	0.215	0.186	0.175	0.175
t-stat	(6.63)	(5.99)	(7.47)	(5.92)	(5.31)	(4.56)
R-square	7.44	7.17	8.35	7.35	7.21	8.03
No. of observations	385	384	383	382	381	380
	France					
n	1	2	3	4	5	6
Coefficient	0.347	0.232	0.192	0.175	0.158	0.135
t-stat	(5.53)	(3.56)	(3.49)	(3.45)	(2.97)	(2.53)
R-square	12.02	6.63	5.28	4.88	4.33	3.36
No. of observations	385	384	383	382	381	380

TABLE 16.3   Predictability of Value at Risk (*Continued*)

	Italy					
n	1	2	3	4	5	6
Coefficient	**0.295**	**0.218**	**0.178**	**0.127**	**0.087**	**0.096**
t-stat	(3.61)	(4.35)	(4.41)	(3.92)	(3.19)	(2.77)
R-square	**8.71**	**6.04**	**4.55**	**2.56**	**1.27**	**1.62**
No. of observations	385	384	383	382	381	380

Note: This table reports the parameter estimates from the regressions of 1-month-ahead realized VaR on the lagged VaR estimates. As described in Section 16.2, realized VaR in a certain period is computed by using the daily returns in that period. *n* denotes the number of months used to compute the VaR estimates. Each month is assumed to have 21 days. For each of the G7 countries, we report the coefficient estimates and the Newey-West (1987) adjusted t-statistics. To correct for the autocorrelation and heteroskedasticity in standard errors, we use the number of overlapping periods plus 1 as the optimal lag in Newey-West estimation. For each estimation, we also report the *R*-squares and the number of observations.

Our first observation is that the persistence in VaR is lower than the persistence in variance across all countries. Hence, as expected, extreme events (or higher-order moments) are harder to predict.

Similar to the variance analysis, the R-squares are lower for the United States: they range from 4.79% to 6.78%. However, in contrast to the variance analysis, Japan stands out in terms of the persistency in tail risk. As shown in Table 16.3, in the case of Japan, the R-squares range from 6.47% to 15.41%. Also, as opposed to Table 16.2, all the parameter estimates are not statistically significant. For example, nonparametric VaR that is estimated using the past 5 to 6 months of daily data cannot significantly forecast the monthly future VaRs.

Finally, although not reported in the chapter, we repeat our analysis in a multivariate setting by using control variables that are related to the U.S. business cycle. After controlling for the term premium, default premium, aggregate dividend yield, and the stochastically detrended riskless rate, we find that our conclusions do not change.

16.5   CONCLUSION

We investigate the predictability of variance and value-at-risk in international stock markets. Monthly variance and nonparametric VaR obtained from the daily stock index returns of the United States, the United Kingdom,

Germany, Japan, Canada, France, and Italy are used in our estimations. We find that for all G7 countries considered in the chapter, persistency in variance is significantly higher than the persistency in VaR. For the United States, we conclude that the realized variance shows statistically significant persistence; however, the economic significance of its persistence is rather low, whereas the variance of the stock market indices for Germany and Italy has the highest persistence, which is both statistically and economically significant. Finally, in contrast to the variance analysis, the strongest predictability of VaR is obtained for Japan. We conclude that although the second moment of the stock return distributions is highly predictable for Germany and Italy, the tails of the distribution are more persistent for Japan.

REFERENCES

Bali, T. G., Cakici, N., Yan, X., and Zhang, Z. (2005). Does idiosyncratic risk really matter? *Journal of Finance* 60:905–29.

Bali, T. G., Demirtas, K. O., and Levy, H. (2007). Is there a relation between downside risk and expected returns? *Journal of Financial and Quantitative Analysis*, forthcoming.

French, K. R., Schwert, G. W., and Stambaugh, R. F. (1987). Expected stock returns and volatility. *Journal of Financial Economics* 19:3–29.

Goyal, A., and Santa-Clara, P. (2003). Idiosyncratic risk matters! *Journal of Finance* 58:975–1008.

Hendricks, D. (1996). Evaluation of value-at-risk models using historical data. *FRBNY Economic Policy Review* 4:39–69.

Newey, W. K., and West, K. D. (1987). A simple, positive semi-definite, heteroskedasticity and autocorrelation consistent covariance matrix. *Econometrica* 55:703–8.

Schwert, G. W. (1989). Why does stock market volatility change over time? *Journal of Finance* 44:1115–53.

CHAPTER 17

Surging OBS Activities and Bank Revenue Volatility

How to Explain the Declining Appeal of Bank Stocks in Canada

Christian Calmès and Raymond Théoret

CONTENTS

17.1 INTRODUCTION

It is widely believed that bank stock is a relatively safe asset from the standpoint of the risk-return trade-off. However, it was also thought that the process of banking deregulation, which began in the 1980s, in both Canada and the United-States,[*] by allowing banks to engage in new activities (such as off-balance sheet (OBS) activities), would give rise to important diversification effects (Rose, 1989; Saunders and Walter, 1994). Indeed, traditional finance predicts that these effects would reduce bank stock volatility. Although these conjectures are at odds with the facts, in both Canada and the United States, researchers have shown that OBS activities triggered a substantial increase in the volatility of bank net operating revenue growth (Stiroh, 2004, 2006a; Stiroh and Rumble, 2006; Calmès and Liu, 2007). However, this volatility surge is not associated with risk-adjusted accounting measures of bank returns (e.g., the return on assets and the return on equity). Incidentally, these measures might have decreased, following the upward trend of the share of noninterest income in banks' net operating revenue. Given the influence of the accounting measures of bank performance on the level and the volatility of bank market returns, these developments are obviously problematic for the investor.

In this chapter, we provide new complementary evidence about the detrimental effects of the increase in the relative importance of noninterest income on the performance of the Canadian banking sector. Our contribution is to demonstrate that the surging volatility of bank revenues has given rise to a risk premium as measured with various accounting returns, as was suggested, but not tested, by DeYoung and Roland (2001).

This chapter is organized as follows. Section 17.2 provides an overview of the existing literature on the effects of the increase of the noninterest income share on banks' net operating revenue. In Section 17.3, we provide some stylized facts related to the surge of OBS activities in the North American banking industry and we run regressions to document the impact of the growing share of noninterest income on Canadian banks' performance. Also in this section we test for the presence of a risk premium related to the increasing volatility of the growth of banks' net operating revenue. In Section 17.4, we formulate a conjecture about the decreasing diversification of the Canadian banking sector, which may be explained by the development of a herding behavior in this industry.

---

[*] For the deregulation process in Canada, see Théoret (1999) and Calmès (2004).

17.2 LITERATURE REVIEW

Financial deregulation in Canada and the United States has allowed banks to move toward more market-based activities (Calmès, 2004). Banks can now underwrite securities for their customers and pool some of their loans for securitization. In addition, deregulation has also allowed Canadian banks to offer fiduciary services and portfolio advice to investors.

New bank activities resulting from the banking deregulation process are mainly classified as OBS activities that generate noninterest income, as opposed to interest income, the revenue associated with the traditional lending activity of banks. Noninterest income is a heterogeneous aggregate that includes different components: trading income, gains (losses) on instruments held for other than trading purposes, fiduciary income, service fees, insurance, other fees, and commissions.

The valuation of OBS activities presents many measurement problems (Calmès, 2004), but we can tackle them by resorting to the method suggested by Boyd and Gertler (1994), who propose to compute an asset-equivalent measure of OBS activities. Let r_{BS} be the mean return on balance sheet activities, A_{BS} be the value of balance sheet assets, and N_{BS} the net revenue associated with balance sheet activities. We have:

$$r_{BS}\, A_{BS} = N_{BS}$$

therefore,

$$A_{BS} = \frac{N_{BS}}{r_{BS}}$$

The balance sheet assets are thus the capitalization, at the r_{BS} rate, of the net revenue generated by these assets. Similarly, we can write

$$A_{OBS} = \frac{N_{OBS}}{r_{OBS}}$$

where A_{OBS} is the asset-equivalent of OBS activities, N_{OBS} is the net revenue associated with OBS activities, and r_{OBS} is the mean return on OBS activities. Assume that

$$r_{BS} = r_{OBS}$$

is the capitalization rate of balance sheet assets and is the same as the capitalization rate of OBS assets. We can thus write

$$A_{OBS} = \frac{N_{OBS}}{N_{BS}} A_{BS} = \frac{N_{OBS}/NOR}{N_{BS}/NOR} A_{BS}$$

where *NOR* stands for net operating revenue. We measure respectively the ratio (N_{OBS}/NOR) by the share of noninterest income and the ratio (N_{BS}/NOR) by the share of net interest income in net operating revenue. We thus arrive at the following measure of OBS activities, used for the eight Canadian domestic banks. For example, for the fourth quarter of 2007, we have

$$A_{OBS} = \frac{snonin}{sni} \times A_{BS} = \frac{0.55}{0.45} 2283 = 2790$$

where *snonin* represents the share of noninterest income, and *sni* the share of net interest income. According to the asset equivalent computation, the assets related to Canadian banks OBS activities are equal to $2,790 billion, or approximately 122% larger than the level of balance sheet assets, but by comparison, they only represented 39% of balance sheet assets in 1988. Similarly to the United States banks, Canadian banks' activities are increasingly dominated by OBS activities.

Figure 17.1 shows the growing importance of the share of noninterest income in bank net operating revenue. The upward trend began in 1992 and lasted until the bursting of the market bubble at the beginning of the second millennium. By 2000, noninterest income accounted for 57% of

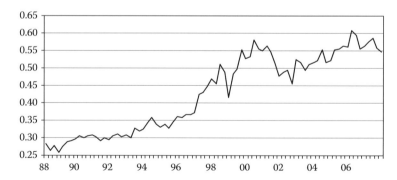

FIGURE 17.1 The growing share of noninterest income in eight Canadian domestic banks from 1988 to 2007.

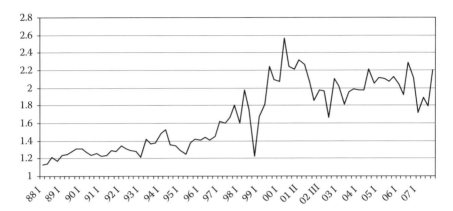

FIGURE 17.2 Noninterest income per $100 of assets for the eight Canadian domestic banks from 1988 to 2007.

net operating revenue, up from only 25% in 1988. This ratio recovered after the market turmoil in the first few years of the millennium and culminated to 60% in the first quarter of 2006, before decreasing thereafter as a result of the weakening of financial markets. Note also that fluctuations of the share of noninterest income are much larger after 1997 than before. Indeed, this share became increasingly sensitive to the fluctuations of financial markets (Calmès, 2004; Calmès and Liu, 2007).

As shown in Figure 17.2, the growing share of noninterest income has boosted the bank ratio of noninterest income to balance sheet assets. Excluding the collapse of this ratio during the 1998 financial crisis (related to the Russian debt), this ratio increased progressively from 1.13% in 1988 to 2.32% in 2001. It decreased steeply during the collapse of the financial markets at the beginning of the second millennium, and it did not recover thereafter, fluctuating around 2%. Similarly to the share of noninterest income, this ratio has also been increasingly dependent on financial markets fluctuations.

Activities related to noninterest income are much more volatile than those associated with net interest income (Stiroh, 2004; Calmès and Liu, 2007). Their direct contribution is to increase the volatility of the bank's net operating revenue growth. There is actually a diversification effect due to the fact that the correlation between interest and noninterest income is less than 1, but this indirect effect is quite low in comparison with the direct effect (Calmès and Liu, 2007). Moreover, the correlation between these two forms of income is quite unstable. Hence, the direct contribution of noninterest income to the volatility of net operating revenue growth dominates largely. Increasing the operating leverage magnifies the volatility of profits growth (De Young and Roland, 2001).

Following Stiroh (2004) and Calmès and Liu (2007), we decompose the net operating revenue growth with a portfolio approach to analyze its volatility with two components: volatility of net interest income growth and volatility of noninterest income growth. The growth of net operating revenue (*NOR*) is computed as

$$d\ln(NOR) = \ln\left(\frac{NOR_t}{NOR_{t-1}}\right) = \ln(NOR_t) - \ln(NOR_{t-1})$$

Its variance may thus be decomposed as

$$\sigma^2_{d\ln(NOR)} = w^2 \sigma^2_{d\ln(NONIN)} + (1-w)^2 \sigma^2_{d\ln(NI)}$$

$$+ 2w(1-w)\text{cov}(d\ln(NONIN), d\ln(NI))$$

where *NONIN* stands for noninterest income, and *NI* for net interest income, and where $w = \frac{NONIN}{NONIN+NI}$, the share of noninterest income in the bank's net operating revenue. The direct contribution of noninterest income to $\sigma^2_{d\ln(NOR)}$ is given by $w^2 \sigma^2_{d\ln(NONIN)}$, while the contribution of net interest income to $\sigma^2_{d\ln(NOR)}$ is equal to $(1-w)^2 \sigma^2_{d\ln(NI)}$. Since noninterest income is more volatile than net interest income, the growing importance of noninterest income in bank net operating revenue directly increases $\sigma^2_{d\ln(NOR)}$. Nonetheless, as long as the correlation between the growth rates of noninterest income and net interest income is not equal to 1, the trade-off between net operating revenue growth and volatility can improve.

Table 17.1 reports the variance decomposition of net operating revenue growth over subperiods ranging from 1988 to 2007 with time intervals corresponding to different legislative periods. In the subperiods 1988–1992 and 1993–1997, noninterest income seems to help reduce net operating revenue variance below what it would have been if banks relied solely on interest income. For example, in the 1988–1992 period, net operating revenue variance was 14.2, which was lower than the 16.9 variance of net interest income. From 1993 to 1997, there were diversification benefits, with net interest income volatility being higher than net operating revenue, and the correlation between the two components of net operating revenue being negative.

TABLE 17.1   Decomposition of the Variance of Net Operating Revenue Growth, Before Provisions, Canadian Banks from 1988 to 2007

	1988–1992			1993–1997		
	Average Share	Variance	Contribution to Variance	Average Share	Variance	Contribution to Variance
Net operating revenue		14.2			9.4	
Net interest income	0.70	16.9	8.4	0.64	9.8	4.0
Noninterest income	0.30	30.2	2.6	0.36	40.4	5.3
Covariance		7.5	3.1		−0.9	−0.4
Correlation		0.33			−0.04	
	1998–2002			2003–2007		
	Average Share	Variance	Contribution to Variance	Average Share	Variance	Contribution to Variance
Net operating revenue		57.4			22.3	
Net interest income	0.49	9.7	2.3	0.45	13.6	2.8
Noninterest income	0.51	212.3	55.9	0.55	75.7	22.6
Covariance		6.1	3.0		−4.2	−2.1
Correlation		0.14			−0.13	

However, the two following subperiods are quite different. During both subperiods, the variance of net operating revenue growth is significantly higher than the variance of net interest income growth, implying that noninterest income growth increased substantially the volatility of net operating revenue growth. The variance of net operating revenue growth has also jumped compared to the previous subperiods. The subperiod 1998–2002, plagued by excessive financial market fluctuations, is particularly symptomatic. The variance of noninterest income growth jumped to 212.3, and was less than 40 prior to this period. During this subperiod, income from trading and investment activities was one of the major contributors to noninterest income volatility both in Canada (Calmès and Liu, 2007) and in the United States (Stiroh, 2004).

During the subperiod 2003–2007, the volatility of net operating revenue growth receded, but it remained much higher than prior to the 1998–2002 financial crisis. In fact, the volatility of noninterest income growth approximately doubled with respect to its level before the 1998–2002 subperiod. However, during this subperiod the correlation between net interest and noninterest income growth became negative, a new situation that contributed to the dampening of the direct pervasive impact of noninterest income on the volatility of net operating revenue growth. In addition, the volatility of noninterest income growth is increasingly related to the one of income from trading and investment activities, the highest among the components of noninterest income, a worrying situation from the standpoint of the risk-return trade-off.

In other respects, even if noninterest income increases the volatility of bank net operating revenue growth, that might be compensated by an increase in expected profitability, as measured by the return on assets (*ROA*) or return on equity (*ROE*). A priori, accounting reasoning suggests that OBS activities should tend to increase these profitability measures. For instance, removing assets from the balance sheet should increase *ROA*. Furthermore, OBS activities are a source of capital relief for a bank and should thus increase *ROE*. We also know that $ROA = ROE\left(\frac{1}{L}\right)$, L being a measure of leverage equal to (A/E), where A denotes the level of balance sheet assets, and E the level of shareholder equity. According to this formula, if OBS activities reduce bank leverage, growing OBS activities should also increase *ROA*. But data reveal that the relationship between OBS activities and leverage is not clear, as banks can use up the leeway resulting from these activities in shifting to riskier mixes of activities instead of holding less equity.

Hence, the usual accounting logic is at odds with the facts. Recent studies (Stiroh, 2004, 2006a; Calmès and Liu, 2007) have shown that an increase in the share of noninterest income tends to depress profitability measures, especially when expressed on a risk-adjusted basis. Besides considerations related to the optimality of bank activities, the higher volume of noninterest income has clear negative implications for supervisors, shareholders, managers, and borrowers, all of whom care about the mean and volatility of bank profits.

Section 17.3 features an empirical model testing the impact of the growing share of noninterest income in net operating revenue on the aggregated measures of performance constructed with a pool of eight Canadian domestic banks. We also run these regressions, individually, on three Canadian banks differing in size.

17.3 EMPIRICAL RESULTS

17.3.1 The Model

To test for the impact of the growing share of noninterest income on bank performance, researchers (Stiroh, 2004; Calmès and Liu, 2007) have resorted to an empirical model taking the following form:[*]

$$y_t = \beta_0 + \beta_1 y_{t-1} + \beta_2 snonin_t + \beta_3 X_t + \varepsilon_t \qquad (17.1)$$

where y_t is a bank performance measure, $snonin_t$ is the share of noninterest income in net operating revenue, X_t is a vector of control variables, and ε_t is the innovation or error term. For instance, the vector X_t may control for bank size, for the riskiness of loans, for asset growth, or for any other factor that may impact on bank performance. Following Calmès and Liu (2007), we retain only the ratio of loan loss provisions to total assets as a control variable because the other ones were found not significant.

After Stiroh (2004) and Calmès and Liu (2007), we also estimate Equation (17.1) on a risk-adjusted basis by dividing y_t by a four-quarter moving average of the standard deviation of y_t. We also introduce a new measure of risk, deflating y_t by its conditional volatility as measured by a GARCH(1,1) model. We tested for other well-known econometric specifications of conditional volatility, like GARCH(p,q), TARCH, EGARCH, and PARCH, using also different distributions for the error term (normal, Student, and generalized error (GED)), but the GARCH(1,1) specification was the best measure of conditional volatility based on traditional measures of econometric model evaluation, such as the Akaike and Schwarz criterions.

Our main contribution is to introduce a risk measure directly in Equation (17.1). Indeed, according to DeYoung and Roland (2001), the surging volatility of bank revenues gave rise to the incorporation of a risk premium in various measures of bank accounting returns. However, the authors did not test this conjecture.

The relationship between expected return and risk is in line with basic finance. Traditional finance establishes a risk-return trade-off such that

$$r_t = \theta_1 + \theta_2 risk_t + \mu_t$$

[*] For an alternative model of bank performance see Théoret (1991).

where r_t stands for return, $risk_t$ is a risk measure, and μ_t is the innovation. We introduce risk in Equation (17.1) by resorting to an ARCH-M model,[*] that is:

$$y_t = \beta_0 + \beta_1 y_{t-1} + \beta_2 snonin_t + \beta_3 X_t + \beta_4 \sigma_{c,t} + \varepsilon_t \qquad (17.2)$$

where $\sigma_{c,t}$, the conditional volatility, is computed using the following equation

$$\sigma_{c,t}^2 = \theta_0 + \theta_1 \sigma_{c,t-1}^2 + \theta_2 \varepsilon_{t-1}^2$$

The ARCH-M procedure is very appealing to estimate the risk premium because it directly incorporates the conditional volatility, chosen as a measure of risk, in the return equation.

17.3.2 Results

Table 17.2 reports the estimation of Equations (17.1) and (17.2) for the pool of the eight major Canadian domestic banks for the period running from the first quarter of 1988 to the fourth quarter of 2007. Data come from the Canadian Bankers Association and the Office of the Superintendent of Financial Institutions (Canada). Unit root tests suggest that all data are stationary, so they are modeled in levels.

Estimation of Equation (17.1) for the ratios ROE and ROA gives very satisfying results in terms of adjusted R^2, which is equal to 0.72 for both ratios. Before adjustment for risk, estimation of Equation (17.1) reveals that the coefficient of the share of noninterest income is significantly negative for both performance ratios. This suggests that OBS activities reduce the performance of Canadian banks, while they also increase the volatility of bank net operating revenue growth. These findings cast doubt on the belief that noninterest income activities can lead to better bank performance through diversification activities (reduction in risk or higher returns). Moreover, consistent with expectations that loan loss provisions lower profits, the coefficient of the ratio of loan loss provisions to total assets is negative in all equations. Since this ratio jumps during recessions, that accentuates the procyclicality of ROE and ROA, which have yet been made more procyclical following the banks increasing involvement in OBS activities.

---

[*] The ARCH-M model is due to Engle et al. (1986).

TABLE 17.2 Profitability of the Eight Canadian Domestic Banks versus Noninterest Income Share, 1988Q1–2007Q4

	ROE(1)	ROE(2)	ROE/$\sigma_{uc,t}$	ROE/$\sigma_{c,t}$	ROA(1)	ROA(2)	ROA/$\sigma_{uc,t}$	ROA/$\sigma_{c,t}$
c	0.24***	0.25***	12.58***	2.94***	1.02***	0.21***	23.71***	5.16***
y_{t-1}	0.15**	-0.01	0.75***	0.60***	0.11*	0.01	-4.72**	0.11
snonin	-0.11**	-0.20***	-16.43***	-1.72	-0.39**	-0.72***	-22.32*	-2.20**
LLP	-0.14***	-0.15***	-7.97***	-1.99***	0.55***	-0.59***	-9.31**	-2.49***
DUM2Q	-0.02	-0.01	1.24	-0.60*	-0.06	-0.03	2.37	-0.37
DUM3Q	-0.02	-0.01	-0.51	-0.20	-0.05	-0.05	3.32	-0.28
DUM4Q	-0.03**	-0.02*	0.32	-0.65**	-0.11**	-0.09**	2.57	-0.55
$\sigma_{c,t}$	—	1.85**	—	—	—	9.78***	—	—
Adjusted R^2	0.72	0.80	0.67	0.68	0.72	0.83	0.15	0.70

Note: Explanatory variables: y_{t-1}, lagged dependent variable; snonin, share of noninterest income in net operating revenue; LLP, ratio of loan loss provisions over total assets; DUMiQ, dummy variable taking the value of 1 for the ith quarter and 0 otherwise; $\sigma_{uc,t}$, unconditional volatility of the dependent variable computed using a rolling window of four quarters; $\sigma_{c,t}$, conditional volatility of the dependent variable using a GARCH(1,1) model. ROE(1) and ROA(1) are models without conditional volatility. ROE(2) and ROA(2) are ARCH-M models incorporating the conditional volatility of the dependent variable. Asterisks indicate the significance levels: *, 10%; **, 5%; and ***, 1%.

Regressing Equation (17.1) using risk-adjusted performance ratios leads to a decrease of adjusted R^2, due to the fact that the scaling factor fluctuates greatly from one period to another. Results tend to improve when using conditional volatility instead of the historical one, to scale the performance ratios, especially for *ROA*, where the adjusted R^2 increases from 0.15 to 0.70 when shifting from historical to conditional volatility. In other respects, the results are similar to those obtained for the regressions without risk adjustment.

We also consider the estimation of Equation (17.2) with the ARCH-M procedure, a new feature for investigating bank performance in this framework. This equation incorporates a risk premium to account for the increasing volatility of bank revenues. We first observe that the introduction of a risk premium in the equations of *ROE* and *ROA* results in a jump of the adjusted R^2. It increases from 0.72 to 0.80 when regressing Equation (17.2) instead of Equation (17.1) using *ROE* as the dependent variable, and from 0.72 to 0.83 when using *ROA* as the dependent variable. We may thus observe that the risk premium has an important impact on *ROE* and *ROA*. Note also that, for both ratios, the risk premium is significant at the 1% level. We thus conclude that banks have reacted to the increasing volatility of their net operating revenue growth by adding a risk premium to the return of their OBS activities, a quite rational, and reassuring, behavior.

We also estimate Equation (17.2) for three individual Canadian banks differing by size: a relatively small-sized bank, the National Bank of Canada (NBC); a medium-sized bank, the Toronto-Dominion Bank (TD); and a large-sized bank, the Royal Bank of Canada (RBC). Let us note that the NBC is very involved in OBS activities in spite of its relatively small size, its share of noninterest income being as high as 0.70 by the end of 2007. Otherwise, the shares of RBC and TD banks were respectively 0.67 and 0.50 for the same period. The share of noninterest income in net operating revenue does not seem to be correlated to bank size in Canada, contrary to what is observed in the United States (Houston and Stiroh, 2006). Figure 17.3 provides a comparison of the evolution of the noninterest income share for the three banks and for the total of the eight Canadian domestic banks. We note that the behavior of the RBC's share is much more stable than that of the other two banks and is in constant progression over the 1988–2007 period, although it has increased at a slower pace since 2003. In addition, RBC has a weight of 26% in Canadian banks total assets, which is a good benchmark to test the impact of the growing share of noninterest income on the accounting measures of bank returns.

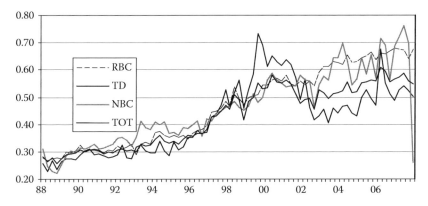

FIGURE 17.3   Share of noninterest income, three Canadian domestic banks from 1988 to 2007.

NBC's and TD's shares have become very volatile since the financial crisis of 1997. While NBC's share has remained on an upward trend before collapsing in the fourth quarter of 2007, TD's share has decreased substantially since 2000. The dispersion between banks' proportions has also greatly increased since 1997.

Table 17.3 provides our results for the chosen Canadian banks. We suspect that the substantial reduction of the R^2 observed with the disaggregation of the sample stems from the presence of a high idiosyncratic risk at the individual level. Being the largest Canadian bank, RBC estimated equations of *ROE* and *ROA* are quite similar to the aggregate, except for

TABLE 17.3   Profitability of Three Canadian Banks versus Noninterest Income Share from 1988 to 2007

	RBC		NBC		TD	
	ROE	*ROA*	*ROE*	*ROA*	*ROE*	*ROA*
c	2.88***	0.98***	0.10***	0.48***	−0.03	0.34***
y_{t-1}	−0.02	−0.07	−0.04	0.06**	0.77***	0.68***
snonin	−0.06***	−0.23**	0.13***	0.12	−0.05***	−0.28***
LLP	−0.06***	−0.37***	−0.08***	−0.55***	−0.02***	−0.23***
DUM2Q	−0.01	0.03	−0.01	−0.03**	−0.02***	−0.08***
DUM3Q	−0.01***	0.02	−0.01	−0.02	−0.02***	−0.04***
DUM4Q	−0.01***	−0.04	−0.01	−0.03**	−0.01***	−0.03***
$\sigma_{c,t}$	0.37***	0.56***	0.75***	1.99***	29.10***	0.38***
Adjusted R^2	0.46	0.39	0.10	0.48	0.10	0.14

the adjusted R^2 which is somewhat smaller, 0.46 for the *ROE* and 0.39 for the *ROA*. These R^2 values remain acceptable, for regressions run on individual bank performance ratios. The share of noninterest income has a significant negative impact on both performance ratios (significant at the 99% confidence level for the *ROE* ratio). On the other hand, the risk premium is relatively high and significant at the confidence level of 99% for both ratios. Finally, the ratio of loan loss provisions impacts negatively and significantly both ratios, especially the *ROA* one.

As expected, the results are not so satisfying for the other two banks retained for this analysis. For NBC, the coefficients associated with the conditional volatility and the ratio of loan loss provisions have the right sign and are significant at the 99% confidence level. But the coefficient of the share of noninterest income is positive. NBC bank thus seems to have benefited from its increasing involvement in OBS activities, which, as mentioned earlier, is much larger than the average for Canadian banks. Perhaps its OBS activities are better priced than those of other banks. Or perhaps there is more synergy between the components of these activities. However, we note that the explanatory power of Equation (17.2), as measured by adjusted R^2, is low for the *ROE* ratio, and moderate for the *ROA* one, the adjusted R^2 for these ratios being, respectively, 0.10 and 0.48.

Finally, the performance of Equation (17.2) is low for explaining the *ROE* and *ROA* ratios of the TD bank, even if the estimated coefficients of the explanatory variables have all the right signs and are, for most of them, significant at the 99% confidence level. The poor performance of the regressions run with the TD sample may be explained by the presence of outliers and idiosyncratic risk. In fact, over the estimation period, the TD bank has been implied in important transactions that had major repercussions on its financial results.

17.3.3 Concluding Remarks

Surging bank OBS activities are associated with an important increase in the ratio of direct to indirect finance, and therefore to more complete financial markets and a more market-oriented financial industry (Calmès, 2004). In fact, Canadian firms increasingly fund their investments by resorting to financial markets, issuing bonds and equity, instead of taking out bank loans. This new financing regime was fostered by the amendments made to the Canadian Bank Act in 1987, which allowed banks to be involved in investment banking activities such as underwriting securities. Instead of issuing loans directly and cashing interest income, banks cash income fees by structuring bonds and stocks issues. These operations allow them to save capital for other purposes. Securitization is another OBS activity used by Canadian banks since the early 1990s, which is also linked to a move toward a more market-based financial system. Securitization can facilitate the trading of previously illiquid loans. Securitization activity has also strengthened the capacity of banks to supply new loans to households and firms for a given amount of funding (Altunbas et al., 2007).

But these developments are obviously very problematic for the protagonists of free markets and for central banks, which aim at preserving financial stability. Indeed, our empirical work shows that the jump in OBS activities increases the volatility of bank net operating revenue growth. Moreover, these activities tend to depress the accounting measures of bank returns. Even if they give rise to a risk premium that is a partial compensation for their increasing volatility, the fact remains that OBS activities, the product of a more market-oriented economy, increase the risk of banks operations—and this was not the expected result associated with increased markets completion.

17.3.3.1 The Credit Channel

Referring to Calmès' paper (2004), Roldos (2006) noted that there were structural breaks in the response of the Canadian economy to monetary shocks in the 1980s and 1990s associated with key changes in the Bank Act. These structural changes gave rise to a weakening of the credit channel, a very important link in the traditional monetary policy transmission mechanism. Monetary policy was perhaps a destabilizing factor when its credit channel was stronger. We know that the financial accelerator was an important link of the credit channel. It is possible that the financial accelerator was amplifying fluctuations of macroeconomic aggregates, even causing overshooting on a large scale. The idiosyncratic shocks associated with the credit channel were probably destabilizing the economy. The financial deepening* resulting from the deregulation process seems to have led to a decrease in the volatility of key macroeconomic variables like GDP or productivity. The Canadian and American economies have become more stable, while the traditional credit channel of monetary policy was losing its steam. For instance, we note an important decrease of output volatility since 1984 in the United States (Stiroh, 2006b; Ozenbas and San Vincente

---

* Associated with an increase of the ratio of direct to indirect finance.

Portes, 2006), a trend also shared by other macroeconomic time series such as inflation. These developments are called the great moderation. In their conclusions, Calmès and Liu (2007) have even made the conjecture that the new trend in dampened inflation volatility might be partly explained by the fact that Canadian firms tend to rely increasingly on financial markets rather than contracting loans. According to this conjecture, inflation control has lessened the contribution of monetary policy.

17.4 REVENUE VOLATILITY, BANK HERDING BEHAVIOR, AND AGGREGATE RISK: A CONJECTURE

The developments related to surging OBS banks' activities are also worrying from another macroeconomic standpoint. According to Houston and Stiroh (2006), aggregate risk has increased in the American banking sector since 1990 while idiosyncratic risk has receded.

> We thus conjecture that net interest income, being related to physical stocks, e.g. loans, would mainly respond to idiosyncratic shocks, like borrower default, whereas noninterest income, being related to flows, e.g. service fees and trading revenues, would respond to aggregate shocks, like unexpected changes in stock market indices and macroeconomic aggregates. Since the former shocks are diversifiable and the latter are not, this conjecture complements the idea that the changing structure of bank revenues is associated with increasingly volatile equity market returns which follows. (Calmès, 2004; Calmès and Liu, 2007)

Following their growing involvement in OBS activities, Canadian banks are thus increasingly exposed to aggregate shocks.

Being exposed to aggregate shocks, banks are also more likely to have similar reactions to economic events, which increase banking risk. Bank herding, i.e., a tendency for banks to move together in periods of economic uncertainty, which has been observed in the United States (Baum et al., 2002) and in Canada (Calmès and Salazar, 2006), seems symptomatic of the greater exposure of banks to aggregate shocks. Incidentally, Quagliariello (2006) notes that, to the best of his knowledge, there were only two papers that investigated the issue of the link between uncertainty regarding future macroeconomic conditions and bank herding behavior (Baum et al., 2005; Calmès and Salazar, 2006). He reports that Canadian intermediaries show herding behavior when they deal with more pronounced aggregate uncertainty. Quagliariello (2006) observes a similar herding behavior for the

Italian banks. His contribution is to distinguish aggregate uncertainty from the idiosyncratic one. In the case of Italian banks, he notes that the herding behavior is at play when macroeconomic or aggregate uncertainty increases. However, when idiosyncratic risk increases, banks behave heterogeneously. According to Quagliariello (2006), this last observation would be due to the competitive advantage of better informed banks behaving in a different way compared to poorly informed intermediaries.

Hence, if aggregate shocks were increasingly important in the Canadian banking system, and since, according to our conjecture, the exposure of banks to those shocks would rise due to the increasing portion of their OBS activities, bank herding could appear more a structural and not just a cyclical phenomenon as previously thought, which would then translate into an increased correlation between banks accounting and equity returns—bad news for the investors in search of portfolio diversification. Indeed, herding is at the antipodes of diversification, and it threatens the stability of any banking system. Traditionally, a portfolio pooling many different bank stocks was seen as a relatively safe investment, but if this conjecture proved to be right, this would no longer be the case because of this systemic herding regime.

17.5 CONCLUSION

The Canadian banking sector, which is traditionally considered relatively safe, is becoming increasingly risky as banks progressively shift their operations toward OBS activities. An important question from the standpoint of the optimality of banking operations remains to be answered: Is this move exogenous or endogenous to the banking sector? It may be viewed as exogenous if we note that the growth of the volume of traditional banking activities such as lending, but also the return or margin on these activities, was steeply trending downward during the 1980s and 1990s due to the exacerbating competition between banks, while the traditional four pillars of the Canadian financial system were eroding. The branch network of Canadian banks was no longer profitable, and the banking system once based on bricks and mortar disappeared. Perfect competition reduces economic profit on traditional activities near zero. In this context, Canadian banks have no choice but to increasingly rely on OBS activities. According to this reasoning, the increasing weight of OBS activities in banks' total operations may be viewed as exogenous.

But this move may also be viewed as endogenous if it was originated by banks themselves. Under this scenario, banks fostered the financial deregulation process by shifting their activities toward a priori more profitable ones like underwriting and securitization. To do so, they would even have encouraged their customers to be more market oriented by substituting direct securities issues to loans. But, if this is the case, the optimality of such a move has to be questioned because according to our empirical work, OBS activities decreased the accounting measures of bank returns and increased the volatility of the growth of net bank operating revenue, a very unfavorable evolution from the standpoint of the risk-return trade-off. However, our contribution in this chapter is to show that such a detrimental evolution gave rise to a partial compensation for banks in the form of the addition of a risk premium to the accounting measures of returns.

Being riskier, the Canadian banking system is now more sensitive to aggregate shocks, which seem to have increased in the financial sector since the beginnings of the 1990s. Bank stock returns could thus be increasingly volatile, bad news for risk-averse investors who consider bank stock a relatively safe value. Regulators should also be more aware of the fact that the risk in the banking sector increases as banks seem increasingly exposed to common aggregate shocks, a conjecture still to be confirmed. This is left for future work.

REFERENCES

Altunbas, Y., Gambacorta, L., and Marques, D. (2007). Securitization and the bank lending channel. Working paper 653, Banca d'Italia, Rome.

Baum, C. F., Caglayan, M., and Ozhan, N. (2002). The impact of macroeconomic uncertainty on bank lending behaviour. Working paper, University of Liverpool, Liverpool.

Baum, C., Caglayn, M., and Ozhan, N. (2005). The second moments matter: The response of bank lending behaviour to macroeconomic uncertainty. Working paper 521, Boston College, Boston.

Boyd, J. H., and Gertler, M. (1994). Are banks dead? Or are the reports greatly exaggerated? *Federal Reserve Bank of Minneapolis Quarterly Review* 18:1–27.

Calmès, C. (2004). Regulatory changes and financial structure: The case of Canada. *Swiss Journal of Economics and Statistics* 140:1–35.

Calmès, C., and Liu, Y. (2007). Financial structure change and banking income: A Canada–U.S. comparison. *Journal of International Financial Markets, Institutions and Money,* in press.

Calmès, C., and Salazar, J. (2006). Variance Macroéconomique Conditionnelle et Mesure de Dispersion des Actifs dans les Portefeuilles Bancaires. In *Finance Computationnelle et Gestion des Risques,* ed. F. E. Racicot and R. Théoret. Ste-Foy, Québec: Presses de l'Université du Québec 22:687–700.

DeYoung, R., and Roland, K. P. (2001). Product mix and earnings volatility at commercial banks: Evidence from a degree of total leverage model. *Journal of Financial Intermediation* 10:54–84.

Engle, R. F., Lilien, D. M., and Robins, R. P. (1986). Estimating time-varying risk premia in the term structure: The ARCH-M model. *Econometrica* 55:391–407.

Houston, J. F., and Stiroh, K. J. (2006). *Three decades of financial sector risk.* Staff Report 248, Federal Reserve Bank of New York.

Ozenbas, D., and San Vincente Portes, L. (2006). On balance sheets, idiosyncratic risk and aggregate volatility: Is firm volatility good for the economy? Working paper, Montclair State University.

Quagliariello, M. (2006). Macroeconomic uncertainty and banks' lending decisions: The case of Italy. Working paper 2006/02, New York University.

Roldos, J. (2006). Disintermediation and monetary transmission in Canada. Working paper, International Monetary Fund, Washington.

Rose, P. S. (1989). Diversification of the banking firm. *Financial Review* 24:251–80.

Saunders, A., and Walter, I. (1994). *Universal banking in the United-States: What could we gain? What could we lose?* New York: Oxford University Press.

Stiroh, K. J. (2004). Diversification in banking: Is noninterest income the answer? *Journal of Money, Credit and Banking* 36:853–82.

Stiroh, K. J. (2006a). A portfolio view of banking with interest and noninterest activities. *Journal of Money, Credit and Banking* 38:1351–61.

Stiroh, K. J. (2006b). *Volatility accounting: A production perspective on increased economic stability.* Staff Report 245, Federal Reserve Bank of New York.

Stiroh, K. J., and Rumble, A. (2006). The dark side of diversification: The case of US financial holding companies. *Journal of Banking & Finance* 30:2131–61.

Théoret, R. (1991). Un Modèle Économétrique des Marges Bénéficiaires des Caisses Populaires Desjardins du Québec et des Banques à Charte Canadiennes. *L'Actualité économique, Revue d'analyse économique* 67:58–79.

Théoret, R. (1999). *Traité de Gestion Bancaire.* Ste-Foy, Québec: Presses de l'Université du Québec.

CHAPTER 18

Usage of Stock Index Options

Evidence from the Italian Market

Rosa Cocozza

CONTENTS

18.1 INTRODUCTION

The usage of the stock index options is increasing at a fast pace in many industrialized countries because of their wide application in many complex portfolios. This trend in Italy had a dramatic peak over the last years (IDEM, 2006). Although it is easy to ascribe this increase to the growing complexity of financial products and to the corresponding increasing involvedness of the management process, the way traders use the stock index option within a specific market has not yet formally been investigated. A very wide and well-known literature deals with the pricing and the value management of option portfolios, starting from the seminal work of Black and Scholes (1973), Merton (1973), and Cox et al. (1979). Another rich area of literature regards the behavior of the traders with special reference to the sentiment of the market as defined by technical analysis practices. Relatively little is known about the trading of this important class of securities. With this respect, an important paper is that by Lakonishok et al. (2004), where a full analysis of the investor behavior was performed with the aim of documenting major empirical facts about the option market activity of different types of investor on the Chicago Board of Trade.

This chapter tries to investigate which kind of strategy is prevailing on the Italian stock exchange as far as the stock index option market is concerned. Given the possibility of using the option for both directional strategies and volatility strategies, the main answer we are looking for concerns firstly the existence of a prevailing behavior on the market and secondly, given a positive answer to the first question, which one is the most popular. The analysis is performed by a graphical analysis and by using an official data set of the clearing house of the Italian stock exchange from December 2007 to May 2008. With respect to the previous literature, our study is aimed at verifying which strategy prevails in the Italian market, since there is a fundamental connection between the exploitation of volatility strategies and the complexity of the reference financial markets. The main finding, at least with reference to the period under observation, confirms the prevalence of directional strategies and of a certain mixture of volatility and directional strategies. Very rarely we found evidence of strong exploitation of volatility trading.

The remainder of the chapter is organized as follows. Section 18.2 introduces the Option Strategy Matrix as an instrument of market analysis. Section 18.3 explains the fundamentals of the reference market data set. Section 18.4 reports the empirical analysis. Section 18.5 concludes.

18.2   RISK AND VALUE DRIVERS IN OPTION TRADING

As known, options are financial instruments that convey the right, but not the obligation, to engage in a future transaction on some underlying security. The holder of a call (put) has the right to buy (sell) a specified quantity of a security at a set strike price at some time on (European option) or before (American option) expiration. Upon the option holder's choice to exercise the option, the party who sold, or wrote, the option must fulfill the terms of the contract.

Therefore, the theoretical value of an option is conceptually the present value of the future cash flow arising from the agreement: for a call option it is positively linked to the difference between the price of the underlying at the exercise time and the strike price ($S_t - s$), while for a put option it is positively related to the opposite ($s - S_t$). As the agreement is not binding for the holder, the payoff is kept positive by setting the maximum value between 0 and the relevant difference. Therefore, the holder of a call (put) option will gain a profit if the price of the underlying rises (falls) with respect to the strike. From the viewpoint of the buyer of the option, once the price of the option is paid, the contract has only upside potential. Conversely, the seller of the option obtains the premium immediately, and then faces the risk of losses upon option exercise. Option contracts are unique insofar as one side (the buyer) has a nonbinding option of going through with a defined transaction, while the other party (the writer) has no such flexibility. This flexibility is valuable: as a consequence, the row differences between the current and the strike price are not able to capture all the elements influencing the value of an option contract. Even on an intuitive basis, the crucial point is the probability of the underlying to reach the strike price. The formalization of the relevant probability distribution was the key to the closed solution provided by Black and Scholes (1973) for the pricing of the European options (plain vanilla). The Black and Scholes formulation provides a pricing formula as a function of the underlying instrument (S_t) and of the valuation time t, given the strike price (k), the spot risk-free rate (r), the volatility of the underlying instrument (σ), and the expiration date (T). Formally, C_t being the price of a call option and P_t being the price of a put option at time t, we have

$$C_t = f(S_t, t \mid k, r, \sigma, T)$$

$$P_t = f(S_t, t \mid k, r, \sigma, T).$$

Both the call and the put options are sensitive to the same explanatory variables; nevertheless, the way the option price reacts to the variation of the independent variables is somehow different for the two contracts. Since in the Black and Scholes environment, the risk-free rate and the volatility of the underlying instrument are assumed to be constant, there are three basic partial derivatives that we can appraise for the sensitivity analysis:

1. the first derivative of the price with respect to the underlying ($\partial C_t / \partial S_t$, $\partial P_t / \partial S_t$)

2. the second derivative of the price with respect to the underlying ($\partial^2 C_t / \partial S_t^2$, $\partial^2 P_t / \partial S_t^2$)

3. the derivative of the price with respect to the time t ($\partial C_t / \partial t$, $\partial P_t / \partial t$)

The first two, respectively delta and gamma,[*] are widely used to set the option strategies and to enhance the potential of option trading. As known, the delta, by measuring the price sensitivity to a change in the value of the underlying instrument, is able to quantify the increase/decrease in the value of the option due to a directional change in the price S_t. In this perspective, it is intuitive to say that the delta of a call option—for the holder of the option—must be positive since the contract gains value when the underlying rises with respect to the strike price; similarly, the delta of a put option—in the holder perspective—must be negative since the contract loses value when S_t rises.[†] The option writer is in the mirror position and face deltas multiplied by −1. The gamma measures the sensitivity of the price to a quadratic change in the price of the underlying instrument and is able to indirectly quantify the increase in the option value—from the holder's standpoint—due to a volatility change of S_t. It is possible to show that the gamma is always positive for the holder of both call and put options, while it is always negative for the writer of the contracts.[‡]

Since any changing parameter in a pricing formula is a risk driver as well as a value driver, the delta and the gamma dynamics confirm that option trading can gain value with reference to both directional and volatility changes of the underlying instrument. With respect to this, if the trader focuses on the directional changes, he or she will go for a long call (or even a short put) in the case of bullish expectations, while he or she will go for a long put (or even a short call) in the case of bearish prospects.[§]

---

[*] The notation in the pricing formula suggests that the partial derivatives themselves are functions of the explanatory variables. Hence, one may envisage some further, high-order partials. The traditional Black and Scholes vanilla option pricing environment uses the three we mentioned in the text. Nevertheless, further partial derivatives are brought into picture as the Black and Scholes assumptions are relaxed gradually. For details see Hull (2006), Neftci (2004), Wilmott (2007).

[†] The delta of a call option takes values between 0 and 1, while the delta of a put option takes values between −1 and 0. The limitation of the value array accounts for the asymmetry of the instrument.

[‡] This property accounts for the convexity of the contract. For details see, among others, Hull (2006).

[§] The asymmetry of the option contract naturally determines a substantial difference between the adoption of a long or a short position. The possibility to consider the short position on a call (put) as an alternative to the long position on a put (call) is linked to two practical observations: the eventual limited variance interval of the underlying and the opportunity to combine this product in a complex portfolio through the call-put parity relationship.

FIGURE 18.1   The Option Strategy Matrix.

As far as the volatility scenario is concerned, the expectation of rising unpredictability forces the trader toward long positions, while the expectation of a decreasing instability drives toward short positions, as shown by the Option Strategy Matrix (OSM) represented by Figure 18.1.

The OSM provides for a complete and useful map of portfolio potential/effective strengths and weaknesses, at least in terms of directional and volatility strategies. The matrix can be used as a strategy selection tool according to the expectations toward price and volatility development, but also as an explanatory map of the market behavior. In this perspective, if the OSM is applied to the option market as a unique entity, the number of contracts in each quadrant offers a depiction of prevailing strategies. As it can be easily appreciated, if the contracts cluster along the vertical axis, there is a considerable usage of directional strategies, according to the market prevailing sentiment about price movements, while if they concentrate along the horizontal axis, there is an extensive implementation of volatility strategies, according to the sentiment regarding volatility expectations. If there is a mixture of the two strategies, the contracts will gather in the center of the map.

18.3   THE REFERENCE MARKET

The analysis applies to the S&P/MIB index that is the benchmark for the Italian market. This index measures the performance of forty stocks listed on the markets organized by the Italian Stock Exchange (Borsa Italiana) and is the only Italian equity index that relies on Standard & Poor's world-renowned approach to index construction. It features free-float adjustment, high liquidity, and broad representation of market performance based on the leading companies in the leading industries.*

Option contracts are the highest-growing products of the Italian Derivatives Market (IDEM). The IDEM is an order-driven market where transactions are anonymous. During continuous trading the execution of contracts occurs, for the quantities available, by automatic matching of proposals from opposite sides present in the market book. Confirmation of all trades is given automatically. Liquidity is supported by the presence of more than twenty market makers quoting continuously or responding to request for quotes. Market makers are granted a reduction in trading fees. They are also provided with a quicker market access and a mass quotation functionality, which allows market makers to send to the market up to 100 quotations simultaneously in just one transaction.

The IDEM is guaranteed by a central counterparty (CCP) guarantee system that takes the counterparty risk starting from the conclusion of the contracts. The CCP service is carried out by CC&G (Cassa di Compensazione e Garanzia), which manages the clearing and guarantee function for the IDEM.

The IDEM currently trades S&P/MIB index options, which were introduced on March 22, 2004, and replaced the MIB30 index options, which were in turn excluded from the list on September 17, 2004. The S&P/MIB index option showed for the current year a total turnover in millions of euro equal to 133,294.5, accounting for approximately 25% of the total IDEM market (Borsa Italiana, 2008).

Index options are European-style options with an underlying notional value equal to the current level of the S&P/MIB index multiplied by

---

* The S&P/MIB index is derived from the universe of stocks trading on the Italian stock exchange main equity market. The index has been created to be suitable for futures and options trading, as a benchmark index for exchange traded funds, and for tracking large capitalization stocks in the Italian market. It is calculated in real time at 30" (09:05–17:31 CET), from the continuous trading phase in the blue-chip segment of electronic shares market (MTA and MTAX), using the last price of each constituent. The S&P/MIB index provides diversity over ten economic sectors by adhering to the Global Industry Classification Standard, or GICS2. Launched in 1999 by Standard & Poor's and Morgan Stanley Capital International (MSCI), GICS has become the industry standard, providing the financial community with one complete set of global sector and industry definitions. The ten GICS sectors that underlie the S&P/MIB index are consumer discretionary, consumer staples, energy, financials, health care, industrials, information technology, materials, telecommunication services, and utilities.

2.5 euro. Many expirations are available, and quite a few number of strikes are quoted. The complete contract specifications are reported in the following box (IDEM, 2006).

THE S&P/MIB INDEX OPTION CONTRACT SPECIFICATIONS

Exercise: S&P/MIB Index Options are exercisable only on their expiration day.

Contract size: Each S&P/MIB Index Option represents a notional value of 2.5 euro per index point. This means that if the index value equals 30,000, each contract has an underlying value of 75,000 euro.

Expirations up to 12 months: Six expirations are always available for trading: four quarterly (March, June, September, and December) expirations, plus the two nearest nonquarterly calendar months.

Expirations beyond 12 months up to 36 months: Four 6-month expirations are always available (June and December). For each maturity up to 12 months (monthly and 3-month maturities), at least fifteen exercise prices shall be traded for both the call and the put series, with intervals of 500 index points. For the four 6-month maturities more than 12 months, at least twenty-one exercise prices shall be listed for both the call and the put series, with intervals of 1,000 index points. But when 6-month maturities fall within the 12 months, new exercise prices shall be introduced with intervals of 500 index points, up to at least fifteen exercises prices shall be traded for both the call and put series, with intervals of 500 index points.

At the end of each trading day, all of the maturities of the following option series shall be excluded from listing:

- call series whose exercise price, with respect to that of the at-the-money series, is higher than the 10th out-of-the-money exercise price or lower than the 10th in-the-money exercise price; or
- put series whose exercise price, with respect to that of the at-the-money series, is lower than the 10th out-of-the-money exercise price or higher than the 10th in-the-money exercise price; and
- for which the following conditions are simultaneously satisfied:
  - there is no open interest
  - the open interest of the put (call) with the same exercise price and maturity is zero
  - the open interest of all the call and put series with exercise price furthest from the at-the-money price, with respect to that of the series to be excluded, is zero

New exercise prices shall be introduced where the reference value of the S&P/MIB Index of the preceding trading day is:

- for call options, higher (lower) than the average of the at-the-money price and the first out-of-(in-)the-money price
- for put options, higher (lower) than the average of the at-the-money price and the first in-(out-of-)the-money price

Borsa Italiana may introduce additional strike prices with respect to those referred to when it is necessary to ensure regular trading, with account taken of the performance of the underlying index. The strike prices will be generated with the interval specified in paragraph 4 or their multiples for call and put options.

Premiums: The premiums of S&P/MIB Index Options are quoted in index points. The level of an Index Option's premium also determines the minimum price movement for the Index Option as follows:

Option Price	Min. Tick Value
Less than 100 index points	1 index point
Between 102 and 500 index points	2 index points
Above 505 index points	5 index points

SOURCE: HTTP://WWW.BORSAITALIANA.IT

18.4 THE EMPIRICAL EVIDENCE

Since the analysis was aimed at evaluating the usage of stock index option, we had to select a definite risk horizon. For the sake of the efficiency of the data set, such risk horizon was set moderately short: the closest expiration contract was selected for each day. Therefore, the maximum risk horizon is no longer than 20 working days. The data derive directly from the Italian CC&G database.[*] The specific information retrieved from the database are:

1. the closing prices of the underlying stock index for the option (index points)

2. the annualized historical volatility calculated as the standard deviation of the daily log returns of the index over the preceding 21 working days multiplied by $\sqrt{252}$

---

[*] The database can be requested directly to the academic service of Borsa Italiana.

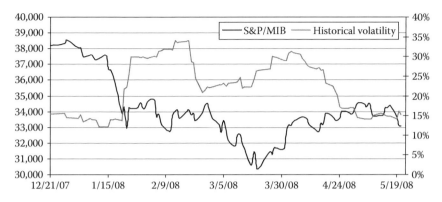

FIGURE 18.2   S&P/MIB and historical volatility (12/21/2007–05/21/2008).

3. the open interests of the next closer maturity put and call contract (all strikes), that is, the total number of options contracts that are not closed or delivered at the end of the day

4. the implied volatility for each contract, as given by the CC&G system[*]

The time period selected is able to cover a full cycle for both S&P/MIB index price and the historical volatility as can be easily appreciated by Figure 18.2, reporting the daily S&P/MIB closing values and the corresponding historical volatility.

For each trading day we built a bubble plot of the implied volatility against the corresponding strike prices for both call and put option, where the size of the single bubble is proportional to the open interest of the selected contract. By this representation it is possible to graphically evaluate the area with higher density. The four quadrants of this coordinate system are made up by the intersection of the day closing S&P/MIB value (index points) with the corresponding historical volatility value. The results of this graphical analysis are reported in the charts in Figures 18.1–18.11, where daily plots were grouped on a monthly basis, with blank spaces for holidays.

---

[*] The implied volatility is calculated by a recursive procedure. The clearing house applies for the risk-free rate, the Euribor rate (Euro Interbank Offered Rate) referred to the same maturity of the option if available, or to an interpolated value between the two closer maturities if not directly available. As far as the dividend correction is concerned, a calibrated estimate of the dividend of the underlying stock basket is applied. Technical details can be obtained directly from CC&G (http://www.ccg.it/) or Borsa Italiana (http://www.borsaitaliana.it/).

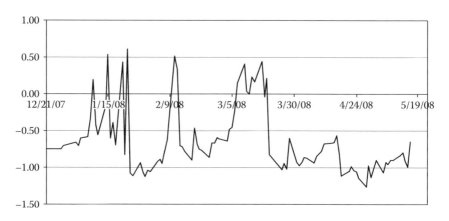

FIGURE 18.3   Strategy Index (21/12/2007 – 16/05/2007).

A synthetic depiction of the prevailing strategy can be gained by the analysis of the ratio of the daily correlation coefficient between the implied volatility and the corresponding strike price to the put-call, i.e. the Strategy Index (SI), given by

$$Strategy\ Index = \frac{Correlation\,(implied\ volatility;\, strike\ price)}{Put\ Open\ Interest/Put\ Open\ Interest}$$

The daily series of the SI is reported by Figure 18.3.

A positive value of the index accounts for a positive relationship between volatility and price expectations. In this case the market will express a preference for call options in the long position and put options in the short position or for short call and long put. Therefore, the positive value of the strategy index is able to signal a prevailing volatility trading. This is the case, for example, of January 23, which exhibits the maximum value of the SI and whose OSM is reported by Figure 18.4.

A negative value of the index accounts for a negative relationship between volatility and price expectations. In this case the market will express a preference for long put and short call or long call and short put. Therefore, the negative value of the strategy index is able to signal a prevailing directional strategy. This is the case, for example, of April 28, which exhibits the minimum value of the SI and whose OSM is reported by Figure 18.5.

A null value of the SI is able to account for no strong relationships between volatility and price expectations. In this case the market will not express a prevailing strategy. This is the case of March 13, which exhibits a null value of the SI and whose OSM is reported by Figure 18.6.

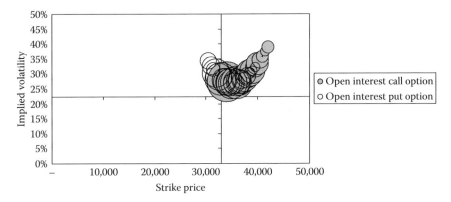

FIGURE 18.4   Option strategy map (01/23/2008).

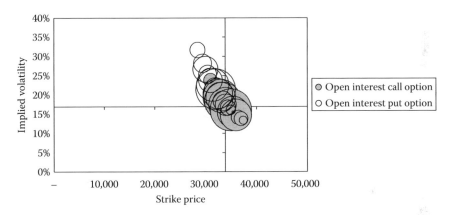

FIGURE 18.5   Option strategy map (04/28/2008).

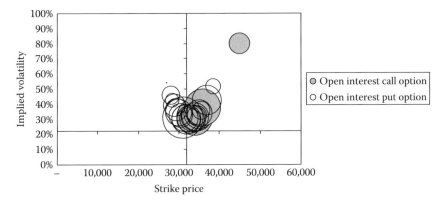

FIGURE 18.6   Option strategy map (03/13/2008).

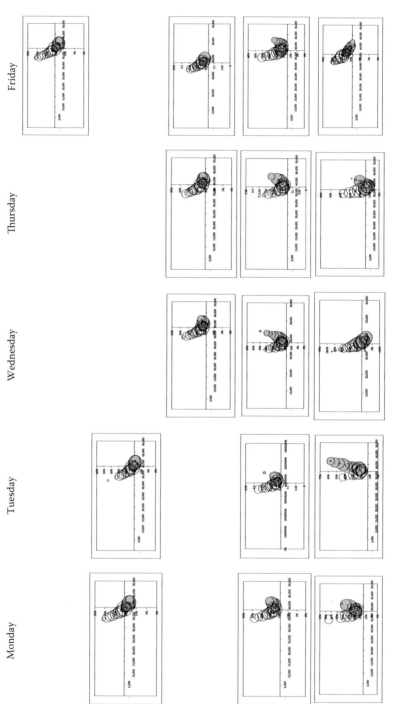

FIGURE 18.7 Option strategy map (12/21/2007–01/18/2008) (gray bubble call option–white bubble put option).

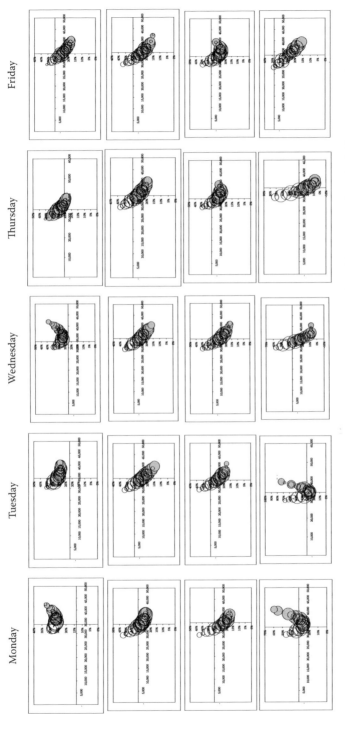

FIGURE 18.8 Option strategy map (01/21/2008–02/15/2008) (gray bubble call option–white bubble put option).

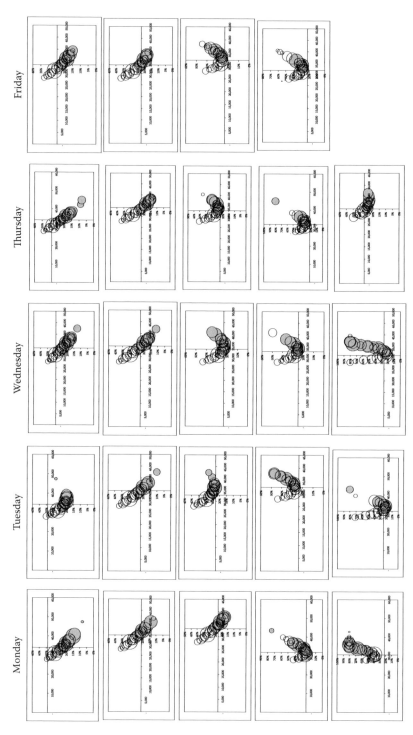

FIGURE 18.9 Option strategy map (02/18/2008–03/20/2008) (gray bubble call option–white bubble put option).

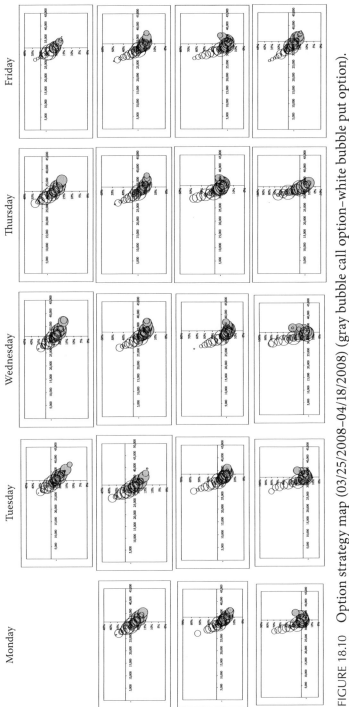

FIGURE 18.10 Option strategy map (03/25/2008–04/18/2008) (gray bubble call option–white bubble put option).

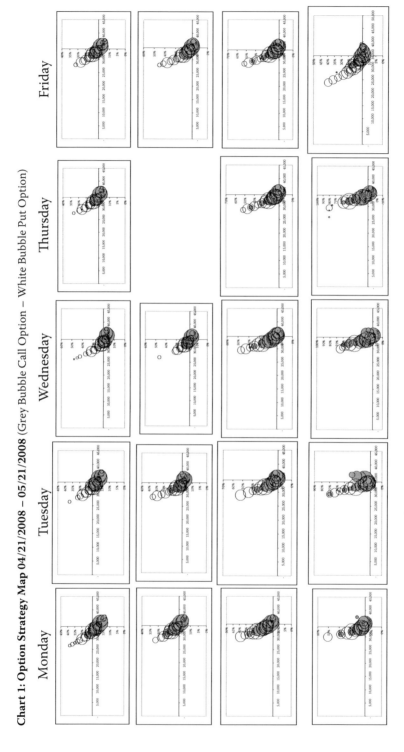

Chart 1: Option Strategy Map 04/21/2008 – 05/21/2008 (Grey Bubble Call Option – White Bubble Put Option)

FIGURE 18.11   Option strategy map (04/21/2008–05/21/2008) (gray bubble call option–white bubble put option).

18.5   CONCLUSION

As it can be easily appreciated by the SI evolution (Figure 18.3), the Italian market for the S&P/MIB stock index option is generally prone, at least with reference to the selected data set, to directional strategies. Very rarely the market shows volatility trading. The results obtained therefore confirm a pragmatic intuition: the usage of the stock index option is mainly related to immunization objectives. Certainly, these results could be different if we extended the time period and the number of expirations under consideration. There is also some seasonal effect in the behavior that could be further exploited, especially with reference to a higher number of expirations for each observation date.

From a methodological standpoint, the Option Strategy Matrix and the Strategy Index appear to be two useful instruments for a general analysis of the market.

REFERENCES

Black, F., and Scholes, M. (1973). The pricing of options and corporate liabilities. *Journal of Political Economy* 8:654–737.

Borsa Italiana. (2008). Monthly key figures. Technical document 11(5), Milan, Italy. Available at http://www.borsaitaliana.it/documenti/statistiche/mediaitaliano/sintesimensili/2008/sintesimensili200805.en_pdf.htm.

Cox, J. C., Ross, S. A., and Rubinstein, M. (1979). Option pricing: A simplified approach. *Journal of Financial Economics* 7:229–63.

Hull, J. C. (2006). *Options, futures, and other derivatives*. 6th ed. Upper Saddle River, NJ: Prentice Hall.

IDEM. (2006). Characteristics and risks of IDEM equity and index option. Technical document. Available at http://www.borsaitaliana.it/quotazioni/derivati/optionsdisclosuredocument_pdf.htm.

Lakonishok, J., Lee, I., and Poteshman, A. M. (2004). Investor behavior and the option market. Working paper, National Bureau of Economic Research, Cambridge, MA. Available at http://www.nber.org/papers/w10264.

Merton, R. C. (1973). The theory of rational option pricing. *Bell Journal of Economics and Management Science* 4:141–83.

Neftci, S. H. (2004). *Financial engineering*. Amsterdam: Elsevier.

Wilmott, P. (2007). *Paul Wilmott introduces quantitative finance*. 2nd ed. Chichester, UK: John Wiley & Sons.

CHAPTER 19

Cross-Sectional Return Dispersions and Risk in Global Equity Markets

Thomas C. Chiang

CONTENTS

19.1 INTRODUCTION

Investors have long considered the behavior of stock return volatility as an important factor in forming their portfolio decisions. For instance, in the mean-variance framework, expectations about the volatility of returns influence portfolio choice through investors' demand for a required rate of return and their attitude toward risk. From a market perspective, the magnitude of stock market volatilities also provides information for assessing the economy, since periods of high volatility tend to coincide with downward market movements (Ang et al., 2006). The financial crisis literature suggests that in an integrated capital market, stock return volatility in one market tends to spill over to other asset markets through a contagion effect, causing widespread financial market instability. Thus, understanding the behavior of volatility and its relationship to stock returns is crucial to investors and government policy makers/regulatory agencies.

In his seminal paper, Merton (1973) presents a theoretical analysis of the expected stock return in relation to risk. He postulates a positive relation between expected excess returns and risk. Following Merton's (1973, 1980) theoretical prediction, voluminous studies have been devoted to investigating this risk-return hypothesis. For instance, French et al. (1987) find evidence that the expected market return is positively related to the predicted volatility of stock returns. Similar findings are shown in the research papers by Baillie and DeGennaro (1990), Ghysels et al. (2005), and Bali and Peng (2006), among others.

However, Glosten et al. (1993) and Koopman and Uspensky (2002) document different results and find evidence of a negative relation between stock return and the predicted volatility. Thus, the empirical evidence on the risk-return trade-off is inconclusive. In reviewing the existing literature, it appears that the sign of the return-risk relation is conditioned on the models being used, the time horizons of the sample under study, and the way risk is being measured, among other factors (Backus and Gregory, 1993; Bali and Peng, 2006).

Since stock returns are observed to be stochastic and the evolution of stock return volatility displays a clustering phenomenon, over the last two decades, the GARCH-type models proposed by Engle and his associates have been considered the most popular approach for modeling the time-varying risk process, including the studies in the above-mentioned literature. The beauty of a GARCH-in-mean model is that it allows researchers to estimate the return and conditional variance simultaneously, so that the link between stock returns and their predicted variance can be established. Using this model, financial economists can conduct empirical analyses for testing the return-risk trade-off hypothesis. Collectively, the work by Engle (1995) and the survey by Bollerslev et al. (1992) summarize the applications of GARCH models in the finance literature.

Despite its powerful capacity for capturing the time-series properties of volatility, the GARCH model fails to capture the risk reflected in
the cross-sectional data. The purpose of this paper is to use the cross-
sectional return dispersion as an incremental variable for the risk factor to explain stock return. We demonstrate that the return dispersions
contain information on excess volume, excess market conditions, and
outliers. Thus, by including the return dispersion, the test equation for
conditional variance is able to outperform a simple conditional time-series variance model.

The remainder of the chapter is organized as follows: Section 19.2
describes the model, sample data, and variable measurements. Section 19.3
presents the empirical evidence on the risk-return relation using a simple GARCH-M model and augmented GARCH-M model incorporating
return dispersions. Section 19.4 presents the information content of cross-
sectional return dispersions. Section 19.5 contains conclusions.

19.2 MODEL, SAMPLE DATA DESCRIPTION, AND VARIABLE MEASUREMENTS

19.2.1 The AGARCH-M Model

It is convenient to start with a simple asymmetric GARCH(1,1)-M model
that will serve as a basis for comparison. The model can be expressed by

$$R_t = \delta + \phi R_{t-1} + \theta \cdot \sqrt{h_t} + \varepsilon_t \tag{19.1}$$

$$h_t = \omega_0 + \omega_1 \varepsilon_{t-1}^2 + \omega_2 I[\varepsilon_{t-1} < 0] \cdot \varepsilon_{t-1}^2 + \omega_3 h_{t-1} \tag{19.2}$$

where R_t is the stock return; h_t is the conditional variance; δ, ϕ, θ, and
ω_i ($i = 0, 1, 2,$ and 3) are constant parameters; I is an indicator variable,
$I = 1$ if $\varepsilon_{t-1} < 0$, and $I = 0$, otherwise; and ε_t is a random error term.

Equation (19.1) is the mean equation, which is assumed to follow an AR(1)
process, reflecting nonsynchronous trading (Lo and MacKinlay, 1990), partial adjustment (Koutmos, 1998), or the presence of positive-feedback trading (Sentana and Wadhwani, 1992; Antoniou et al., 2005). The term $\sqrt{h_t}$ is
the conditional standard error, which is used to proxy for measuring risk.

Thus, a positive value and statistical significance on θ from a regression test would be evidence supporting the return-risk trade-off hypothesis.[*]

Equation (19.2) describes an evolutionary process of conditional variance, h_t, which is assumed to be dependent on past shocks squared and an AR(1) term of variance. We assume that the conditional variance follows a GARCH (1,1) process based on the parsimonious principle popularized by Bollerslev et al. (1992). In this model, we also include an indicator variable, I, $I = 1$ if $\varepsilon_{t-1} < 0$, and 0 otherwise, to reflect the asymmetric responses of the conditional variance to previous shocks (Glosten et al., 1993; Chiang and Doong, 2001).[†] It follows that good news, $\varepsilon_{t-1} > 0$, and bad news, $\varepsilon_{t-1} < 0$, have differential effects on the conditional variance; good news has an impact of ω_1, while bad news has an impact of $\omega_1 + \omega_2$. If $\omega_2 > 0$, the evidence suggests that bad news aggravates volatility. Therefore, if $\omega_2 = 0$ is rejected from a regression estimation, we would conclude that the impact of news is asymmetric. Thus, this AGARCH model is appealing, since the return and variance processes are estimated jointly and the variance is characterized by time-varying and asymmetric responses to previous shocks.

19.2.2 The Data

To estimate the model, we use daily stock price indices for five major markets from January 4, 1990 through December 31, 2006. The data consist of both sector stock indices and market price indices. The samples cover Hong Kong (HK), Japan (JP), Germany (GR), the United Kingdom (UK), and the United States (US). The industrial data set for each market contains 156 sectors. Following the conventional approach, we calculate stock return as $R_t = (p_t - p_{t-1}) \times 100$, where p_t and p_{t-1} are the natural logarithms of a stock index for each market or sector at time t and $t - 1$. All of the data were taken from Datastream International.

19.2.3 Evidence of the AGARCH-M Model

Table 19.1 reports the regression estimates of Equations (19.1) and (19.2). Consistent with the partial adjustment hypothesis (Koutmos, 1998) or the

---

[*] In the literature, both the standard error of stock returns and the variance of stock returns have been used to proxy for risk (see French et al., 1987). Here we report only the results from using the standard error to save space.

[†] A number of conditional variance models in the GARCH family have been proposed in the literature. We refer to Engle (1995) or the paper by Cappiello et al. (2006) for details.

TABLE 19.1 Estimates of Stock Return and Conditional Volatility Using the AGARCH(1,1)-M Model

Index	δ	ϕ	θ	ω_0	ω_1	ω_2	ω_3	$-LL$
HK	0.063	0.067***	-0.017	0.078***	0.049***	0.121***	0.840***	6,606
	(1.25)	(4.89)	(0.35)	(6.66)	(3.09)	(5.78)	(51.15)	
JP	-0.073	0.088***	0.075	0.042***	0.030***	0.134***	0.866***	5,892
	(1.44)	(5.54)	(1.34)	(6.90)	(3.73)	(9.98)	(75.18)	
GR	0.042	0.109***	0.006	0.020***	0.039***	0.112***	0.866***	4,704
	(1.27)	(6.73)	(0.11)	(8.28)	(3.52)	(7.39)	(74.53)	
UK	0.039**	0.113***	0.013	0.013***	0.053***	0.111***	0.873***	3,941
	(1.87)	(8.94)	(0.33)	(4.58)	(3.23)	(5.12)	(52.80)	
US	-0.003	0.086***	0.060	0.013***	0.017**	0.116***	0.911***	4,994
	(0.07)	(5.49)	(1.23)	(6.36)	(2.28)	(10.44)	(110.51)	

Note: The statistics in this table are based on the following regression system:

$$R_t = \delta + \phi R_{t-1} + \theta \cdot \sqrt{h_t} + \varepsilon_t$$

$$h_t = \omega_0 + \omega_1 \varepsilon_{t-1}^2 + \omega_2 I[\varepsilon_{t-1} < 0] \cdot \varepsilon_{t-1}^2 + \omega_3 h_{t-1}$$

where I is an indicator variable, $I = 1$ if $\varepsilon_{t-1} < 0$, and 0 otherwise, reflecting asymmetric responses of the conditional variance to previous shocks. Values in parentheses are the absolute z-statistics. ***, **, and * denote significance at the 1%, 5%, and 10% levels, respectively. $-LL$ denotes minus the log-likelihood function.

presence of nonsynchronous trading (Lo and MacKinlay, 1990), all of the coefficients of the AR(1) term are positive, lying between zero and unity, and statistically significant.

While testing the return-risk trade-off hypothesis, we cannot find evidence that the estimated coefficient on the conditional standard error term, θ, is statistically significant. This indicates the weakness of the return-risk trade-off hypothesis, since none of the markets offer supportive evidence in their daily data. Nevertheless, we should be more prudent in drawing a conclusion based on this preliminary model. Two related issues are worth considering. First, it is important to ask whether the mean equation is correctly specified, since a misspecification of the mean equation could generate a misleading error term, which could in turn distort the measure of the conditional variance. Second, the variance in Equation (19.2) may well capture its time-series pattern based on a specific dynamic process. However, other information, such as macroeconomic risk factors or cross-sectional variations revealed in the market data, is excluded from the model. Since most macroeconomic fundamental variables are not available on a daily basis, this leads us to focus on the second issue and ask whether the conditional variance equation is correctly specified. We shall return to this issue later.

Checking the variance equation, the evidence shows that all of the coefficients in the GARCH(1,1) equations are statistically significant, indicating that stock-return volatilities are characterized by a heteroskedastic process with a clustering phenomenon. Since the sum of the coefficients on the variance equation is close to unity, we find that the volatility is highly persistent. A special feature emerging from the variance equation is that the coefficient of the asymmetric term, ω_2, is positive and highly significant. This holds true for all of the markets. We can conclude that the asymmetric effect is present in all of the stock return series, demonstrating that volatility is higher in a falling market than it is in a rising market.

19.3 THE ROLE OF CROSS-SECTIONAL RETURN DISPERSION

19.3.1 Stock Returns and Cross-Sectional Return Dispersion

As we argued in the previous section, the failure to confirm the return-risk trade-off hypothesis using data from the five markets mentioned previously may be due to the fact that conditional variance based on a univariate GARCH specification may not sufficiently capture the information on risk, and hence lacks the power to measure risk. As noted by Pagan and Ullah (1988), in the GARCH-M model, the estimates of the parameters in the conditional mean equation are not asymptotically independent of the estimates of the parameters in the conditional variance; hence, any misspecification in the variance equation generally leads to biased and inconsistent estimates of the parameters in the mean equation. In this section, we shall use the cross-sectional return dispersion (CRD) as an incremental variable to explain stock returns.

In the literature, the dispersion of stock returns may project a depressed state of the economy, signifying higher risk for holding stocks in a portfolio. For instance, Loungani et al. (1990) examine data on U.S. monthly stocks and discover that stock return dispersion leads to unemployment. Christie and Huang (1995) find that the dispersion is higher during recessions and positively covaries with the yield spread between high- and low-rated corporate bonds. Connolly and Stivers (2006) find that the daily return dispersion contains information about the future volatility of portfolio returns. In general, higher return dispersion may represent less agreement among investors concerning the outcomes of market returns (Connolly and Wang, 2003). It is in this sense that return dispersion reflects different beliefs, indicating more risk aversion toward the stock market.

Chang et al. (2000) and Duffee (2001) use the cross-sectional absolute deviation to measure return dispersion because it is less sensitive to outliers. Specifically, CRD_t is defined as

$$CRD_t = \frac{1}{N} \sum_{i=1}^{N} |R_{i,t} - R_{m,t}| \qquad (19.3)$$

where CRD_t is measured by the average of the cross-sectional absolute deviation from the market return on a particular trading day, $R_{i,t}$ is the stock return for sector i ($i = 1, 2, 3, \ldots, 156$), and $R_{m,t}$ is the mean value of the 156 industrial stock returns for each national market. As we argued earlier, CRD_t may have substantial information content for measuring risk. A simple way to test its significance is to add the CRD_t term to the right-hand side of the mean equation as expressed by

$$R_t = \delta + \phi R_{t-1} + \theta \cdot \sqrt{h_t} + \eta \cdot CRD_t + \varepsilon_t \qquad (19.4)$$

where CRD_t can be viewed as an incremental variable for explaining stock returns. Equations (19.4) and (19.2) are estimated jointly, and the results are reported in Table 19.2.[*]

The estimates in Table 19.2 suggest that the coefficients of the variance equation and the AR(1) term produce results very similar to those we achieved earlier. Although some of the coefficients on θ are statistically significant, the sign is negative. Thus, no evidence is found to support the return-risk trade-off hypothesis. However, there is an interesting finding for the coefficient of CRD_t. That is, the coefficient of η shows a positive value and is statistically significant for all of the markets. From this perspective, we find some evidence of a positive relationship between stock returns and risk. The evidence tends to point in the direction of CRD_t having more information content for explaining stock returns.

19.3.2 Conditional Variance and Cross-Sectional Return Dispersion

Instead of placing CRD_t in the mean equation, we attempt to use CRD_t to model the conditional variance. In particular, we write:

$$R_t = \delta + \phi R_{t-1} + \theta_F \cdot \sqrt{h_{F,t}} + \varepsilon_{it} \tag{19.5}$$

$$h_{it} = \varpi + \alpha \varepsilon_{i,t-1}^2 + \beta h_{i,t-1} + \eta \varepsilon_{i,t-1}^2 I_{t-1} \tag{19.6}$$

where $h_{F,t}$ in Equation (19.6) is the full information estimator of variance based on the conditional variance h_t in Equation (19.2) plus the risk information denoted by CRD_t and CRD_{t-1}. Thus, the risk factor in the mean Equation (19.5) as measured by $\sqrt{h_{F,t}}$ contains the information derived from both the time-series pattern of the variance and the cross-sectional dispersions. The estimations of Equations (19.5) and (19.6) are reported in Table 19.3.

By checking the estimates in Table 19.3 and comparing them with those reported in Table 19.1, where the information contained in CRD_t was excluded from the conditional variance, we find that the coefficients

---

[*] Alternatively, we can use the cross-sectional standard deviation (CSSD) to calculate return dispersion (Christie and Huang, 1995), expressed as $CSSD_t = (\Sigma_{i=1}^{N}(R_{i,t} - R_{m,t})^2/(N-1))^{1/2}$, where N is the number of sectors in the portfolio, $R_{i,t}$ is the observed stock return of industry i at time t, and $R_{m,t}$ is the cross-sectional average stock of N returns in the portfolio at time t. Since the results are similar, we do not report them in Table 19.2.

TABLE 19.2   Estimates of Stock Return Based on Time-Series Volatility and Cross-Sectional Return Dispersions

Index	δ	ϕ	θ	η	ω_0	ω_1	ω_2	ω_3	$-LL$
HK	-0.047	0.086***	-0.138**	0.239***	0.089***	0.028***	0.155***	0.823***	7,918
	(0.83)	(5.84)	(2.49)	(11.44)	(12.87)	(2.79)	(12.95)	(86.04)	
JP	-0.107***	0.083***	-0.155***	0.335***	0.030***	0.021***	0.163***	0.867***	7,011
	(2.63)	(5.66)	(3.13)	(9.04)	(8.09)	(3.19)	(17.11)	(121.56)	
GR	0.031	0.111***	-0.111*	0.119***	0.033	0.034***	0.155***	0.830***	5,447
	(0.82)	(6.73)	(1.84)	(5.24)	(11.98)	(3.31)	(12.12)	(75.43)	
UK	0.046*	0.161***	-0.066	0.053**	0.016***	0.064***	0.107***	0.844***	4,808
	(1.64)	(10.55)	(1.24)	(2.52)	(10.51)	(6.83)	(10.79)	(94.33)	
US	-0.018	0.083***	0.018	0.073**	0.019***	0.015**	0.122***	0.894***	6,178
	(0.49)	(5.57)	(0.34)	(2.05)	(9.47)	(2.01)	(12.46)	(131.74)	

Note: The statistics in this table are based on the following regression system:

$$R_t = \delta + \phi R_{t-1} + \theta \cdot \sqrt{h_t} + \eta \cdot CRD_t + \varepsilon_t$$

$$h_t = \omega_0 + \omega_1 \varepsilon_{t-1}^2 + \omega_2 I[\varepsilon_{t-1} < 0] \cdot \varepsilon_{t-1}^2 + \omega_3 h_{t-1}$$

where I is an indicator variable, $I = 1$ if $\varepsilon_{t-1} < 0$, and 0 otherwise, reflecting asymmetric responses of the conditional variance to previous shocks. CRD_t is the cross-sectional return dispersions on the industrial sectors. Values in parentheses are the absolute z-statistics. ***, **, and * denote significance at the 1%, 5%, and 10% levels, respectively. $-LL$ denotes minus the log-likelihood function.

TABLE 19.3 Estimates of Stock Return and Conditional Volatility Based on AGARCH-M and Return Dispersion

Index	δ	ϕ	θ_F	ω_0	ω_1	ω_2	ω_3	ω_4	ω_5	$-LL$
HK	-0.210***	0.074***	0.266***	0.051***	0.034*	0.125***	0.589***	1.107***	-0.770***	6,344
	(5.07)	(5.08)	(6.13)	(2.94)	(1.76)	(5.25)	(15.93)	(27.74)	(65.10)	
JP	-0.218***	0.064***	0.235***	0.045***	0.011	0.077***	0.793***	1.970***	-1.736***	5,892
	(6.72)	(4.35)	(5.96)	(5.03)	(1.34)	(5.12)	(76.71)	(104.58)	(85.17)	
GR	-0.010	0.106***	0.089**	0.013***	0.000	0.153***	0.666***	0.578***	-0.419***	4,353
	(0.45)	(7.27)	(2.34)	(2.11)	(0.03)	(5.78)	(14.55)	(25.88)	(13.90)	
UK	-0.067***	0.129***	0.019***	0.408***	0.800	0.794***	0.256***	17.614***	-5.598***	5,229
	(6.39)	(59.85)	(5.60)	(3.13)	(0.93)	(3.07)	(5.54)	(5.28)	(3.54)	
US	-0.054*	0.070***	0.130***	-0.001	0.000	0.065***	0.925***	0.927***	-0.897***	4,994
	(1.67)	(4.60)	(2.78)	(0.30)	(0.03)	(6.54)	(80.52)	(79.60)	(255.28)	

Note: The statistics in this table are based on the following regression system:

$$R_t = \delta + \phi R_{t-1} + \theta_F \cdot \sqrt{h_{F,t}} + \varepsilon_t$$

$$h_{F,t} = \omega_0 + \omega_1 \varepsilon_{t-1}^2 + \omega_2 I[\varepsilon_{t-1} < 0] \cdot \varepsilon_{t-1}^2 + \omega_3 h_{F,t-1} + \omega_4 CRD_t + \omega_5 CRD_{t-1}$$

where I is an indicator variable, $I = 1$ if $\varepsilon_{t-1} < 0$, and 0 otherwise, reflecting asymmetric responses of the conditional variance to previous shocks. CRD_t is the cross-sectional return dispersions in the industrial sectors. Values in parentheses are the absolute z-statistics. ***, **, and * denote significance at the 1%, 5%, and 10% levels, respectively. $-LL$ denotes minus the log-likelihood function.

of the AR(1) term in the mean equation, θ, and in the variance equation, ω_3, are highly significant. This is understandable, since the lagged dependent variable has the capacity to summarize the historical information, including the missing values for longer lags of the explanatory variables. The statistics from Table 19.3 also support the hypothesis that volatility responds asymmetrically to previous shocks, as evidenced by the statistical significance of ω_2.

The results shown in Table 19.3 demonstrate three significant changes compared with those in Table 19.1. First, none of the lagged shock squared terms in the variance equation are found to be statistically significant at the 5% level, which means that the information content in the ARCH component has been vanishing. Second, the coefficients for both CRD_t and CRD_{t-1} are highly significant, suggesting that the ARCH effect has been overtaken by the cross-sectional return dispersion. Third, and more important, the estimated values on θ_F are all positive and statistically significant at the 1% level. This finding suggests that higher return is associated with a higher risk, supporting the return-risk trade-off hypothesis. The validity of this finding is rooted in the measure of the risk variable, $h_{F,t}$, which is not only evolving with its time-series dynamics, but also influenced by the cross-sectional dispersions. The latter appears to be more significant. The next question then is: What information lies behind CRD_t? We shall answer this in the next section.

19.4 WHAT EXPLAINS THE CROSS-SECTIONAL RETURN DISPERSIONS?

It has been observed that fear rises and risk increases when the market undergoes extreme movements. This phenomenon often shows up in excess trading volumes. Clark (1973) documents that the variances of stock returns and trading volumes are both driven by the same latent variable. In their study of U.S. stocks, Lamoureux and Lastrapes (1990) demonstrate that trading volume has significant explanatory power for stock returns. They show that when the volume variable is included in the estimated equation, the GARCH effect is weakened. Wagner and Marsh (2005) extend the Lamoureux and Lastrapes (1990) model by considering an asymmetric GARCH-in-mean specification. They contend that surprise volume (unexpected above-average trading volume) provides a superior model fit and helps to explain the persistence of volatility as well as excess kurtosis. Following this line of argument, we set up the cross-sectional return dispersion as a function of excess trading volume and market stock return squared. Specifically, we write:

$$CRD_t = \beta_0 + \beta_1 U_t + \beta_2 R^2_{m,t} + \sum_{i=1}^{5} \lambda_i D_i + \varepsilon_i \qquad (19.7)$$

where U_t is excess trading volume, which is calculated by subtracting normal volume from actual volume. Following Wagner and Marsh's (2005) model, normal volume is obtained using the Hodrick-Prescott (1997) filtering method.[*,†] The $R^2_{m,t}$ term is the market return squared to reflect extreme market conditions (Chang et al., 2000). In Equation (19.7), we also include five dummy variables to insulate the contamination of outliers that may distort the estimated results (Tsay, 1988; Peña, 2001).

Table 19.4 reports the regression estimates of the cross-sectional return dispersion equation. Two significant results are worth noting. First, with the exception of the UK market, the excess volume variable is highly significant, meaning that the return dispersion reflects the market activity of excess trading. Second, the return dispersion is positively correlated with the market return squared and all of the coefficients are significant at the 1% level. This finding suggests that when the market undergoes extreme movements, the market return dispersions are expected to be more diverse, which may reflect more profound fear and risk aversion. Given the information content of Equation (19.7), we can link it to Equation (19.6), which helps us to understand the predictive power of return dispersion in the variance equation. Because of this information content, we find evidence to support the return-risk trade-off hypothesis.

19.5   CONCLUSION

This chapter examines the relation between stock return and risk by applying the data from five major global markets: Hong Kong, Japan, Germany, the United Kingdom, and the United States. Testing the hypothesis by using a standard asymmetric GARCH(1,1) model, we cannot find any

---

[*] The natural logarithm of trading volume is used to restrict the volume to be nonnegative.

[†] As expounded by Longin (1997), return volatility, volume, and liquidity are all positively related to each other, although these variables may be associated with different trading processes. To some extent, the volume can be set up as a proxy of liquidity, which has the advantage of being easy to measure. Based on the information we observed, it is appealing to use excess trading volume to explain CRD_t.

TABLE 19.4 Estimates of the Cross-Sectional Return Dispersion Based on the Excess Trading Volume and Stock Return Squared

Index	β_0	β_1	β_2	λ_1	λ_2	λ_3	λ_4	λ_5	$\bar{R}^2$
HK	1.216***	0.007**	0.043***	4.570***	7.559***	5.737***	5.2968**	4.954***	0.28
	(64.13)	(2.02)	(8.61)	(52.95)	(51.197)	(65.79)	(73.518)	(57.46)	
JP	0.819***	97.707***	0.036**	2.479***	1.978***	2.041***	0.767***	2.371***	0.19
	(56.77)	(6.80)	(7.58)	(151.39)	(53.67)	(35.53)	(5.085)	(85.85)	
GR	0.893***	0.019***	0.085***	1.595***	0.840***	1.773***	1.566***	2.130***	0.22
	(44.18)	(2.63)	(3.61)	(15.08)	(26.18)	(18.66)	(105.32)	(25.07)	
UK	0.704***	0.005	0.141***	-1.504***	2.037	2.274***	0.912***	1.678***	0.34
	(61.68)	(1.34)	(11.72)	(5.11)	(1.39)	(118.83)	(29.36)	(65.79)	
US	0.776***	3.231***	0.048***	1.871***	1.405***	0.591***	1.917***	0.979***	0.27
	(61.23)	(9.65)	(6.94)	(72.92)	(17.39)	(25.89)	(36.77)	(3.89)	

Note: The statistics in this table are based on the following regression using a Newey-West (1987)–consistent estimator:

$$CRD_t = \beta_0 + \beta_1 U_t + \beta_2 R_{m,t}^2 + \sum_{i=1}^{5} \lambda_i D_i + \varepsilon_t$$

where $CRD_t = \frac{1}{N}\sum_{i=1}^{N}|R_{i,t} - R_{m,t}|$ is the dependent variable, $R_{m,t}$ is the value of an equally weighted realized return of all industry indexes on date t, $R_{m,t}^2$ is the square term for each market, and U_t is the excess trading volume. The reported coefficient of U_t is obtained by multiplying by 10^3. D_i is the dummy variable; it takes a value of unity for present outlier, otherwise 0. Five dummy variables are included in the test equation to capture unusual market disturbances. The dates of dummy variables for each market are identified as follows:

HK: $(D_1, D_2, D_3, D_4, D_5) = (5/7/93, 3/10/00, 3/20/00, 6/27/01, 7/12/01)$
JP: $(D_1, D_2, D_3, D_4, D_5) = (1/5/00, 1/11/00, 3/13/00, 4/17/00, 4/21/00)$
GR: $(D_1, D_2, D_3, D_4, D_5) = (3/10/98, 6/24/99, 12/3/99, 2/28/00, 1/2/01)$
UK: $(D_1, D_2, D_3, D_4, D_5) = (9/17/92, 4/2/99, 3/16/00, 7/5/01, 10/2/02)$
US: $(D_1, D_2, D_3, D_4, D_5) = (4/19/99, 3/15/00, 1/3/01, 1/4/01, 9/17/01)$

The daily data span the period from January 4, 1990 to December 31, 2006 for Hong Kong (HK), Japan (JP), Germany (GR), the United Kingdom (UK), and the United States (US). All of the daily data are taken from the database of Datastream International. ***, **, and * denote significance at the 1%, 5%, and 10% levels, respectively.

supporting evidence for the return-risk trade-off hypothesis, although a substantial clustering of volatility is present.

We then construct the cross-sectional return dispersions (CRD) based on 156 sectors for each market and use CRD as an independent argument for explaining the stock return for each market. It turns out that there is a significant relation between stock return and CRD. This finding suggests that using a univariate specification of the AGARCH model may not sufficiently capture all of the information pertinent to explaining the variance evolution. Our empirical test indicates that by including cross-sectional return dispersions in the prediction of the conditional variance, we find evidence supporting the return-risk trade-off hypothesis.

Further testing the information content underlying cross-sectional return dispersions, we find evidence to suggest that the CRD is positively correlated with excess trading volume and market return squared. This implies that when the market experiences excessive trading volume and extreme market conditions, risk tends to be higher, which in turn gives rise to a higher return to compensate investors.

REFERENCES

Ang, A., Hodrick, R., Xing, Y., and Zhang, X. (2006). The cross-section of volatility and expected returns. *Journal of Finance* 61:259–99.

Antoniou, A., Koutmos, G., and Pericli, A. (2005). Index futures and positive feedback trading: Evidence from major stock exchanges. *Journal of Empirical Finance* 12:219–38.

Backus, D., and Gregory, A. (1993). Theoretical relations between risk premiums and conditional variance. *Journal of Business and Economic Statistics* 11:177–85.

Baillie, R., and DeGennaro, R. (1990). Stock returns and volatility. *Journal of Financial and Quantitative Analysis* 25:203–14.

Bali, T., and Peng, L. (2006). Is there a risk-return tradeoff? Evidence from high-frequency data. *Journal of Applied Econometrics* 21:1169–98.

Bollerslev, T., Chou, R. Y., and Kroner, K. F. (1992). ARCH modeling in finance: A review of the theory and empirical evidence. *Journal of Econometrics* 52:5–59.

Cappiello, L., Engle, R. F., and Sheppard, K. (2006). Asymmetric dynamics in the correlations of global equity and bond returns. *Journal of Financial Econometrics* 4:537–72.

Chang, E. C., Cheng, J. W., and Khorana, A. (2000). An examination of herd behavior in equity markets: An international perspective. *Journal of Banking and Finance* 24:1651–79.

Chiang, T. C., and Doong, S. C. (2001). Empirical analysis of stock returns and volatilities: Evidence from seven Asian stock markets based on TAR-GARCH model. *Review of Quantitative Finance and Accounting* 17:301–18.

Christie, W. G., and Huang, R. D. (1995). Following the pied piper: Do individual returns herd around the market? *Financial Analysts Journal* 51:31–37.

Clark, P. K. (1973). A subordinated stochastic process model with finite variance for speculative prices. *Econometrica* 41:135–56.

Connolly, R., and Stivers, C. (2006). Information content and other characteristics of the daily cross-sectional dispersion in stock returns. *Journal of Empirical Finance* 13:79–112.

Connolly, R. A., and Wang, F. A. (2003). International equity market comovements: Economic fundamentals or contagion? *Pacific-Basin Finance Journal* 11:23–43.

Duffee, G. R. (2001). Asymmetric cross-sectional dispersion in stock returns: Evidence and implications. Working paper, Federal Reserve Bank of San Francisco.

Engle, R. T. (1995). *ARCH: Selected readings.* New York: Oxford University Press.

French, K., Schwert, W., and Stambaugh, R. (1987). Expected stock returns and volatility. *Journal of Financial Economics* 19:3–29.

Ghysels, E., Santa-Clara, P., and Valkanov, R. (2005). There is a risk-return tradeoff after all. *Journal of Financial Economics* 76:509–48.

Glosten, L., Jagannathan, R., and Runkle, D. (1993). On the relation between the expected value and the volatility of the nominal excess return on stocks. *Journal of Finance* 48:1779–802.

Hodrick, R. J., and Prescott, E. C. (1997). Postwar U.S. business cycles: An empirical investigation. *Journal of Money, Credit and Banking* 29:1–16.

Koopman, S., and Uspensky, E. (2002). The stochastic volatility in mean model: Empirical evidence from international stock markets. *Journal of Applied Econometrics* 17:667–89.

Koutmos, G. (1998). Asymmetries in the conditional mean and the conditional variance: Evidence from nine stock markets. *Journal of Economics and Business* 50:277–90.

Lamoureux, C. G., and Lastrapes, W. D. (1990). Heteroskedasticity in stock return data: Volume versus GARCH effects. *Journal of Finance* 45:221–29.

Lo, A., and MacKinlay, C. (1990). An econometric analysis of nonsynchronous trading. *Journal of Econometrics* 45:181–211.

Longin, F. (1997). The threshold effect in expected volatility: A model based on asymmetric information. *Review of Financial Studies* 10:837–69.

Loungani, P., Rush, R., and Rave, W. (1990). Stock market dispersion and unemployment. *Journal of Monetary Economics* 25:367–88.

Merton, R. (1973). An intertemporal capital asset pricing model. *Econometrica* 41:867–87.

Merton, R. (1980). On estimating the expected return on the market: An exploratory investigation. *Journal of Financial Economics* 8:323–61.

Newey, W. K., and West, K. (1987). A simple positive semi-definite, heteroskedasticity and autocorrelation consistent covariance matrix. *Econometrica* 55:703–8.

Pagan, A. R., and Ullah, A. (1988). The econometric analysis of models with risk terms. *Journal of Applied Econometrics* 3:87–105.

Peña, D. (2001). Outliers, influential observations, and missing data. In *A course in time series analysis*, ed. D. Peña, G. C. Tiao, and R. S. Tsay. Hoboken, NJ: John Wiley & Sons.

Sentana, E., and Wadhwani, S. (1992). Feedback traders and stock return auto-correlations: Evidence from a century of daily data. *Economic Journal* 102:415–35.

Tsay, R. (1988). Outliers, level shifts, and variance changes in time series. *Journal of Forecasting* 7:1–20.

Wagner, N., and Marsh, T.A (2005), Surprise volume and heteroskedasticity in equity market returns. *Quantitative Finance* 5:2, 153–168.

CHAPTER 20

News, Trading, and Stock Return Volatility

Vladimir Zdorovtsov

CONTENTS

20.1 INTRODUCTION

Numerous studies find that the return variances over periods when the exchanges are open significantly exceed those over periods when the exchanges are closed (e.g., Fama, 1965; Oldfield and Rogalski, 1980; French and Roll, 1986; Barclay et al., 1990; Stoll and Whaley, 1990). Three potential explanations for the phenomenon have been offered in the literature: (1) more public information reaches the marketplace during normal business hours; (2) the trading activity of informed investors reveals their private information, inducing greater return variance; and (3) the process of trading itself introduces noise into stock prices and returns as investors overreact to other's trades, leading to more volatile returns over trading periods. The literature generally concludes that although there is some evidence of noise-induced trading return volatility (e.g., French and Roll (1986) offer an estimate of 4% to 12% of the daily return variance), the bulk of the difference between variances of trading and nontrading windows is attributable to the trading of informed market participants.

I show that the natural experiment approach utilized in the extant studies to control for public information may not be appropriate to the extent that information arrival itself is a function of trading. I provide a direct empirical test of the competing hypotheses by analyzing the volatility of close-to-open and open-to-close returns for NASDAQ securities with and without active extended-hours trading, while jointly and explicitly controlling for the firm-specific contemporaneous public information flow. My methodology disentangles the effects of noise, public information, and private information on stock return volatility. I also contribute to the burgeoning literature analyzing trading activity and return characteristics in the quickly growing extended-hours market.

By comparing the variances over multiday windows spanning days when the exchanges are closed, with single-day close-to-close variances, existing studies make inferences about the volatilities over trading and nontrading periods. For example, French and Roll (1986) investigate return behavior around weekends and business days when the NYSE and AMEX were closed. Barclay et al. (1990) examine returns on weeks when the Tokyo Stock Exchange was open on Saturdays. By assuming that the characteristics of the flow of public information on a business day when the exchanges are closed or on a Saturday when they are open are similar to those of a typical business day or typical Saturday, respectively, the authors draw conclusions about the impact of public information flow on return variances. However, an inevitable assumption in these studies is that the incidence of news releases is exogenously determined. While seemingly innocuous, one can offer several likely scenarios of how this conjecture might be violated. For example, a number of theoretical and empirical studies indicate that corporations strategically time information releases conditionally upon the presence of trading, as opposed to merely based on the business hours cycle.[*] By obtaining a comprehensive measure of firm-specific time-stamped information releases over the concurrent time window, I am able to control directly for the effects of public information flow, and for the possible information endogeneity issues, avoiding such potentially biasing assumptions.

The literature on return volatility largely ignores trades that take place outside regular trading hours (currently 9:30 a.m. to 4:00 p.m. Eastern Time.) Yet, a number of studies suggest that trading activity in after-hours and especially in premarket sessions, although low in volume, is dominated by informed participants (e.g., Barclay and Hendershott, 2003; Chan, 2002). Thus, I posit that stocks with more active extended-hours trading will, *ceteris paribus*, have greater overnight return variances. Furthermore, if after-hours and premarket trading convey private information, a shift in the timing of price discovery will occur, reducing the volatility of the subsequent open-to-close returns.[†] Alternatively, if trading only introduces additional noise, an increase in extended-hours volume will lead to greater overnight volatility and will not affect that of the subsequent regular trading session.[‡]

I find that the effects of after-hours and premarket trading on return volatility are markedly different. The less informed order flow in after-hours sessions is associated with little price impact and appears to be greatest on low information asymmetry days. The volatilities of close-to-open and open-to-close returns are negatively related to after-hours volume, and the volatility ratio is unaffected by such volume.

---

[*] For example, see Patell and Wolfson (1982), Gennotte and Trueman (1996), Baginski et al. (1996), Bushee et al. (2004), and Libby et al. (2002).

[†] Although the terms *after-hours* and *extended-hours* are sometimes used interchangeably, formally, the extended-hours window encompasses all transactions outside of the regular 9:30 a.m.–4:00 p.m. session and can be broken into *after-hours* (the period starting at 4:00 p.m. and generally extending until 8:00 p.m.) and *premarket* (generally accepted as the 7:00–9:30 a.m. period.) See http://www.nasdaq.com/reference/glossary.stm.

[‡] Utilizing extended-hours trades also allows me to avoid potential confounding effects of the home bias likely present in the analyses of internationally listed securities.

Conversely, higher trading volume in the premarket session, typically composed predominantly of anonymous information-based trades, is associated with greater overnight and lower subsequent regular session volatility, indicating that price discovery shifts toward the premarket hours. Consequently, the volatility ratios decrease in premarket trading volume.

Unlike the existing studies, I offer evidence in support of the public information hypothesis. Greater flow of public information over trading (nontrading) hours increases the open-to-close (close-to-open) return volatility, and the ratio of return volatilities is directly related to the news flow differential.

The rest of the chapter is organized as follows: Section 20.2 offers an overview of the existing literature and the development of hypotheses. Section 20.3 presents the data and methodology. Section 20.4 contains the empirical results, and Section 20.5 concludes.

20.2 LITERATURE REVIEW AND HYPOTHESIS DEVELOPMENT

20.2.1 Variance Ratios, Information Flow, and Trading Noise

Several studies find that stock returns are more volatile over exchange trading hours than they are over nontrading periods. French and Roll (1986) analyze equity return behavior around business days when NYSE and AMEX were closed. The authors assume that the flow of public information is not affected by exchange closures but is rather a by-product of the business hours activities. Since private information is conveyed through trading of the informed investors, and assuming this trading occurs only during the regular trading session hours, French and Roll (1986) conclude that it is the trading of the informed investors that leads to the bulk of variance differences. Similarly, Barclay et al. (1990) investigate equity returns during the period when the Tokyo Stock Exchange (TSE) was open on Saturdays. The authors assume that by analyzing weeks with and without Saturday trading, the effects of the flow of public information are held constant. Their analysis shows that during weeks with Saturday trading, weekend variance almost doubles, weekly volume goes up, but weekly variance is unaffected. The higher weekend variance is offset by lower variances on subsequent days as informed traders accelerate their trading. The study concludes that the results support the rational trading models based on private information and are inconsistent with public information or noise hypotheses.

The existing analyses lack a direct test of the effects of public information. While indeed some public information may be a by-product of business activities, and its arrival would thus largely coincide with the timing of the exchange operations, other research shows that the news release policy frequently contains an element of strategic timing, one of the critical parameters of which is the presence (or absence) of trading.* For example, Baginski et al. (1996) find that, consistent with voluntary disclosure predictions of Diamond (1985) and King et al. (1990), management strategically releases larger earnings surprises outside of the regular trading session hours. Similarly, Gennotte and Trueman (1996) suggest that management will prefer to issue negative information in extended hours and positive news during normal trading hours. Consistent with this, Patell and Wolfson (1982) and Francis et al. (1992) demonstrate that negative announcements tend to cluster outside of the normal exchange trading hours. Libby et al. (2002) present evidence of overnight news releases being more significant. Bushee et al. (2004) show that, subsequent to the Regulation Fair Disclosure (FD) requiring equal investor access to material information, firms tend to host their conference calls in extended hours to discourage trading by the less sophisticated investors during the calls, and thereby lower the excess volatility it induces. Lastly, the Securities and Exchange Commission appears to exhibit a preference that firms make corporate announcements during periods without an available trading venue.†

To the extent that the arrival of firm-specific information releases is potentially conditional on the presence of trading, the natural experiment approach of earlier studies may underestimate the effects of public information. It is important to control directly for the flow of contemporaneous firm-specific news in testing the effects of public information on equity return variances.

20.2.2 Evidence of Informed Trading in Extended Hours

Rational trading models (e.g., Kyle, 1985; Admati and Pfleiderer, 1988), predict that it is optimal for traders with private information to trade when the liquidity traders are most active. However, such models assume that the informed agents have a sufficiently low information decay rate. In many instances, the informational advantage is short-lived. The preponderance

---

* Note also that the overlap between business and trading hours will be a function of the time zone.

† Special study: Electronic Communication Networks and After-Hours Trading, Division of Market Regulation, June 2000. See http://www.sec.gov/news/studies/ecnafter.htm.

of traders participating in the extended-hours sessions plausibly either have or believe to have such a short-lived advantage. Indeed, given significantly greater extended-hours transactions costs, agents with long-lived information would likely delay their trades until the more liquid regular sessions.

Barclay and Hendershott (2003) present compelling evidence that such trades are substantially more informed and lead to significant price discovery, despite considerably higher spreads and generally low extended-hours volume. The average trade size is two to three times larger, due to lack of retail orders outside of the regular session.[*] A related study (SEC, 2000) finds that although the extended-hours session is more a market of stocks than a stock market, trading is relatively active for stocks subject to major corporate news announcements issued outside of the regular session hours.

The extant rational trading models suggest that prices will be most informative at times of high trading volume due to the high numbers of privately informed traders. However, although the absolute number of informed participants is likely lower in extended-hours trading, their relative number is potentially substantially higher, as discretionary transaction-cost-elastic liquidity traders opt to defer their trades until the less costly regular session. The lower proportion of liquidity and retail traders and, consequently, a greater ratio of informed to uninformed participants will result in a more informed order flow. Barclay and Hendershott (2003) demonstrate that while volatility per unit of time is generally lower in extended hours than it is during the trading day, volatility per trade is higher. The authors conclude that when trading is conducted by the most informed market participants, significant price discovery can occur even on low trading volume.[†]

---

[*] Until fairly recently, extended-hours trading was available almost exclusively only to institutional and professional traders. Although the advent and expansion of electronic communication networks (ECNs) has enabled individual investors to place anonymous orders eligible for execution in extended hours, the existing empirical and anecdotal evidence suggests their activities outside of the regular session hours remain immaterial. Under the existing NASD rules 2110 and 2210, member firms have an obligation to disclose the material risks of extended-hours trading to their retail customers before permitting customers to engage in this activity.

[†] Similarly, Barclay et al. (2001, 2002) develop and empirically confirm a theoretical model that predicts a higher percentage of informed traders on ECNs. They show that although ECN trading volume is lower, it has a substantially greater permanent price impact and explains approximately two-thirds more price volatility than market-maker trades.

20.2.3 Overnight Price Discovery

Greene and Watts (1996) and Masulis and Shivakumar (1999) show that stock price reactions to overnight earnings news and seasoned equity offerings, respectively, are significantly faster on NASDAQ. Bacidore and Lipson (2001) find that the overnight price discovery for NASDAQ securities is much larger than it is for securities listed on the NYSE, and that this difference appears to be an increasing function of firm size. They also find that a greater percentage of the daily volume is executed at the open on NYSE compared to NASDAQ. It is reasonable to hypothesize that this is attributable to substantially greater volume of extended-hours trading in NASDAQ securities.*

Hong and Wang (2000) develop a theoretical model that shows how the incidence of periodic market closures alone can generate empirical patterns, including higher volatility over trading periods than over nontrading periods, even assuming constant information flow. Thus, insofar as extended-hours trading diminishes this closure effect, one can expect the disparity in volatilities of returns over close-to-open and open-to-close windows to be smaller for stocks with more active trading outside of the regular session, *ceteris paribus*.

20.2.4 Hypotheses

If return variances are caused by the arrival of public information, then:

> *Public information hypothesis:* The close-to-open and open-to-close return volatilities will be positively related to the public information flow over the respective time periods.

If return variances are caused by trading of informed market participants, the volatility of overnight returns will increase in extended-hours trading volume. Furthermore, if the rise in close-to-open variance is due to the greater amount of private information impounded through such trading, the

---

* For example, Barclay and Hendershott (2001) find that such volume accounts for almost 4% of daily trading volume on NASDAQ and only 0.5% on NYSE, and that it is positively related to daily volume (and therefore firm size). Indeed, the authors note that given this difference in extended-hours volume, the studies investigating the speed with which information is incorporated into the opening prices across markets are problematic.

timing of price discovery will shift and the volatility of open-to-close returns will correspondingly decline. More formally:

> *Private information hypothesis:* The close-to-open (open-to-close) return volatility will be increasing (decreasing) in the extended-hours trading volume.

Barclay and Hendershott (2003) and Chan (2002) show that whereas most order flow in after-hours is relatively uninformed and represents position adjustment and hedging, the premarket session trading is primarily information based. Consequently, the hypothesized private information effects are expected to be especially prominent for premarket trading volume.

If, on the other hand, trading only induces noise as investors overreact to each other's actions, no shift in the timing of price discovery will occur. The extended-hours trading will cause additional overnight return variance and will not affect the open-to-close returns. Thus:

> *Noise hypothesis:* The close-to-open (open-to-close) return volatility will be increasing in (independent of) the extended-hours trading volume.

20.3 DATA AND METHODOLOGY

20.3.1 Sample Selection

I start with all NASDAQ securities covered by the Center for Research in Security Prices (CRSP) during the 2000–2001 period. I limit my sample to NASDAQ securities for a number of reasons. Many existing studies show that the trading mechanism has significant effects on stock return behavior (e.g., Amihud and Mendelson, 1987; Miller, 1989; Stoll and Whaley, 1990; Bacidore and Lipson, 2001). Thus, by restricting the sample to NASDAQ, I avoid the potential confounding effects caused by the institutional and procedural differences. Second, the volume and cross-sectional variation of extended-hours trading in NYSE securities is relatively small. Third, and perhaps most important, the premarket trades for NYSE securities are not captured by the NYSE Trade and Quote (TAQ) database during my sample period.

The sample is then restricted to stocks that never trade at prices below $5 per share during this period, yielding 1,571 securities. I leave out penny stocks due to their extreme percentage price swings in extended hours (e.g., see SEC, 2000). I further require at least ten trades on at least 250 days over the 2-year window spanning exactly 500 trading days. This screen reduces the sample to 1,094 firms. I am able to locate TAQ data for 1,001 of these firms. Data on capitalization are obtained from CRSP as of the last trading day of 2001. I impose the usual screens for out-of-sequence, nonstandard delivery, and erroneous trade prints and obtain all TAQ transactions for each sample firm-day over the 7:30 a.m.–7:00 p.m. window.*

To obtain a proxy for public information flow, I use a Web crawler to search CBS.MarketWatch.com and its twenty news sources for all firm-specific information released over the 2000–2001 window. The list of news providers contains Reuters, BusinessWire, PR Newswire, Market Wire, Edgar Online, CNET News.com, CBS News, Knight Ridder, $ TheStreet.com, RealTime Headlines, TV & Radio, New York Times, FT.com, Market Pulse, and United Press Intl., among others, and represents a broad array of coverage sources. Conducting the search electronically allows me to have a substantially larger sample and a much more extensive list of news providers than prior studies analyzing the effects of public information flow. I download up to the last 100 news releases going back from December 31, 2001. The number of news items per firm is bounded from above at 100 due to search constraints imposed by CBS.MarketWatch.com. This constraint is binding for 132 companies.† For 949 companies, I am able to locate the ticker in the CBS.MarketWatch.com database. For eleven of these companies, not a single news release is located. For the remainder of the sample the search generates 40,694 news items time-stamped to the minute. The mean (median) number of news releases per company is 43.38 (33).

20.3.2 Calculation of Variance and Information Flow Ratios

I compute return moments for the following intervals: close-to-close, close-to-open, and open-to-close. Not all stocks in the sample necessarily trade every day when the exchange is open. In computing the moments listed above, I omit the days where the exchange is open but the stock does not trade according to TAQ. The classification for such windows as trading or nontrading is at best ambiguous, and therefore, I opt to omit them from the

---

* I require TAQ correction codes of 1 or 0, condition of Regular Way (Blank or *) or T for extended-hours trades, and trade size and price above zero.

† Admittedly, this proxy for public information flow is not perfect (e.g., I expect some news releases to be stale or noninformative). However, these criticisms plague most investigations dealing with news flow data, and insofar as they equally apply to releases made during and outside of the regular trading session, no bias is expected for my results. Also note that to the extent the truncated 100 releases obtained are representative of the full population (within and outside of the normal trading hours), no bias is introduced by this constraint for the 132 firms for which is it binding. The results are not sensitive to exclusion of these firms.

analysis. Instead, nontrading period returns are computed as the change in price from the close of a trading day with executed transactions to the open of the next such adjacent trading day without any trading days with zero trades in between.[*] The volatility ratio on a per-hour basis is calculated as follows:

$$\sigma^2_{RATIO\,j} = \frac{\sigma^2_{OpCl\,j} / \overline{HoursTr_j}}{\sigma^2_{ClOp\,j} / \overline{HoursNTr_j}} \tag{20.1}$$

where $\sigma^2_{OpCl\,j}$ is the time-series variance of open-to-close returns for security j; $\sigma^2_{ClOp\,j}$ is the time-series variance of close-to-open returns for security j; and $HoursTr_j$ and $HoursNTr_j$ are the average time lengths of the open-to-close and close-to-open periods in hours for security j.[†]

To explicitly control for the flow of public information over close-to-open and open-to-close periods, I allocate all news releases into these two groups for each security according to their time stamps. A news flow ratio is calculated as follows:

$$PerHourNewsRatio_j = \frac{NewsTr_j / \overline{HoursTr_j}}{NewsNTr_j / \overline{HoursNTr_j}} \tag{20.2}$$

where $NewsTr_j$ and $NewsNTr_j$ represent the number of news items released over the open-to-close periods and close-to-open periods for security j, and $HoursTr_j$ and $HoursNTr_j$ are as defined above.

20.4 EMPIRICAL ANALYSIS

20.4.1 Intraday Dynamics of Trading Activity

Figure 20.1 presents an intraday distribution of the trading volume. Consistent with prior literature (e.g., Foster and Viswanathan, 1993; Harris, 1986), I find evidence of a U-shaped pattern in trading volume during the regular trading session both in dollar terms and in the number of trades. As in Barclay and Hendershott (2003), the bulk of extended-hours trading volume occurs around the opening and closing of the regular trading session. I find that after-hours volume substantially exceeds

---

[*] All but 266 sample firms trade every day. The results are not sensitive to exclusion of these firms.

[†] The average lengths of these periods are calculated to account for their variation across securities. For example, due to more omitted trading days with no trades for some sample firms, they may have relatively more Friday-close-to-Monday-open nontrading returns or a different number of shortened trading days (e.g., due to exchange holidays).

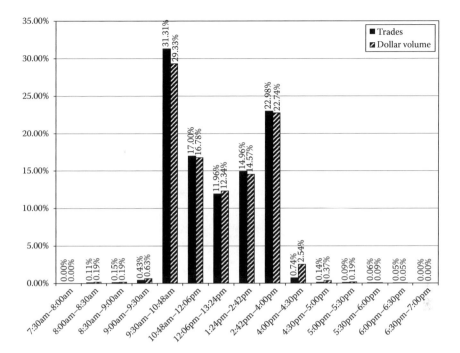

FIGURE 20.1 Intraday Distribution of trading volume. This graph demonstrates the dynamics of the number of trades and dollar volume over extended-hours and regular trading session subintervals.

premarket volume: while the total extended-hours dollar volume represents 4.24% of the aggregate daily volume, more than three-quarters of it is transacted in after-hours.

Figure 20.2 shows the dynamics of the mean and median trade size across intraday time increments in dollar terms. The average trade size in extended hours is significantly higher than that in normal trading hours. The average trade size in the premarket window starts out at a level comparable to that during the regular session and then dramatically surges during the 8:00–8:30 a.m. interval, far exceeding the levels during the rest of the day. This spike is most likely attributable to the fact that although some ECNs begin operating as early as 7:00 a.m., the majority of brokers offer premarket trading starting at 8:00 a.m. Thus, this time effectively represents the first opportunity to act on new private or public information for the bulk of traders. The trade size abruptly rises after the end of the regular session and peaks around 5:00 p.m.

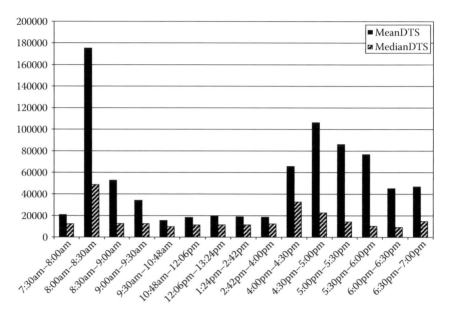

FIGURE 20.2   Intraday distribution of trade size. DTS is the dollar size of trade.

20.4.2   Analysis of Public Information Flow

Table 20.1 presents descriptive statistics on the flow of public information. The average number of informational releases per firm is 43.38, out of which 13.84 and 29.54 occur during and outside of the regular session trading hours, respectively. Unlike Berry and Howe (1994), Patell and Wolfson

TABLE 20.1   Information Flow Statistics

	N_Firms	Mean	Median	Std Dev	Min	Max
NofNewsAll	938	43.38	33.00	33.77	1.00	100.00
NewsTr	938	13.84	10.00	12.25	0.00	77.00
NewsNTr	938	29.54	22.00	24.46	0.00	90.00
NewsRatio	938	0.35	0.32	0.20	0.00	1.00
PerHourNewsRatio	918	3.04	1.98	4.37	0.00	61.41

Note: This table summarizes average per-firm per-day news release statistics. Time-stamped news releases are obtained by a computerized search of cbs.marketwatch.com database. *NofNewsAll*, *NewsTr*, and *NewsNTr* denote all news releases, trading, and nontrading news releases, respectively; *NewsRatio* is the ratio of the number of releases made over trading hours to the total. *PerHourNewsRatio* is the ratio of the per-hour number of news releases over trading hours to the per-hour news releases over nontrading hours.

(1982), and Francis et al. (1992), I find that in aggregate, there are fewer information releases during trading hours than outside of them—35% versus 65%. Berry and Howe (1994) document that for the universe-aggregated information flow, the per-hour volume of releases made during normal trading hours exceeds that outside such hours by a factor of 3. In my case, using the aggregated procedure of Berry and Howe yields a ratio of only 1.26. The discrepancy appears to indicate a general shift of public information flow toward nontrading hours.

Interestingly, however, the mean (median) per-hour news ratio across firms is considerably higher at 3.04 (1.98), indicating that the aggregated results obscure the effects of less informationally intensive firms. While the overall volume of releases for these companies is low (thus having a minor effect on the aggregated ratio), such firms appear to have relatively more news during trading hours.

In results not reported here, I show that there appears a clear upward trend in the amount of available public information, starting with a spike in October 2000.[*] Interestingly, this coincides with the passage of the Securities and Exchange Commission Regulation Fair Disclosure. While this evidence is intended as suggestive only, it indicates that contrary to the suggestion of the opponents of Regulation FD that the quantity of information reaching the market will decline, companies appear to have substituted public communication channels for private venues and the flow of public information has increased since October 2000.[†]

Berry and Howe (1994) document that on a typical day, information flow (as proxied by Reuters News Service) begins to substantially increase around 8:30 a.m., continues to build until noon, and then shows a lull. The flow then rises again during the remainder of the trading session and peaks between 4:30 and 5:00 p.m.

Figure 20.3 plots the intraday distribution of the number of news releases in 30-minute increments. Panel 1 examines the flow of information on trading days and shows a pattern generally resembling that found in Berry and Howe (1994) with several key differences. First, the flow of information begins to rise considerably earlier. There is a sharp surge in the number of news items starting at 6:00 a.m. The volume of information continues to climb steeply until the beginning of regular session trading at

---

[*] This result is not sensitive to exclusion of the 132 firms for which the number of news is truncated at 100.

[†] This is consistent with the findings of Heflin et al. (2003), who show that the quantity of firms' voluntary disclosures increased post–Regulation FD.

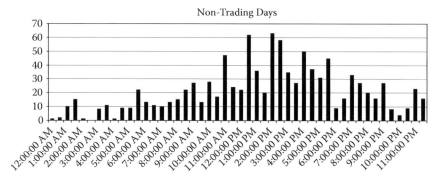

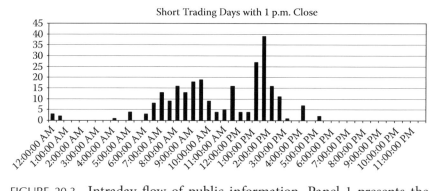

FIGURE 20.3 Intraday flow of public information. Panel 1 presents the number of news releases in 30-minute intraday increments for all trading days during the 2000–2001 period. Panel 2 displays the releases for all nontrading days in the sample. Panel 3 plots the intraday releases for 5 trading days with shortened regular trading sessions (1:00 p.m. close).

9:30 a.m. Information flow begins to abate thereafter, gradually diminishing until the end of regular session trading at 4:00 p.m. However, immediately after the end of normal trading hours, the rate of information arrival more than quadruples, peaking at 3,224 releases and declining monotonically until 9:00 p.m. Interestingly, no similar patterns are observed for nontrading days, where the rate of information arrival generally increases with time until midday and declines thereafter (Panel 2). Examining the flow of news announcements on 5 trading days in my sample when the U.S. exchanges close at 1:00 p.m. (Panel 3), it becomes clear that the flow of public information is indeed closely linked to exchange operating hours. The sharp increase in news volume that is observed at 4:00 p.m. for normal trading days shifts to 1:00 p.m. for trading days when the exchanges close at this hour. The effect is statistically significant at conventional levels.

Several conclusions can be drawn from this evidence. First, compared to the results of the earlier studies, there seems to have been a general shift of information flow away from regular trading hours. More importantly, the pattern of information arrival is clearly tied to the boundaries of the normal exchange trading hours, casting doubt on the assumption that information flow is not a function of trading activity, implicitly used in prior return volatility studies, and further corroborating the need for direct control for the effects of public information.

20.4.3   Univariate Variance Analysis

Table 20.2 summarizes the effects of news flow on return volatility and variance ratios. Several results stand out. Consistent with the public information hypothesis, overnight return volatility for companies with lower per-hour news ratio (i.e., greater flow of news overnight relative to that during regular session hours) significantly exceeds that of firms with the news ratio above the sample median. Open-to-close and close-to-close variances are not affected, and the variance ratios are consequently lower. These results appear independent of the potentially related volume effects, since the premarket, after-hours, and regular session volumes are not significantly different across the high and low news ratio subsets.

Results of the univariate effects of extended-hours volume are presented in Table 20.3. The average per-hour volatility ratio is 15.95 and is comparable to those of prior studies (e.g., Oldfield and Rogalski (1980), French and Roll (1986), and Stoll and Whaley (1990) report average ratios of 12.78, 13.20, and 16.20, respectively). Consistent with both the private information and the noise hypotheses, the overnight variance increases

TABLE 20.2   Effects of News on Return Volatility

	Low News Ratio	**High News Ratio**	*p*-value
PerHourNewsRatio	1.1720	4.8991	0.0011
VAR_Ratio	15.7304	16.6056	0.0693
ClClVar, %	0.2585	0.2450	0.2402
ClOpVar, %	0.0708	0.0671	0.0000
OpClVar, %	0.2161	0.2022	0.1305
NofNewsAll	47.8388	40.5926	0.0011
NewsTr	10.7015	17.3551	0.0000
NewsNTr	37.1373	23.2375	0.0000
RELEH, %	0.0702	0.0628	0.1848
RELAH, %	0.0561	0.0496	0.1143
RELPM, %	0.0141	0.0131	0.5740
RELREG, %	1.4861	1.3492	0.3716
N	459	459	

Note: *Var_Ratio* is the per-hour ratio of open-to-close variances to close-to-open variances; *ClClVar*, *ClOpVar*, and *OpClVar* denote close-to-close, close-to-open, and open-to-close variances, respectively; *NofNewsAll*, *NewsTr*, and *NewsNTr* denote all news releases, trading, and nontrading news releases, respectively; *RELEH*, *RELAH*, *RELPM*, and *RELREG* represent extended-hours, after-hours, premarket, and regular session volume scaled by firm capitalization. Average numbers are given for sample subsets composed of stocks with the per-hour ratios of the number of news items over trading periods to number of news items over nontrading periods above and below the median level, respectively.

in relative extended-hours dollar volume. The open-to-close and close-to-close variances are also higher for firms with greater overnight dollar volume. Although this result appears to be contrary to the predictions of the private information hypothesis, it can be potentially attributable to the fact that companies with greater relative extended-hours volume also generally have greater regular session volume (as is clearly seen in Table 20.6), and the latter in turn leads to greater open-to-close variances. Furthermore, since the extended-hours volume is composed of the after-hours and premarket volume, and to the extent that these sessions appear to exhibit markedly different trading processes with regard to the informativeness of the order flow (e.g., see Barclay and Hendershott, 2003; Chan, 2002), one needs to examine their effects separately.

Several additional interesting results are worth noting. First, consistent with Patell and Wolfson (1982) and Francis et al. (1992), the skewness of overnight returns is negative and further declines in extended-hours volume, indicating that the information made public or revealed through trading during this window tends to be of a negative nature. Second, stocks

TABLE 20.3   Effects of Extended-Hours Volume

	All	**Low RELEH**	**High RELEH**	*p*-value
VAR_Ratio	15.95	14.75	17.16	0.00
ClClVar, %	0.25	0.14	0.35	0.00
ClOpVar, %	0.07	0.05	0.09	0.00
OpClVar, %	0.21	0.13	0.26	0.00
SkewnessClOp	−2.10	−1.09	−3.12	0.00
SkewnessOpCl	0.46	0.48	0.44	0.38
PerHourNewsRatio	3.04	3.83	2.31	0.00
NewsTr	13.84	7.43	19.93	0.00
NewsNTr	29.54	14.71	43.64	0.00
RegNofTrades	1,477	211	2,741	0.00
AHNofTrades	16.31	3.16	29.43	0.00
PMNofTrades	10.43	1.19	19.66	0.00
Cap (thousands)	2,346,305	1,425,569	3,265,203	0.05
RELEH, %	0.06	0.02	0.11	0.00
RELAH, %	0.05	0.02	0.09	0.00
RELPM, %	0.01	0.00	0.02	0.00
RELREG, %	1.36	0.30	2.43	0.00
OpenTime	9:42:53	9:53:53	9:31:54	0.00
CloseTime	15:46:05	15:35:52	15:56:17	0.00
OpenTradeSize	434.60	496.65	372.67	0.00
CloseTradeSize	856.49	831.60	881.33	0.03
N	1001	500	501	

Note: The averages are presented for firms with relative extended-hours volume below and above the sample median and for the overall sample.

with more active extended-hours trading sessions open and close the regular sessions considerably closer to the official 9:30 a.m.–4:00 p.m. bounds. This effect is significant at conventional levels and holds after controlling for firm size (not reported). Stoll and Whaley (1990) examine NYSE stocks for years 1982–1986 and show that for more (less) actively traded firms a delay at the open leads to greater (lower) overnight volatility. The authors suggest that opening delays for actively traded stocks imply large order imbalances at the open, whereas for less active stocks the delays merely denote absence of orders. In results available upon request, I show that the opening delays are associated with lower overnight variances regardless of firm size. However, it should be noted that overnight trading is largely nonexistent during the sample period analyzed in Stoll and Whaley (1990) and is smaller for NYSE securities than for NASDAQ stocks in general. Consequently, if trading in after-hours and premarket sessions alleviates potential imbalances at the open and helps establish the new opening price (see, e.g., Chan, 2002), then one can argue that stocks with inflows of significant private or public overnight information that would have had large opening imbalances, longer opening delays, and greater overnight variances in years without a relatively active extended-hours market, will now have greater extended-hours volume, greater overnight volatility, and shorter opening delays. Yielding additional support to this argument is the fact that the average size of the first regular session trade decreases in extended-hours volume.

In results available upon request, I find that the effects of the extended-hours dollar volume on the close-to-open, open-to-close, and close-to-close return variances, and on the variance ratios, are not monotonic. As the extended-hours dollar volume as a fraction of firm capitalization increases to about the sample median level, there appears to be no effect on the overnight (i.e., close-to-open) return variance. On the other hand, the open-to-close variance steadily increases, and as a result, the close-to-close variance and the per-hour variance ratios go up. Increases in the extended-hours volume beyond the median level lead to greater overnight, open-to-close, and close-to-close variances. The rising overnight volatility more than offsets increasing open-to-close volatility and the variance ratios steeply decline. This indicates that, consistent with Barclay et al. (1990), a significant volume of extended-hours trading needs to exist before overnight variances are affected.[*] A likely explanation for the rising open-to-close volatility is that, since extended-hours volume is strongly correlated with open-to-close volume, increases in the former (while not necessarily sufficient to induce a noticeable effect on the overnight variance) are related to increases in the latter, which in turn result in greater open-to-close and close-to-close variances and higher variance ratios. Thus, it is critical to control for the related effects of regular session volume in analyzing those of the extended-hours volume.

Table 20.4 examines the effects of premarket, after-hours, and total extended-hours relative volume on variance ratios by regular session volume quintiles. The effects of the premarket and after-hours volume appear to be different. Within the extended-hours volume, it is the premarket volume that tends to lead to greater overnight variances and lower variance ratios. The result is less significant for lower regular volume quintiles and is reversed for the lowest quintile. Again, to the extent that premarket volume

---

[*] Similarly, Forster and George (1995) find that for cross-listed stocks, foreign trading facilitates price discovery if there is sufficient trading volume in the foreign market.

TABLE 20.4   Effects of Extended-Hours Volume on Variance Ratios by Regular Session Volume

RelRegQuintile	RELPM			RELAH			RELEH		
	Low	High	p-value	Low	High	p-value	Low	High	p-value
Lowest	9.97	13.39	0.00	9.74	13.63	0.00	9.66	13.71	0.00
2	16.12	15.69	0.72	13.22	18.59	0.00	13.71	18.10	0.00
3	20.08	18.47	0.30	18.80	19.75	0.54	18.66	19.89	0.43
4	19.68	16.25	0.01	17.39	18.54	0.39	17.89	18.04	0.91
Highest	16.92	12.99	0.00	15.65	14.25	0.14	16.42	13.49	0.00

Note: *RELEH, RELAH, RELPM*, and *RELREG* represent extended-hours, after-hours, pre-market, and regular session volume scaled by firm capitalization.

is correlated with regular session volume, this indicates that substantial trading activity needs to exist for the overnight variances to be affected.

20.4.4   Regression Analysis

The evidence presented above indicates that the effects of private information, public information, and noise need to be analyzed jointly. Also, since open-to-close volatility is linked to open-to-close volume (e.g., see Stoll and Whaley, 1990), and because regular session volume and extended-hours volume are correlated, one needs to control for the effects of the regular session volume in determining those of the extended-hours volume.

I address these concerns within a two-way fixed effects OLS regression framework and estimate the following models:[*]

$$LogAbsClOp_{j,t} = \beta_0 + \beta_1 LAGNewsTr_{j,t} + \beta_2 NewsNTr_{j,t} + \beta_3 LAGRELAH_{j,t}$$
$$+ \beta_4 RELPM_{j,t} + \beta_5 LAGRELREG_{j,t} + v_j + u_t + \varepsilon_{j,t} \qquad (20.3)$$

$$LogAbsOpCl_{j,t} = \beta_0 + \beta_1 NewsTr_{j,t} + \beta_2 NewsNTr_{j,t} + \beta_3 LAGRELAH_{j,t}$$
$$+ \beta_4 RELPM_{j,t} + \beta_5 RELREG_{j,t} + v_j + u_t + \varepsilon_{j,t} \qquad (20.4)$$

$$LogRatio_{j,t} = \beta_0 + \beta_1 LAGRELAH_{j,t} + \beta_2 RELPM_{j,t} + \beta_3 RELREG_{j,t}$$
$$+ \beta_4 NewsDiff_{j,t} + v_j + u_t + \varepsilon_{j,t} \qquad (20.5)$$

---

[*] Random effects estimations yield similar results. Fixed effects results are presented based on the Hausman specification test.

where

$$LogAbsClOp_{j,t} = Ln \left| \frac{ClOp_{j,t}}{ClOpHours_{j,t}} \right| \qquad (20.6)$$

Similarly, $LogAbsOpCl_{j,t}$ is the natural logarithm of the absolute value of the per-hour open-to-close return; $LogRatio_{j,t}$ is the difference between $LogAbsOpCl_{j,t}$ and $LogAbsClOp_{j,t}$; $NewsTr_{j,t}$ and $NewsNTr_{j,t}$ are the numbers of trading and nontrading hours news items, respectively; $LAGNewsTr_{j,t}$ is the number of news releases over the regular trading hours of the preceding trading day; $LAGRELAH_{j,t}$ is the prior trading day's after-hours dollar volume scaled by capitalization; $RELPM_{j,t}$ and $RELREG_{j,t}$ are the premarket and regular session dollar volumes scaled by capitalization; $LAGREREG_{j,t}$ is the prior trading day's regular session dollar volume scaled by capitalization; $NewsDiff_{j,t}$ is the difference between $NewsTr_{j,t}$ and $NewsNTr_{j,t}$;* v_j, u_t, and $\varepsilon_{j,t}$ are the error terms; and $j \in [1,1001]$ and $t \in [1,499]$ denote the firm and the trading day, respectively.

Table 20.5 summarizes the results. Several key conclusions emerge. Consistent with prior literature and the results shown earlier, I find that the after-hours and premarket trading volume exhibit different effects. Specifically, greater premarket volume leads to substantially higher (lower) close-to-open (open-to-close) return volatility and to considerably lower volatility ratios. This result supports the private information hypothesis and shows that indeed the premarket trading volume is largely composed of information-motivated trades. Greater informed trading in the premarket session shifts price discovery toward the close-to-open period, increasing volatility of the overnight returns, reducing that of the open-to-close returns, and leading to lower volatility ratios.

Conversely, the after-hours volume is negatively related to the overnight and open-to-close volatility. This evidence is consistent with the suggestion of Barclay and Hendershott (2003) that large liquidity-motivated after-hours trades are more likely to execute on low information asymmetry days and are associated with little price impact. Unlike the premarket volume, after-hours volume is not significantly related to volatility ratios.

---

* Given the high frequency of zero news volume, using a ratio leads to a considerable reduction in sample size.

TABLE 20.5   Regression Analysis

Intercept	LogAbsClOp		LogAbsOpCl		LogRatio	
	Estimate	p-value	Estimate	p-value	Estimate	p-value
	−9.438	0.000	−5.683	0.000	3.792	0.0001
NewsTr			0.099	0.000		
LAGNewsTr	−0.023	0.001				
NewsNTr	0.196	0.000	0.057	0.000		
LAGRELAH	−1.573	0.003	−4.140	0.000	−0.868	0.209
RELPM	39.790	0.000	−14.386	0.000	−42.994	0.000
RELREG			3.117	0.000	1.342	0.000
LAGRELREG	0.232	0.000				
NewsDiff					0.075	0.000
N	400,857		429,387		383,188	
Adj-Rsqr	0.335		0.164		0.187	

Note: The following two-way fixed effects models are estimated by OLS:

$$LogAbsClOp_{j,t} = \beta_0 + \beta_1 LAGNewsTr_{j,t} + \beta_2 NewsNTr_{j,t} + \beta_3 LAGRELAH_{j,t}$$

$$+ \beta_4 RELPM_{j,t} + \beta_5 LAGRELREG_{j,t} + v_j + u_t + \varepsilon_{j,t}$$

$$LogAbsOpCl_{j,t} = \beta_0 + \beta_1 NewsTr_{j,t} + \beta_2 NewsNTr_{j,t} + \beta_3 LAGRELAH_{j,t}$$

$$+ \beta_4 RELPM_{j,t} + \beta_5 RELREG_{j,t} + v_j + u_t + \varepsilon_{j,t}$$

$$LogRatio_{j,t} = \beta_0 + \beta_1 LAGRELAH_{j,t} + \beta_2 RELPM_{j,t} + \beta_3 RELREG_{j,t}$$

$$+ \beta_4 NewsDiff_{j,t} + v_j + u_t + \varepsilon_{j,t}$$

where $LogAbsClOp_{j,t}$ and $LogAbsOpCl_{j,t}$ are the natural logarithms of the absolute values of the close-to-open and open-to-close returns on a per-hour basis, respectively; $LogRatio_{j,t}$ is the difference between $LogAbsOpCl_{j,t}$ and $LogAbsClOp_{j,t}$; $NewsTr_{j,t}$ is the number of news releases over regular trading session hours; $NewsNTr_{j,t}$ is the number of news items released between the end of the previous trading day's regular session and the beginning of the current trading day's regular session; $LAGNewsTr_{j,t}$ is the number of news releases over the regular trading hours of the preceding trading day; $LAGRELAH_{j,t}$ is the prior trading day's after-hours dollar volume scaled by capitalization; $RELPM_{j,t}$ and $RELREG_{j,t}$ are the premarket and regular session dollar volumes scaled by capitalization; $LAGRELREG_{j,t}$ is the prior trading day's regular session dollar volume scaled by capitalization; $NewsDiff_{j,t}$ is the difference between $NewsTr_{j,t}$ and $NewsNTr_{j,t}$; v_j, u_t, and $\varepsilon_{j,t}$ are the error terms; and $j \in [1,1001]$ and $t \in [1,499]$ denote the firm and the trading day, respectively.

Contrary to the conclusions of the earlier studies, the evidence on the effects of news flow yields credence to the public information hypothesis. Higher volume of public information released outside regular trading hours leads to higher close-to-open return volatility. Consistent with overnight releases being more significant, there is evidence of a spillover of volatility into the subsequent regular session.[*] Greater volume of public information reaching the market during regular trading hours leads to higher open-to-close volatility. Interestingly, not only is there no spillover of volatility into the subsequent overnight period, but the volatility of the latter appears to decline. This indicates that the typically less influential daytime releases are completely priced in during normal trading hours, causing the degree of information asymmetry and price uncertainty to decline. Lending further support to the public information hypothesis is the significant positive relation between the volatility ratio and the news flow differential. In other words, greater flow of public information over trading hours versus nontrading hours is associated with higher ratios of open-to-close to close-to-open volatilities.

A theoretical model developed in Holden and Subrahmanyam (1992) shows that aggressive competition among the informed traders leads to faster revelation of their information. Thus, if this competition is greater in extended hours due to a higher proportion of informed agents, trading during this period will impound information into prices faster. This prediction, combined with the fact that there are fewer short-selling restrictions in extended hours, leads one to expect the link between information and volatility to be stronger for overnight return windows.[†] The results in Table 20.5 are generally consistent with this conjecture. Indeed, while the rate of overnight information arrival is strongly related to overnight return variability, the link is more economically pronounced over close-to-open periods.

One potential criticism of the news flow data is the possible presence of redundant news releases merely reiterating the subject matter of an earlier story from a different (or the same) source. To check the sensitivity of the above results to such noninformative releases, I repeat the estimations with the news volume variables replaced by dummy variables equal to 1 for windows with one or more releases and 0 otherwise. The news differential in these specifications is computed as the difference between the values of such dummy variables. The results (not reported) are qualitatively and quantitatively similar.

---

[*] Note that this is contrary to He and Wang (1995), who develop a rational expectations model predicting that public information has a rather short-lived effect and leads to trading only in the contemporaneous period.

[†] Although some ECNs do not allow short-sale transactions at prices below the close of the previous regular session, the NASD's short-sale rule is not applicable outside of regular market hours during my sample window (NASD Notice to Members 94–68).

To further check the robustness of the preceding analysis, I also estimate cross-sectional models relating time-series return variances to measures of information flow and trading volume aggregated at the firm level.
The qualitative results available upon request remain largely unchanged.
Not surprisingly, the statistical significance declines as power is lost in the
aggregation process.[*]

20.4.5   Within-Firm Effects of Trading Volume and Information Flow

One advantage of the natural experiment approach employed in the prior
studies is the implicit control for the firm-specific characteristics that can
potentially affect return volatility, since the same securities are investigated across different time periods. To examine the sensitivity of the above
results to the effects of potentially omitted variables, the following procedure is performed: I locate firms that trade every day and have 100 news
releases with at least 100 days between the dates of the first and last news
item. These selection screens yield a sample of 107 companies. For each
firm, the trading days spanned by the news data are subdivided into high
premarket volume, low premarket volume, high after-hours volume, low
after-hours volume, high news difference, and low news difference, based
on the respective mean levels. Table 20.6 reports average ratios of the absolute value of the per-hour open-to-close return to the absolute value of
the per-hour close-to-open return for the corresponding subsamples, the
number of firms for which the difference in average ratios across such subsamples is positive and negative, as well as the number of firms for which
such differences are significant at the 10% level.

The findings are in agreement with the conclusions of the preceding
analysis. Specifically, consistent with the private information hypothesis,
days with the premarket dollar volume above the mean level have lower
volatility ratios for 92 of the 107 examined firms. For the overwhelming
majority of such companies the difference is significant at the 10% level.
Unlike the premarket volume, the less informed after-hours dollar volume
exhibits no clear link to the volatility ratios. Consistent with the public
information hypothesis, days with greater arrival of news over regular trading hours versus nontrading hours are accompanied by higher volatility

---

[*]   Because the dependent variables in these estimations (close-to-open or open-to-close variance) have nonnegative domains, I repeat the analysis using Tobit regressions as well as using
OLS after taking the natural logarithm of the respective dependent variables. The results are
qualitatively and quantitatively unchanged.

TABLE 20.6 Within-Firm Effects of Premarket Volume, After-Hours Volume, and News Differential on Volatility Ratios

Panel A: Effects of Premarket Volume

$\dfrac{\sum_{j=1}^{107} \overline{\text{Ratio}}_{\text{PMHigh}\,j}}{107}$	$\dfrac{\sum_{j=1}^{107} \overline{\text{Ratio}}_{\text{PMLow}\,j}}{107}$	$(\overline{\text{Ratio}}_{\text{PMHigh}\,j} - \overline{\text{Ratio}}_{\text{PMLow}\,j}) < 0$		$(\overline{\text{Ratio}}_{\text{PMHigh}\,j} - \overline{\text{Ratio}}_{\text{PMLow}\,j}) > 0$	
		N	N_Significant	N	N_Significant
25.23	39.81	92	70	15	0

Panel B: Effects of After-Hours Volume

$\dfrac{\sum_{j=1}^{107} \overline{\text{Ratio}}_{\text{AHHigh}\,j}}{107}$	$\dfrac{\sum_{j=1}^{107} \overline{\text{Ratio}}_{\text{AHLow}\,j}}{107}$	$(\overline{\text{Ratio}}_{\text{AHHigh}\,j} - \overline{\text{Ratio}}_{\text{AHLow}\,j}) < 0$		$(\overline{\text{Ratio}}_{\text{AHHigh}\,j} - \overline{\text{Ratio}}_{\text{AHLow}\,j}) > 0$	
		N	N_Significant	N	N_Significant
38.81	35.80	49	15	58	13

Panel C: Effects of News Difference

$\dfrac{\sum_{j=1}^{107} \overline{\text{Ratio}}_{\text{NewsDiffHigh}\,j}}{107}$	$\dfrac{\sum_{j=1}^{107} \overline{\text{Ratio}}_{\text{NewsDiffLow}\,j}}{107}$	$(\overline{\text{Ratio}}_{\text{NewsDiffHigh}\,j} - \overline{\text{Ratio}}_{\text{NewsDiffLow}\,j}) < 0$		$(\overline{\text{Ratio}}_{\text{NewsDiffHigh}\,j} - \overline{\text{Ratio}}_{\text{NewsDiffLow}\,j}) > 0$	
		N	N_Significant	N	N_Significant
38.19	33.03	30	2	77	36

Note: The sample consists of 107 firms that trade every day and have at least 100 news items with at least 100 days between the first and the last. For each firm j, the days are subdivided into "high" and "low" groups based on premarket volume, after-hours volume, and news difference (*PMHigh, PMLow, AHHigh, AHLow, NewsDiffHigh, NewsDiffLow*, respectively), using the corresponding company mean levels. The table reports average ratios of the absolute value of the per-hour open-to-close return to the absolute value of the per-hour close-to-open return, and the number of firms for which the difference in means across the subsamples is positive and negative. *N_Significant* refers to the number of firms for which the above difference is significant at the 10% level.

ratios for seventy-seven of the firms. The relation is significant in thirty-six cases. In only two instances the difference is significantly negative.

20.5 CONCLUSION

I reexamine the puzzling phenomenon of greater stock return volatility over trading periods versus nontrading periods. Data on order flow in the after-hours, premarket, and regular trading sessions, along with a unique extensive data set covering the concurrent firm-specific public information flow for a large sample of NASDAQ securities over the 2000–2001 period, allow me to carry out a direct test of the competing hypotheses and to offer new evidence on the determinants of return volatility.

Consistent with the existing studies, my results support the private information hypothesis. Higher trading volume in the premarket session, composed predominantly of anonymous information-based ECN trades, is associated with greater overnight return volatility and lower regular session volatility, indicating that price discovery shifts toward the premarket hours. Consequently, the volatility ratios decrease in premarket trading volume. Consistent with previous research, I show that the volume in after-hours is associated with little price impact and appears to be greatest on low information asymmetry days. The volatility of close-to-open and open-to-close returns is negatively related to after-hours volume, and volatility ratios are unaffected by such volume.

Unlike the existing studies, however, I also offer evidence consistent with the public information hypothesis. Greater flow of public information over trading (nontrading) hours increases the open-to-close (close-to-open) return volatility, and the ratio of return volatilities increases in the news flow differential.

The evidence on information spillover effects confirms the findings of prior studies that public information released outside regular trading hours tends to be of greater economic significance.

The analysis also presents new evidence on the trading processes in the rapidly growing extended-hours session and on the dynamics of public information flow.

ACKNOWLEDGMENTS

The views and opinions expressed do not represent State Street Global Advisors.

I thank seminar participants at the University of South Carolina, Xavier University, 2004 European Summer Symposium in Financial Markets,

2004 Financial Management Association, SSgA Advanced Research Center, 2005 Multinational Finance Society, 2005 Australasian Finance and Banking Conference, and 2006 European Financial Management meetings, and an anonymous referee for helpful comments.

REFERENCES

Admati, A., and Pfleiderer, P. (1988). A theory of intraday patterns: Volume and price variability. *Review of Financial Studies* 1:3–40.

Amihud, Y., and Mendelson, H. (1987). Trading mechanisms and stock returns: An empirical investigation. *Journal of Finance* 42:533–53.

Bacidore, J., and Lipson, M. (2001). The effects of opening and closing procedures on the NYSE and Nasdaq. Working paper, University of Virginia, Charlottesville.

Baginski, S., Hassell, J., and Pagach, D. (1996). Further evidence on nontrading period information release. *Contemporary Accounting Research* 12:207–21.

Barclay, M., and Hendershott, T. (2003). Price discovery and trading after hours. *Review of Financial Studies* 16:1041–73.

Barclay, M., Hendershott, T., and McCormick, D. (2001). Electronic communications networks and market quality. Working paper, University of Rochester, Rochester, New York.

Barclay, M., Hendershott, T., and McCormick, D. (2002). Information and trading on electronic communications networks. Working paper, University of Rochester, Rochester, New York.

Barclay, M., Litzenberger, R., and Warner, J. (1990). Private information, trading volume, and stock-return variances. *Review of Financial Studies* 3:233–53.

Berry, T., and Howe, K. (1994). Public information arrival. *Journal of Finance* 49:1331–46.

Bushee, B., Matsumoto, D., and Miller, G. (2004). Managerial and investor responses to disclosure regulation: The case of Reg FD and conference calls. *Accounting Review* 79:617–43.

Chan, Y. (2002). Volatility, volume and pricing efficiency in the stock index futures market when the underlying cash market does not trade. Working paper, Hong Kong Polytechnic University, Kowloon, Hong Kong.

Diamond, D. (1985). Optimal release of information by firms. *Journal of Finance* 40:1071–94.

Fama, E. (1965). The behavior of stock market prices. *Journal of Business* 38:34–105.

Forster, M., and George, T. (1995). Trading hours, information flow and international cross-listing. *International Review of Financial Analysis* 4:19–34.

Foster, F., and Viswanathan, S., (1993). Variations in trading volume, return volatility, and trading costs: Evidence on recent price formation models. *Journal of Finance* 48:187–211.

Francis, J., Pagach, D., and Stephan, J. (1992). The stock market response to earnings announcements released during trading versus nontrading periods. *Journal of Accounting Research* 30:165–84.

French, K., and Roll, R. (1986). Stock return variances: The arrival of information and the reaction of traders. *Journal of Financial Economics* 17:5–26.

Gennotte, G., and Trueman, B. (1996). The strategic timing of corporate disclosures. *Review of Financial Studies* 9:665–90.

Greene, J., and Watts, S. (1996). Price discovery on the NYSE and the Nasdaq: The case of overnight and daytime news releases. *Financial Management* 25:19–42.

Harris, L. (1986). A transaction data study of weekly and intradaily patterns in stock returns. *Journal of Financial Economics* 16:99–117.

He, H., and Wang, J. (1995). Differential information and dynamic behavior of stock trading volume. *Review of Financial Studies* 8:919–72.

Heflin, F., Subramanyam, K., and Zhang, Y. (2003). Regulation FD and the financial information environment: Early evidence. *Accounting Review* 78:1–37.

Holden, C., and Subrahmanyam, J. (1992). Long-lived private information and imperfect competition. *Journal of Finance* 47:247–70.

Hong, H., and Wang, J. (2000). Trading and returns under periodic market closures. *Journal of Finance* 55:297–354.

King, R., Pownall, G., and Waymire, G. (1990). Expectations adjustments via timely management forecasts: Review, synthesis, and suggestions for future research. *Journal of Accounting Literature* 9:113–44.

Kyle, A. (1985). Continuous auctions and insider trading. *Econometrica* 53:1315–35.

Libby, T., Mathieu, R., and Robb, S. (2002). Earnings announcements and information asymmetry: An intraday analysis. *Contemporary Accounting Research* 19:449–72.

Masulis, R., and Shivakumar, L. (1999). Intraday market response to equity offering announcements: A NYSE/AMEX-NASDAQ comparison. Working paper, Vanderbilt University, Nashville, TN.

Miller, E. (1989). Explaining intraday and overnight price behavior. *Journal of Portfolio Management* 15:10–17.

Oldfield, G., and Rogalski, R. (1980). A theory of common stock returns over trading and non-trading periods. *Journal of Finance* 35:729–51.

Patell, J., and Wolfson, M. (1982). Good news, bad news, and the intraday timing of corporate disclosures. *Accounting Review* 57:509–27.

Securities and Exchange Commission. (2000). Special study: Electronic communication networks and after-hours trading. Available at http://www.sec.gov/news/studies/ecnafter.htm#exec.

Stoll, H., and Whaley, R. (1990). Stock market structure and volatility. *Review of Financial Studies* 3:37–71.

CHAPTER 21

The Correlation of a Firm's Credit Spread with Its Stock Price

Evidence from Credit Default Swaps

Martin Scheicher

CONTENTS

21.1 INTRODUCTION

The market developments surrounding the declining credit quality of General Motors (GM) highlight the interdependence between equity and corporate debt markets. In March 2003, when GM was still an investment-grade debtor, its stock cost US$34 and the quoted premium on credit default swaps (CDS) was 365 basis points. Two years later, market participants' increasing concerns about GM's financial situation had raised the premium to more than 550 basis points, while the stock price had declined to below US$30. In November 2005, following successive downgrades by the major rating agencies, the market quote for the CDS premium now exceeded 1,000 basis points and the stock price was around US$23. This simultaneous comovement of the stock price and the CDS premium raises the following question: What are the general nature and determinants of the correlation between the equity and corporate debt market?

This chapter conducts an empirical analysis of the linkages between the equity and corporate debt markets. These are the two markets that firms use for raising capital. So far, there is little empirical evidence on linkages between the two markets. In the model of Merton (1974), corporate debt and equity represent alternative claims on a firm's assets. The two securities' common dependence on the firm asset value may create measurable linkages between the market prices of a firm's stock and its corporate bonds.

The strong growth of credit derivatives in the last few years has significantly simplified the trading of credit risk.[*] The most commonly used credit derivative is the credit default swap, which functions like a traded insurance contract against the losses arising to its creditors from a firm's default. Standardized contracts, low transaction costs, and a large and heterogeneous set of market participants have helped credit default swaps to hold the benchmark function for the price discovery process in the corporate debt markets.

The use of CDS–stock price pairs reduces differences in the information content of the two market prices. In particular, the CDS market quote is the cleanest available measure for the risk premium that investors require to bear corporate default risk. Historically, the diversity of individual bond features such as seniority, coupon structure, and embedded options, and the fact that many investors follow a buy-and-hold strategy, all have contributed to comparatively low liquidity in the corporate bond market. In contrast, the homogeneity and standardization of CDS contracts have supported the development of an active market, therefore reducing the liquidity premia observed in corporate bond spreads.[†] In addition, using CDS data removes the need to specify a risk-free term structure in order to calculate credit spreads.

The correlation of stock prices and credit spreads is an important variable in corporate finance decision making and in banks' risk management models. In a corporate finance context, the comovement of the two variables affects industrial firms' cost of capital, and thus influences how firms

---

[*] In this chapter, default risk is defined as the exposure to losses arising from a borrower's default, whereas credit risk also captures the losses arising from a borrower's downgrading.

[†] Blanco et al. (2005) document that CDS premia lead bond spreads and are taking an increasingly important role in the price discovery process.

choose the mix of equity and debt financing in order to optimize their capital structure. In a risk management context, this chapter's methodology is specifically relevant for the recently popular trading strategy of capital structure arbitrage. This strategy relies on relative pricing differences between a firm's debt and equity and is commonly used by hedge funds. Value-at-risk (VaR) modeling for this trading strategy requires an analysis of the CDS-equity comovement.

Few papers have so far directly studied the comovement of the market prices of stocks and corporate debt. For a sample from 1986 to 1990, Kwan (1996) finds significant negative unconditional correlations of stock returns and the changes in corporate bond yields. Norden and Weber (2008) show that in the period from 2000 to 2002, stock returns have led CDS premia. However, they use a rather restricted methodology that does not allow for time variation, and they do not study the determinants of the linkages. Schaefer and Strebulaev (2004) study the linkage between stocks and corporate bonds in order to evaluate the hedging performance of structural models. They find that the Merton model provides a good prediction of the sensitivity of corporate bond returns to stock returns. Another related study is Acharya and Johnson (2007) on insider trading in the CDS market. Using news reflected in equity prices as a benchmark for public information, they find incremental information revelation in the CDS market, providing evidence consistent with the occurrence of insider trading.

This chapter estimates the correlation of the log differences of stock prices and CDS premia of 240 major European and North American companies. The sample consists of CDS-stock price pairs for actively traded firms for the period from March 2003 to November 2005. I measure the linkage between the CDS premium and the stock price by means of the firm-specific conditional correlation in log differences, because a panel approach cannot simultaneously capture the considerable time variation (e.g., during the market turbulence in May 2005) as well as the cross-sectional variation from AA-rated firms to those in Chapter 11. I estimate the conditional correlation by means of a simplified bivariate generalized autoregressive conditional heteroskedasticity (GARCH) model.

My main finding is a statistically significant negative linkage between log changes in individual stock prices and CDS premia. Correlations are largest for high-yield firms, but the difference to the investment-grade category is weaker in Europe than North America. Furthermore, conditional correlations are characterized by sizable time variation, and among individual firms there is limited homogeneity. These empirical results illustrate the advantages of a conditional firm-specific correlation model because only this methodology can detect periods of significant comovement of stock returns and log differences of CDS premia such as the market turbulence in May 2005.

The rest of this chapter is organized as follows. In Section 21.2, I describe the mechanism of credit default swaps, the sample, and the correlation model. Section 21.3 details the empirical analysis. Section 21.4 concludes the chapter by summarizing the main results.

21.2 SAMPLE AND METHODOLOGY

21.2.1 Sample Construction

Credit default swaps are the most commonly traded credit derivatives. They transfer the risk that a certain individual entity defaults from the "protection buyer" to the "protection seller" in exchange for the payment of a premium. Commonly, CDS have a maturity of 1 to 10 years, with most of the liquidity concentrated on the 5-year horizon (see Longstaff et al., 2005).

A major step in the evolution of the credit risk transfer market has been the launch of harmonized CDS indices. In June 2004 a new family of indices was introduced, namely iTraxx in Europe and Asia and CDX in North America. This harmonization has led to generally accepted benchmarks for the credit market, therefore increasing market transparency and market liquidity.

The composition of this index family provides the basis for the selection of my sample. In the investment-grade corporate segment, the indices contain the equally weighted CDS premia of the 125 most liquid firms. Selection of index constituents is based on a semiannual poll of the main CDS dealers, which then leads to an update of the index composition in March and September of each year.

My analysis comprises individual European and U.S. firms in both the investment-grade and high-yield segments. The sample is designed to be representative across ratings and across industry sectors, covering financial firms as well as industrial firms. The starting point for the firm selection is the set of firms in the iTraxx Europe and Dow Jones CDX NA investment-grade index and the iTraxx Europe Crossover* and Dow Jones CDX NA high-yield index,† with the composition as of October 2005.

---

* This index contains the thirty most liquid nonfinancial names from Europe that are rated Baa3 or lower and are on a negative outlook.

† This index contains the 100 most liquid nonfinancial North American names that are rated Baa3 or lower and are on a negative outlook.

To construct the sample I match the CDS data and the stock price data of the iTraxx and CDX member firms in Bloomberg. Using a weekly frequency and a sample period start of March 2003, my sample consists of 111 European and 129 North American companies. The sample is diversified across sectors, as it contains energy firms, industrial entities, consumer cyclical and noncyclical firms, insurance companies, banks, telecoms, as well as automobile firms. The ratings at the end of the sample range from AA to D, therefore covering the entire spectrum of credit quality. There are two defaulted firms. Delphi filed for bankruptcy on October 8, 2005. Furthermore, Dana, another manufacturer of automobile components, went into the Chapter 11 procedure on March 3, 2006, i.e., shortly after the end of the sample period. Overall, most of the observations come from the rating categories between AA and BB.

21.2.2 Sample Description

The sample period is characterized by a steady decline in CDS premia and by a period of market turmoil. Figure 21.1 plots the time series of weekly premia.

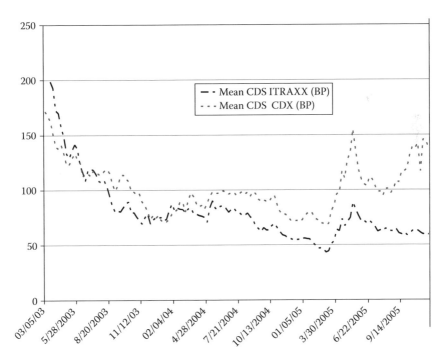

FIGURE 21.1 Time-series plots of CDS. (Data: Bloomberg, author's calculations.)

A downward trend in the CDS premia is common for both regions. Among the main factors behind this decline in risk premia was a benign macroeconomic environment, low equity market volatility, and the "hunt for yield." This phenomenon describes institutional investors' strong demand for higher-yielding assets in the aftermath of the collapse of stock prices, which had started in March 2000.[*] This search for higher-yielding assets manifested itself in many asset classes. In the credit markets, this demand pressure together with low default rates and the steadily declining equity market volatility contributed to a sharp decline in credit risk premia. For instance, in March 2003, the median CDS premium for European firms was around 80 basis points, whereas it measured around 35 basis points in November 2005.

An upward jump in CDS premia is observed in May 2005 after S&P downgraded Ford and General Motors to the high-yield segment of the credit market. The market turbulence following this announcement drove CDS premia up for a limited period. The market turmoil at that time had an adverse impact on the functioning of the credit derivatives market, reportedly causing large losses among some hedge funds.

In both samples, the median premium is around 50 basis points and the standard deviation equals 140 basis points. As regards the extreme values, the maximum of 2,500 basis points is observed in the iTraxx sample for Corus in March 2003, and the minimum of 7.75 basis points is recorded for Barclays Bank in September 2004. In the CDX data set, the maximum is observed for Delphi in October 2005 with 3,144 basis points, and the minimum is recorded for Wal-Mart in March 2005.

21.2.3 The Estimation of the Conditional Correlation

Before estimating the firm-specific correlations, I need to decide whether to use the levels of the variables or the first log differences of CDS premia or stock prices. In order to find the appropriate specification of the variables I apply an augmented Dickey-Fuller test with five lags and an intercept. At the 5% level, the null hypothesis of a unit root cannot be rejected for 95 European and 100 North American CDS premia. Therefore, all further analysis is based on weekly log differences of CDS premia and stock prices.

Among the alternative parameterizations for the conditional covariance matrix offered in the literature (see Bauwens et al. (2006) for a survey on GARCH-based conditional correlation models), I choose the exponentially

---

[*] See Chapter VI in BIS (2004) for a discussion of the search for yield.

weighted moving average (EWMA) model. This model, introduced by JP Morgan in its RiskMetrics methodology, combines a flexible parameterization of the second moments with a comparatively low computational burden. These properties are important criteria in model selection, as the sample composition requires estimating 240 bivariate GARCH models.

The EWMA specification describes the time-varying behavior of second moments by means of a simple specification, where the same parameter determines the persistence of both variances and the covariance. Motivated by computational tractability, this approach relies on the assumption that there are no cross-sectional differences in the determinants of time variation in individual firms' second moments:

$$\begin{pmatrix} \Delta \log P_t \\ \Delta \log CDS_t \end{pmatrix} = \begin{pmatrix} \mu_1 \\ \mu_2 \end{pmatrix} + \begin{pmatrix} \varepsilon_{1t} \\ \varepsilon_{2t} \end{pmatrix}$$

$$\varepsilon_t \,|\, I_{t-1} \sim N(0, H_t)$$

$$H_t = \begin{pmatrix} h_{11t} & h_{12t} \\ h_{12t} & h_{22t} \end{pmatrix}$$

(21.1)

$$h_{iit} = 0.94 h_{iit-1} + 0.06 \varepsilon_{iit-1}^2$$

$$h_{12t} = 0.94 h_{12t-1} + 0.06 \varepsilon_{12t-1}$$

where P_t is the stock price and CDS_t is the premium on the credit default swap.

The selection of the EWMA is supported by the empirical results of Ferreira and Lopez (2005), who find that its performance is not necessarily inferior relative to more complex multivariate GARCH models.

21.3 EMPIRICAL RESULTS

Before presenting the correlation estimates I discuss the estimates of the two conditional volatilities. Figure 21.2 plots the aggregate time series of EWMA volatilities. The graphs show that stock return volatility has declined strongly in the period from March 2003 to November 2005, whereas the downward trend in the CDS volatility is considerably weaker. The levels of the two volatilities are quite different. On average, median stock return volatility is 25%, about half the CDS volatility of 50%. Furthermore, stock return volatility never exceeds CDS volatility. During the market turbulence in May 2005, the CDS volatility almost doubled, whereas the return volatility did not change to such a large extent.

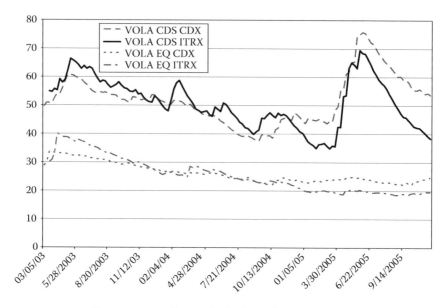

FIGURE 21.2 Time-series plots of CDS and equity volatilities. (Data: Bloomberg, author's calculations.)

Summary measures of the conditional correlations are given in Table 21.1, and Figure 21.3 plots the time series of the cross-sectional averages. As a categorization scheme for the total sample, I group the 240 firms according to their geographic region and their credit rating in November 2005 (obtained from the Fitch Ratings database).

Four observations emerge from the time-series plots and the descriptive statistics. First, across rating categories the mean correlations between stock returns and CDS changes range between −0.2 and −0.4, indicating an almost continuous decline in the correlation from A-rated firms[*] to the high-yield segment of the market. Relying on the credit rating as a proxy for firms' default risk,[†] this result indicates that declining credit quality is indeed linked to a higher correlation in absolute value. Overall, the rating category with the strongest correlation (mean value of −.048) is the North American D segment, which comprises Delphi and Dana. This value is three times the correlation in the CDX A-rated segment.

---

[*] Among the A-rated iTraxx segment, Siemens, Deutsche Bank, and Muenchner Rueckversicherung all have a mean correlation of around −0.4.

[†] This analysis neglects rating migrations (see Longstaff et al. (2005) for a similar approach).

TABLE 21.1   Descriptive Statistics of Estimated Correlations

			Europe			
Rating	**AA**	**A**	**BBB**	**BB**	**B-C**	
Mean	−0.26	−0.20	−0.22	−0.29	−0.31	
Min	−0.68	−0.87	−0.88	−0.71	−0.68	
Max	0.39	0.51	0.92	0.60	0.17	
SD	0.20	0.21	0.22	0.21	0.19	
N	1,846	5,538	6,532	1,420	426	
			North America			
Rating	**AA**	**A**	**BBB**	**BB**	**B-C**	**D**
Mean	−0.16	−0.19	−0.17	−0.29	−0.43	−0.48
Min	0.22	0.57	0.88	0.34	0.48	−0.18
Max	−0.82	−0.88	−0.85	−0.86	−0.89	−0.86
SD	0.18	0.19	0.22	0.19	0.24	0.15
N	661	5114	7647	1638	1086	249

Source: Bloomberg, author's calculations.

Note: This table reports the descriptive statistics of the conditional correlations estimated from the EWMA model across rating categories (with the Fitch rating as of November 2005). SD is the standard deviation and N is the number of observations in each rating category. The sample contains weekly observations from March 2003 to November 2005 for the 111 iTraxx firms and the 129 CDX firms.

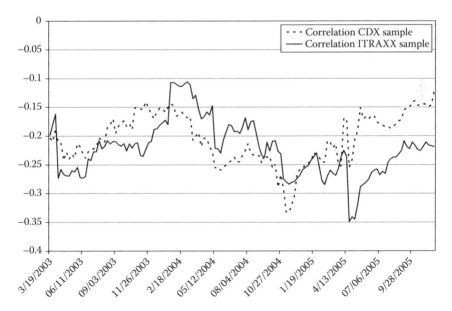

FIGURE 21.3   Time series of conditional correlations. (Data: Bloomberg, author's calculations.)

Second, the time series of correlations are characterized by sizable variation. For example, for the European data set, annual average conditional correlations are strongest in 2004, with a value of -0.27. In 2003, the average conditional correlations were around -0.16, and in 2005 they were around -0.21. This pattern is valid for all rating categories and particularly strong for the BB-rated segment, where the value for 2004 is recorded at -0.38.

Third, the standard deviation of the conditional correlation is larger than 0.18 for all rating categories, but there are no clear common features in the movement across rating categories. As regards the extreme values, the range of correlations increases with declining ratings. In particular, the maximum and minimum are, e.g., 0.39 and -0.68 for European A-rated firms but 0.60 and -0.71 for European BB-rated firms.

Finally, the turmoil in the credit market in May 2005 significantly affected the movement of the conditional correlations. In absolute value, the correlation increased during May 2005. As an example for this change, for A-rated iTraxx firms the mean correlation declined from -0.25 to -0.40. During this episode of market turbulence, comovement between the stock market and the credit market hence became stronger. Thus, a market participant with a long position in both risk categories would have seen an increase in aggregate portfolio risk.

My finding of stronger linkages between debt and equity for firms with lower credit quality has been documented by other papers. In a regression framework, Huang and Kong (2003) find for bond-based credit spreads that the sensitivity to the stock price increases with a firm's credit risk.

There is little evidence of common patterns among individual firms' correlation series. As an example, Figure 21.4 plots the correlations for four firms: Deutsche Bank, France Telecom, GM, and Delphi. These firms are chosen to represent the diversity in sectors as well as in credit quality, which is present in my sample. Figure 21.4 shows that the range of the correlation estimates differs across the firms, with France Telecom also recording positive correlations in the first half of the sample. Thus, there was an episode where both the CDS premium and the stock price rose. This situation could arise potentially due to an increase in leverage, which raises the profitability of the firm as well as its credit risk.

For GM and Delphi, the strongest correlations are observed during May 2005, with a value around -0.9. Given the close linkages between the two firms, the time-series movement of their correlations is similar. Before Delphi entered the Chapter 11 procedure, its correlation amounted

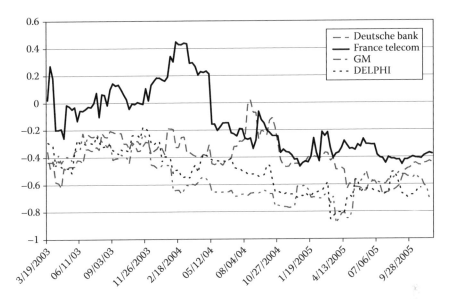

FIGURE 21.4   Conditional correlations of Deutsche Bank, France Telecom, GM, and Delphi. (Data: Bloomberg, author's calculations.)

to −0.62. At the same time, the correlation for GM was −0.55. During October 2005, it then fell to end the month also at −0.62.

Standard deviations of the correlation vary considerably across firms. Computing individual firms' t-statistics, I find that for 39 iTraxx and 36 CDX firms, the mean firm-specific conditional correlation of returns and first differences of CDS premia is significantly different from zero at the 5% level. In all of these cases, the mean correlation is negative.

In summary, individual firms' conditional correlations are quite volatile over time, mostly negative, and show limited homogeneity in the cross section.

21.4   CONCLUSION

For a sample of 240 firms, this chapter conducts a study of the linkages between stock returns and the changes in CDS premia. Using a bivariate EWMA model, I find that average conditional correlations are around −0.2. In absolute terms, the correlations strongly increase during periods of market turbulence, such as May 2005. Among individual firms, there is considerable volatility and little homogeneity in the correlation variation. The estimation results for GM and Delphi provide an additional perspective on the strength of the CDS-equity comovement. For these two firms, correlations are around −0.9 during May 2005. Comparing results for the North American and European samples, I find that the differences are rather limited.

Overall, the results indicate a negative linkage between individual stock returns and first differences of CDS premia. Thus, I confirm the results of Kwan (1996), who documented significantly negative linkages between stock prices and corporate bonds. My results also suggest that the linkage between stock returns and first differences of CDS premia is clearly measurable, but it is characterized by sizable time variation. Furthermore, this linkage is existent for both the investment-grade as well as the high-yield segment of the credit market. In particular, the existence of credit-equity linkages also for the upper segment of the credit market has so far not been documented in the literature.

These results have implications for risk modeling, in particular with respect to the interaction of credit and market risk in risk models. According to the empirical findings presented here, the linkages between the two risk categories increase in times of market turbulence, such as in May 2005. For the modeling of portfolio risk, this finding implies that diversification benefits between the two markets may be limited in volatile periods. For a bank using separate value-at-risk models to measure market and credit risk, the comovement documented here may lead to a misspecification of her overall VaR model and, in this manner, affect the accuracy of her estimates for overall required capital.

The empirical findings are particularly relevant for the recently popular trading strategy of capital structure arbitrage, where debt and equity are traded in relative value terms. Another application is the trading of convertible bonds, where again both credit risk and market risk components are present in the risk profile.

ACKNOWLEDGMENTS

The opinions in this chapter do not necessarily reflect those of the ECB. I am grateful to Til Schuermann and seminar participants at the European Central Bank, Deutsche Bundesbank, and the CREDIT 2006 conference for helpful comments.

REFERENCES

Acharya, V., and Johnson, T. (2007). Insider trading in credit derivatives. *Journal of Financial Economics* 84:110–41.

Bank for International Settlements. (2004). 74th annual report. Basel, Switzerland.

Bauwens, L., Laurent, S., and Rombouts, J. (2006). Multivariate GARCH models: A survey. *Journal of Applied Econometrics* 21:79–109.

Blanco, R., Brennan, S., and Marsh, I. (2005). An empirical analysis of the dynamic relationship between investment-grade bonds and credit default swaps. *Journal of Finance* 60:2255–81.

Ferreira, M., and Lopez, J. (2005). Evaluating interest rate covariance models within a value-at-risk framework. *Journal of Financial Econometrics* 3:126–68.

Huang, J., and Kong, W. (2003). Explaining credit spread changes: Some new evidence from option-adjusted spreads of bond indices. *Journal of Derivatives* 56:30–44.

Kwan, S. (1996). Firm-specific information and the correlation between individual stocks and bonds. *Journal of Financial Economics* 40:63–80.

Longstaff, F., Mithal, S., and Neis, E. (2005). Corporate yield spreads: Default risk or liquidity? New evidence from the credit default swap market. *Journal of Finance* 60:2213–53.

Merton, R. (1974). On the pricing of corporate debt: The risk structure of interest rates. *Journal of Finance* 29:449–70.

Norden, L., and Weber, M. (2008). The comovement of credit default swap, bond and stock markets: An empirical analysis. Working paper, University of Mannheim, Germany.

Schaefer, S., and Strebulaev, I. (2004). Structural models of credit risk are useful: Evidence from hedge ratios on corporate bonds. Working paper, London Business School, London.

CHAPTER 22

Modeling the Volatility of the FTSE100 Index Using High-Frequency Data Sets

David E. Allen and Marcel Scharth

CONTENTS

22.1 INTRODUCTION

The availability of ultra-high-frequency stock market data and the subsequent introduction of realized volatility measures enabled the development of new econometric models of volatility (Andersen and Bollerslev, 1998) featuring more precise parametric models of time-varying volatility. One such measure, the realized variance, is defined as the sum of squared intraday returns sampled at a sufficiently high frequency, consistently approximating the integrated variance over the fixed interval where the observations are summed. Realized volatility is the squared root of the realized variance. In practice, high-frequency measures do suffer some contamination from microstructure noise such as bid-ask bounce, etc. (see Biais et al., 2005). This *ex post* volatility measure can be modeled as an observable variable (see Andersen et al. (2003) and Barndorff-Nielsen and Shephard (2002) for the theoretical foundations of realized volatility (RV)). Several recent papers have proposed corrections to estimation of RV in order to take the microstructure noise into account (see McAleer and Medeiros (2008b) and Gatheral and Oomen (2007) for reviews).

In this chapter we refer to realized volatility as a consistent estimator of the squared root of the integrated variance and model the volatility of the FTSE100 index using high-frequency data sets in a period including the onset of the subprime mortgage debacle in the United States. We show that the presence of high and time-varying volatility of volatility is a fundamental stylized fact of stock market volatility, bringing additional uncertainty in the tails of the distribution of asset returns, explaining why events of several standard deviations may be observed, and rendering point forecasts of realized volatility a very poor measure of risk during critical moments of the financial crisis. We argue that higher moments of returns should be modeled to deal with this problem and show that the volatility of volatility is subject to strong leverage effects and is strongly and positively related to the level of volatility. In this chapter, we give a brief introduction on how this can be done within a realized volatility framework and explain how the daily distribution of returns (from which value-at-risk, expected shortfall, and other measures of interest can be extracted) can be forecasted from the model.

Our results suggest that the use of point forecasts of volatility is insufficient for obtaining adequate coverage and systematically underestimates the VaR intervals, but the introduction of a Monte Carlo method–based density forecast based on a specification that takes into account the volatility of volatility corrects this failure in the lower tail. Moreover, results are improved when intraday volatility feedback effects, which skew the *ex ante* distribution of the returns, are taken into account. In the case of expected shortfall, our method significantly improves forecasts and the results strongly favor specifications with time-varying volatility of volatility and asymmetric effects. These results are stronger than the ones obtained by Corsi et al. (2008) in that we argue that ignoring time-varying volatility of volatility and intraday leverage effects renders risk measures strongly biased and density forecasts inaccurate.

22.2   REVIEW OF PRIOR WORK

Empirical work on the characteristics of asset returns suggested that both fractional integration and structural changes can describe the volatility of asset returns (Lobato and Savin, 1998; Martens et al., 2004; Beltratti and Morana, 2006; Morana and Beltratti, 2004; Hyung and Franses, 2002). Some researchers have applied simpler time-series models that are consistent with high persistence in relevant horizons, even though they do not rigorously exhibit long memory (hence their label as quasi-long-memory models). Some examples are the mixed data sample (MIDAS; see, for example, Ghysels et al., 2007) and heterogeneous autoregressive (HAR; Corsi, 2004) models, both of which explore data sampled at different frequencies, and the unobserved ARMA component (UC) of Koopman et al. (2005). Other contributions to the realized volatility modeling and forecasting literature are by Martens et al. (2004), who develop a nonlinear (ARFIMA) model to accommodate level shifts, day-of-the-week effects, leverage effects, and volatility level effects; Andersen et al. (2007); and Tauchen and Zhou (2005), who argue that the inclusion of jump components significantly improves forecasting performance. McAleer and Medeiros (2008a) extend the HAR model to account for nonlinearities, while Hillebrand and Medeiros (2007) also consider nonlinear models and evaluate the benefits of bootstrap aggregation (bagging) for volatility forecasting. Ghysels et al. (2007) argue that realized absolute values outperform square return-based volatility measures in predicting future increments in quadratic variation. Scharth and Medeiros (2006) introduce multiple regime models linked to asymmetric effects.

22.3   OUR APPROACH TO MODELING VOLATILITY

Given the variety of approaches evident in the literature, we commence by considering what the characteristics of realized volatility series are that can have significant impact for risk management and other applications.

We argue that a basic property of the observed realized volatility series is that they exhibit a very large degree of volatility themselves, and that this volatility of volatility is itself time varying; periods of stable and more predictable volatility alternate with episodes where the series display large swings and assume values within a potentially broad range. This has important implications for the tails of the distribution of returns.

We will give next a qualitative discussion of this claim. First, the presence of high and time-varying volatility of volatility means high uncertainty in the tails of the distribution of daily returns. If the volatility in the next day is relatively unpredictable, then a conditional expectation of this variable will not contain much information about what might happen in terms of very negative or very positive returns. Even though returns standardized by (*ex post*) quadratic variation measures are nearly Gaussian, returns standardized by fitted or predicted values of time-series volatility models are far from normal. Given the uncertainty in volatility, this is expected and should not be seen as evidence against those models; explicitly modeling the higher moment is necessary. Second, forecasting improvements brought by the body of work discussed previously are marginal and swamped by the size of the volatility of volatility. In an extreme example, Scharth and Medeiros (2006) calculate that even a simple exponentially weighted moving average (EWMA) of realized volatility delivers predictions that are very close to ARFIMA and HAR specifications; it is crucial to account for the fact that the series is highly persistent, but the way this is done has very little economic relevance. Hence, it is very easy to predict the level of volatility in relation to the history of the series, but there is not much relevant information to it given the uncertainty in the variable. We thus take the view that even though point forecasts have been the main output from which volatility models in general have been evaluated (extensive comparisons of forecasting performance like the one performed by Hansen and Lunde (2005) are common in the literature), those statistics do not necessarily convey much information about the relative economic significance of the volatility models (see Fleming et al., 2001, 2003; Chan and Kalimipalli, 2006). In particular, and perhaps not surprisingly, small and possibly statistically insignificant forecasting performance differences may overshadow important relative modeling qualities.

When high-frequency data were not widely available, the volatility of volatility could not be observed, and latent volatility models were the only available option, the typical solution of the literature for volatility models that did not generate normally distributed standardized returns was to assume an *ad hoc* distribution for returns conditional on volatility that would sufficiently inflate the tails. The implications of the volatility of volatility for the tails of returns were understood, and this modification could account in part for the mixing properties of volatility and returns. However, this is not entirely satisfactory when the volatility of volatility is time varying with nonconstant size of the tails. Hence, we begin by analyzing the time-series properties of the volatility of volatility of the FTSE100 index. To do so, we use the concept of realized quarticity (see Barndorff-Nielsen and Shephard, 2002, 2004, 2006; Andersen et al., 2007), which can be seen as an estimate for the variance of the return variation, suggesting that the volatility of volatility is characterized by long-memory properties, strong leverage effects, short-lived explosive regimes, and high correlation with the level of volatility. To the best of our knowledge, the last three of these characteristics have not been documented in the literature so far.

The next step is to directly model both the volatility and the volatility of volatility. We propose that the informative but noisy realized quarticity series be combined with the latent variable approach implemented by Corsi et al. (2008), the first to consider the volatility of volatility and who extended the framework for modeling volatility by specifying a GARCH process to allow for clustering in the squared residuals of those realized volatility models and assumed a normal inverse Gaussian (NIG) distribution to accommodate fat tailedness and skewness in the distribution of the residuals. We suggest, however, that this approach should be extended to make use of the greater volume of information available in the context of high-frequency data. In particular, modeling the (possibly nonlinear) relation between the volatility of volatility and the level of volatility seems to improve the model. With the assumptions for the realized volatility process and following the evidence in assuming that returns conditional volatility are normally distributed, we have a mixing hypothesis that will enable us to propose a straightforward two-step Monte Carlo method for calculating value-at-risk, expected shortfall, and other density-related measures for returns. The procedure consists in first simulating realized volatility and then using each of these simulated values to simulate returns. The empirical distribution function of the simulated returns can then be used for obtaining a prediction of value-at-risk and other density-related measures of interest.

22.4 EMPIRICAL SETUP

22.4.1 Data and Realized Volatility Measurement

The empirical analysis focuses on the realized volatility of the FTSE100 index, which is plotted on Figure 22.1. The raw intraday quote data were obtained from the TaqTiq/SIRCA (Securities Industry Research Centre of Asia-Pacific) database. The period of analysis starts on January 2, 1996, and ends on December 28, 2007, providing a total of 3,001 trading days. We start by removing nonstandard quotes, computing mid-quote prices, filtering possible errors, and obtaining 1-second returns for trading hours. Following the results of Hansen and Lunde (2006), we adopt the previous tick method for determining prices at precise time marks.

To measure the realized volatility, we turn to the theory developed by Barndorff-Nielsen et al. (2005, 2007a, 2007b) and implement a subsampled

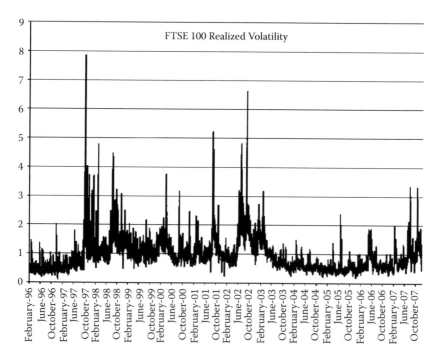

FIGURE 22.1   FTSE100 realized volatility.

realized kernel estimator based on 1-minute returns and the modified Tukey-Hanning kernel, which is consistent in the presence of microstructure noise.

22.4.2 Modeling Realized Volatility

Our general specification is given by

$$\phi_p(L)(1-L)^d(RV_t - \mu(W_t)) = X_t\lambda + h_t(h_{t-1}, \varepsilon_{t-1}, Z_t)\varepsilon_t \qquad (22.1)$$

where here d denotes the fractional differencing parameter, L the lag operator, ε_t is independent and identically distributed (i.i.d.) with $E(\varepsilon_t) = 0$, ϕ_p (L) is a polynomial of order L, W_t is a vector of variables affecting the mean of RV_t, X_t is a vector of explanatory variables, and Z_t is a vector of variables that may affect the volatility of volatility. We discuss below the specification for the conditional mean of volatility and the distribution of the errors; we postpone the analysis of heteroskedasticity for subsequent subsections.

22.4.3 Autoregressive Fractionally Integrated Specification

When $0.5 < d < 0.5$, we have a stationary autoregressive fractionally integrated model for the realized volatility. After running a battery of specification tests centered on the Schwarz information criterion, we set $\phi_p(L) = (1 - \phi_p L)$ (that is, an ARFIMA(1, d, 0) model) for all our estimations. Such models have been extensively estimated for realized volatility, for example, in Andersen et al. (2003), Areal and Taylor (2002), Beltratti and Morana (2005), Deo et al. (2006), Martens et al. (2004), and Thomakos and Wang (2003), among others.

22.4.4 Heterogenous Autoregressive (HAR) Specification

The heterogeneous autoregressive (HAR) model proposed by Corsi (2004) is an unfolding of the heterogeneous ARCH (HARCH) model developed by Müller et al. (1997). It is specified as a multicomponent volatility model with an additive hierarchical structure, leading to an additive time-series model of the realized volatility that specifies the volatility as a sum of volatility components over different horizons.

Turning to our general specification, let $d = 0$ (underlining the view that the model does not genuinely exhibit long memory) and

$$\phi_p P(L) = 1 - \phi_p L - \phi_p \sum_{i=1}^{5} L^i - \phi_3 \sum_{i=1}^{22} L_i$$

Furthermore, consider the notation $RV_{t,j}=\Sigma^t_{i=t-j+1}RV_{t,j}/j$, which will be used extensively in this chapter. We can then write our HAR model with daily, weekly, and monthly components as

$$RV_t=\mu(W_t)+\phi_1 RV_{t-1}+\phi_2 RV_{t-2}+\phi_3 RV_{t-1,5}+\phi_4 RV_{t-1,22}+X_t\beta+h_t(Z_t)\varepsilon_t$$

$$(22.2)$$

We can see that the HAR specification is an AR, Equation (22.2), model rendered parsimonious by several parameter restrictions. Simulations reported in Corsi (2004) show that the generous number of autoregressive lags renders the HAR model capable of reproducing the observed hyperbolic decay of the sample autocorrelations of realized volatility series over not too long horizons. Moreover, the model displays forecasting performance that is similar to that of ARFIMA models, which is generally true for any model that exhibits high persistence (and not necessarily authentic long-memory properties). For its estimation simplicity, the HAR-RV has been commonly favored in the high-frequency econometrics literature (e.g., Andersen et al., 2007). Nevertheless, it is difficult to further justify the HAR model. One of its drawbacks is that it tends to estimate parameters that are generally inconsistent with each other when different direct forecasting estimations are implemented.

22.4.5 Asymmetric Effects and Jumps

Bollerslev et al. (2006) and Scharth and Medeiros (2006) highlight the impact of leverage effects on the dynamics of realized volatility. The latter argue for the existence of regime switching behavior in volatility, with large falls (rises) in prices being associated with persistent regimes of high (low) variance in stock returns. The authors show that the incorporation of cumulated daily returns as a explanatory variable brings modeling advantages by capturing this effect, which can be quite large; after analyzing certain stocks in the Dow Jones Index, the authors document that falls in the horizon of less than 2 months are associated with volatility levels that are up to 60% higher than the average of periods with stable or rising prices. We estimate models with and without such effects. Moreover, we consider jump components that have been receiving growing attention in the realized volatility literature. Building on theoretical results for bipower variation measures, articles such as Andersen et al. (2007), Tauchen and Zhou (2005), and Barndorff-Nielsen and Shephard (2006) established related frameworks for the nonparametric estimation of the jump component in asset return volatility by explicitly considering the presence of less persistent elements in the volatility of stocks in contrast with the smooth and very slowly mean-reverting part associated with long-memory properties. In this chapter, we follow Ghysels et al. (2007) and take the realized absolute variation (denoted RAV_t), calculated as the sum of intraday absolute returns as a more robust measure of the persistent component in volatility, thus separating the effect of jumps. We find only the first lag of this variable to be significant, yielding in the least parsimonious case:[*]

$$X_t\lambda = \lambda_1 I(r_{t-1} < 0)r_{t-1} + \lambda_2 I(r_{t-1} > 0)r_{t-1} + \lambda_3 I(r_{t-1,5} < 0)r_{t-1,5}$$

$$+ \lambda_4 I(r_{t-1,5} > 0)r_{t-1,5} + \lambda_5 I(r_{t-1,22} < 0)r_{t-1,22} + \lambda_6 I(r_{t-1,22} > 0)r_{t-1,22}$$

$$+ \lambda_7 \sqrt{RAV_{t-1}} + I(r_t < 0)(r_t/RV_t) + I(r_t > 0)(r_t/RV_t) \qquad (22.3)$$

where the indicator functions have been included to reinforce the asymmetry between the effect of positive and negative returns and r_t/RV_t is to be interpreted as an exogenous shock following the standard normal distribution.

22.4.6 The Distribution of ε_t

To account for the non-Gaussianity of the error terms we follow Corsi et al. (2008) and assume that the (unconditional) i.i.d. innovations ε_t are distributed normal inverse Gaussian (NIG), which is flexible enough to allow for excessive kurtosis and skewness and reproduce a number of symmetric and asymmetric distributions (including the normal itself). The density of the NIG distribution is given by

$$f(x,\alpha,\beta,\mu,\delta) = \frac{\alpha}{\pi} \frac{k_1\left(\alpha\delta\sqrt{1+\left(\frac{x-\mu}{\delta}\right)^2}\right)}{\sqrt{1+\left(\frac{x-\mu}{\delta}\right)^2}} \exp\left\{\delta\left(\sqrt{\alpha^2-\beta^2} + \beta\left(\frac{x-\mu}{\delta}\right)\right)\right\}$$

$$(22.4)$$

where $K_i(x)$ is the modified Bessel function of the second kind with index i, $\mu \in \mathfrak{R}$ denotes the location parameter, $\delta > 0$ the scale, $\alpha > 0$ the

---

[*] In what follows, we will make use of the abbreviations AE, VF, and RAV in the acronyms of the models that contain asymmetric effects, intraday volatility feedback effects, and jumps, respectively (for example, an HAR/AE model).

shape, and $\beta \in (-\alpha, \alpha)$ the skewness parameter. Mean and variance are given by

$$E(x)=\mu+\frac{\alpha}{\alpha^2-\beta^2}\ Var(x)=\frac{\alpha^2}{\sqrt{\alpha^2-\beta^2}^3} \tag{22.5}$$

So that the distribution is standardized by setting

$$\mu=-\frac{\alpha^2}{\sqrt{\alpha^2-\beta^2}^3} \quad \text{and} \quad \delta=\frac{(\alpha^2-\beta^2)^{3/2}}{\alpha^2}$$

22.4.7 Estimation

The parameters are estimated by maximizing the log-likelihood function:

$$\ell(\hat{\varphi},\hat{\lambda},\hat{\alpha},\hat{\beta}, RV_{1...T}, X_{1...T}, Z_{1...T})=T\log(\hat{\alpha})-T\log(\pi)$$

$$+\sum_{t=1}^{T}\log\left[K_1\left(\hat{\alpha},\hat{\delta}\left(1+\hat{y}_t^2\right)^{1/2}\right)\right]$$

$$-0.5\sum_{t=1}^{T}\log\left(1+\hat{y}_t^2\right)+T\hat{\delta}(\hat{\alpha}^2-\hat{\beta}^2)^{1/2}$$

$$+\hat{\delta}\hat{\beta}\sum_{t=1}^{T}\hat{y}_t-0.5\sum_{t=1}^{T}\log\left(\hat{h}_t^{1/2}\right) \tag{22.6}$$

In the case of the ARFIMA model, we employ a two-step estimation. In the first we apply the widely used log periodogram estimator (GPH) of Geweke and Porter-Hudak (1983) to filter the data. The number of ordinates used in each regression is selected by the plug-in method of Hurvich and Deo (1999). We then apply the maximum likelihood estimator above for the filtered series.

22.5 EMPIRICAL RESULTS AND INTRODUCING TIME-VARYING VOLATILITY OF VOLATILITY

22.5.1 The *Ex Ante* Distribution of the FTSE100 Returns

With our basic time-series model for the realized volatility defined, we are ready to state the empirical problem at the center of our analysis. Even though it has been long recognized that the distributions of the stock returns scaled by realized standard deviations are approximately

TABLE 22.1 FTSE100 Daily Returns Standardized by the Fitted Values of a Typical Realized Volatility Time-Series Model (1996–2007)

Mean	−0.001
Median	0.05
Maximum	3.88
Minimum	−4.60
Standard deviation	1.03
Skewness	−0.23
Kurtosis	3.89

Gaussian (e.g., Andersen et al., 2001), Table 22.1 and Figure 22.2 (which illustrate how the point forecasts tend to fall short of the realized volatility by as far as 50% exactly on the riskiest days) reveal that this is far from the case when we scale returns by the in-sample predicted values of our best-fitting model in terms of in-sample forecasts (the HAR model with leverage effects and the square root of the realized absolute variation as explanatory variables).

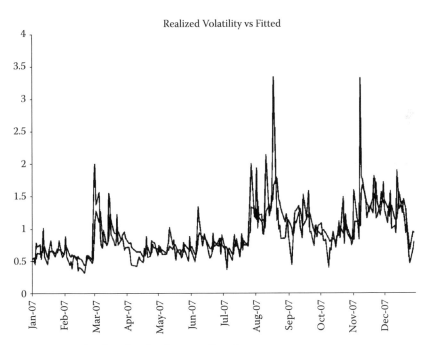

FIGURE 22.2 Realized volatility vs. fitted.

Despite the sizable forecasting gains made possible by volatility models based on high-frequency data, our descriptive results can be directly related to the failure of GARCH volatility models to completely account for the excess kurtosis of returns (see, for example, Malmsten and Teräsvirta, 2004, Carnero et al., 2004). The researcher or practitioner interested in evaluating the density of returns from the perspective of a time-series model still lives in a fat-tailed world, and models of the conditional expectation of volatility have little to say about it. In this chapter, we do not interpret those facts as evidence against those models, but as a consequence of high day-to-day unpredictability of the shocks that affect the volatility (excessive kurtosis) and the intraday correlation between those shocks and returns (negative skewness). We argue that an adequate volatility model for return density forecasting and risk management in this setting should illuminate the dynamics of the higher moments. To pursue this objective, we will turn to the idea of time-varying volatility of realized volatility (Corsi et al., 2008), which will allow for time-varying kurtosis on the general model introduced previously.

22.5.2 Incorporating Time-Varying Volatility of Volatility into the Time-Series Model

To account for the dynamic properties of the volatility of volatility, we follow Corsi et al. (2008) and initially specify a GARCH(1,1) model for the conditional heteroskedasticity of the realized volatility. Recalling Equation (22.1), we have that in this case the conditional variance of the realized volatility follows:

$$h_t = \theta_0 + \theta_1 h_{t-1} + \theta_2 \varepsilon_{t-1}^2 \tag{22.7}$$

However, our empirical estimations (which we omit here for conciseness) indicated that there is a substantial degree of positive correlation between the level of realized volatility and the volatility of volatility, so that a more empirically relevant and better-fitting specification is given by

$$h_t = \theta_0 + \theta_1 h_{t-1} + \theta_2 \varepsilon_{t-1}^2 + \theta_3 \tilde{R}V_t + \theta_4 \tilde{R}V_t^2 \tag{22.8}$$

where RV_t denotes the conditional mean of the realized volatility. Surprisingly, the presence of this variable renders the GARCH coefficients almost insignificant. The estimated volatility of volatility for the FTSE data using Equation (22.8) is displayed by Figure 22.3. It can be seen that

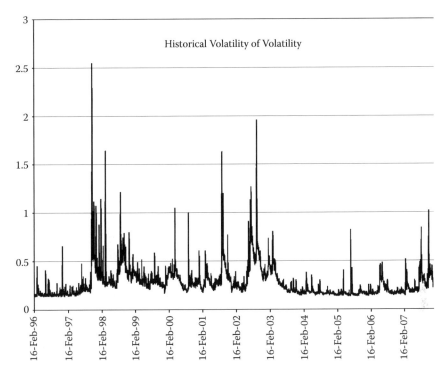

FIGURE 22.3   Historical volatility of volatility.

in line with the realized volatility the variable displays a marked crisis behavior during the Asian, Russian, Internet, 2002, and subprime crises and other episodes. As we shall illustrate, this stylized fact has important consequences for issues of risk management.

22.5.3   Density Forecasting

A density forecast for the stock index returns, which can be used for calculating a number of risk measures, can be calculated from our model by Monte Carlo as follows:

1. In the first step, the functional form of Equation (22.1) is used for the evaluation of predictions of the realized volatility and the volatility conditional on past realized volatility observations, returns, the estimated volatility of volatility series and shocks, and other variables.

2. We randomly generate n shocks distributed as the standardized NIG with the parameters estimated from the data, which multiplied

by $h_t^{1/2}$ and added to RV_t originate a vector of n simulated realized volatilities for day t, where $\tilde{x}_t$ denotes a prediction of a generic variable x for day t.

3. Under the hypothesis that standardized returns are normally distributed, we employ each of these n simulated volatilities to simulate m associated returns. The empirical density function of the set of $n \times m$ simulated returns yield our final density forecast.

22.5.4   Some Illustrative Out-of-Sample Results

Even though in the introduction to this chapter we argued that point forecasting may not be a powerful tool for comparing volatility models, predictions have been the main basis of comparison in the volatility literature and are the subject of extensive analysis. Therefore, we start by analyzing the performance of the models described previously in this aspect as a first layer of comparison. The evaluation of forecasts is based on the mean absolute error (MAE), the root mean squared error (RMSE), and the R^2 of a regression of the observed realized volatility on the forecasts. A formal test of the forecasting differences is given by the superior predictive ability (SPA) test developed by Hansen (2005). The null hypothesis is that a given model is not inferior to any other competing models in terms of a given loss function.

The forecasting statistics are displayed in Table 22.2. The results suggest that whenever cumulated returns are included in the conditional mean specification, the respective model fares unambiguously better than

TABLE 22.2   Point Forecasts for 2007

	R-squared	RMSE	MAE	SPA in R^2	SPA in MSE
HAR	0.438	0.295	0.210	0.002	0.104
HAR/AE	0.484	0.272	0.191	0.832	0.787
HAR-GARCH	0.413	0.296	0.209	0.001	0.027
HAR/AE-GARCH	0.447	0.283	0.200	0.005	0.032
HAR/AE/RAV-GARCH	0.451	0.284	0.201	0.000	0.022
ARFIMA	0.341	0.303	0.198	0.004	0.006
ARFIMA/AE	0.384	0.291	0.187	0.059	0.016
ARFIMA-GARCH	0.340	0.303	0.198	0.009	0.009
ARFIMA/AE-GARCH	0.367	0.296	0.194	0.037	0.022
HAR/AE/VF-GARCH	0.455	0.273	0.173	0.876	0.036
HAR/AE-Aug.GARCH	0.447	0.275	0.174	0.762	0.216

its counterpart without leverage effects (on all criteria). Nevertheless, the results support our previous claims: the differences in performance are economically small and swamped by the size of the volatility of volatility; the benefits of the increased efficiency of alternative conditional variance specifications fail to materialize in this context.

We now proceed to analyze the ability of different models and the Monte Carlo method in forecasting adequate quantiles over the lower tail of the return distribution and the whole density. The first is illustrated by the proportion of returns that exceed the 2.5% value-at-risk forecast (we use this interval and not the more traditional 1% one so as to have more violations to examine), which is analyzed by means of the likelihood ratio tests for unconditional coverage (UC) developed by Christoffersen (1998). To evaluate the accuracy of the whole density forecasts we rely on the theory of density evaluation developed by Diebold et al. (1998). Below the Monte Carlo method is compared with the results obtained when only the point forecasts are taken into account in predicting the densities.

The results are organized in Table 22.3. As expected, the method of calculating VaR values based only on the point forecast of volatility is strongly biased toward underestimating the value-at-risk, failing to provide

TABLE 22.3   Value-at-Risk and Density Forecasting Results for 2007

	Point Forecast			Monte Carlo Method		
	2.5% VaR Failures	UC p-value	p-value for Density Test	2.5% VaR Failures	UC p-value	p-value for Density Test
HAR	0.036	0.306	0.000	0.028	0.781	0.183
HAR/AE	0.040	0.168	0.000	0.032	0.510	0.203
HAR-GARCH	0.040	0.168	0.000	0.040	0.168	0.251
HAR/AE-GARCH	0.044	0.086	0.000	0.032	0.510	0.247
HAR/AE/ RAV-GARCH	0.044	0.086	0.000	0.032	0.510	0.249
ARFIMA	0.075	0.000	0.000	0.064	0.001	0.000
ARFIMA/AE	0.075	0.000	0.000	0.064	0.001	0.000
ARFIMA-GARCH	0.075	0.000	0.000	0.056	0.007	0.000
ARFIMA/ AE-GARCH	0.068	0.000	0.000	0.060	0.003	0.000
HAR/AE/ VF-GARCH	0.060	0.003	0.000	0.032	0.510	0.210
HAR/AE-Aug. GARCH	0.060	0.003	0.000	0.032	0.510	0.233

adequate coverage. For this reason, the whole density forecasts are also strongly rejected in this case (for example, the point forecast method indicates that the probability of the two lowest returns on the data set—August 10, 2007, 3.55%, and August 16, 2007, 3.82%—was less than 1 in 20,000). On the other hand, the table provides evidence that taking the time-varying volatility of volatility into account importantly mitigates or eliminates this problem. Even though a complete and more rigorous analysis is out of the scope of this chapter, these results illustrate our main point: it is fundamental to take into account the high unpredictability in the volatility in performing this type of risk analysis. In doing so, we rule out implausible results for our data set.

22.6   CONCLUSION

We have used higher moments of volatility and leverage effects to demonstrate that it is possible to achieve more accurate estimates of relevant *ex ante* risk measures based on realized volatility by incorporating these components into our model.

REFERENCES

Andersen, T., and Bollerslev, T. (1998). Answering the skeptics: Yes, standard volatility models do provide accurate forecasts. *International Economic Review* 39:885–906.

Andersen, T., Bollerslev, T., and Diebold, F. (2007). Roughing it up: Including jump components in the measurement, modeling and forecasting of return volatility. *Review of Economics and Statistics* 89:701–20.

Andersen, T., Bollerslev, T., Diebold, F. X., and Ebens, H. (2001). The distribution of realized stock return volatility. *Journal of Financial Economics* 61:43–76.

Andersen, T., Bollerslev, T., Diebold, F. X., and Labys, P. (2003). Modeling and forecasting realized volatility. *Econometrica* 71:579–625.

Areal, N., and Taylor, S. R. (2002). The realized volatility of FTSE-100 futures prices. *Journal of Futures Markets* 22:627–48.

Barndorff-Nielsen, O., Hansen, P., Lunde, A., and Shephard, N. (2005). Realised kernels can consistently estimate integrated variance: Correcting realised variance for the effect of market frictions. Discussion paper, Nuffield College, Oxford.

Barndorff-Nielsen, O., Hansen, P., Lunde, A., and Shephard, N. (2007a). Designing realised kernels to measure the ex-post variation of equity prices in the presence of noise. Unpublished manuscript, Stanford University.

Barndorff-Nielsen, O., Hansen, P., Lunde, A., and Shephard, N. (2007b). Subsampling realized kernels. Working paper, Stanford University.

Barndorff-Nielsen, O., and Shephard, N. (2002). Econometric analysis of realized volatility and its use in estimating stochastic volatility models. *Journal of the Royal Statistical Society B* 64:253–80.

Barndorff-Nielsen, O., and Shephard, N. (2004). Power and bipower variation with stochastic volatility and jumps. *Journal of Financial Econometrics* 2:1–37.

Barndorff-Nielsen, O., and Shephard, N. (2006). Econometrics of testing for jumps in financial economics using bipower variation. *Journal of Financial Econometrics* 4:1–30.

Beltratti, A., and Morana, C. (2005). Statistical benefits of value-at-risk with long memory. *Journal of Risk* 7:47–73.

Beltratti, A., and Morana, C. (2006). Breaks and persistency: Macroeconomic causes of stock market volatility. *Journal of Econometrics* 131:151–77.

Biais, B., Glosten, L., and Spatt, C. (2005). Market microstructure: A survey of microfoundations, empirical results, and policy implications. *Journal of Financial Markets* 8:217–64.

Bollerslev, T., Litvinova, J., and Tauchen, G. (2006). Leverage and volatility feedback effects in high-frequency data. *Journal of Financial Econometrics* 4:353–84.

Carnero, M. A., Pena, D., and Ruiz, E. (2004). Persistence and kurtosis in GARCH and stochastic volatility models. *Journal of Financial Econometrics* 2:319–42.

Chan, W. H., Jha, R., and Kalimipalli, M. (2006). The economic value of trading with realized volatility in the SP 500 index options market. Working paper, School of Business & Economics, Wilfrid Laurier University, Waterloo, Ontario.

Christoffersen, P. (1998). Evaluating interval forecasts. *International Economic Review* 39:841–62.

Corsi, F. (2004). A simple long memory model of realized volatility. Manuscript, University of Southern Switzerland.

Corsi, F., Kretschmer, U., Mittnik, S., and Pigorsch, C. (2008). The volatility of realized volatility. *Econometric Reviews* 27:46–78.

Deo, R., Hurvich, C., and Lu, Y. (2006). Forecasting realized volatility using a long-memory stochastic volatility. *Journal of Econometrics* 131:29–58.

Diebold, F. X., Gunther, T. A., and Tay, A. S. (1998). Evaluating density forecasts with applications to financial risk management. *International Economic Review* 39:863–83.

Fleming, J., Kirby, C., and Ostdiek, B. (2001). The economic value of volatility timing. *Journal of Finance* 56:329–52.

Fleming, J., Kirby, C., and Ostdiek, B. (2003). The economic value of volatility using realized volatility. *Journal of Financial Economics* 67:473–509.

Gatheral, J., and Oomen, R. (2007). Zero intelligence variance estimation. Working paper, Warwick Business School.

Geweke, J., and Porter-Hudak, S. (1983). The estimation and application of long memory time series models. *Journal of Time Series Analysis* 4:221–38.

Ghysels, E., Sinko, A., and Valkanov, R. (2007). MIDAS regressions: Further results and new directions. *Econometric Reviews* 26:53–90.

Hansen, P. (2005). A test for superior predictive ability. *Journal of Business Economic Statistics* 23:365–80.

Hansen, P. R., and Lunde, A. (2005). A forecast comparison of volatility models: Does anything beat a GARCH(1,1) model? *Journal of Applied Econometrics* 20:873–89.

Hansen, P. R., and Lunde, A. (2006). Realized variance and market microstructure noise (with discussion). *Journal of Business and Economic Statistics* 24:127–218.

Hillebrand, E., and Medeiros, M. (2007). Forecasting realized volatility models: The benefits of bagging and nonlinear specifications. Discussion paper, Pontifical Catholic University of Rio de Janeiro.

Hurvich, C., and Deo, R. (1999). Plug-in selection of the number of frequencies in regression estimates of the memory parameter of a long-memory time series. *Journal of Time Series Analysis* 20:331–41.

Hyung, N., and Franses, P. (2002). *Inflation rates: Long-memory, level shifts, or both?* Report 2002–08, Econometric Institute, Erasmus University, Rotterdam.

Koopman, S., Jungbacker, B., and Hol, E. (2005). Forecasting daily variability of the SP 100 stock index using historical, realised and implied volatility measurements. *Journal of Empirical Finance* 23:445–75.

Liu, C., and Maheu, J. (2007). Forecasting realized volatility: A Bayesian model averaging approach. Department of Economics, University of Toronto.

Lobato, I. N., and Savin, N. E. (1998). Real and spurious long-memory properties of stock-market data. *Journal of Business and Economic Statistics* 16:261–68.

Malmsten, H., and Teräsvirta, T. (2004). *Stylized facts of financial time series and three popular models of volatility.* Working Paper Series in Economics and Finance 563, Stockholm School of Economics, Stockholm.

Martens, M., Van Dijk, D., and De Pooter, M. (2004). *Modeling and forecasting S&P 500 volatility: Long memory, structural breaks and nonlinearity.* Discussion Paper 04-067/4, Tinbergen Institute, Rotterdam.

McAleer, M., and Medeiros, M. (2008a). A multiple regime smooth transition heterogeneous autoregressive model for long memory and asymmetries. *Journal of Econometrics*, (147) 1:104–115.

McAleer, M., and Medeiros, M. (2008b). Realized volatility: A review. *Econometric Reviews* 27:10–45.

Morana, C., and Beltratti, A. (2004). Structural change and long range dependence in volatility of exchange rates: Either, neither or both? *Journal of Empirical Finance* 11:629–58.

Müller, U., Dacorogna, M., Dav, R., Olsen, R., Pictet, O., and von Weizscker, J. (1997). Volatilities of different time resolutions—Analyzing the dynamics of market components. *Journal of Empirical Finance* 4:213–40.

Scharth, M., and Medeiros, M. (2006). Asymmetric effects and long memory in the volatility of Dow Jones stocks. Working paper, Department of Economics, Pontifical Catholic University of Rio de Janeiro.

Tauchen, G., and Zhou, H. (2005). Identifying realized jumps on financial markets. Working paper, Department of Economics, Duke University, Durham.

Thomakos, D., and Wang, T. (2003). Realized volatility in the futures market. *Journal of Empirical Finance* 10:321–53.

Weizsacker, J. E. (1997). Volatilities of different time resolutions—Analysing the dynamics of market components. *Journal of Empirical Finance* 4:213–39.

IV

Emerging Market Volatility

Economic Integration on the China Stock Market, before and after the Asian Financial Crisis

Jack Penm and R. D. Terrell

CONTENTS

23.1 INTRODUCTION

Since 1980, China's economy has gradually developed under the influence of Deng Xiaoping's philosophy. Following a series of reforms that opened its socialist market economy, China has steadfastly developed foreign trade and actively attracted foreign investment. China's trade value ranks fifth in the world. Its foreign capital attraction has been measured as the foremost among all the developing countries, with annual foreign direct investment (FDI) inflows of about US$72.41 billion and US$69.47 billion in 2005 and 2006, respectively. Reform and opening up have not only promoted the sustained, swift, and sound development of China's national economy, but also helped restructure its economic system.

The economy of modern China has been well integrated into the global economy, especially with the United States, United Kingdom, Japan, Taiwan, Hong Kong, and Singapore. For example, bilateral trade turnover (the sum of exports and imports) between China and the United States reached US$389.7 billion in 2007, over 150 times that of 1979. In 2007, exports from the UK to China were less than US$18.5 billion, and the imports from China were US$31.7 billion, while the exports and imports between China and Japan were US$102.1 billion and US$134.0 billion, respectively. In 2007, Taiwan's exports to and imports from Mainland China were US$101 billion and US$24.5 billion respectively. At the same time, Hong Kong's exports to and imports from Mainland China were US$12.8 billion and US$184.4 billion, respectively. On the other hand, in 2007 the bilateral trade volume between China and Singapore was US$47.2 billion.

Chinese stock markets have developed rapidly especially in recent years. By 2007, the total market capitalization of the Shanghai Stock Exchange (SHSE) was US$2.38 trillion. The Shanghai Composite Index was compiled to reflect the stock price movements in SHSE. It is a weighted average stock price index, with the weighting being determined by the number of shares issued by all listed companies. It used December 19, 1990, as the base day and has been officially published since July 15, 1991 (*SHSE 2007 Fact Book*).

Having entered the World Trade Organization (WTO), China has further opened its capital markets. Whether the degree of correlation between the returns of market indices in China and major developed countries or regional economies will correctly reflect the integration between economies, and whether investment in Chinese stock markets can diversify the risk of portfolio investment, will attract global investors' attention. Our hypothesis is that Chinese stock markets may have low-correlation relationships with major developed countries and regional counterparts. Hence, investment in Chinese stock markets can represent a feasible element of their portfolio to enhance the reward-to-volatility ratio even though China is interacting into the global economy gradually. Findings of a differential impact of the Chinese stock markets on the other six stock markets can lead to further insights into socioeconomic connections. Specifically, the investigation of co-movement relationships will provide useful information for both domestic and foreign investors.

The benefit of international diversification, however, is limited when national equity markets are cointegrated, because the presence of common factors limits the amount of independent variation. Cointegration among national equity markets implies that there are fewer assets available to investors for portfolio diversification than a simple count of the number of stocks. Moreover, cointegration would also mean Granger causality in levels, and hence would be suggestive of inefficiency in the market (Penm et al., 2003).

This chapter contributes to the literature by investigating whether stock co-movements exist between China and the three world leaders, the United States, UK, and Japan, and between China and three regional counterparts, Taiwan, Hong Kong, and Singapore, before and after the Asian financial crisis. It is observed that before the crisis is the most volatile period in the stock markets involved, and markets became less volatile after the crisis. We first examine co-movements between China and these six stock markets by employing the Engle-Granger (1987) two-step cointegration technique. Next, we employ the minimum final prediction error criterion to determine the optimum lag structures (Hsiao, 1981). Finally, we use the error correction model (ECM) or vector autoregressive (VAR) models to find the causal relationship between the Chinese stock market and the other six stock markets. Our results provide evidence that only Japan and Taiwan were cointegrated with the Chinese stock market before the crisis. However, after the crisis, all six stock markets became cointegrated with the Chinese stock market. Moreover, the Chinese stock market is more cointegrated with the three regional counterparts than the three world leaders. Their close socioeconomic, trade, and cultural relationships with China emphasize this important feature for global investors.

The rest of the chapter is organized as follows. Section 23.2 provides a literature review. Section 23.3 presents the data and methodology used. Section 23.4 discusses the empirical findings and interpretation of the results. Section 23.5 provides a conclusion.

23.2 LITERATURE REVIEW

A number of studies have examined co-movements of international stock markets. A considerable amount of work has been done on the interrelationships among the world equity markets, especially focusing on the major developed markets like the United States and Japan. The presence of strong co-movements among national stock markets limits the benefit of international diversification.

The performance of developed markets was the focus of world attention before and after the crash of 1987. The crash of October 1987 made people realize that most national equity markets are closely integrated.

The developed markets, notably that of the United States, exert a strong influence on other, smaller markets. Lee and Kim (1994), using a correlation approach, examine the effect of the October 1987 crash on the co-movements among national stock markets. "They find that national stock markets became more interrelated after the crash, and the strengthening co-movements among national stock markets continued for a long period after the crash. In addition, it is shown that the co-movements among national stock markets were stronger when the U.S. stock markets became more volatile" (Lee and Kim, 1994). There is also focus in the literature on price discovery in world markets in investigating international co-movements. Naturally, cointegration and error correction modeling provide a useful framework for analyzing price adjustments in internationally linked markets.

In recent years new capital markets have emerged in many parts of the world, and foreign capital controls have also been relaxed to a certain extent. With this relaxation of capital controls there has been an increase in investors' interest in international diversification, as it allows investors to have a larger basket of foreign securities to choose from and to add to their portfolio assets to diversify investment risk. A number of studies have examined co-movements in stock returns with reference to the expected return and diversification benefits of emerging-market investments.

Asian capital markets are key players among the emerging markets. Many studies have been done in the 1990s and thereafter to study the co-movements between Asian markets and the stock markets in developed countries. Kwan et al. (1995) study the stock markets of Australia, Hong Kong, Japan, Singapore, South Korea, Taiwan, the UK, the United States, and Germany, and suggest that these markets are not weak form efficient, as they find significant lead-lag relationships between these equity markets. Palac-McMiken (1997) employ monthly ASEAN market indices (Indonesia, Malaysia, Philippines, Singapore, and Thailand) from the 1987–1995 period and discover that with the exception of Indonesia, all markets are connected with one another and jointly they are not efficient. The author proposes that there still exists some leeway for diversification across these markets even though there is proof of interdependence among ASEAN stock markets. Johnson and Soenen (2002) examine the equity market integration of the Japanese stock market and twelve other equity markets in Asia. The authors conclude that equity markets in Australia, Hong Kong, Malaysia, New Zealand, and Singapore are extremely integrated with the Japanese stock market. It appears that previous empirical studies on the relationship between world stock markets do not provide consistent results. The reasons for the inconsistent results are numerous, including the choice of markets, different sample periods, different frequency of observations, and the different methodologies employed. Looking at the increasing importance and integration of the Chinese economy in the world economy, this study takes China into account, where it has not previously been examined. The purpose of assessing co-movements between China and the six stock markets is unique to this study.

23.3 DATA AND METHODOLOGY

We use weekly stock indices of the major stock exchanges in China (Shanghai Composite), the United States (S&P 500), the UK (FTSE 100), Japan (Nikkei 225 Stock Average), Taiwan (Taiwan SE Weighted), Hong Kong (Hang Seng), and Singapore (Straits Times Index). All the indices are expressed in terms of local currencies and obtained from DataStream. Our sample covers the period from January 1, 1991 through December 31, 2007. Weekly indices are used to avoid representation bias from some thinly traded stocks, i.e., the problem of nonsynchronous trading. In addition, we use Wednesday indices to avoid the day-of-the-week effect of stock returns (Lo and MacKinlay, 1988). We divided our sample into two periods to look at the effect of the Asian financial crisis: January 1, 1991–December 31, 1996 (before crisis) and January 1, 1997–December 31, 2007 (after crisis).

To examine the co-movements between the stock indices in China and the six markets with three developed world leaders—the United States, the UK, and Japan—and three regional counterparts—Taiwan, Hong Kong, and Singapore—we study the relationship

$$Y_t = a + bX_t + e_t \qquad (23.1)$$

where Y_t denotes the Chinese stock index; X_t denotes the index of any of the six other countries (the United States, the UK, Japan, Taiwan, Hong Kong, or Singapore); and e_t denotes the error term. As the stock indices are likely to be nonstationary, the statistical concept of cointegration plays a major role in determining the validity and reliability of the relationship.

Cointegration tests, which are important in determining the presence and nature of an equilibrium economic relation, were first introduced by Granger (1981) and later developed by Granger (1987). A detailed description of cointegration can be found in Brailsford et al. (2008). Before testing for cointegration, a unit root test has to be performed to test for nonstationarity for both endogenous and exogenous variables.

Cointegration tests in this paper consist of two steps. The first step is to examine the stationarity properties of the various stock indices in our study. If a series, say y_t, has a stationary, invertible, and stochastic ARMA representation after differencing d times, it is said to be integrated of order d, and denoted as $y_t = I(d)$. To test the null hypothesis H_0: $y_t = I(1)$ versus the alternative hypothesis H_1: $y_t = I(0)$, we apply the Dickey-Fuller (1981) unit root test procedure based on the OLS regression

$$\nabla y_t = b_0 + a_0 t + a_1 y_{t-1} + \varepsilon_t \qquad (23.2)$$

or apply the augmented Dickey-Fuller (ADF) test based on the OLS regression

$$\nabla y_t = b_0 + a_0 t + a_1 y_{t-1} + \sum_{i=1}^{p} b_i \nabla y_{t-i} + \varepsilon_t \qquad (23.3)$$

where $\Delta y_t = y_t - y_{t-1}$ and y_t can be Y_t or X_t as defined in Equation (23.1). The regressions in Equations (23.2) and (23.3) allow for a drift term (b_0) and a deterministic trend (a_0). The regression in Equation (23.3) allows a stochastic structure in the error term, ε_t, while p is chosen in Equation (23.3) to achieve white noise residuals. Testing the null hypothesis of the presence of a unit root in y_t is equivalent to testing the hypothesis that $a_1 = 0$ in Equations (23.2) and (23.3). If a_1 is significantly less than zero, the null hypothesis of a unit root is rejected. When $p = 0$, the test is known as the Dickey-Fuller (DF) test. This test assumes that the residuals ε_1 are independently and identically distributed. If serial correlation exists in the residuals, then $p > 0$ and the ADF test must be applied.

In addition, we test the hypothesis that y_t is a random walk with drift, i.e., $(b_0, a_0, a_1) = (b_0, 0, 0)$, and y_t is random walk without drift, $(b_0, a_0, a_1) = (0, 0, 0)$, using Equation (23.2). The test statistics are the likelihood ratios Φ_3 or Φ_2 found in Dickey and Fuller (1981). Following Wong et al. (2005a), if the hypotheses that $a_1 = 0$, $(b_0, a_0, a_1) = (b_0, 0, 0)$, or $(b_0, a_0, a_1) = (0, 0, 0)$ are accepted, we can conclude that y_t is $I(1)$. If we cannot reject the hypotheses that y_t is $I(1)$, we need to further test the null hypothesis H_0: $y_t = I(2)$ versus the alternative hypothesis H_1: $y_t = I(1)$. Note that most nonstationary series are integrated of order one. If both Y_t and X_t are of the same order, say $I(d)$, with $d > 0$, we then estimate the cointegrating parameter of (1) by OLS regression. If the residuals of Equation (23.1) are stationary, the two series, Y_t and X_t, are said to be cointegrated. Otherwise, Y_t and X_t are not cointegrated. The most common tests for stationarity of estimated residuals are Dickey-Fuller (CRDF) and augmented Dickey-Fuller (CRADF) tests based on the OLS regression

$$\nabla \hat{e}_t = \gamma \hat{e}_{t-1} + \sum_{i=1}^{p} \gamma_i \nabla \hat{e}_{t-1} + \xi_t \qquad (23.4)$$

should be employed, where $\hat{e}_t$ are residuals from the cointegrating regression in Equation (23.1) and p is chosen to achieve empirical white noise residuals.

> Engle and Granger (1987) pointed out that when a set of variables is cointegrated, a vector autoregression in first differences will be misspecified, and any potentially important long-term relationship between the variables will be unclear. Thus, inferences based on vector autoregression in first differences may lead to incorrect conclusions (Sims et al., 1990). However, there exists an alternative representation, an error correction representation of such variables, which takes account of a short- and long-run equilibrium relationship shared by those variables. (Wong et al., 2005b)

Once the cointegration relationship between the Chinese stock market and the markets of other countries has been decided, we can adopt the bivariate VAR model to test for Granger causality. If the cointegration does not exist between the two markets, following Granger et al. (2000), we employ

$$\nabla Y_t = a_0 + \sum_{i=1}^{n} a_{1i} \nabla Y_{t-i} + \sum_{j=1}^{m} a_{2j} \nabla X_{t-j} + u_{1t}$$

$$\nabla X_t = b_0 + \sum_{i=1}^{n} b_{1i} \nabla X_{t-i} + \sum_{j=1}^{m} b_{2j} \nabla Y_{t-j} + u_{2t}$$

(23.5)

where Y_t and X_t represent the indices of the Chinese stock market and any of these six stock markets, respectively, n and m are the optimum lags, and u_{1t} and u_{2t} are the error terms. We test the null hypothesis, H_0: $a_{21} = a_{22} = \cdots = a_{2m} = 0$, which implies that any of these six stock markets do not Granger-cause the Chinese stock market. Similarly, we test H_0: $b_{21} = b_{22} = \cdots = b_{2m} = 0$ to confirm that the Chinese stock market does not Granger-cause any of these six stock markets.

If the series are cointegrated, there is a long-term, or equilibrium, relationship among the series. Their dynamic structure can be exploited for further investigation. An error correction model (ECM) abstracts the short- and long-run information in the modeling process. The ECM proposed by Engle and Granger (1987) corrects for disequilibrium in the short run. Engle and Granger (1987) show that cointegration is implied by the existence of an error correction representation of the indices involved. An important theorem, known as the Granger representation theorem, states that if two variables Y and X are cointegrated, then their relationship can be expressed as an ECM (Wong et al., 2005a, p. 8).

In this situation, an error correction term $(e_{t-1} = Y_{t-1} - \delta X_{t-1})$ is added to the equation to test for Granger causality.

$$\nabla Y_t = a_0 + a e_{t-1} + \sum_{i=1}^{n} a_{1i} \nabla Y_{t-i} + \sum_{j=1}^{m} a_{2j} \nabla X_{t-j} + u_{1t}$$

$$\nabla X_t = b_0 + b e_{t-1} + \sum_{i=1}^{n} b_{1i} \nabla X_{t-i} + \sum_{j=1}^{m} b_{2j} \nabla Y_{t-j} + u_{2t}$$

(23.6)

The existence of cointegration implies causality among the set of variables as manifested by $|a| + |b| > 0$; a and b denote speeds of adjustment (Engle and Granger, 1987). If we do not reject H_0: $a_{21} = a_{22} = \cdots = a_{2m} = 0$ and $a = 0$, then none of these six stock markets (either the United States, the UK, Japan, Taiwan, Hong Kong, or Singapore) Granger-cause the Chinese stock market. Similarly, not to reject H_0: $b_{21} = b_{22} = \cdots = b_{2m} = 0$ and $b = 0$ suggests that the Chinese stock market does not Granger-cause any of the other six stock markets individually (Granger et al., 2000).

To test the hypothesis H_0: $a_{21} = a_{22} = \cdots = a_{2m} = 0$, we find the sum of square of residuals for both the full regression, SSE_F, and the restricted regression, SSE_R, in Equation (23.6) and apply the F test:

$$F = \frac{(SSE_R - SSE_F)/m}{SSE_F/(N-m-n-2)}$$

where N is the number of observations, and n and m are defined in Equation (23.5) or Equation (23.6). If H_0 is true, F is distributed as $F(m, N - m - n - 2)$. Hence, the hypothesis H_0 is rejected at the α level of significance if $F > F(\alpha; m, N - m - n - 2)$ and the reduced model is accepted if H_0 is not rejected. Similarly, we can test for the hypothesis $H_0: b_{21} = b_{22} = \cdots = b_{2m} = 0$ and then make decisions on causality. We apply the usual simple t statistics to test $H_0: a = 0$ and $H_0: b = 0$.

The minimum final prediction error criterion (Hsiao, 1981) is used to determine the optimum lag structures in the regressions of Equations (23.5) and (23.6), where n and m are the maximum lags of the corresponding variables to be used in the right-hand side of Equations (23.5) and (23.6), and u_{1t} and u_{2t} are disturbance terms obeying the assumptions of the classical linear regression model. The final prediction error statistic of ∇Y_t, with n lags of ∇Y_t and m lags of ∇X_t, is

$$FPE_{\nabla Y_t}(n,m) = \frac{(N+n+m+1)\Sigma(\nabla Y_t - \nabla \hat{Y}_t)^2}{(N-n-m-1)N}$$

where N is the number of observations. The FPE statistic for ∇X_t is found in the same way. To determine the minimum $FPE_{\Delta Y_t}$, the first step is to run the regression in the first part of Equation (23.5), excluding ∇X_t, and with only lags of ∇Y_t to be included. We start from $m = 0$ and $n = 1$ and calculate $FPE_{\Delta Y_t}(1,0)$. We proceed with the same step until $n = n^*$, where FPE is minimized for $m = 0$. Then by holding $n = n^*$, we systematically lag m until $m = m^*$ minimizes the FPE. The same procedure is repeated with the second part of Equation (23.5), where $n = n^{**}$ and $m = m^{**}$ minimize $FPE_{\Delta X_t}(n,m)$.

23.4 EMPIRICAL RESULTS AND INTERPRETATION

Table 23.1 shows the results of testing the order of integration of the seven series for the two periods before and after the Asian financial crisis. We do not reject that all seven stock indices are $I(1)$ in our sample at the 5% significance level for both periods. Our findings show that there exists no strong lead-lag effects in testing lead-lag effects by using both daily and weekly data. Thus, the results originally obtained are a good measure for testing co-movement between stock indices between China and other markets.

Having established that the stock indices in our study are all $I(1)$, we then estimate the cointegrating Equation (23.1). We conduct the unit root test on the residuals from Equation (23.4) to test the cointegration. Panel A of Table 23.2 shows that in the period before the Asian financial crisis

TABLE 23.1   Unit Root Test Results for the Weekly Indices in the Chinese Stock
Market and the Other Six Stock Markets

Variable	Period	DF	ADF	Φ_2	Φ_3
China	1991–1996	−1.82	−1.82	0.92	2.28
Shanghai SE Composite	1997–2007	−1.89	−1.89	0.99	2.80
United States	1991–1996	−1.74	−1.74	4.62	1.64
S&P 500 Composite	1997–2007	−2.00	−2.00	1.35	3.74
UK	1991–1996	−3.34	−3.34	2.00	5.60
FTSE 100	1997–2007	−2.10	−2.02	1.25	3.80
Japan	1991–1996	−2.14	−2.14	0.28	2.46
Nikkei 225 Stock Average	1997–2007	−2.00	−2.00	0.93	2.07
Taiwan	1991–1996	−2.10	−2.10	0.30	2.31
Taiwan SE Weighted	1997–2007	−2.68	−2.68	0.16	3.71
Hong Kong	1991–1996	−2.41	−2.41	3.64	3.25
Hang Seng	1997–2007	−1.65	−1.79	0.12	1.38
Singapore	1991–1996	−2.08	−2.08	2.05	2.70
Straits Times	1997–2007	−1.62	−1.66	0.21	1.33

Note: DF is the Dickey-Fuller t-statistic; ADF is the augmented Dickey-Fuller statistic. Φ_2
and Φ_3 are the Dickey-Fuller likelihood ratios. *, $p < 0.05$; **, $p < 0.01$.

(from 1991 to 1996), only Japan and Taiwan are cointegrated with the
Chinese stock market at the 5% significance level, and Japan is also cointegrated with China at the 1% significance level. However, in the period after
the Asian financial crisis (from 1997 to 2007), Panel B of Table 23.2 shows
that all six stock markets are cointegrated with the Chinese stock market
at the 5% significance level. Furthermore, the three regional counterparts,
Taiwan, Hong Kong, and Singapore, are cointegrated with China at the
1% significance level. This suggests that since the Asian financial crisis,
the Chinese stock market has been more closely integrated into the global
economy, especially with its regional counterparts. These closer relationships could be related to geographical proximity, partnerships in trade,
and cultural and historical similarities. Yang et al. (2003) point out that
the Asian financial crisis altered the degree of market integration in the
region over time, though China was not included in their study, and our
findings extend their approach to include China.

One possible explanation for the existence of cointegration between the
Chinese stock market and all others is that it is the outcome of Chinese
economic reform and its open-door policy. Since 1997, the Chinese
economy has been gradually integrated into the global economy, especially with world leaders like the United States, the UK, and Japan, and

TABLE 23.2   Cointegration Results for the Chinese Stock Market and the Other Six Stock Markets

A. Before Crisis: 1991–1996				
Country	**Model**	**R^2**	**CRDF**	**CRADF**
United States	Y(China) = −4.6332 + 1.7756Y(US) (−4.63) (10.94)	0.2780	−1.74	−1.74
UK	Y(China) = −11.3909 + 2.2106Y(UK) (−7.39) (11.49)	0.2980	−1.76	−1.76
Japan	Y(China) = 38.959 − 3.2994Y(Japan) (18.94) (−15.87)	0.4475	−2.61**	−2.61**
Taiwan	Y(China) = 6.1897 + 0.0147Y(Taiwan) (3.87) (0.08)	0.0000	−2.13*	−2.13*
Hong Kong	Y(China) = −4.4922 + 1.2157Y(HK) (−7.70) (18.55)	0.5252	−1.91	−1.91
Singapore	Y(China) = −4.3212 + 1.45Y(Singapore) (−4.23) (10.41)	0.2582	−1.74	−1.74
B. After Crisis: 1997–2007				
Country	**Model**	**R^2**	**CRDF**	**CRADF**
United States	Y(China) = 3.4549 + 0.5497Y(US) (9.18) (10.25)	0.2377	−2.06*	−2.06*
UK	Y(China) = 4.6906 + 0.3053Y(UK) (8.40) (4.69)	0.0613	−2.16*	−2.16*
Japan	Y(China) = 9.3015 − 0.2086Y(Japan) (23.83) (−5.10)	0.0717	−2.19*	−2.42*
Taiwan	Y(China) = 10.1419 − 0.3224Y(Taiwan) (28.69) (−8.01)	0.1600	−2.55*	−2.71**
Hong Kong	Y(China) = 4.1844 + 0.333Y(HK) (8.92) (6.67)	0.1165	−2.69**	−2.69**
Singapore	Y(China) = 5.4315 + 0.2536Y(Singapore) (14.44) (5.00)	0.0690	−2.67**	−2.67**

Note: CRDF is the cointegration regression Dickey-Fuller statistic for stationarity of the estimated residuals. CRADF is the comparable test statistic for the augmented Dickey-Fuller. *, $p < 0.05$, **, $p < 0.01$.

the regional counterparts such as Taiwan, Hong Kong, and Singapore. China's comprehensive national purchasing power has been remarkably strengthened, with high GDP growth. China is now the seventh largest trading nation in the world. The Chinese government has reduced the tariff rate and expanded the opening of trade in goods and services. China is progressively liberalizing its service sectors, like finance, insurance, telecommunication, transportation, and tourism. With greater economic integration with the world, the Chinese stock market cannot be isolated from world stock markets. All these fundamental economic factors are reflected in the performance of the Chinese stock market.

We further test the Granger causality relationship among the seven countries. Since all the other six markets are cointegrated with China after the crisis, ECM is employed to test for the Granger causality in the period after the crisis. ECM only applied to Japan and Taiwan in the period before the crisis due to their cointegration relationship. The other markets are tested by VAR as they are not cointegrated. The significance of $a \neq 0$ in Table 23.3 leads us to reject the null hypotheses that any other stock market does not Granger-cause the Chinese stock market. However, the results lead us to accept the null hypotheses that the Chinese stock market does not Granger-cause any other stock market. Therefore, the outcomes indicate that unilateral causality arises from any of these six stock markets, except Japan, to the Chinese stock market, especially in the period after the crisis. Several explanations may account for the causal relationships between any two equity markets. They include economic relationships, regulatory structures, exchange rate policies, and trade flows.

The above outcomes reflect the existence of a Granger causality among the five stock markets other than Japan and the Chinese stock market, but this does not explain why the direction is unilateral. Our results in Panel B of Table 23.3 show that we fail to reject H_0: $b_{21} = b_{22} = \cdots = b_{2m} = 0$ and $b = 0$ for the second period, except for Japan. This implies that the Chinese stock market does not Granger-cause any of these stock markets. One possible reason is that the Chinese stock market is a policy market. According to a study conducted in China (Wang et al., 2006), nearly 50% of the significant market movements were caused by changes in trading rules or changes in policies. Changes in rules in China could influence the movement of the Chinese stock market, but should not have any effect on its influence upon other markets. If this view is correct, the Chinese stock market does not Granger-cause other stock markets. Also, there is evidence of a disconnection between stock returns and real economic growth in China. For instance, the annual return of the Shanghai Composite Index was −4% in 1998, while the GDP growth rate for China for that year was 7.8%. In 2000, due to a policy shift that was favorable to the stock market, the annual return of the index achieved a staggering 52%, even though the GDP growth for that year was 8%. In 2001, the return of the index swung to −21%, when the government sold a huge number of state-owned shares in the secondary

TABLE 23.3   Granger Causality Results for the Chinese Stock Market
and the Other Six Stock Markets

	A. Before Crisis: 1991–1996					
Country	**Granger Cause**	**N**	**m**	**p-values[a]**	**p-values[b]**
United States (us)	us → cn	1	1	0.3777	n.a.
	cn → us	6	1	0.4077	n.a.
UK (uk)	uk → cn	1	2	0.2165	n.a.
	cn → uk	2	6	0.0326*	n.a.
Japan (jp)	jp → cn	1	1	0.1381	0.0385*
	cn → jp	1	3	0.0259*	0.4765
Taiwan (tw)	tw → cn	1	1	0.4489	0.0376*
	cn → tw	6	1	0.9294	0.3985
Hong Kong (hk)	hk → cn	1	2	0.0355*	n.a.
	cn → hk	1	1	0.4569	n.a.
Singapore (sg)	sg → cn	1	3	0.0361*	n.a.
	cn → sg	5	1	0.6111	n.a.

	B. After Crisis: 1997–2007					
Country	**Granger Cause**	**N**	**m**	**p-values[a]**	**p-values[b]**
United States (us)	us → cn	6	1	0.9444	0.0086**
	cn → us	1	1	0.3155	0.7970
UK (uk)	uk → cn	6	1	0.6545	0.0070**
	cn → uk	4	1	0.3337	0.2238
Japan (jp)	jp → cn	6	1	0.6805	0.0140*
	cn → jp	1	3	0.0043**	0.7686
Taiwan (tw)	tw → cn	6	1	0.0762	0.0122*
	cn → tw	1	1	0.6683	0.5896
Hong Kong (hk)	hk → cn	6	1	0.3382	0.0006**
	cn → hk	4	1	0.1183	0.5979
Singapore (sg)	sg → cn	6	1	0.2163	0.0007**
	cn → sg	5	1	0.5607	0.3510

Note: n.a. means ECM not applicable in the model due to no cointegration between the
two variables. → implies Granger cause, e.g., us → cn implies the United States
Granger-causes China. *, $p < 0.05$; **, $p < 0.01$.

[a] p-values of F test on H_0: $a_{21} = a_{22} = \cdots = a_{2m} = 0$ or H_0: $b_{21} = b_{22} = \cdots = b_{2m} = 0$.

[b] p-values of t test on H_0: $a = 0$ or H_0: $b = 0$ in ECM model.

market, while the GDP was 7.3% in that time. Further, in 2006 the return
of the index surged 130%, while the GDP rose 10.7%. Another likely reason
is the speculative nature of the Chinese stock market. Stock prices often do
not really reflect the underlying assets of the firms. Thus, the Chinese stock

market is still in the developing stage, and is not yet sufficiently mature to Granger-cause other markets in the world.

23.5 CONCLUSION

The economies of China and the United States, the UK, Japan, Taiwan, Hong Kong, and Singapore have become increasingly integrated with growing bilateral trade and direct investments. The purpose of this study is to see whether growing economic integration is reflected in the stock price movements between the Chinese stock market and these six markets. Our results show that cointegration exists significantly between the Chinese stock market and each of these six stock markets, between China and the three regional markets together, and the three world markets together, and all six markets together after the Asian financial crisis. Our results also show that cointegration does not exist between the Chinese stock market and the markets of the United States, the UK, Hong Kong, and Singapore, but does exist between China and the three regional markets together, and the three world leaders together, and all six markets together before the Asian financial crisis. This implies that economic integration has been incorporated into the performance of stock markets in the long term, especially after the Asian financial crisis. It also confirms that higher levels of cointegration are typically associated with lower levels of market volatility. Furthermore, we find that the three regional counterparts, Taiwan, Hong Kong, and Singapore, are more cointegrated with China after the Asian financial crisis. This significant closer relationship could be related to geographical proximity, partnerships in trade, and cultural and historical similarities. Our results are robust to the exchange rate effect when the Chinese Yuan is pegged to the U.S. dollar. We also find that all these stock markets, except Japan, Granger-cause the Chinese stock market, but not vice versa. This unilateral causal relationship may be due to the economic relationships, regulatory structures, exchange rate policy, and trade flows between the countries. In addition to their reflection in policy or regulatory changes, the Chinese stock market is still in the developing stage in terms of Granger-causing others in the world.

REFERENCES

Brailsford, T. J., O'Neill, T. J., and Penm, J. (2008). A new approach for estimating relationships between stock market returns: Evidence of financial integration in the Southeast Asian region. *International Finance Review* 8:17–38.

Dickey, D., and Fuller, W. (1981). Likelihood ratio statistics for autoregressive time series with unit roots. *Econometrica* 49:1057–72.

Engle, R., and Granger, C. W. J. (1987). Cointegration and error correction: Representation, estimation and testing. *Econometrica* 55:251–76.

Granger, C. W. J. (1981). Some properties of time series data and their use in economic model specification. *Journal of Econometrics* 29:121–30.

Granger, C. W. J. (1987). Developments in the study of cointegrated economic variables. *Bulletin of Economics and Statistics* 48:213–28.

Granger, C. W. J., Huang, B. N., and Yang, C. W. (2000). A bivariate causality between stock prices and exchange rates: Evidence from recent Asian flu. *Quarterly Review of Economics and Finance* 40:337–54.

Hsiao, C. (1981). Autoregressive modelling of money income causality detection. *Journal of Monetary Economics* 7:85–106.

Johnson, R., and Soenen, L. (2002). Asian economic integration and stock market co-movement. *Journal of Financial Research* 25:141–57.

Kwan, A. C. C., Sim, A. B., and Cotsomitis, J. A. (1995). The casual relationship between equity indices on world exchanges. *Applied Economics* 27:33–37.

Lee, S. B., and Kim, K. J. (1994). Does the October 1987 crash strengthen the co-movements among national stock markets? *Review of Financial Economics* 3:89–102.

Lo, A. W., and MacKinlay, A. C. (1988). Stock market prices do not follow random walks: Evidence from a simple specification test. *Review of Financial Studies* 1:41–66.

Palac-McMiken, E. (1997). An examination of ASEAN stock markets: A cointegration approach. *ASEAN Economic Bulletin* 13:299–311.

Penm, J., Penm, J. H. C., and Terrell, R. D. (2003). *Collaborative research in quantitative finance and economics*. ACT, Australia: Evergreen Publishing.

Sims, C. A., Stock, J. H., and Watson, M. W. (1990). Inference in linear time series models with some unit roots. *Econometrica* 58:113–44.

Wang, X. L., Shi, K., and Fan, H. X. (2006). Psychological mechanisms of investors in Chinese stock markets. *Journal of Economic Psychology* 27:762–80.

Wong, W.-K., Agarwal, A., and Du, J. (2005a). Financial integration for India stock market, a fractional cointegration approach. Working paper, Department of Economics, National University of Singapore.

Wong, W.-K., Khan, H., and Du, J. (2005b). *Money, interest rate and stock prices: New evidence from Singapore and the United States*. Graduate School, Universitas 21 Global, Singapore.

Yang, J., Kolari, J. W., and Min, I. (2003). Stock market integration and financial crises: The case of Asia. *Applied Financial Economics* 13:477–86.

CHAPTER 24

Do Tigers Care about Dragons? Spillovers in Returns and Volatility between Chinese Stock Markets

Bartosz Gebka[*]

CONTENTS

24.1 INTRODUCTION

The emergence and rapid development of the Chinese capital market has attracted considerable interest from investors, policy makers, and academics alike. Among other issues, the integration of China into the world financial

---

[*] I thank Lynne Evans for her helpful suggestions. All remaining errors are my own responsibility.

system and the efficiency of its domestic stock markets have been intensively investigated and discussed. This chapter contributes to the knowledge of the functioning of the Chinese capital market, as well as that of investors operating on the Shanghai and Shenzhen stock exchanges. Special attention is given to the issue of informational efficiency of these two trading venues, the behavior of domestic and foreign investors in China, and the lessons policy makers and investors can learn from our observations.

For a long time a distinctive feature of the Chinese capital market was the existence of two types of stocks: class A stocks, available exclusively to domestic investors, and class B stocks, which could be traded only by foreigners. Given the existence of two trading locations for stocks in China, i.e., Shanghai and Shenzhen, this results in four different stock markets: class A and B stock markets, in both Shanghai and Shenzhen. Consequently, information about the Chinese corporate sector relevant for asset pricing could reach investors through one of these four markets. This would result in an immediate price reaction in one of them and a subsequent adjustment in the remaining ones, i.e., in information spillovers, or causality, across markets. Even if financial system reforms in the 2001–2002 period relaxed some restrictions by allowing some domestic (foreign) investors to trade in B (A) stocks, a partial separation still exists: trading in A (B) stocks is dominated by domestic (foreign) investors. Hence, investigating spillovers between A and B markets (as well as between Shanghai and Shenzhen) enables us to draw conclusions concerning the differences in efficiency across markets and in informativeness of trades by domestic and foreign investors.

The issue of cross-market differences in efficiency can be approached by analyzing information spillovers in returns and volatility. Spillovers in returns (in volatility) take place if returns on stocks traded in one market change as a reaction to changing returns (volatility) in another market. The existence of return spillovers implies predictability in returns, which could be exploited by traders to realize abnormal profits. In addition, the existence of linkages between markets in the form of return spillovers could be utilized to benefit from portfolio diversification.

Causality in volatility captures another dimension of cross-market efficiency. From a theoretical perspective, changes in volatility have been argued to reveal arrivals of news and its assimilation by market participants (Ross, 1989; Engle et al., 1990). In the case of agents holding heterogeneous beliefs, their updating in response to news can generate correlation in volatility over time (Shalen, 1993). Hence, the cross-market causality in volatility indicates information transmission between markets, especially for markets with different sets of investors, i.e., with different information endowment and beliefs. Knowledge about the existence of causality in volatility between markets can be utilized by investors to better understand the issue of risk (Merton, 1980), i.e., to obtain superior estimates and forecasts of risks of assets. It can be used in the valuation of financial products such as derivatives, and in hedging techniques (Ng, 2000). The linkages in volatility can also allow for more accurate estimates and forecasts of variance-covariance matrices, benefiting investors seeking diversification opportunities.

Given the theoretical and practical relevance of spillovers, it should be no surprise that there exists a vast number of studies on spillovers between national capital markets. Emerging markets are generally found to be driven by developed ones (e.g., Bekaert et al., 2005), but some studies indicate that spillovers in returns from emerging to developed countries exist during the turbulent periods (e.g., Gebka and Serwa, 2006). Spillovers in volatility seem to be a rarer phenomenon; e.g., Bekaert and Harvey (1997) and Ng (2000) find the volatility in Pacific Basin countries to be only weakly, but increasingly, dependent on volatility in mature markets, and Kim (2005) argues that strong spillovers from the United States to Asia emerged only after the 1997 crisis.

In addition to spillovers from mature markets, several studies report linkages in returns and volatilities between emerging markets, located both in one region and on different continents. The general conclusions are that substantial spillovers in returns, and to a smaller extent in volatility, can exist between emerging markets located in the same region. Linkages between countries from different regions, however, are found to be substantially weaker (Edwards and Susmel, 2001; Sola et al., 2002; Fuji, 2005; Gebka and Serwa, 2007).

This chapter is organized as follows. We describe the history and institutional features of the Chinese markets in Section 24.2. Section 24.3 reviews the hypotheses about the causality between Chinese markets, and previous empirical findings are reported in Section 24.4. Methodology to test for causality is described in Section 24.5, and data used in this study in Section 24.6. We proceed with the presentation of results and their interpretation in Section 24.7. Section 24.8 concludes.

24.2 THE CHINESE STOCK MARKETS

Stock exchanges in China opened in the early 1990s in two locations: Shanghai Stock Exchange (SSE) was established in the largest Chinese city and an important commercial center in December 1990, followed by the Shenzhen Stock Exchange (SZSE), which opened in the first special economic zone in July 1991. SSE started its operations with eight stocks, with an annual turnover of US$857 million, and market capitalization of US$552 million in 1991. On the SZSE, six companies were initially listed, with a total market value of US$1,495 millions, and annual value of trades of US$667 million in 1991. Shares of Chinese companies can be issued in two forms: nonnegotiable, nontradable shares and negotiable ones, tradable on a stock exchange. A substantial fraction of shares are nonnegotiable (59% of total equity as of August 2007[*]). The nonnegotiable shares can be further divided into sponsor shares (46% of total equity), shares held by the legal persons, and shares owned by the employees (marginal fractions). The largest part of nonnegotiable shares is sponsor shares, the government being the largest shareholder.

As for the tradable shares, they can be issued as class A, B, or H shares. The latter category encompasses shares issued by Chinese companies and traded in Hong Kong from 1993 onward. Class A shares are ordinary equity shares that until 2002 were available exclusively to Chinese investors and traded mostly by individuals in Shanghai or Shenzhen. Companies going public are required to issue at least 25% of their equity as tradable A shares (Sun and Tong, 2000). B shares are traded either on the SSE (denominated in U.S. dollars) or on the SZSE (in HK dollars) and until 2001 could have been traded exclusively by foreign inventors. The rights and obligations of holders of A and B class shares are identical, as explicitly recognized by the passage of the Securities Law of China on June 1, 1999 (Karolyi and Li, 2003).

In addition, Chinese companies can issue class S shares tradable in Singapore and class N shares tradable at the New York Stock Exchange (Chen et al., 2006). Lastly, there are partially state-owned companies incorporated and listed in Hong Kong, with the bulk of their businesses based in China, referred to as the red chips (Sun and Tong, 2000).

On February 19, 2001, a major reform of the Chinese capital market took place, whereby the Securities Regulatory Commission announced that Chinese domestic investors with foreign currency accounts would be allowed to invest in B stocks previously available only to foreign traders. Following a suspension of trading and reopening on February 28, a surge in the number of investors resulted, thus increasing the trading volume and prices of class B shares (Karolyi and Li, 2003). Initially, domestic

---

[*] The latest figures are available from the China Securities Regulatory Commission website.

TABLE 24.1   Shanghai (SSE) and Shenzhen (SZSE) Stock Exchange in 2006

	SSE			SZSE		
	Total	**A**	**B**	**Total**	**A**	**B**
No. stocks listed	1,028,393.00	1,012,427.92	15,965.49	586,129.22	568,433.40	17,695.27
No. stocks traded (in millions)	725,225.34	718,056.78	7,168.56	409,575.59	401,042.34	8,533.13
Value of trades in stocks (in USD millions)	206,069.56	199,867.68	6,201.88	107,564.80	97,806.54	9,758.26
Market value of tradable stocks (in USD millions)	898,273.38	892,071.46	6,201.88	223,168.77	209,427.22	9,978.48
Market value of all stocks (in USD millions)	1,028,393.00	1,012,427.92	15,965.49	586,129.22	568,433.40	17,695.27

Source: Shenzhen Stock Exchange Handbook 2006 and Shanghai Stock Exchange Handbook 2006, www.sse.com.cn.

investors were allowed to buy B shares only if they had had accounts denominated in foreign currencies before the opening announcement was made. However, this restriction was temporary and was lifted on June 1, 2001. Due to the nonconvertibility of the Chinese national currency, Renminbi, the availability of class B shares traded in foreign currencies to Chinese investors is still partially limited. China also introduced the Qualified Foreign Institutional Investor (QFII) system in November 2002. Effective from December 1, 2002, approved foreign institutional investors are allowed to trade on China's domestic capital markets, including A markets (Chen et al., 2006).

In 2006, there were 886 and 621 companies listed on the SSE and SZSE, respectively (see Table 24.1). Class B stocks constituted only 1.5% (3%) of this number for the SSE (SZSE). The total value of companies listed in Shanghai was around US$900 billion, of which US$200 billion was tradeable. The fraction of B shares in capitalization was 0.7% and

3%, respectively. For the SZSE, the total capitalization amounted to US$223 billion, with half of it being the market value of tradable shares. The fraction of B shares was higher, at 4.5% and 9%, respectively. The number of stocks traded in Shanghai was around 1,000 billion (Shenzhen: 586 billion), the majority of trades carried out in A stocks (98.5% for SSE and 97% for SZSE). The total value of trades was US$5,780 billion in Shanghai and US$3,265 billion in Shenzhen, mostly in A stocks (99% and 98%).

24.3 REASONS FOR SPILLOVERS BETWEEN A AND B MARKETS

The literature identifies several differences between A and B markets that potentially result in different speeds of price reactions to news. First, as stated above, A shares are traded by Chinese domestic, mostly individual, investors, whereas the majority of trades in B shares are conducted by foreign, institutional investors. Second, trading volume in B shares is a fraction of the A market's volume, making differences in efficiency more likely. In addition, trading location can matter, as stocks traded in Shanghai are issued by large companies with the bulk of their business conducted in mainland China, whereas the B market is dominated by small, export-oriented companies (Wang et al., 2004). These differences could result in different patterns of price behavior across trading locations.

There is no agreement in the literature on which market, A or B, is more efficient. Some studies argue that foreign investors may receive news about the Chinese economy faster than domestic Chinese investors due to the information barriers that exist within China (Chui and Kwok, 1998; Xu, 2000). This would imply quicker adjustment of B prices to news and an unidirectional causality from B to A markets. In addition, as trading in A (B) stocks is dominated by individual (institutional) investors and stocks with high (low) institutional ownership have been shown to lead those with high individual ownership (Badrinath et al., 1995), B stocks would be expected to lead A stocks.

On the other hand, various studies document the existence of a discount in prices of B shares, especially prior to the 2001 reform (Karolyi and Li (2003) report it to be over 75% before February 2001 and to have declined to 8% a few months later). It prevailed despite the fact that for many Chinese companies listed simultaneously as both class A and B shares, both shareholder groups had equal rights to the future cash flows. Various explanations have been given for this phenomenon. Foreign investors were argued to have an information disadvantage over Chinese firms, relative to local investors, due to language barriers, different accounting standards, and a lack of reliable information sources (Brennan and Cao, 1997). Chakravarty et al. (1998) offer a theoretical framework that explains the existence of the B price discount with segmentation of A and B markets and the information advantage of domestic Chinese investors. In line with theoretical predictions, recent empirical studies document the information asymmetry between A and B markets, with trades in A stocks being more likely to be driven by information (Karolyi and Li, 2003; Chan et al., 2008). This would imply causality from A to B stock returns and volatilities. Lastly, whereas foreign investors with access to the world market are well diversified, portfolios of domestic investors contain a large fraction of A shares. Consequently, developments on the A markets are more important to the total wealth of Chinese investors, which generates a stronger incentive for them to obtain and quickly react to news. This further suggests higher information content of A markets.

As for the liquidity aspect, there is evidence that stocks with higher trading volume react more quickly to market-wide information, in both mature (Chordia and Swaminathan, 2000) and emerging markets (Gebka, 2008). Applied to Chinese data, this effect would be expected to result in information spillovers from highly liquid A to less liquid B markets, and from SSE to SZSE. On the other hand, higher trading volume can be generated by investors trading on noise rather than information, potentially resulting in lower information content of trades on the market with higher liquidity and in causality from the less (class B or SZSE) to more (class A or SSE) liquid market. Indeed, studies on the behavior of domestic Chinese investors dominating the A market show that they engage in short-term speculations, lack investment knowledge and skills, and consequently underestimate risk and suffer from overconfidence (Sun and Tong, 2000; Chen et al., 2004; Mei et al., 2005; Wang et al., 2006).

The differences in capitalization can also generate spillovers between Shanghai and Shenzhen. Large cap stocks have been shown to lead small cap stocks in other countries due to their quicker adjustment to market-wide news (Chordia and Swaminathan, 2000; Gebka, 2008) and to higher quality of cash flow information (Yu and Wu, 2001). Hence, we could expect returns and volatility of relatively larger Shanghai-listed companies to have higher information content and to lead stocks issued by smaller companies listed in Shenzhen.

24.4 PREVIOUS STUDIES ON CAUSALITY IN CHINA

Depending on the sample period, studies on spillovers between Chinese markets fall into three categories. Interestingly, most of the empirical evidence comes from studies investigating an early period of the Chinese market development, i.e., 1990–1997. Most authors report the markets for B shares to be more informative and lead the A shares in returns (Chui and Kwok, 1998; Sjoo and Zhang, 2000; Laurence et al., 1997; Kim and Shin, 2000). However, the overall evidence is far from conclusive, with returns on Shanghai A stocks found to contain information on future returns on other stocks (Laurence et al., 1997; Lee and Rui, 2000), especially before 1996 (Kim and Shin, 2000). This leadership of one market suggests an informational disadvantage of investors trading on a foreign market. The evidence on the relative informativeness of the trading locations is also mixed: Laurence et al. (1997), Lee and Rui (2000), and Kim and Shin (2000) suggest that Shanghai was leading Shenzhen, but Fung et al. (2000) report the opposite. As for the volatility spillovers in this early period, Su and Fleisher (1999) find that A markets receive more news and react more strongly and more persistently to it than B markets. Hence, the higher informational content of trading in A shares could imply spillovers in volatility from A to B markets. However, empirical studies report no spillovers in volatility between A and B stocks: Lee and Rui (2000) find bidirectional volatility spillovers only between B shares traded in Shanghai and Shenzhen, and in A stocks from Shanghai to Shenzhen. Hu et al. (1997) investigate A shares only and find volatilities in both locations to be contemporaneously correlated, but no evidence for lagged spillovers between trading venues.

Another set of studies focuses on an extended sample period, i.e., ending after 2000. Chiang et al. (2008) report spillovers in returns from A to B markets weaken after 2001, and Wang and Di Iorio (2007) find that A and B markets become increasingly integrated after 2000. These results suggest that contemporaneous linkages between A and B stocks increased but lagged spillovers became weaker, implying an increase in the efficiency. Qiao et al. (2008) also report lagged causality to be less pronounced after the 2001 reform. The finding of decreasing informational advantage of the A market is in line with the findings of Mei et al. (2005), who suggest that domestic Chinese investors trading in A shares speculate more; consequently, their trades are less informative. Chelley-Steeley and Qian (2005) find volatilities of A and B stocks to be contemporaneously correlated, but little evidence for integration of A and B markets in general. The evidence on volatility spillovers between Shanghai and Shenzhen is mixed: Fabozzi et al. (2004) find volatilities across locations to be independent from each other, whereas Zhu et al. (2004) report a causal relation in volatilities in both directions. Qiao et al. (2008) observe nonlinear causality in returns from B to A markets before 2001 and in the opposite direction after liberalization.

Lastly, studies describing the Chinese markets from the late 1990s onward find domestic Chinese investors trading in A shares to be better informed and A markets to be more informationally efficient (Fifield and Jetty, 2008; Karolyi and Li, 2003), implying that spillovers from A to B shares should be observed. However, Yang (2003) reports the B stocks traded in Shanghai lead all other markets in returns, whereas spillovers in returns from A stocks are limited to the Shanghai market influencing the A market in Shenzhen. Accordingly, He et al. (2003) find higher volatility of B stocks to be a result of higher informed trading in B stocks.

In summary, the results on spillovers between Chinese markets (A and B shares and Shanghai and Shenzhen) are inconclusive. However, a conclusion can be drawn that spillovers in returns are observed more frequently than in volatilities. This is a surprising finding, given that the theory indicates that volatility should spill over due to differences in information availability and prior beliefs. The findings reported in previous studies further seem to depend on the methodology used and the sample investigated. The latter might suggest that the informational leadership can be time varying. In addition, the increasing efficiency of the Chinese markets could have resulted in the disappearance of lagged linkages and an increase in the contemporaneous linkages between returns and volatility of stocks traded in different locations and by different investors, as their prices adjust to news instantaneously rather than with a lag.

24.5 METHODOLOGY

To test for causality in returns and volatility between Chinese markets, we employ the procedure of Cheung and Ng (1996). This test is conducted in two steps. First, returns on each Chinese market are modeled separately. Second, the causality statistics are calculated. Index returns for each market (A and B markets in Shanghai and Shenzhen) are modeled as an autoregressive process with autoregressive conditional heteroskedasticity

(ARCH) in disturbances (Engle, 1982).[*] Additionally, a variable X_t representing external shocks is included into the mean and variance equation. Hence, the model is

$$R_t = \alpha_0 + \sum_{i=1}^{m} \alpha_i R_{t-i} + \sum_{i=0}^{n} \phi_i X_{t-i} + \lambda C_{t-1} + z_t \qquad (24.1)$$

$$z_t = \varepsilon_t h_t^{1/2}, \varepsilon_t \sim NID(0, 1) \qquad (24.2)$$

$$h_t = \beta_0 + \sum_{i=1}^{q} \delta_i \varepsilon_{t-i}^2 + \sum_{i=0}^{v} \eta_i X_{t-i}^2 \qquad (24.3)$$

where R_t is the log index return on a Chinese market and X_t is a measure of external shocks. These shocks are proxied by index returns from a developed market, the choice of which is discussed in the next section. Sluggish adjustment to past shocks originating at home is accounted for by lagged values of the local stock index returns, R_{t-i}. Tests of Engle and Granger (1987) and Johansen (1991) show that variables R_t and X_t are cointegrated for all markets—hence we include an error correction term, C_{t-1}, into Equation (24.1). The conditional variance of the error term, h_t, is described by an autoregressive conditional heteroskedastic (ARCH) process. The error term z_t represents the component of index returns independent of past shocks originating on this market and abroad. They are assumed to be independently, identically, and normally distributed with zero mean and unit variance.[†]

The second part of this procedure is a test of causality (spillovers, linkages) between returns on indices from Chinese markets. For each of the four markets, we estimate the model—Equations (24.1) to (24.3). Next, standardized innovations, ε_t, are derived. For two markets with index returns $R_{1,t}$ and $R_{2,t}$, innovations $\varepsilon_{1,t}$ and $\varepsilon_{2,t}$ capture unsystematic return components (i.e., independent from past shocks and external effects) and are denoted ξ_t and ζ_t, respectively. They are further used to test for

---

[*] Engle and Ng (1993) tests reject the hypothesis of asymmetric impact of news on volatility for all markets analyzed.

[†] The maximum-likelihood estimates are asymptotically consistent even for nonnormal standardized innovations (Bollerslev et al., 1994). Moreover, the causality test applied here is robust against the nonnormality of error terms.

causality in returns between markets. The test of Cheung and Ng (1996) employed here utilizes the estimations of the cross-correlation functions, denoted r, for standardized residuals. Under the null hypothesis no causality between the two markets is present. In particular, there is no causality in returns when residuals from the first market, ξ_t, are uncorrelated with residuals from the second market, ζ_{t-i}, at all leads and lags ($i = 0, \pm 1, \pm 2, \ldots$). The test for the hypothesis of no causality at all lags from j to k is performed using the test statistic

$$S_{\xi\zeta}(j,k) = T \sum_{i=j}^{k} r_{\xi\zeta}(i)^2 \qquad (24.4)$$

The alternative hypothesis is of causality in returns at some lag i. T is the sample size and $r_{\xi\zeta}(i)$ and $r_{UV}(i)$ are proxied by the sample cross-correlation functions. Under the null hypothesis of no causality Equation (24.4) has asymptotic χ^2 distribution with $(k - j + 1)$ degrees of freedom. Additionally, the statistic $S_{\xi\zeta}(i,i)$, i.e., $S(j,k)$ for $j = k = i$, can be calculated to test for causality at a certain lag i. The hypothesis of causality in returns at the selected lag i is that market 1 ($R_{1,t-i}$) causes market 2 ($R_{2,t}$) or market 2 ($R_{2,t-i}$) causes market 1 ($R_{1,t}$) (for $i > 0$).

To test for causality in volatility, Equations (24.1) to (24.3), are reestimated, with lagged returns from another Chinese market as additional explanatory variables in Equation (24.1) if the test for causality in returns reveals the existence of such causality at any lag i. Next, squared innovations for both markets are calculated as $U_t \equiv \xi_t^2$ and $V_t \equiv \zeta_t^2$. Causality in volatility is not present when squared shocks, U_t and V_{t-i}, are uncorrelated at all leads and lags ($i = 0, \pm 1, \pm 2, \ldots$). The null hypothesis of no causality in volatility at all lags from j to k is tested using the statistic

$$S_{UV}(j,k) = T \sum_{i=j}^{k} r_{UV}(i)^2 \qquad (24.5)$$

against the alternative hypothesis of causality in volatility at some lag i. Under the null hypothesis, Equation (24.5) follows asymptotic χ^2 distribution with $(k - j + 1)$ degrees of freedom. Additionally, to test for causality at a certain lag i, statistics $S_{UV}(i,i)$, i.e., $S(j,k)$ for $j = k = i$, are used. If the statistic $S_{UV}(i,i)$ exceeds its critical value for a selected lag i, then market 1 ($R_{1,t-i}$) causes market 2 ($R_{2,t}$) or market 2 ($R_{2,t-i}$) causes market 1 ($R_{1,t}$) in volatility (for $i > 0$).

The causality in returns (in volatility) reveals that movements in returns (volatility of returns) on one market are transferred to returns (volatility) on another market. For both tests, contemporaneous (lagged) causality occurs when the alternative hypothesis is true for $i = 0$ ($i \neq 0$). The approach of Cheung and Ng (1996) was demonstrated to have good empirical power and size properties. The testing procedure is independent from the specification of the model and is robust to asymmetric and leptokurtic errors. Therefore, it is more robust against misspecification problems that reduce the power of tests based on multivariate GARCH models (Hafner and Herwartz, 2004). Hong (2001) argued that the Cheung and Ng approach suffers from lower power for large lag values i. However, we tested for a maximum of five lags, which were shown to reduce the test's power only marginally. Also, it has been argued (Van Dijk et al., 2005) that the presence of specific breaks in volatility decreases the power of the Cheung-Ng test. To account for this effect, we use time-varying conditional volatility h_t and the moving window estimates with window-specific volatilities.

24.6 DATA

We perform our analysis of daily data on indices for four Chinese stock markets, class A and B stocks listed in Shanghai and Shenzhen, obtained from Datastream. The sample period is January 1997 to March 2008 and excludes the early, most volatile period in the history of these markets. Daily index returns are computed as differences in log index values. All indices are measured in a common currency (USD). Econometric tests reveal that all returns series are stationary.[*] To obtain an optimal measure of global shocks, X_t, we estimate Equations (24.1) to (24.3) for each Chinese market using different proxies of X_t: returns on Nikkei 500, Hang Seng, S&P500, and the MSCI WORLD index (the two latter ones with a lag of 1 day). Based on the values of the log-likelihood functions, we infer that the Hang Seng index best captures the global shocks in returns and volatilities; hence, it is used as variable X_t in further analysis. The number of lags in Equations (24.1) and (24.3) is set to be equal to 5, i.e., $m = n = q = v = 5$, to allow for sluggish adjustment to news and to account for day-of-the-week effects.

---

[*] We use the augmented Dickey and Fuller (1979) test, the Philips and Perron (1988) test, and the augmented weighted symmetric test (Pantula et al., 1994) to test for the presence of unit roots.

24.7   RESULTS

Results from tests for causality in returns after controlling for the impact of global shocks X_t are reported in Table 24.2, Panel A. The numbers reported indicate the lag at which causality is present, e.g., for spillovers from Shanghai A to Shenzhen B, numbers 0, 1, and 4 indicate that there were significant links between these markets at lag 0, and lagged causality from Shanghai A to Shenzhen B with a 1- and 4-day delay (test statistic given by Equation (24.4) is significant at lags $i(=j=k)$ equal to 0, 1, and 4). In addition, the notation 1–5 indicates that the test for joint lagged causality in returns (lags 1 to 5) reports a significant result (test statistic given by Equation (24.4) is significant at lags $j = $ 1 to $k = 5$), and hence lagged causality from Shanghai A to Shenzhen B exists.

TABLE 24.2   Results from Causality in Returns and Causality in Variance Tests

Panel A: Whole Sample

		Spillovers to:			
		Shanghai A	Shanghai B	Shenzhen A	Shenzhen B
	Shanghai A		0***, 1*, 3** 1–5*	0***	0***, 1**, 4* 1–5*
Spillovers from:	Shanghai B	0***		0***	0***, 3*, 4** 1–5*
	Shenzhen A	0***	0***, 3*		0***, 1*
	Shenzhen B	0***	0***	0***	

Panel B: Before the 2001 Reform

		Spillovers to:			
		Shanghai A	Shanghai B	Shenzhen A	Shenzhen B
	Shanghai A		0***, 3*	0***	0***
Spillovers from:	Shanghai B	0***		0***	0***, 3**, 5* 1–5*
	Shenzhen A	0***	0***, 3*		0***
	Shenzhen B	0***, 4*	0***	0***	

Panel C: After the 2001 Reform

		Spillovers to:			
		Shanghai A	Shanghai B	Shenzhen A	Shenzhen B
	Shanghai A		0***	0***	0***, 1*
Spillovers from:	Shanghai B	0***		0***	0***, 4**
	Shenzhen A	0***	0***		0***
	Shenzhen B	0***	0***	0***	

Note: ***, **, and * indicate significance at the 1%, 5%, and 10% levels, respectively.

For the whole sample, significant contemporaneous causality (i.e., at lag $i = 0$) in returns is present for all market pairs. However, the direction of causality cannot be detected here, as it might indicate reactions of one market to shocks originating in another, or from bidirectional causality. Significant contemporaneous causality could have been expected as prices of Chinese companies, regardless of their listing venue and form (A- or B-type share), should react to the same macroeconomic news about the Chinese economy. It is also possible that the intraday adjustment speed to news differs across markets, but it cannot be investigated using daily data.

Turning our attention to return spillovers at higher lags, the following observations can be made. First, there are several instances of causality from A to B markets, but no lagged causality in the opposite direction. This finding can imply that the A markets are more informationally efficient than the B ones and adjust quickly to new information, whereas prices of B stocks require several days to incorporate the same news. Hence, the lead-lag relationship between A and B stocks emerges. Second, instances of news originating in Shanghai and spilling over to Shenzhen with a time-lag seem to be more numerous than instances of news being transmitted from Shenzhen to Shanghai. In particular, the highest number of significant spillovers originates on the Shanghai A market, whereas B stocks listed in Shenzhen seem to be most sensitive to spillovers from other markets. In addition, there is significant causality from B stocks traded in Shanghai to B stocks in Shenzhen, indicating that the Shanghai market is more efficient than the Shenzhen one.

We also compare the causality results for two subsamples, for the periods prior to and following the reform of February 19, 2001, when domestic investors were allowed to trade in B shares in addition to A shares. The results are reported in Table 24.2, Panels B and C. For the period prior to the reform, we find spillovers in returns from A to B shares to be more numerous than in the opposite direction, suggesting greater informational efficiency of the A markets. Furthermore, the Shenzhen traded stocks appear to be slightly more informative for the Shanghai traded ones than vice versa. The Shanghai B segment is sensitive to changes in stock prices elsewhere on A markets, but leads the Shenzhen B market itself.

After the 2001 reform, several changes in cross-market causality occurred. First, after allowing domestic Chinese investors to trade in B shares, the causality from A to B markets seems to have become less pronounced: Shanghai B market is not driven by the A markets and Shenzhen B stocks started to react to returns on the SSE A index instead.

This may be due to the fact that trades by Chinese investors, which are more informative, have partially moved from A to B markets, so that A markets lost some of the informational advantage over B markets that they used to have when domestic and foreign investors were separated. Another change is that the informational leadership of Shanghai over Shenzhen seems to have risen, as there are no lagged spillovers from Shenzhen to Shanghai following the 2001 reform. The overall evidence also suggests that efficiency increased in the 2001 onward period, as indicated by the smaller number of lagged linkages between the markets. A possible interpretation of this finding is that each market reacts to news originating on other markets within a day rather than with a time lag, and hence hardly any causality beyond the contemporaneous linkages can be observed.

We also conduct tests for spillovers in volatility between Chinese markets, while controlling for spillovers in returns as identified before, and for the impact of global shocks in both returns, X_t, and volatility, X_t^2. We find significant contemporaneous linkages in volatility for all market pairs. For lagged causality in volatility, however, no significant results can be observed. These patterns of causality are further identical for the pre- and postreform period. This finding corresponds to previous studies reporting significant lagged spillovers in returns but failing to find significant cross-market lagged causality in volatility. The overall evidence may indicate that information transmitted via volatilities is accommodated quickly by other markets, but it can take more than one day for those aspects of news carried over by returns to be priced.

To further investigate the changes in causality in returns and volatility between the Chinese stock markets, we conduct the analysis using the Cheung and Ng (1996) methodology for a moving annual estimation window. Specifically, we estimate Equations (24.1) to (24.3) and test for causality in both returns and volatility for a window of 250 days (corresponding to 1 year), and moving this window by 21 days (roughly 1 month) in each step. This procedure results in a time series representing the values of Cheung and Ng's (1996) test statistics given by Equations (24.4) and (24.5). Specifically, we are interested in the contemporaneous causality (captured by $S_{\xi\zeta}(0,0)$ and $S_{UV}(0,0)$) and joint lagged causality (captured by $S_{\xi\zeta}(1,5)$ and $S_{UV}(1,5)$) for each pair of the Chinese markets. To formally analyze the changes of causality over time, we estimate two types of regressions:

$$S_t = \alpha_0 + \alpha_1 D_t + \varepsilon_t \qquad (24.6)$$

and

$$S_t = \beta_0 + \beta_1 D_t + \beta_2 t + \beta_3 tD_t + \varepsilon_t \qquad (24.7)$$

where S_t is one of the causality statistics, i.e., $S_t \in \{ S_{\xi\zeta}(0,0), S_{UV}(0,0), S_{\xi\zeta}(1,5), S_{UV}(1,5) \}$, and captures the strength of the contemporaneous or lagged causality (as measured by the statistics $S(0,0)$ and $S(1,5)$, respectively). D_t is a dummy variable that equals 0 prior to the February 19, 2001, reform and 1 thereafter, and t is the time trend. Hence, Equation (24.6) measures the difference in the average strength of causality between each market prior to and following the reform: $\hat{\alpha}_0$ describes the average causality strength before the reform, $\hat{\alpha}_0 + \hat{\alpha}_1$ thereafter, and $\hat{\alpha}_1$ indicates how the causality changed due to the reform. Equation (24.7) allows the causality to evolve over time, and allows this evolution to change around the reform date. The trend before the reform is described by $\hat{\beta}_0 + \hat{\beta}_2 t$, with $\hat{\beta}_2$ being the estimate of the speed in changes of causality between two markets. After the reform, this trend is allowed to change to $(\hat{\beta}_0 + \hat{\beta}_1) + (\hat{\beta}_2 + \hat{\beta}_3)t$, with the postreform speed being equal to $\hat{\beta}_2 + \hat{\beta}_3$ and $\hat{\beta}_3$ capturing the impact of the reform on the speed of changes in causality.

Given the increasing volume of trading and quality of information available to more sophisticated investors, one could expect the market efficiency to improve over time. Specifically, the strength of contemporaneous causality, $S(0,0)$, should increase over time, hence $\hat{\alpha}_1$ and both $\hat{\beta}_2$ and $\hat{\beta}_2 + \hat{\beta}_3$ would be positive and significant. Further, lagged causality, $S(1,5)$, could have decreased or increased. The former effect potentially occurs if information adjustment speeds up and causality recorded at higher lags at the beginning of the sample tends to occur during the same day as time proceeds. The latter is possible if causality is slow at first and tends to occur at lags higher than 5 but speeds up later on and is captured by the statistic for lags 1 to 5.

In addition, if the reform had improved informational efficiency of one market (e.g., B markets due to increased trading by better informed domestic Chinese investors), we would expect the informational content of, and hence causality by, other markets to be less pronounced in the period after February 19, 2001. Hence, a shift in causality from lagged to a contemporaneous one should be observed, resulting in positive $\hat{\alpha}_1$ in regressions with $S(0,0)$ as the dependent variable. Further, if the reform increased at the speed at which the markets' efficiency increases, $\hat{\beta}_3$ in regressions with $S(0,0)$ should be positive as well.

The results for changes in contemporaneous causality in returns and volatility over time are reported in Table 24.3. In panel A, the results for

TABLE 24.3 Estimation Results for Equation (24.6): Causality in Returns (Panel A) and Volatility (Panel B) at Lag 0 (S(0,0)) and 1–5 (Y to X and X to Y)

X Y	SHA SHB		SHA SZA		SHA SZB		SHB SZA		SHB SHB		SZA SZB	
	$\hat{\alpha}_0$	$\hat{\alpha}_1$	$\hat{\alpha}_0$	$\hat{\alpha}_1$	$\hat{\alpha}_0$	$\hat{\alpha}_1$	$\hat{\alpha}_0$	$\hat{\alpha}_1$	$\hat{\alpha}_0$	$\hat{\alpha}_1$	$\hat{\alpha}_0$	$\hat{\alpha}_1$
Panel A: Causality in Returns												
S(0,0)	54.58***	68.85***	221.93***	1.36	55.34***	68.97***	50.62***	78.91***	129.70***	35.13***	54.69***	76.47***
Y to X	2.52***	0.50	1.25***	-0.14	4.00***	-0.50	6.10***	-3.33***	3.75***	-1.47***	4.31***	-1.32
X to Y	6.24***	-2.85**	1.32***	-0.04	4.24***	-1.45*	2.90***	0.02	4.23***	-1.95***	3.93	-1.36
Panel B: Causality in Volatility												
S(0,0)	26.42***	33.55**	207.51***	5.36	25.47***	44.86***	21.47***	44.68***	90.01***	39.29***	27.49***	49.22***
Y to X	3.28***	0.18	11.36***	-2.81	4.56***	2.25	9.88***	-1.46	17.75***	-4.75	3.58***	3.26***
X to Y	9.24***	-4.10	13.34***	-7.20*	7.13***	-1.94	3.19***	2.96*	18.56***	-9.14**	7.17***	-0.71

Equation (24.6) for causality in returns at lag 0 (denoted S(0,0)) and 1–5 (denoted X to Y and Y to X) are presented. A and B markets in Shanghai and Shenzhen are denoted as SHA, SZA, SHB, and SZB, respectively. For the contemporaneous causality in returns, all but one $\hat{\alpha}_1$ parameter are positive and significant, indicating that the strength of contemporaneous linkages increased for all market pairs following the reform of 2001. Hence, Chinese markets seem to have become more informationally efficient. For causality at lags 1–5, only five results are significant, and all of them negative. This indicates that lagged causality decreased, which can be interpreted as evidence of improving efficiency. The results further suggest that returns on SSE A index partially lost their information content (as other markets depend on them to a smaller extent), and that SSE B market's efficiency increased (as it depends less on other markets). This is in line with a decreasing information advantage of A markets as reported in Table 24.2.

Table 24.4, Panel A reports estimation results for Equation (24.7) for causality in returns at lags 0 and 1–5. For the prereform era, we can observe an upward trend in the strength of contemporaneous causality between Chinese markets, as indicated by the positive and significant $\hat{\beta}_2$. However, negative and significant $\hat{\beta}_3$ indicates that this effect became less pronounced after the reform. Hence, although the reform was aimed at increasing the efficiency of the Chinese stock markets, the growth rate in efficiency actually diminished. However, this result might be due to the fact that the efficiency level was relatively high prior to 2001, so that additional improvement was more difficult to achieve. With no reform, the growth in efficiency could have been even slower. Further, even if weaker, the postreform growth in efficiency is still positive in most cases ($\hat{\beta}_2 + \hat{\beta}_3 > 0$). When looking at changes in efficiency at lags 1–5, however, we discover five significant cases of increasing lagged adjustment speed ($\hat{\beta}_3 > 0$). Combined with decreasing speed at lag 0, this finding could suggest that the overall efficiency of mostly B markets started improving in the postreform period, as they were increasingly more able to predict A markets at lags higher than 0. However, given the fact that most lagged causalities are insignificant in the postreform era, this effect does not seem to be of any economic relevance. To summarize, the findings indicate slower but still positive improvements in efficiency of Chinese markets following the 2001 reform, especially of class B stocks.

We also analyze changes in causality in variance over time, with results for Equation (24.6) estimated for causality in variance at lags 0 and 1–5

TABLE 24.4 Estimation Results for Equation (24.7): Causality in Returns (Panel A) and Volatility (Panel B) at Lag 0 (S(0,0)) and 1–5 (Y to X and X to Y)

Panel A: Causality in Returns

X Y	SHA SHB				SHA SZA				SHA SZB			
	$\hat\beta_0$	$\hat\beta_1$	$\hat\beta_2$	$\hat\beta_3$	$\hat\beta_0$	$\hat\beta_1$	$\hat\beta_2$	$\hat\beta_3$	$\hat\beta_0$	$\hat\beta_1$	$\hat\beta_2$	$\hat\beta_3$
S(0,0)	4.45	114.80**	2.45***	−2.40***	204.49***	57.24***	0.85***	−1.31***	6.90	92.77**	2.36***	−2.07***
Y to X	4.32***	2.41	−0.09***	0.04**	1.03***	−0.06	0.01	−0.01	6.31***	−0.02	−0.11*	0.08
X to Y	6.47***	−8.41**	−0.01	0.07	1.68***	−1.69***	−0.02	0.03**	2.62***	−1.45	0.08***	−0.06**

X Y	SHB SZA				SHB SZB				SZA SZB			
	$\hat\beta_0$	$\hat\beta_1$	$\hat\beta_2$	$\hat\beta_3$	$\hat\beta_0$	$\hat\beta_1$	$\hat\beta_2$	$\hat\beta_3$	$\hat\beta_0$	$\hat\beta_1$	$\hat\beta_2$	$\hat\beta_3$
S(0,0)	3.86	108.64**	2.28***	−2.08***	111.72***	85.17***	0.88*	−1.26*	7.24	88.30**	2.31***	−1.89**
Y to X	5.87***	−5.74***	0.01	0.02	4.84***	−5.64***	−0.05**	0.09***	7.05***	−1.36	−0.13*	0.10*
X to Y	5.61***	1.00	−0.13**	0.09**	3.48**	−0.07	0.04	−0.05	0.86***	0.86	0.15***	−0.14***

Panel B: Causality in Volatility

X Y	SHA SHB				SHA SZA				SHA SZB			
	$\hat\beta_0$	$\hat\beta_1$	$\hat\beta_2$	$\hat\beta_3$	$\hat\beta_0$	$\hat\beta_1$	$\hat\beta_2$	$\hat\beta_3$	$\hat\beta_0$	$\hat\beta_1$	$\hat\beta_2$	$\hat\beta_3$
S(0,0)	16.37*	71.26	0.49	−0.82	189.69***	51.80***	0.87***	−1.21***	6.41	57.72	0.93*	−0.86
Y to X	2.98***	0.66	0.01	−0.02	18.56***	−17.1***	−0.35**	0.44**	4.97***	−9.52***	−0.02	0.15*
X to Y	4.17	9.10**	0.25	−0.34*	23.72***	−21.4***	−0.51***	0.55***	7.15***	−1.55	0.00	0.00

X Y	SHB SZA				SHB SZB				SZA SZB			
	$\hat\beta_0$	$\hat\beta_1$	$\hat\beta_2$	$\hat\beta_3$	$\hat\beta_0$	$\hat\beta_1$	$\hat\beta_2$	$\hat\beta_3$	$\hat\beta_0$	$\hat\beta_1$	$\hat\beta_2$	$\hat\beta_3$
S(0,0)	8.70	71.98	0.62*	−0.79	111.95***	35.61	−1.07	0.85	9.62	49.79	0.87**	−0.67
Y to X	3.73	6.27	0.30*	−0.32*	17.19***	5.45	0.03	−0.14	4.63***	−5.11**	−0.05*	0.14***
X to Y	1.97***	−3.19	0.06**	0.03	13.24*	4.34	0.26	−0.36	6.28***	2.08	0.04	−0.07

shown in Table 24.3, Panel B. Contemporaneous causality between all but one market pair improved after the 2001 reform, as indicated by positive and significant parameters $\hat{\alpha}_1$. As for the differences in lagged causality, the results for most market pairs are insignificant. Hence, the overall evidence is that of increased efficiency of the Chinese markets, as they rely less on the lagged volatility of their counterparts.

Table 24.4, Panel B reports estimation results for Equation (24.7) for causality in volatility at lags 0 and 1–5. In the prereform era, most of the $\hat{\beta}_2$ estimates in regressions with S(0,0) as a dependent variable are positive and significant, suggesting that there was an increase in the strength of contemporaneous causality in volatility over time. Further, this trend seems to have continued after February 19, 2001, as all but one $\hat{\beta}_3$ estimate are insignificant. Only for the causality between Shanghai and Shenzhen A markets can we observe a decline in contemporaneous linkages and an increase in lagged volatility causality following the reform, suggesting deteriorating efficiency of these two biggest Chinese markets. A closer look at the time series behavior of test statistics S(0,0) and S(1,5) reveals that the increase in the latter is due to its high values at the sample's end. This might be due to the arrival of uninformed speculators and a loss in informational content of prices resulting therefrom. However, the overall lagged causality in volatility is not significant, as reported above, so these effects are of marginal magnitude. Overall, the evidence suggests that linkages in volatility at lag 0 have been improving over time, but the 2001 reform had no noticeable impact on this process.

24.8 CONCLUSION

The results presented in this chapter allow us several insights into the functioning and evolution of the Chinese stock markets. First, there is substantial evidence of the causality in returns, with A markets, dominated by domestic individual Chinese investors, being more informationally efficient than their B-type counterparts, which are dominated by foreign institutional investors. In addition, the overall evidence suggests that the efficiency of both market types is improving over time. Further, B markets seem to be losing their informational disadvantage as compared to the A markets, which coincides with the reform of 2001 allowing Chinese investors to trade in B-type shares. The location of trades also seems to matter, with returns on stocks traded at the SSE having more predictive power for the SZSE-listed stocks than vice versa in the post-2001 reform era (and the opposite effect before). Lastly, the results indicate that causality in volatility takes place within 1 trading day for all market pairs and is improving over time.

These findings indicate that the Chinese domestic investors trading predominantly in class A stocks have had an informational advantage over foreigners trading in B stocks, as the returns on the former lead those on the latter. This is in line with findings by Karolyi and Li (2003), Chan et al. (2008), and others. Even if foreign investors are mostly institutions and domestic ones are individuals, and institutional investors have been shown to be better informed (Badrinath et al., 1995), these results indicate that the location of an investor and knowledge of local language can also matter. Further, markets with higher capitalization and liquidity (class A and the SSE) lead those with a lower total value of assets and trading, despite documented irrationality of Chinese traders. Hence, the findings reported from other countries that size and liquidity improve efficiency are confirmed for the Chinese markets. Moreover, the predictive power of Shanghai over Shenzhen-listed companies could be due to the fact that the latter are export oriented and less sensitive to the domestic events, and hence the fact that their holders pay less attention to China-specific news and react to it with a time lag. Lastly, only contemporaneous causality in volatility was found, suggesting that traders acting on different markets differ in their beliefs and knowledge but learn quickly from other investors' reactions to news.

The existence of spillovers in returns, mostly from the SSE-listed and class A stocks, indicates that returns of B shares and those traded in Shenzhen are partially predictable. This finding could be of interest to portfolio investors, as it creates potential for obtaining excess returns. Also, these slower than instantaneous adjustments of asset prices across markets suggest that return correlation is less than perfectly positive. Hence, when investing in Chinese companies, potential diversification benefits are possible by spreading the capital widely across class A and B stocks and the SSE and SZSE, rather than investing in only one type of shares listed on one stock exchange. As the linkages in volatility are contemporaneous, no predictability is possible, although causality at higher frequencies (intradaily) could still be possible. However, the knowledge of significant correlations in volatilities across markets implies the existence of commonalities in risk, and could be utilized for improved estimation of asset risk, to be used in the valuation of assets themselves, e.g., in the CAPM type of models, and of derivatives. Further, it could help in the estimation of variance-covariance matrices used in statistical testing procedures. However, investors should also be aware of the time-varying nature of the causality among the Chinese markets.

The results show that the efficiency of the Chinese markets has been increasing over time, which can be at least partially attributed to the liberalization process, including the 2001 reform. This confirms that the legislative changes have been heading in the right direction, as they have improved the allocation of scarce financial resources within the Chinese economy. The existence of lagged causality in returns suggests that the adjustment of prices to news is still sluggish, and hence further reforms are necessary. Educational measures to increase investors' knowledge, and hence reduce noise trading, could also contribute to the goal of improved market efficiency.

REFERENCES

Badrinath, S. G., Kale, J. R., and Noe, T. H. (1995). Of shepherds, sheep, and cross-autocorrelations in equity returns. *Review of Financial Studies* 8:401–30.

Bekaert, G., and Harvey, C. R. (1997). Emerging equity market volatility. *Journal of Financial Economics* 43:29–77.

Bekaert, G., Harvey, C. R., and Ng, A. (2005). Market integration and contagion. *Journal of Business* 78:32–63.

Bollerslev, T., Engle, R. F., and Nelson, D. B. (1994). ARCH models. In *Handbook of econometrics*, ed. R. F. Engle and D. L. McFadden. (4) 2959–3038. Amsterdam: Horth-Holland.

Brennan, M., and Cao, H. (1997). International portfolio investment flows. *Journal of Finance* 52:1851–80.

Chakravarty, S., Sarkar, A., and Wu, L. (1998). Information asymmetry, market segmentation and the pricing of cross-listed shares: Theory and evidence from Chinese A and B shares. *Journal of International Financial Markets, Institutions and Money* 8:325–56.

Chan, K., Menkveld, A. J., and Yang, Z. (2008). Information asymmetry and asset prices: Evidence from the China foreign share discount. *Journal of Finance* 63:159–96.

Chelley-Steeley, P., and Qian, W. (2005). Testing for market segmentation in the A and B share markets of China. *Applied Financial Economics* 15:791–802.

Chen, D.-H., Blenman, L. P., Bin, F.-S., and Chen, J. (2006). The effects of open market reforms on the behavior of China's stock prices. *International Research Journal of Finance and Economics* 5:95–110.

Chen, G.-M., Kim, K. A., Nofsinger, J. R., and Rui, O. M. (2004). Behavior and performance of emerging market investors: Evidence from China. Working paper, Washington State University, Pullman.

Cheung, Y., and Ng, L. K. (1996). A causality-in-variance test and its application to financial market prices. *Journal of Econometrics* 72:33–48.

Chiang, T. C., Nelling, E., and Tan, L. (2008). The speed of adjustment to information: Evidence from the Chinese stock market. *International Review of Economics and Finance* 17:216–29.

Chordia, T., and Swaminathan, B. (2000). Trading volume and cross-autocorrelations in stock returns. *Journal of Finance* 55:913–35.

Chui, A. C. W., and Kwok, C. C. Y. (1998). Cross-autocorrelation between A shares and B shares in the Chinese stock market. *Journal of Financial Research* 21:333–35.

Dickey, D., and Fuller, W. (1979). Distribution of the estimators for autoregressive time series with a unit root. *Journal of American Statistical Association* 74:427–31.

Edwards, S., and Susmel, R. (2001). Volatility dependence and contagion in emerging equity markets. *Journal of Development Economics* 66:505–32.

Engle, R. F. (1982). Autoregressive conditional heteroskedasticity with estimates of the variance of the U.K. inflation. *Econometrica* 50:987–1008.

Engle, R. F., and Granger, C. (1987). Co-integration and error correction: Representation, estimation and testing. *Econometrica* 55:251–76.

Engle, R. F., Ito, T., and Lin, W. L. (1990). Meteor showers or heat waves? Heteroscedastic intra-daily volatility in the foreign exchange market. *Econometrica* 58:525–42.

Engle, R. F., and Ng, V. K. (1993). Measuring and testing the impact of news on volatility. *Journal of Finance* 48:1749–78.

Fabozzi, F. J., Tunaru, R., and Wu, T. (2004). Modeling volatility for the Chinese equity markets. *Annals of Economics and Finance* 5:79–92.

Fifield, S. G. M., and Jetty, J. (2008). Further evidence on the efficiency of the Chinese stock markets: A note. *Research in International Business and Finance*, 22: 351–361.

Fujii, E. (2005). Intra and inter-regional causal linkages of emerging stock markets: Evidence from Asia and Latin America in and out of crisis. *Journal of International Financial Markets, Institutions and Money* 15:315–42.

Fung, H., Lee, W., and Leung, W. K. (2000). Segmentation of the A- and B-share Chinese equity markets. *Journal of Financial Research* 23:179–95.

Gebka, B. (2008). Volume- and size-related lead-lag effects in stock returns and volatility: An empirical investigation of the Warsaw Stock Exchange. *International Review of Financial Analysis* 17:134–55.

Gebka, B., and Serwa, D. (2006). Are financial spillovers stable across regimes? Evidence from the 1997 Asian crisis. *Journal of International Financial Markets, Institutions and Money* 16:301–17.

Gebka, B., and Serwa, D. (2007). Intra- and inter-regional spillovers between emerging capital markets around the world. *Research in International Business and Finance* 21:203–21.

Hafner, C. M., and Herwartz, H. (2004). *Testing for causality in variance using multivariate GARCH models.* Econometric Institute Report 2004-20, Erasmus University Rotterdam, Rotterdam, The Netherlands.

He, Y., Wu, C., and Chen, Y.-M. (2003). An explanation of the volatility disparity between the domestic and foreign shares in the Chinese stock markets. *International Review of Economics and Finance* 12:171–86.

Hong, Y. (2001). A test for volatility spillover with application to exchange rates. *Journal of Econometrics* 103:183–224.

Hu, J. W., Chen, M., Fok, R. C. W., and Huang, B. (1997). Causality in volatility and volatility spillover effects between U.S., Japan and four equity markets in the South China growth triangular. *Journal of International Financial Markets, Institutions and Money* 7:351–67.

Johansen, S. (1991). Estimation and hypothesis testing of cointegration vectors in Gaussian vector autoregressive models. *Econometrica* 59:1551–80.

Karolyi, G. A., and Li, L. (2003). A resolution of the Chinese discount puzzle. Working paper, Ohio State University, Columbus.

Kim, S.-J. (2005). Information leadership in the advanced Asia-Pacific stock Markets: Returns, volatility and volume information spillovers from the U.S. and Japan. *Journal of Japanese and International Economies* 19:338–65.

Kim, Y., and Shin, J. (2000). Interactions among China-related stocks. *Asia-Pacific Financial Markets* 7:97–115.

Laurence, M., Cai, F., and Qian, S. (1997). Weak-form efficiency and causality tests in Chinese stock markets. *Multinational Finance Journal* 1:291–307.

Lee, C. F., and Rui, O. M. (2000). Does trading volume contain information to predict stock returns? Evidence from China's stock markets. *Review of Quantitative Finance and Accounting* 14:341–60.

Mei, J., Scheinkman, J., and Xiong, W. (2005). *Speculative trading and stock prices: Evidence from Chinese A-B share premia.* Working Paper 11362, NBER, Cambridge, MA.

Merton, R. C. (1980). On estimating the expected returns on the market: An explanatory investigation. *Journal of Financial Economics* 8:323–61.

Ng, A. (2000). Volatility spillover effects from Japan and the U.S. to the Pacific-Basin. *Journal of International Money and Finance* 19:207–33.

Pantula, S. G., Gonzales-Farias, G., and Fuller, W. A. (1994). A comparison of unit-root test criteria. *Journal of Business and Economic Statistics* 12:449–59.

Philips, P., and Perron, P. (1988). Testing for a unit root in time series regression. *Biometrika* 75:335–46.

Qiao, Z., Chiang, T. C., and Wong, W. K. (2008). Long-run equilibrium, short-term adjustment, and spillover effects across Chinese segmented stock markets. *Journal of International Financial Markets, Institutions and Money* 18:276–89.

Ross, S. (1989). Information and volatility: The no-arbitrage martingale approach to timing and resolution irrelevancy. *Journal of Finance* 44:1–17.

Shalen, C. T. (1993). Volume, volatility, and the dispersion of beliefs. *Review of Financial Studies* 6:405–34.

Sjoo, B., and Zhang, J. (2000). Market segmentation and information diffusion in China's stock markets. *Journal of Multinational Financial Management* 10:421–38.

Sola, M., Spagnolo, F., and Spagnolo, N. (2002). A test for volatility spillovers. *Economics Letters* 76:77–84.

Su, D., and Fleisher, B. M. (1999). Why does return volatility differ in Chinese stock markets? *Pacific-Basin Finance Journal* 7:557–86.

Sun, Q., and Tong, W. H. S. (2000). The effect of market segmentation on stock prices: The China syndrome. *Journal of Banking and Finance* 24:1875–902.

Van Dijk, D., Osborn, D. R., and Sensier, M. (2005). Testing for causality in variance in the presence of breaks. *Economics Letters* 89:193–99.

Wang, P., Liu, A., and Wang, P. (2004). Return and risk interactions in Chinese stock markets. *Journal of International Financial Markets, Institutions and Money* 14:367–83.

Wang, X. L., Shi, K., and Fan, H. X. (2006). Psychological mechanisms of investors in Chinese stock markets. *Journal of Economic Psychology* 27:762–80.

Wang, Y., and Di Iorio, A. (2007). Are the China-related stock markets segmented with both world and regional stock markets? *Journal of International Financial Markets, Institutions and Money* 17:277–90.

Xu, C. K. (2000). The microstructure of the Chinese stock market. *China Economic Review* 11:79–97.

Yang, J. (2003). Market segmentation and information asymmetry in Chinese stock markets: A VAR analysis. *Financial Review* 38:591–609.

Yu, C.-H., and Wu, C. (2001). Economic sources of asymmetric cross-correlation among stock returns. *International Review of Economics and Finance* 10:19–40.

Zhu, H., Lu, Z., Wang, S., and Soofi, A. (2004). Causal linkages among Shanghai, Shenzhen, and Hong Kong stock markets. *International Journal of Theoretical and Applied Finance* 7:135–49.

CHAPTER 25

Optimal Settlement Lag for Securities Transactions

An Application to Southeast Stock Exchanges

Marco Rossi and Raphael W. Lam

CONTENTS

25.1 INTRODUCTION

An asset is considered liquid either if it trades in a market with a sufficient number of participants to allow purchases and sales on short notice at prices near the contemporaneous equilibrium value of the instrument, or if the asset's equilibrium value is unlikely to change substantially over a given time interval. Trading costs potentially affect the investor's capacity to convert a certain asset, as cheaply and risklessly as possible, into a means of payment whose wide acceptability can ensure finality of payment. In short, these costs reduce, *ceteris paribus*, assets' liquidity, and can therefore substantially affect an investment strategy, including the optimal portfolio turnover and trade location.

The microstructure of equity markets is key in assessing the size of trading costs and determining the liquidity of a given asset. This chapter focuses on one specific feature of the microstructure of equity transactions: settlement.* With the surge in trading activities, including cross-border, since the early 1980s, increasing attention was paid to managing risks in clearing and settlement systems, which had become more complex, particularly in the aftermath to the October 1987 market break. Several recommendations—most notably by the Group of Thirty (G-30, 1989)—started to be implemented in the early 1990s to facilitate the setting of industry standards, including guidelines about position limits, collateral and mark-to-market requirements, netting procedures, borrowing and lending facilities, guarantee/clearing funds, and finality of transaction. Technological advances and the recognition that the efficiency and security of the clearing and settlement mechanism affect the attractiveness of national financial centers and their ability to compete with other centers in the region have moved stock exchanges toward settlement systems, which incorporate many of the recommended features and, in particular from this chapter's perspective, the requirement that delivery and payment occur simultaneously and with short delays, generally a few days.

Immediacy, however, while an effective risk reduction tool, may also entail costs, as a participant's ability to settle depends on its liquidity position at any point in time, which in turn is affected by its trading activities, the liquidity of the local interbank market, and its ability to tap it efficiently. This chapter develops an analytical framework to assess the optimal delay to settle securities transactions, where *optimal* means that a longer (shorter) delay would generate too high risks (costs). It points out that credit and liquidity risks, stock price volatility, and money markets liquidity are all elements of a cost-risk trade-off. As an application of the framework to the specifics of local markets, the chapter reports illustrative numerical examples for several stock exchanges in the Asian and Pacific region (Table 25.1).

The chapter is organized as follows. Section 25.2 discusses the various risks associated with securities trading, and the costs involved in mitigating these risks. Section 25.3 presents an analytical framework to consider the trade-off between risks and costs and shows that the optimal settlement

---

* Dealing costs, information disclosure requirements, research, and other services are important factors in attracting securities transactions to a particular stock exchange.

TABLE 25.1 Settlement Systems in Selected Asia-Pacific Stock Exchanges

	Stock Exchange	Rolling Settlement System[a]
Australia	Australian Securities Exchange (ASX)	T + 3
China	Shanghai Stock Exchange (SSE)	T + 3
	Shenzhen Stock Exchange	T + 3
Hong Kong SAR	Hong Kong Stock Exchange (HKSE)	T + 2
India	Bombay Stock Exchange Limited (BSE)	T + 3
	National Stock Exchange of India (NSE)	T + 3
Indonesia	Indonesia Stock Exchange (IDX) or Jakarta Stock Exchange	T + 4
Korea	Korea Stock Exchange (KRX)	T + 2
Malaysia	Malaysia Exchange (MYX) or Kuala Lumpur Stock Exchange	T + 3
Singapore	Singapore Exchange (SGX)	T + 3
Taiwan, Province of China	Taiwan Stock Exchange (TSE)	T + 2
Thailand	Stock Exchange of Thailand (SET)	T + 3
United Kingdom	London Stock Exchange	T + 3
United States	New York Stock Exchange (NYSE)	T + 3
	Nasdaq Stock Exchange	T + 3
Japan	Tokyo Stock Exchange	T + 3

Source: Rhee (2000); various stock exchanges.

[a] T refers to the date of the transaction.

lag for securities transactions ultimately depends on the parameters that characterize that trade-off. The numerical examples reported in Section 25.4 show that different settlement lags across stock exchanges are warranted by the specifics of the local capital market. Section 25.5 concludes.

25.2 RISKS AND COSTS IN SECURITIES SETTLEMENT SYSTEMS

Trading is a process that involves a series of specific risks and costs. The process is relatively straightforward. First, two counterparts, willing to trade, agree on the terms of the transaction. Then, trade matching and its confirmation by market makers' clients prepare the ground to trade clearance, and the respective obligations of counterparts to deliver the asset or settle the transaction on a certain date are determined. Finally, the discharge of those obligations occurs with the final transfer of securities from the seller to the buyer (delivery) and the final transfer of funds from the buyer to the seller (payment). Settlement lag refers to the time interval between when the trade is confirmed and when it is settled.

Settling securities transactions takes time, and therefore involves risks between the moment the transaction is effected and the moment it is settled.[*] Only when delivery and payment have occurred is the trading process completed, the uncertainty about the "good end" of the transaction resolved, and risks eliminated. These risks include operational risk (the risk of a technical break-down), third-party credit risk (the risk that a third party to a transaction—a settlement bank or agent—fails during the settlement process), and counterparty risk (the risk of a party to the transaction to fail to deliver either the securities or the funds at time of settlement). This latter is the focus of this chapter, as counterparty risk is what risk management systems are designed to contain more specifically in view of its potential systemic consequences.[†]

The types and sources of risks to counterparts in securities transactions are very much the same as those arising from foreign exchange trades. If an obligation is not settled for full value, the counterparts to the transaction incur a credit loss. Credit risk in settlement systems comprises (1) the risk of loss on unrealized gains (replacement cost risk) and (2) the risk of loss of securities delivered or funds paid to the defaulting party just before the failure is detected (principal risk).

Replacement cost risk refers to the possibility that a defaulting participant can produce losses to his counterpart by forcing the latter to sell or buy securities at the market price instead of at the contract price previously agreed. For example, the seller is exposed to a replacement cost loss if, in case the original transaction fails to settle properly, the market price at which he or she would need to sell the securities is lower than the original contract price. In this example, the buyer would profit as the market has moved in his or her favor since the original transaction failed to settle. Losses on unrealized gains on unsettled transactions clearly depend on the behavior of market prices (price volatility), and therefore the time necessary to complete settlement.

Principal risk is the risk that a buyer makes the payment but does not receive delivery of the securities, or that a seller delivers the securities but does not receive payment for them. In other words, it stems from unsynchronized payment and delivery. Failure to settle would imply the loss of the full value of securities or funds (value of the transaction) that have been transferred to the defaulting party.[‡]

---

[*] See Bank for International Settlements (1992) for a discussion of the types and sources of risks in settling securities transactions. Also see Bank for International Settlements (2006).

[†] See Organization for Economic Cooperation and Development (1991).

[‡] Analogous to principal risk is cross-currency settlement risk in foreign exchange settlement, or Herstatt risk from the failure of Bankhaus Herstatt in 1974.

Reasons may vary as to why a party may not be able to settle an agreed transaction fully at the agreed time. It could be the case that a party's inability to settle is technical and temporary (liquidity shortage) or permanent (solvency). Irrespective of the reason, unwinding a transaction is costly.* Even in the arguably unlikely situation in which it is possible to clearly and immediately determine that illiquidity is the cause of the failed transaction (hence, avoiding systemic consequences), traders would still need to restart the process, that is, to find a third party and complete the intended transaction on newly agreed terms. As this takes time and the result is uncertain, traders—at least one side to the trade—are likely to bear some costs, including replacement and liquidity costs, the latter referring to the fact that the liquidity expected to be obtained through the failed trade would need to be raised elsewhere.

One way to reduce these risks and costs has been to shorten the settlement lag, which, however, imposes other costs on traders. As settlement of securities trades implies a transfer of funds and securities from one account to another, there is a need for traders to maintain cash and securities balances, or to be able to tap a well-functioning local interbank market efficiently. An effort to reduce the opportunity cost of maintaining high cash/securities balances to support a given level of transactions or an illiquid interbank market could result in higher rates of failed transactions. Transactions failures would, in turn, increase replacement cost and liquidity risks by randomizing the expected funds/securities balances at the end of the settlement process. A trade-off between the costs of reducing the settlement lag and those implied by a settlement failure clearly emerges.

In addition to the risks and costs for the individual trader, one party's failure to settle a transaction may generate risks and costs of a systemic proportion as the other party to the transaction may fail to settle other transactions and trigger a typical domino reaction. Below, the focus is on the risks and costs of settling transactions for an individual party rather than for the system as a whole.

25.3 THE ANALYTICAL FRAMEWORK

This section, which builds on Rossi (1994), brings together the various elements of risks and costs involved in settling securities transactions into an analytical framework that could be used to derive the optimal settlement

---

* Some stock exchanges require all transactions to be unwound in case of a settlement failure.

lag, that is, the time interval that minimizes the costs associated with the settlement process.

The first step is to define the probability, $Q(T)$, that a party to a transaction fails to settle—becoming illiquid or insolvent—after a transaction has been agreed, since most of the costs discussed above are contingent on such an event. This probability is likely increasing in the settlement lag and can be expressed as

$$Q(T) = 1 - (1 - q)^T \qquad (25.1)$$

where $0 \leq q \leq 1$ is the instantaneous probability of default. The longer the settlement lag, the higher the probability that a default event occurs stemming from a party's other trading and financial activities or from the impact of other parties' defaults.

The replacement cost is computed by comparing the contract price of a security with its market price. Losses on unrealized gains depend on the behavior of spot prices during the settlement period. As typical in the literature, market prices are assumed to follow a geometric Brownian motion with drift:

$$dS = \mu S dt + \sigma S dz \qquad (25.2)$$

where dz is a Wiener process. It is possible to show that $S(T)$ is described by a lognormal distribution whose expected value and variance are, respectively:

$$E[S(T)] = S\, e^{\mu T} \qquad (25.3)$$

$$Var[S(T)] = S^2 e^{2\mu T}(e^{\sigma^2 T} - 1) \qquad (25.4)$$

where μ, the expected rate of return, and σ, the standard deviation, are both constant.

The replacement cost is measured as the variance of the difference between the contract price $C(0)$ at the time, $t = 0$, the trader decides to effect the transaction and the market spot price $S(T)$ realized on that trade at the time of settlement T.[*]

$$Var[S(T) - C(0)] \qquad (25.5)$$

Given Equation (25.4), the replacement cost is

$$RC(T) = Q(T)[S^2 e^{2\mu T}(e^{\sigma^2 T} - 1)] \qquad (25.6)$$

---

[*] T indicates the length of time between execution and settlement of the transaction, $T = 0$ being the execution date.

The cost associated with principal risk can be measured by

$$PRC = Q(T)\, P \tag{25.7}$$

where P is the value of the transaction.[*]

The cost associated with credit risk is the sum of replacement cost and the cost associated with principal risk:

$$CRC(T) = Q(T)[P + S^2 e^{2\mu T}(e^{\sigma^2 T} - 1)] \tag{25.8}$$

which is an increasing function of the settlement lag T, both directly and through the probability of default $Q(T)$.

To reduce the probability of being unable to settle a transaction, traders maintain cash and securities balances at a cost (opportunity cost). The more quickly traders are able to tap the money and securities lending markets at reasonable terms, the weaker the need to maintain such balances, hence reducing the transaction cost associated with settlement. The longer the settlement lag, the easier liquidity can be obtained even in relatively illiquid money markets, and the lower the cost of holding reserves.

The cost of holding reserves can therefore be expressed as a declining function of the settlement lag:

$$C(T) = \frac{A}{1+T} \tag{25.9}$$

where A is a constant term.

The optimal settlement lag for securities transactions can be derived by minimizing the total cost of settlement—obtained as the sum of the various cost items discussed above (Panel 1 in Figure 25.1)—with respect to T.

$$Min_T\, CT(T) = Min_T\, [1-(1-q)^T][P + S^2 e^{2rT}(e^{\sigma^2 T} - 1)] + \frac{A}{1+T} \tag{25.10}$$

The first-order condition with respect to T is

$$\frac{dCT}{dT} = S^2 e^{2rT}[1-(1-q)^T][2re^{\sigma^2 T} + \sigma^2 e^{\sigma^2 T} - 2r] \qquad (a) > 0$$

$$-(1-q)^T \ln(1-q)[S^2 e^{2rT}(e^{\sigma^2 T} - 1) + P] \qquad (b) > 0 \tag{25.11}$$

$$-\frac{A}{(1+T)^2} \qquad (c) < 0$$

---

[*] See Angelini and Giannini (1993) and Stevens (1998).

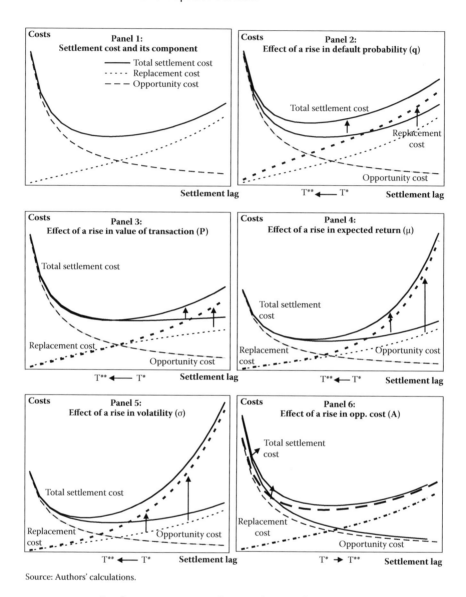

FIGURE 25.1 Settlement costs, settlement lag, and comparative statics.

The sign of this partial derivative changes over the domain for T. There exists a trade-off between the credit risk and liquidity risk components. For small T, the opportunity cost effect (c) offsets the credit risk cost (a) and (b), implying a negative relationship between the total cost and the settlement lag. However, as T increases, the opportunity cost effect (c) becomes more negligible and dominated by the rising credit risk cost, implying the marginal settlement cost is rising with the settlement dates. The optimal settlement lag T^* may be solved by setting Equation (25.11) equal to zero.

A rise in the value of transaction (P), probability of default (Q), stock price (S), expected return (r), and the spot price volatility σ increases, *ceteris paribus*, the cost components of the total settlement cost related to credit risk. An increase in A, on the other hand, increases the opportunity cost component of the total settlement cost. This is shown by the first partial derivatives:

$$\frac{dCT}{dP} = [1-(1-q)^T] > 0$$

$$\frac{dCT}{dq} = [T(1-q)^{T-1}][P+S^2 e^{2rT}(e^{\sigma^2 T}-1)] > 0$$

$$\frac{dCT}{dS} = [1-(1-q)^T][2Se^{2rT}(e^{\sigma^2 T}-1)] > 0 \qquad (25.12)$$

$$\frac{dCT}{dr} = [1-(1-q)^T][S^2 e^{2rT}(e^{\sigma^2 T}-1)2T] > 0$$

$$\frac{dCT}{d\sigma^2} = [1-(1-q)^T][S^2 e^{2rT}e^{\sigma^2 T}T] > 0$$

$$\frac{dCT}{dA} = \frac{1}{1+T} > 0$$

The impact of the probability of default (q), the value of the transaction (P), the expected rate of return (r), and the spot price volatility (σ) on the optimal settlement lag (T^*) and the settlement lag (T) can be obtained using the implicit function theorem. These comparative statics effects are shown in Panels 2–6 in Figure 25.1. An increase in q, P, r, and σ will increase credit risk, and hence the total settlement cost. As long as the marginal increase in credit risk is increasing with settlement time, it reduces the optimal settlement lag from T^* to T^{**} (the total cost function shifts leftward). By contrast, an increase in A lengthens the optimal lag from T^* to T^{**} (the total cost function shifts rightward).

25.4 AN APPLICATION TO STOCK EXCHANGES IN THE ASIAN AND PACIFIC REGION

The surge in cross-border transactions and integration of financial markets across the Asian and Pacific region have made investors more aware that operational support systems, like clearing and settlement mechanisms, form a critical part of the effective capital market.

Rhee (2000) provides a comprehensive survey of the risk management systems of clearing and settlement mechanisms in equity markets across the Asian and Pacific region, but does not provide an analytical framework in addressing the optimal settlement process across stock exchanges.

Braeckevelt (2006) observes that the clearing and settlement infrastructure in Asia is very fragmented. The infrastructure is not cost-efficient and does not mitigate risks in the settlement process in a comprehensive manner. He also suggests that given capacity constraints, Asian stock markets may need to introduce a gradual rationalization of the domestic and regional infrastructure to develop fully integrated U.S.- or European-style clearing and settlement systems.

The analytical framework proposed in the previous section is used here to provide optimal settlement lags for twelve stock exchanges in the region, including both industrialized economies and emerging markets (Table 25.2). The numerical examples illustrate that the optimal settlement lags in these stock exchanges may vary depending on the volatility of stock prices, trading costs, and market liquidity.

Data to estimate the few parameters that are necessary to produce these numerical examples are collected from CEIC, Datastream, and the International Financial Statistics. The data on stock exchanges contain market indices, market capitalization, trading volumes, and the number of existing and newly listed securities over the sample period from January 1990 through March 2008. The data indicate that average returns and volatility vary remarkably across stock exchanges. For example, stock markets in Shanghai and Shenzhen recorded an average return of over 25% during 2000–2007, which was more than double the average returns for the industrialized economies. The average annual growth in market capitalization exceeded 17%, with the value of transactions rising at a similar pace. The default or fail transactions rate has remained low during the sample period, at less than 5% on average, according to the BIS (2002, 2006).

In the numerical examples, a probability of default of 1% for industrialized countries and 5% for emerging market economies are used. Parameters on return (μ), volatility (σ), and the value of transaction (P) are proxied by averages calculated for each stock exchange. The opportunity cost of holding cash/securities balances (A) is proxied by the average deposit rate.[*] This

---

[*] See Bekaert et al. (2007) and Aitkena and Comerton-Forde (2003) on different measures of liquidity.

TABLE 25.2   List of Selected Asia-Pacific Stock Exchanges

	Stock Exchanges	**Major Market Indices**
Australia	Australian Securities Exchange (ASX)	S&P/ASX 200
China	Shanghai Stock Exchange (SSE)	Shanghai Composite Index
	Shenzhen Stock Exchange	Shenzhen blue-chip composite index
Hong Kong SAR	Hong Kong Stock Exchange (HKSE)	Hang Seng Index
India	Bombay Stock Exchange Limited (BSE)	BSE SENSEX
	National Stock Exchange of India (NSE)	Standard & Poor's CRISIL NSE Index 50 (S&P CNX Nifty)
Indonesia	Indonesia Stock Exchange (IDX) or Jakarta Stock Exchange	JSX Composite
Korea	Korea Stock Exchange (KRX)	Korean Composite Stock Price Index (KOSPI)
Malaysia	Malaysia Exchange (MYX) or Kuala Lumpur Stock Exchange	Kuala Lumpur Composite Index (KLCI)/FTSE Bursa Malaysia Index
Singapore	Singapore Exchange (SGX)	Straits Times Index
Taiwan, Province of China	Taiwan Stock Exchange (TSE)	Taiwan Capitalization Weighted Stock Index
Thailand	Stock Exchange of Thailand (SET)	Stock Exchange of Thailand (SET) Index

Source: Various stock exchanges.

rate remained low at about 2.2%, except in India and Indonesia, where it is above 6%. The interbank rates at short maturity across economies were about 3%, where the interbank rate for India has been at 7% on average. For each stock exchange, Table 25.3 reports the estimates that are used to proxy the parameters of the analytical framework.

The numerical results, reported in Table 25.4, suggest that the optimal settlement lags for securities transactions at these twelve stock exchanges in the Asian and Pacific region are broadly in line with international best practices. In particular, the optimal settlement lags range from 2.7 to 4.6 days, with generally longer lags for securities transactions affected in emerging markets economies. These illustrative results show that local practices—most of the twelve stock exchanges have already implemented the T + 3 or less settlement practice—may be more ambitious than

TABLE 25.3   Parameters Used in the Numerical Example[a]

	Default Probability[b]	Expected Return (annualized)	Volatility (standard deviation)	Average Transaction (in thousands USD)	Deposit Rate (in percent)
	q	μ	σ	P	r_d
Stock Exchanges					
Australia	0.01	9.61	11.87	1.85	n.a.
China—Shanghai	0.05	28.85	75.54	2.12	1.58
China—Shenzhen	0.05	26.64	65.49	0.91	1.58
Hong Kong SAR	0.01	13.43	27.72	n.a.	2.71
India—Bombay	0.05	16.16	41.90	2.57	8.83
India—National	0.05	15.13	39.51	2.19	8.83
Indonesia	0.05	15.16	26.92	n.a.	6.02
Korea	0.01	11.85	40.15	3.13	2.68
Malaysia	0.05	7.02	30.27	0.86	3.18
Singapore	0.01	9.00	28.31	2.75	1.66
Taiwan, Province of China	0.01	5.07	28.06	n.a.	1.50
Thailand	0.05	5.63	30.10	0.89	n.a.
Industrialized economies	0.01	9.17	28.20	—	2.27
Emerging markets	0.05	16.82	44.50	—	4.42

Source: CEIC, Datastream, and BIS.

[a] Industrialized economies include Australia, Hong Kong SAR, Korea, Singapore, and Taiwan, Province of China. The rest are considered emerging markets.

[b] Estimates of default probability range from 1% to over 5% (BIS, 2002).

warranted by the reality of local markets and the trade-off between credit risk and opportunity cost (Panel 1 in Figure 25.2).

The numerical results also suggest that, across stock exchanges, credit risk is a significant component of the total settlement cost, while the opportunity cost accounts only for between 10% (industrialized economies) and 26% (emerging markets economies) of the total settlement cost, as the former have generally more liquid money markets.

Despite the relatively small share of opportunity costs in total settlement costs, shortening the settlement lag appears optimal only if it is accompanied by a reduction in the cost of liquidity. In particular, Panel 2 in Figure 25.2 shows that higher opportunity costs lengthen the optimal settlement lags, *ceteris paribus*. Stock exchanges in locations with limited

TABLE 25.4   The Optimal Settlement Lag: Numerical Examples[a]

	Optimal Settlement Lag (in days)	Replacement Risk Cost (% of total cost)	Opportunity Cost (% of total cost)
Stock Exchanges			
Australia	3.33	85.63	14.37
China—Shanghai	3.40	82.34	17.66
China—Shenzhen	4.18	76.52	23.48
Hong Kong SAR	2.82	88.88	11.12
India—Bombay	3.20	83.71	16.29
India—National	3.88	78.81	21.19
Indonesia	4.55	73.74	26.26
Korea	2.57	90.36	9.64
Malaysia	2.88	85.94	14.06
Singapore	2.97	87.97	12.03
Taiwan, Province of China	4.09	80.01	19.99
Thailand	3.88	78.81	21.19
Industrialized economies	2.69	89.67	10.33
Emerging markets	3.62	80.72	19.28

Source: Authors' calculations.

[a] Industrialized economies include Australia, Hong Kong SAR, Korea, Singapore, and Taiwan, Province of China. The rest are considered emerging markets.

liquidity, especially during periods of market distress, are likely to see a surge in the risk of a settlement failure. Panel 3 in Figure 25.2 shows a sharp rise in the rate of settlement failures if settlement is shortened, as the potential benefits from a reduction in the replacement risk cost are relatively small.

Finally, the optimal settlement lag is inversely related to stock prices volatility (Panel 4 in Figure 25.2), as an increase in σ would increase replacement cost risk, albeit at a decreasing rate. A closer look at stock price data shows that volatility has decreased over time in most of the stock exchanges included in the sample (at 5% statistical significance level), suggesting that, *ceteris paribus*, more mature markets may afford relatively longer settlement lags. The data also show that, as one can expect, volatility at times of financial turmoil is higher, underscoring the potential for settlement failure for given settlement mechanisms during such times.

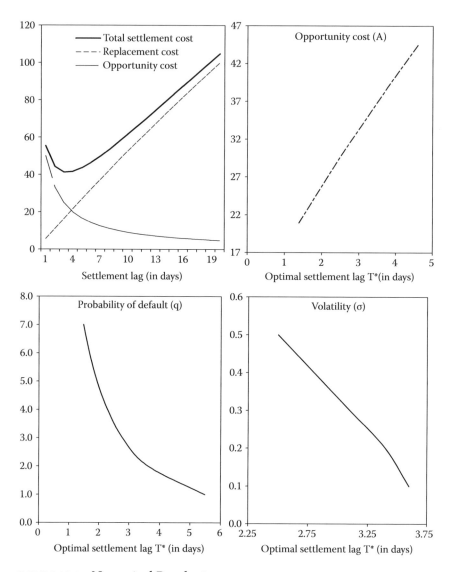

FIGURE 25.2 Numerical Results 1.

25.5 CONCLUSION

Transaction costs can substantially impact actual portfolio performance and, more generally, an investment strategy, including the optimal portfolio turnover and trade location. The microstructure of equity markets is key in assessing the size of trading costs and determining the liquidity of a given asset. This chapter focuses on one feature of the microstructure of equity transactions: settlement.

The rise of emerging markets, and increasing integration of financial markets across the globe, has made the settlement process considerably more complex. Investors and supervisors have become more aware of the potential disruption stemming from settlement failures, and of the need to develop strong support mechanisms to foster the effectiveness of the international financial system. In the aftermath of the October 1987 market break, the Group of Thirty recommended the reduction of the settlement lag for securities transactions to 3 days.

While the recommendation has been supported both by traders and by stock exchanges and central banks, less attention has been paid to its practicality, particularly in markets where deep money and securities markets have not yet fully developed. This chapter proposes an analytical framework to derive the optimal settlement lag, that is, the optimal time period between the moment a transaction is effected and the moment it is settled, taking into account risks and costs involved in the settlement process.

The framework identifies the main risks and costs involved in settling securities transactions and shows that the optimal settlement lag for securities transactions depends on a series of parameters that characterize the local financial market, such as stock price volatility, rate of return, money and securities markets liquidity, and the probability of a credit event. The numerical examples for twelve stock exchanges in the Asian and Pacific region illustrate that the optimal settlement lag, albeit varying across stock exchanges, is broadly in line with international best practices. They also show, however, that the specifics of the local financial market matter when establishing settlement practices, as there is likely a trade-off between shortening the settlement lag and the opportunity cost of ensuring that settlement failure is averted.

ACKNOWLEDGMENTS

The views expressed herein are those of the authors and should not be attributed to the IMF, its executive board, or its management.

REFERENCES

Aitkena, M., and Comerton-Forde, C. (2003). How should liquidity be measured? *Pacific-Basin Finance Journal* 11:45–59.

Angelini, P., and Giannini, C. (1993). *On the economics of interbank payment systems*. Temi di discussione 193, Banca d'Italia, Rome.

Bank for International Settlements. (1992). *Delivery versus payment in securities settlement systems*. Report by the Committee on Payment and Settlement Systems of the Central Banks of the Group of Ten Countries, Basel.

Bank of International Settlements. (2002). *Assessment methodology for recommendations for securities settlement systems.* Report 51 by the Committee on Payment and Settlement Systems, Basel.

Bank for International Settlements. (2006). *General guidance for national payment system development.* Report 70 by the Committee on Payment and Settlement Systems, Basel.

Bekaert, G., Harvey, C., and Lundblad, C. (2007). Liquidity and expected returns: Lessons from emerging markets. *Review of Financial Studies* 20:1783–831.

Braeckevelt, F. (2006). Clearing, settlement and depository issues: Investor services product management in Asia. Bank of International Settlement Papers No. 30, Bank of International Settlements.

Group of Thirty. (1989). *Clearance and settlement systems in the world's securities markets.* Washington, DC: Author.

Organization for Economic Cooperation and Development. (1991). Systemic risks in securities markets. OECD Ad Hoc Group of Experts on Securities Markets, Paris.

Rhee, S. G. (2000). Risk management systems in clearing and settlement: Asian and Pacific equity markets. *Asian Development Review* 18:94–119.

Rossi, M. (1994). *Settling securities transactions.* Mimeo. London: Bank of England.

Stevens, E. (1998). Risk in large-dollar transfer systems. *Federal Reserve Bank of Cleveland Economic Review* 2:2–16.

CHAPTER 26

Seasonality and the Relation between Volatility and Returns

Evidence from Turkish Financial Markets

Oktay Taş, Cumhur Ekinci, and Zeynep İltüzer Samur

CONTENTS

26.1 INTRODUCTION

Over the past decades many studies have provided evidence of seasonal anomalies in asset returns in stock, fixed-income, or foreign exchange markets. The anomalies mentioned in the literature mainly are periodical movements like the time-of-the-day, the day-of-the-week, and the month-of-the-year effects. The day-of-the-week effect has been investigated by Cross (1973), French (1980), Gibbons and Hess (1981), Jaffe and Westerfield (1985), Aggarwal and Rivoli (1989), Keim and Stambaugh (1984), Lakonishok and

Levi (1982), and Lakonishok and Smidt (1988), among others. These empirical papers mostly find that Monday returns are significantly negative and Friday returns are higher than on the other days of the week. However, they differ in the time periods and the stock markets they cover as well as the number of firms and their characteristics. For example, Aggarwal and Rivoli studied the turn-of-the-year effect and the weekend effect in four emerging markets and provided evidence of high January returns and low Monday returns, similarly to the findings on developed countries' equity markets. While Cross (1973), French (1980), Gibbons and Hess (1981), and Keim and Stambaugh (1984) studied the day-of-the-week anomalies in U.S. stock markets and found that Friday return is the highest and Monday return is the lowest, Jaffe and Westerfield (1985) examined the U.S., UK, Canada, Japan, and Australia stock markets and documented that Thursday return is the lowest in Japan and Australia stock markets. After an analysis on the Center for Research in Security Prices (CRSP) equally weighted and value-weighted indices for the period of 1964–1974, Lakonishok and Levi (1982) concluded Monday and Friday effects disappeared by the mid-1970s. On the other hand, Lakonishok and Smidt (1988) examined the seasonal anomalies in Dow Jones Industrial Average by using 90-year daily data and found significantly negative Monday returns.

Monthly seasonality has also drawn a great deal of attention. Keim (1983) provided evidence that small firms showed higher returns in January than large firms do, and that the majority of high returns in January occur during the first week. Ariel (1987) studied the seasonalities within a month. Accordingly, mean return is positive in the first half of the month and insignificant the rest of the month. Jones et al. (1987) questioned whether the underlying cause of the January effect is the tax-motivated transactions. That is, investors seek to reduce their tax expenses by closing their bad positions (realizing losses) at the end of the year, which implies a decline in stock price. However, they found that the January effect is not related to these sales for tax advantage, but rather to the small-firm effect, as in Keim. Lakonishok and Smidt (1984) documented the turn-of-the-year seasonality in small firms. Gültekin and Gültekin (1983) studied the monthly seasonality for the major equity markets and stated that the January effect is common to all except the UK market, in which April is the month with the highest returns. Kato and Schallheim (1985) studied the month-of-the-year effect in the Japanese stock market and reported the January and June anomalies. They grounded their findings on the fact that the Japanese employees used to invest in equities the bonuses they receive in June.

As clear from the literature review, almost all the studies provide evidence of seasonal anomalies, though for different countries, time periods, or firm sizes. Findings vary across studies, whereas high January returns and negative Monday returns remain common. As a result, we observe that except the two anomalies above, the nature of the seasonal anomalies depends on the country or the market.

A parallel research area that reports more heterogeneous results involves the relation between stock returns and their volatility. For instance, French et al. (1987) reported a positive relation between stock returns and volatility. In contrast, Nelson (1991) and Glosten et al. (1993) found a negative relation. Baillie and Degennaro (1990) studied the relation between the returns and volatility of a stock portfolio and, interestingly, could not find any relation. They concluded that traditional two-parameter model relating volatility and return is inappropriate and other measures of risk are needed. By the same token, Theodossiou and Lee (1995) and Corhay and Rad (1994) reported an insignificant relation between returns and their volatility for ten and three countries, respectively.

The analysis of seasonal anomalies in volatility also receives attention. Balaban et al. (2001) studied the relation between returns and volatility of nineteen countries by adding the day-of-the-week effect in the model. They reported that only three of the nineteen countries show a positive relation between return and volatility, while the rest show a zero relation. Moreover, seven countries show the day-of-the-week effect in mean returns, six countries show the day-of-the-week effect in conditional volatility, and two countries show the day-of-the-week effect in both return and volatility. The nature and direction of the effect are not the same for all the countries that show seasonal anomalies. Glosten et al. (1993) developed a new way by integrating seasonality in volatility estimation in an application on CRSP value-weighted index. They reported significant October and January effects in volatility.

For the Turkish stock market, there are also a few papers studying seasonalities on a varying basis. Demirer and Karan (2002) reported a significant Friday effect but no clear evidence of a Monday effect in the Istanbul Stock Exchange (ISE) returns for the 1988–1996 period. Karan and Uygur (2001) investigated the day-of-the-week effect in portfolios composed of stocks selected according to market value for the 1991–1999 period. They reached the same conclusion as Demirer and Karan. Additionally, they found a January effect for ISE stocks. Bildik (2004) examined the daily seasonality in stock market and found that returns are significantly higher in the second part of the week (especially on Friday) and lower in the first two days of the week (especially on Tuesday) by using 11-year index data from 1988 to 1999. The study also revealed that the volatility of the ISE100 index was highest on Monday and lowest on Friday. It also supported the evidence of the January effect found by Karan and Uygur. Oğuzsoy and Güven (2003) investigated the daily seasonality of the ISE100 index as well as the ISE30 index firms for the same period as in Bildik. They found the same evidence for the index. As to stock-based analysis, they confirmed that most index stocks show high (low) returns on Friday (Monday and Tuesday) and that volatility is highest on Monday.

This chapter first tries to detect some daily and monthly anomalies in returns and volatilities of the ISE100 index and the dollar/new Turkish lira exchange rate (US$/TRY), and then seeks whether there remains any relation between volatility and returns when these are deseasonalized from the calendar effects. This also makes sense in that it contributes to our understanding of the relation between risk and return in the context of asset pricing. To the best of our knowledge, this is the first ever study to include a seasonality analysis on conditional variance in addition to mean returns in Turkish financial markets. Our results show some calendar anomalies, but do not confirm a relation between volatility and returns.

The remainder of the chapter proceeds as follows. Section 26.2 describes the data and the methodology, Section 26.3 explores empirical results, and Section 26.4 concludes.

26.2 DATA AND METHODOLOGY

Our data set includes the daily ISE100 index obtained from the ISE website and the US$/TRY exchange rate obtained from the Central Bank of Turkey website for a period of January 3, 2001 to March 1, 2007. This makes 1,731 (1,770) working days for the index (exchange rate). Based on these series, we calculate logarithmic returns.

To test for difference between the mean rates of return across the days of the week and the months of the year, we perform the following Ordinary Least Squares (OLS) regressions as widely used in literature.

$$r_{dt} = \phi_{d1} \cdot D_{1t} + \phi_{d2} \cdot D_{2t} + \phi_{d3} \cdot D_{3t} + \phi_{d4} \cdot D_{4t} + \phi_{d5} \cdot D_{5t} + \varepsilon_{dt} \qquad (26.1)$$

$$r_{mt} = \phi_{m1} \cdot M_{1t} + \phi_{m2} \cdot M_{2t} + \cdots + \phi_{m12} \cdot M_{12t} + \varepsilon_{mt} \qquad (26.2)$$

where r_{dt} (r_{mt}) is the rate of return of either ISE100 or US\$/TRY on day t (month m), and D_{it} (M_{jt}), $i = 1,2\ldots5$ ($j = 1,2\ldots12$), is a dummy variable that equals 1 if trading day t is a Monday, Tuesday, Wednesday, Thursday, or Friday (trading month m is January, February, ..., December), respectively, and 0 otherwise.

For the analysis of the relation between volatility and returns, we use a GARCH-M model. This model adds extra parameters, like seasonal effects in our case, to the dependent variable, to the explanatory variables, or both. Although each variable can be best modeled by different lag values, GARCH (1,1) is predominantly the most widely used. Consequently, as a first step, we define the relation between volatility and returns as in Model 1 below:

$$r_t = c + \lambda f\left(\sigma_t^2\right) + \sum_k \gamma_k DUMMY_{kt} + \varepsilon_t$$
$$\varepsilon_t \approx N\left(0, \sigma_t^2\right) \qquad (26.3)$$
$$\sigma_t^2 = c_0 + \beta_1 \sigma_{t-1}^2 + \sum_k \delta_k DUMMY_{kt} + \alpha_1 \varepsilon_{t-1}^2$$

where $DUMMY$ is the dummy for either the days of the week (D_{it}, $i = 1,2\ldots5$) or the months of the year (M_{jt}, $j = 1,2\ldots12$), f is a function of variance,[*] and λ, called the risk premium, explains the relation between conditional volatility and returns, both refined from seasonal effects.

Model 1 is the standard model used to eliminate seasonal anomalies (see, for instance, Balaban, 2001; Seyyed et al., 2005). Nevertheless, it does not take into account the asymmetry in return innovation. Therefore, as an alternative method, we run the regressions in Model 2 that assume positive and negative return innovations can affect the conditional volatility differently:

$$r_t = c + \lambda f\left(\sigma_t^2\right) + \sum_k \gamma_k DUMMY_{kt} + \varepsilon_t$$
$$\varepsilon_t \approx N\left(0, \sigma_t^2\right)$$
$$\sigma_t^2 = c_0 + \beta_1 \sigma_{t-1}^2 + \sum_k \delta_k DUMMY_{kt} + \alpha_1 \varepsilon_{t-1}^2 + \alpha_2 \varepsilon_{t-1}^2 I_{t-1}$$
$$(26.4)$$

where, in addition to the variables defined above, the dummy variable I_{t-1} takes the value of 1 if the return innovation ε_{t-1} is negative, and 0 if the return innovation ε_{t-1} is positive.

---

[*] We assume three versions of volatility funtion, i.e., f can be equal to σ_t, σ_t^2, or $\ln(\sigma_t^2)$.

26.3 EMPIRICAL RESULTS

This section gives our empirical results in two main subsections. First, we present the evidence for daily and monthly seasonal anomalies. Then, we determine how effective volatility is on returns, after refined from seasonality.

26.3.1 Seasonal Anomalies

In the search of seasonal anomalies, we refer to the descriptive statistics and regression statistics of daily returns and volatility across the days of the week (Table 26.1) and the months of the year (Table 26.2). In Table 26.1, we observe that both ISE100 and US$/TRY mean returns are negative on Monday. They become positive in the last two days of the week for ISE100 and in the middle days for US$/TRY. An interesting observation is that on Friday, standard deviations are strikingly lower for ISE100 and higher for US$/TRY.

A more pertinent analysis can be done by looking at the p-values of the regression coefficients given by Equation 26.1. In what follows, ISE100 returns are significantly negative on Monday and positive on Thursday and Friday at 1%. By comparison, US$/TRY returns are negative on Monday and positive on Tuesday and Thursday, but only at a 10% level of significance. As to a joint test, both parametric (F) and nonparametric (KW) tests strongly reject the null hypothesis that all the days of the week have the same returns, though somewhat less obvious for US$/TRY (F = 2.32) than ISE100 (F = 6.42).

A third analysis is on the anomalies in volatility. For this purpose, we investigate the equality of variances through Bartlett (B), Levene (L), and Brown-Forythe (BF) tests. Under the null hypothesis, each day of the week has the same variance. Table 26.1 reveals that we reject the null for ISE100 by all the tests at 5% and for US$/TRY by Bartlett test (B). However, one can argue that the B-test is not so powerful since it tests whether the variance-covariance matrix equals the identity matrix. That is why we rather refer to the other two test statistics. These cannot reject the null for US$/TRY.

Analogously to Table 26.1, Table 26.2 shows the statistics of daily returns and volatility for the months of the year. Accordingly, April and October (February and March) have significantly higher rates of return than other months for ISE100 (US$/TRY). However, the evidence is weaker than in the previous results since p-values are lower than 0.05 for all the months except October for ISE100. Moreover, F and KW statistics indicate that the null hypothesis of no significant month-of-the-year effect cannot be rejected at 5%, but only at 10% and uniquely for US$/TRY.

TABLE 26.1   Daily Percentage Returns According to the Days of the Week

Panel A: Continuously Compounded Daily Percentage Return on the ISE100 Index				
	Monday	Tuesday	Wednesday	Thursday	Friday
Nb Obs	335	350	348	353	345
Max	10.09	17.77	17.09	11.79	12.68
Min	−15.80	−9.44	−19.97	−10.37	−9.44
Mean	−0.56	−0.04	0.00	0.45	0.37
Std Dev	2.85	2.59	2.78	2.78	2.33
p-value	0.000	0.727	0.954	0.002	0.008

	Equality of Coefficients		Equality of Variances		
Test	F	KW	B	L	BF
Stat	6.42	31.79	17.38	2.66	2.60
p-value	0.000	0.000	0.002	0.031	0.035

Panel B: Continuously Compounded Daily Percentage Return on US$/TRY Exchange Rate				
	Monday	Tuesday	Wednesday	Thursday	Friday
Nb Obs	350	354	355	357	354
Max	11.34	8.82	10.05	6.00	33.47
Min	−5.38	−12.56	−4.33	−8.36	−2.89
Mean	−0.12	0.13	0.08	0.12	0.04
Std Dev	1.14	1.26	1.12	1.01	1.98
p-value	0.083	0.057	0.210	0.082	0.549

	Equality of Coefficients		Equality of Variances		
Test	F	KW	B	L	BF
Stat	2.32	30.13	227.68	0.63	0.65
p-value	0.041	0.000	0.000	0.639	0.626

Notes: ISE100 (US$/TRY) data were obtained from www.imkb.gov.tr (www.tcmb.gov.tr). Returns are calculated according to the following formula: $R_t = \ln(P_t/P_{t-1})$, where P is either ISE100 or US$/TRY and ln is natural log.

Tests of equality of means: Fisher (F), Kruskal-Wallis (KW).

Tests of equality of variances: Bartlett (B), Levene (L), Brown-Forsythe (BF).

Contrary to returns, volatility is more dispersed across the months of the year. For instance, standard deviations are particularly high in February and December and low in August for ISE100, while February appears to be the only month where volatility is particularly higher than the others for US$/TRY. One can notice that the B, L, and BF tests all very strongly reject the null hypothesis of equal variances.

TABLE 26.2 Daily Percentage Returns According to the Months of the Year

Panel A: Continuously Compounded Daily Percentage Return on the ISE100 Index

	Jan	Feb	Mar	Apr	May	Jun	Jul	Aug	Sep	Oct	Nov	Dec
Nb Obs	151	131	144	142	148	150	153	151	145	146	135	135
Max	10.98	9.42	11.38	12.68	6.61	5.90	9.64	4.51	7.53	10.09	11.79	17.77
Min	-8.35	-19.97	-13.34	-6.30	-8.67	-6.61	-9.44	-4.48	-6.43	-3.70	-9.44	-10.37
Mean	0.11	-0.14	-0.17	0.43	-0.33	-0.12	0.05	0.05	0.09	0.57	0.00	0.26
Std Dev	2.77	3.54	3.26	2.62	2.45	2.17	2.53	1.66	2.05	2.39	3.14	3.29
p-value	0.611	0.532	0.439	0.057	0.131	0.566	0.808	0.800	0.670	0.010	0.969	0.262

Equality of Coefficients

Test	F	KW
Stat	1.31	10.60
p-value	0.206	0.477

Equality of Variances

Test	B	L	BF
Stat	929.84	3.29	2.89
p-value	0.000	0.000	0.000

Panel B: Continuously Compounded Daily Percentage Return on US$/TRY Exchange Rate

	Jan	Feb	Mar	Apr	May	Jun	Jul	Aug	Sep	Oct	Nov	Dec
Nb Obs	150	130	144	142	148	150	152	151	145	145	135	134
Max	2.55	33.47	6.26	10.05	4.77	4.20	8.14	4.98	3.77	2.30	1.86	2.32
Min	-2.00	-12.56	-2.18	-5.38	-2.15	-5.22	-8.36	-2.58	-3.02	-2.89	-2.19	-1.48
Mean	-0.05	0.23	0.20	0.09	0.08	0.16	0.01	0.02	0.08	0.07	-0.14	-0.09
Std Dev	0.64	3.37	1.16	1.65	1.05	1.24	1.37	0.86	0.83	0.69	0.62	0.73
p-value	0.628	0.053	0.068	0.398	0.430	0.149	0.887	0.856	0.478	0.483	0.232	0.428

Equality of Coefficients

Test	F	KW
Stat	1.15	19.10
p-value	0.311	0.059

Equality of Variances

Test	B	L	BF
Stat	143.72	4.41	4.29
p-value	0.000	0.000	0.001

Notes: See Table 26.1.

TABLE 26.3   Day and Month Dummies Used in GARCH Equations

Day Dummies				Month Dummies			
ISE100		**US$/TRY**		**ISE100**		**US$/TRY**	
Mean Eq	**Var Eq**	**Mean Eq**	**Var Eq**	**Mean Eq**	**Var Eq**	**Mean Eq**	**Var Eq**
Mon	Mon	Mon			Feb		Feb
Thu		Tue			Dec		
Fri		Thu					

26.3.2   The Relation between Volatility and Returns

This section gives the empirical results about the relation between volatility and returns based on the models given by Equations 26.3 and 26.4. Dummy variables in GARCH equations are selected according to the results of the previous subsection. More explicitly, for mean equations, we selected as dummy the variables with 10% significant coefficients if the null hypothesis was rejected by both F and KW tests, and for variance equations, the outliers if the null hypothesis was rejected by the B, L, and BF tests. These are summarized in Table 26.3.

The mathematical representation of the models can be found in the notes below Table 26.4 to Table 26.7. Table 26.4 reports the parameter estimates for the models, including the day-of-the-week effects for ISE100. Accordingly, for all regressions in the table, the coefficient c is insignificant, indicating no systematic movement in daily returns; γ_2 is positive and significant at 1%, while γ_1 is negative and significant only at 10%, and γ_3 is insignificant, revealing a strong Thursday effect and weak negative Monday effect in ISE100 returns. If λ is negative, it is far from being significant. Hence, we fail to reject the idea that volatility has an impact on returns. In the variance equation, constant term c_0 is insignificant, so we cannot determine any given level of daily volatility. Other estimates on the variance equation are more significant and homogeneous across the models. For instance, β_1 is 0.9, and this is a sign of high positive autocorrelation in volatility, i.e., volatility clusters. α_1 is also positive and quite significant, so we deduce residuals are heteroskedastic and innovation (news) has an impact on volatility. The coefficient α_2, too, is positive and significant at 1%. This means that a decrease in prices (bad news) affects the market more than an increase in prices (good news). Finally, δ_1 is also positive and partially significant at 5% or 10%. This is weak evidence of high volatility on Monday compared to other days.

TABLE 26.4 Parameter Estimates of GARCH(1,1)-M with the Day-of-the-Week Effect for ISE100

	c	λ	γ_1	γ_2	γ_3	c_0	β_1	α_1	α_2	δ_1	Log-Likelihood
Model 1	.0005		−.0025	**.0026**	.0022	0000	**.9039**	**.0858**		**.00006**	4050
	(.446)		(.066)	(.034)	(.101)	(.447)	(.000)	(.000)		(.026)	
Model 1 SQRT(GARCH)	.0014	−.0415	−.0025	**.0026**	.0021	0000	**.9009**	**.0895**		**.00005**	4049
	(.393)	(.586)	(.069)	(.038)	(.111)	(.611)	(.000)	(.000)		(.049)	
Model 1 GARCH	.0005	.0311	−.0025	**.0026**	.0022	0000	**.9001**	**.0901**		**.00005**	4049
	(.547)	(.982)	(.065)	(.037)	(.104)	(.580)	(.000)	(.000)		(.041)	
Model 1 LOG(GARCH)	−.0062	−.0008	−.0024	**.0025**	.0021	0000	**.9010**	**.0896**		.00005	4049
	(.374)	(.324)	(.074)	(.039)	(.118)	(.702)	(.000)	(.000)		(.068)	
Model 2	.0003		−.0025	**.0025**	.0023	0000	**.9017**	**.0620**	**.0488**	.00005	4054
	(.677)		(.064)	(.046)	(.092)	(.779)	(.000)	(.000)	(.001)	(.054)	
Model 2 SQRT(GARCH)	.0016	−.0693	−.0024	.0024	.0022	0000	**.8998**	**.0642**	**.0524**	.00004	4053
	(.303)	(.357)	(.070)	(.050)	(.101)	(.988)	(.000)	(.000)	(.000)	(.117)	
Model 2 GARCH	Not converged till the 500th iteration										
Model 2 LOG(GARCH)	−.0084	−.0011	−.0024	.0024	.0022	0000	**.9004**	**.0642**	**.0525**	.00003	4054
	(.227)	(.203)	(.075)	(.051)	(.104)	(.901)	(.000)	(.000)	(.000)	(.164)	

Model 1: $r_t = c + \lambda f(\sigma_t^2) + \gamma_1 MONDAY + \gamma_2 THURSDAY + \gamma_3 FRIDAY + \varepsilon_t$

$$\sigma_t^2 = c_0 + \beta_1 \sigma_{t-1}^2 + \alpha_1 \varepsilon_{t-1}^2 + \delta_1 MONDAY$$

Model 2: $r_t = c + \lambda f(\sigma_t^2) + \gamma_1 MONDAY + \gamma_2 THURSDAY + \gamma_3 FRIDAY + \varepsilon_t$

$$\sigma_t^2 = c_0 + \beta_1 \sigma_{t-1}^2 + \alpha_1 \varepsilon_{t-1}^2 + \alpha_2 \varepsilon_{t-1}^2 I_{t-1} + \delta_1 MONDAY$$

Notes: Models 1 and 2 follow Equations 26.3 and 26.4, respectively. γ_i (δ_i) is the coefficient for seasonal dummy in returns (volatility). SQRT(GARCH), GARCH, and LOG(GARCH) show how volatility function f is defined: by σ_t, σ_t^2, and In (σ_t^2), respectively. The numbers in parentheses show the p-values. Coefficients significant at 5% (1%) are shown in bold (italic bold).

TABLE 26.5 Parameter Estimates of GARCH(1,1)-M with the Day-of-the-Week Effect for US$/TRY

Model	c	λ	γ_1	γ_2	γ_3	c_0	β_1	α_1	α_2	Log-Likelihood
Model	-.0011		.0008	.0018	.0059	.00001	.2653	1.9209		5516
	(.000)		(.018)	(.000)	(.000)	(.000)	(.000)	(.000)		
Model 1 SQRT(GARCH)	-.0036	.2567	.0008	.0020	.0059	.00001	.2687	1.7609		5543
	(.000)	(.000)	(.029)	(.000)	(.000)	(.000)	(.000)	(.000)		
Model 1 GARCH	-.0012	.5067	.0008	.0018	.0059	.00001	.2677	1.8977		5516
	(.000)	(.443)	(.016)	(.000)	(.000)	(.000)	(.000)	(.000)		
Model 1 LOG(GARCH)	.0178	.0019	.0000	.0012	.00568	.00001	.2490	1.6357		5556
	(.000)	(.000)	(.962)	(.001)	(.000)	(.000)	(.000)	(.000)		
Model 2	-.0017		.0004	.00135	.00603	.00003	.08304	.8886	2.6675	5534
	(.000)		(.290)	(.000)	(.000)	(.000)	(.000)	(.000)	(.000)	
Model 2 SQRT(GARCH)	-.0006	-.1041	.0000	.00085	.00564	.00003	.0583	.6907	3.4725	5536
	(.134)	(.002)	(.866)	(.025)	(.000)	(.000)	(.000)	(.000)	(.000)	
Model 2 GARCH	Not converged till the 500th iteration									
Model 2 LOG(GARCH)	Not converged till the 500th iteration									

Notes: Model 1: $r_t = c + \lambda f(\sigma_t^2) + \gamma_1 MONDAY + \gamma_2 TUESDAY + \gamma_3 THURSDAY + \varepsilon_t$

$$\sigma_t^2 = c_0 + \beta_1 \sigma_{t-1}^2 + \alpha_1 \varepsilon_{t-1}^2$$

Model 2: $r_t = c + \lambda f(\sigma_t^2) + \gamma_1 MONDAY + \gamma_2 TUESDAY + \gamma_3 THURSDAY + \varepsilon_t$

$$\sigma_t^2 = c_0 + \beta_1 \sigma_{t-1}^2 + \alpha_1 \varepsilon_{t-1}^2 + \alpha_2 \varepsilon_{t-1}^2 I_{t-1}$$

For other explanations, see the notes in Table 26.4.

TABLE 26.6 Parameter Estimates of GARCH(1,1)-M with the Month-of-the-Year Effect for ISE100

	c	λ	c_0	β_1	α_1	α_2	δ_1	δ_2	Log likelihood
Model 1	.0012		8.91E-06	.8880	.0999		2.60E-05	-1.09E-05	4049
	(0.011)		(.000)	(.000)	(.000)		(.000)	(.019)	
Model 1 SQRT(GARCH)	.0028	-.0803	8.97E-06	.8870	.1016		2.62E-05	-1.19E-05	4048
	(.042)	(.243)	(.000)	(.000)	(.000)		(.000)	(.007)	
Model 1 GARCH	Not converged till the 500th iteration								
Model 1 LOG(GARCH)	-.0105	-.0014	9.20E-06	.8868	.1015		2.60E-05	-1.36E-05	4050
	(.052)	(.027)	(.000)	(.000)	(.000)		(.000)	(.001)	
Model 2	.0009		9.41E-06	.8881	.0729	.0503	2.40E-05	-9.79E-06	4053
	(.052)		(.000)	(.000)	(.000)	(.001)	(.000)	(.039)	
Model 2 SQRT(GARCH)	.0029	-.1010	9.14E-06	.8877	.0737	.0534	2.42E-05	-1.10E-05	4052
	(.031)	(.137)	(.000)	(.000)	(.000)	(.001)	(.000)	(.013)	
Model 2 GARCH	Not converged till the 500th iteration								
Model 2 LOG(GARCH)	-.0115	-.0015	9.21E-06	.8882	.0739	.0529	2.40E-05	-1.27E-05	4051
	(0.033)	(0.019)	(.000)	(.000)	(.000)	(.001)	(.000)	(.002)	

Notes: Model 1: $r_t = c + \lambda f(\sigma_t^2) + \varepsilon_t$

$$\sigma_t^2 = c_0 + \beta_1 \sigma_{t-1}^2 + \alpha_1 \varepsilon_{t-1}^2 + \delta_1 FEBRUARY + \delta_2 DECEMBER$$

Model 2: $r_t = c + \lambda f(\sigma_t^2) + \varepsilon_t$

$$\sigma_t^2 = c_0 + \beta_1 \sigma_{t-1}^2 + \alpha_1 \varepsilon_{t-1}^2 + \alpha_2 \varepsilon_{t-1}^2 I_{t-1} + \delta_1 FEBRUARY + \delta_2 DECEMBER$$

For other explanations, see the notes in Table 26.4.

TABLE 26.7 Parameter Estimates of GARCH(1,1)-M with the Month-of-the-Year Effect for US$/TRY

	c	λ	c_0	β_1	α_1	α_2	δ_1	Log-Likelihood
Model	-4.98E-05		1.56E-06	.7885	.2011		.00014	5924
	(.731)		(.000)	(.000)	(.000)		(.000)	
Model 1 SQRT(GARCH)	-.000110	.0079	1.64E-06	.7810	.2091		.00015	5921
	(.763)	(.896)	(.000)	(.000)	(.000)		(.000)	
Model 1 GARCH	-8.96E-05	.6152	1.64E-06	.7811	.2087		.00015	5921
	(.593)	(.800)	(.000)	(.000)	(.000)		(.000)	
Model 1 LOG(GARCH)	-7.97E-05	-.0008	1.66E-06	.7795	.2108		.00015	5921
	(.693)	(.710)	(.000)	(.000)	(.000)		(.000)	
Model 2	4.19E-05		2.01E-06	.7750	.2643	-.1208	.00015	5930
	(.787)		(.000)	(.000)	(.000)	(.000)	(.000)	
Model 2 SQRT(GARCH)	Not converged till the 500th iteration							
Model 2 GARCH	-2.26E-05	1.3145	2.10E-06	.7686	.2758	-.1308	.00016	5928
	(.587)	(.900)	(.000)	(.000)	(.000)	(.000)	(.000)	
Model 2 LOG(GARCH)	.0008	7.80E-05	2.10E-06	.7676	.2760	-.1290	.00016	5927
	(.726)	(.733)	(.000)	(.000)	(.000)	(.000)	(.000)	

Notes: Model 1: $r_t = c + \lambda f(\sigma_t^2) + \varepsilon_t$

$$\sigma_t^2 = c_0 + \beta_1 \sigma_{t-1}^2 + \alpha_1 \varepsilon_{t-1}^2 + \delta_1 FEBRUARY$$

Model 2: $r_t = c + \lambda f(\sigma_t^2) + \varepsilon_t$

$$\sigma_t^2 = c_0 + \beta_1 \sigma_{t-1}^2 + \alpha_1 \varepsilon_{t-1}^2 + \alpha_2 \varepsilon_{t-1}^2 I_{t-1} + \delta_1 FEBRUARY$$

For other explanations, see the notes in Table 26.4.

The interpretation of the parameters for US$/TRY in Table 26.5 is analogous. Contrary to ISE100, c is significantly negative, which means that, if we ignore other effects, US$/TRY slightly decreases (TRY appreciates) day by day. The results about λ are diverse. λ is significantly positive in the SQRT(GARCH) and the LOG(GARCH) versions of Model 1, while significantly negative in the SQRT(GARCH) version of Model 2 and insignificant for the GARCH version of Model 1.[*] So, there seems to be a negative relation between volatility and returns, but we are unable to generalize the findings with certainty. γ_2 and γ_3 are positive and quite significant, while γ_1 is positive and partially significant at 5%. These indicate strong Tuesday and Thursday effects and a weak Monday effect in US$/TRY returns. Unlike for the ISE100 analysis above, c_0 is significantly positive. This shows that the FX market has a certain intrinsic volatility. The signs of β_1, α_1, and α_2 are positive, like the ones for ISE100. However, an interesting remark is that both α_1 and α_2 are larger than in the ISE100 analysis, while β_1 is smaller. This means there is a weaker autocorrelation in exchange rate volatility, but the impact of innovation is larger.

Table 26.6 presents the results of the regressions, including the month-of-the-year effect for ISE100. According to this table, the results about c are quite heterogeneous across the models. For instance, c is positive for the SQRT(GARCH) model and negative for the LOG(GARCH) model at nearly a 5% level of significance. λ is negative and partially significant at 5%. This is weak evidence of a negative relation between volatility and returns. c_0, β_1, α_1, α_2, and δ_1 are all positive and significant at 1%, and δ_2 is negative and significant at 5%. Positive c_0 signifies that there is a certain level of daily volatility in the stock market. Similarly to the results in Table 26.4, volatility clustering is high with a β_1 of nearly 0.89. Both α_1 and α_2 are significantly positive, implying that innovation has an impact on volatility and that negative returns (bad news) have a larger impact than positive returns (good news). Significantly positive δ_1 (negative δ_2) indicates that daily ISE100 volatility is higher in February (lower in December) than in other months.

Finally, Table 26.7 gives the parameter estimates of the same regressions, but for US$/TRY. In what follows, c is slightly negative but insignificant. λ, too, is insignificant, so that we cannot find a relation between volatility and returns. However, all the remaining coefficients except α_2 are positive and significant at 1%, providing results similar to the previous

---

[*] In fact, we cannot rely on the results provided by Model 1 since, technically speaking, the model requires $\beta_1 + \alpha_1 < 1$. However, this is not satisfied in the first four rows of Table 26.5.

TABLE 26.8 Summary Results

	c	λ	γ_1	γ_2	γ_3	c_0	β_1	α_1	α_2	δ_1	δ_2
ISE100 D-O-W-E	IC	IS	− S-10%	+ S-5%	+ IS	+ IS	+ S-1%	+ S-1%	+ S-1%	+ PS-10%	
US$/TRY D-O-W-E	− S-1%	IC	+ PS-5%	+ S-1%	+ S-1%	+ S-1%	+ S-1%	+ S-1%	+ S-1%		
ISE100 M-O-Y-E	IC	− PS-5%				+ S-1%	+ S-1%	+ S-1%	+ S-1%	+ S-1%	− S-5%
US$/TRY M-O-Y-E	− IS	+ IS				+ S-1%	+ S-1%	+ S-1%	− S-1%	+ S-1%	

Notes: The first (last) two rows give the estimates of the regressions with the day-of-the-week effect (D-O-W-E) (month-of-the-year effect, M-O-Y-E); γ_i and δ_i refer to different days or months in each model; $+/-$ indicate the sign of the coefficient; and S, PS, IS, and IC indicate significant, partially significant, insignificant, and inconclusive coefficients with percentages showing the significance level, respectively.

ones. For instance, positive c_0 reveals that a certain level of intrinsic volatility exists; a β_1 of nearly 0.8 indicates a high autocorrelation in volatility; a positive α_1 shows innovation has an impact on volatility; and a positive δ_1 implies US\$/TRY volatility is higher in February than in other months. By contrast, α_2 is significantly negative. This means an innovation due to negative returns (appreciation of TRY) has a lower impact on volatility than an innovation due to positive returns (depreciation of TRY).

Based on the findings above, we can summarize and generalize our results, as given in Table 26.8, in the following way:[*]

c: Somewhat negative daily returns in US\$/TRY, inconclusive for ISE100.

λ: Mostly negative for ISE100, although the evidence is quite weak and insignificant for US\$/TRY.

γ: There are positive Thursday returns in ISE100 and positive Tuesday and Thursday returns in US\$/TRY.

c_0: There is a given level of volatility in the market.

β_1: There is a strong positive autocorrelation in volatility in both markets, i.e., volatility persists.

α_1: There is a strong heteroskedasticity in both ISE100 and US\$/TRY returns.

α_2: Bad news affects volatility more than good news for ISE100.

δ: There is strong evidence of higher volatility in February for both ISE100 and US\$/TRY and lower volatility in December for ISE100.

26.4 CONCLUSION

Based on daily data, we analyze in this paper the day-of-the-week and month-of-the-year effects in Turkish financial markets, more specifically in ISE100, the main index in Istanbul Stock Exchange and in US\$/TRY exchange rate, in both mean and conditional variance. If these effects are well understood, they can help investors in deciding the timing of their investments[†] or in developing appropriate pricing methods. Our main

---

[*] For simplicity, the parameter for all the seasonal dummies in returns (conditional volatility) is indicated by $\gamma(\delta)$.

[†] Certainly, one should make several other analyses like the one on transaction costs and liquidity conditions before such a decision.

focus, however, is to investigate, with a GARCH-M (1,1) model and various versions of volatility function (logarithmic, exponential, etc.), whether there is a relation between volatility and returns after filtering from seasonal effects rather than explaining the reasons for seasonality. To the best of our knowledge, this is the first ever study of this kind on Turkish financial markets, in particular, to filter from seasonality both mean and conditional volatility.

Our results can be summarized as follows: We find significantly negative returns on Monday and positive returns on Thursday and Friday for ISE100 and, to a lesser extent (at 10%), negative returns on Monday and positive returns on Tuesday and Thursday for US$/TRY. On the other hand, we cannot find any significant difference in daily returns across the months of the year. However, volatility is particularly high in February for both ISE100 and US$/TRY and low in August for ISE100. Interestingly, no January effect, as mentioned in the literature, was detected. With the GARCH specification, however, we can only confirm higher Thursday returns for ISE100 and higher Tuesday and Thursday returns for US$/TRY. In addition, higher February volatility is obvious for both ISE100 and US$/TRY and lower December volatility is detected for ISE100.

A relation between volatility (i.e., conditional variance) and returns after controlling for day-of-the-week or month-of-the-year effects is not obvious. We can only partially find some evidence of a negative relation in ISE100.

Secondary but more evident findings include a high level of volatility as well as volatility clustering in both ISE100 and US$/TRY; the heteroskedastic nature of the error terms; and asymmetry in positive and negative shocks. More precisely, the latter indicates negative shocks in ISE100 have a higher impact on volatility than positive shocks. This asymmetry is in line with previous findings in the literature and shows the market is more sensitive to risky situations. This idea works for US$/TRY; i.e., negative shocks in US$/TRY (in cases where TRY depreciates vis-à-vis US$)[*] have a higher impact on volatility only in the specification with monthly dummies.

---

[*] This is a little tricky because a negative shock in US$/TRY return normally is a downward movement in the exchange rate, i.e., an appreciation of TRY vis-à-vis US$. However, for most market participants, a negative shock in the exchange rate designates a depreciation of TRY. That is why here we define a negative shock as a situation where US$/TRY suddenly rises. This distinction is important for the interpretation of the results concerning α_2.

Our results partially confirm previous studies on the Istanbul Stock Exchange in that we find negative returns on Monday and positive returns on Thursday and Friday in ISE100. Nevertheless, we cannot find any significant January effect. This may be due to the newer data set we use. Alternatively, one can argue that investors have learned about these calendar anomalies and these effects have been disappearing through time.

This research can be extended, first by adding a third major instrument, namely, a fixed-income indicator such as a bond market index. Second, these three can be linked to each other in a vector autoregression (VAR) setup. These constitute potential research topics.

REFERENCES

Aggarwal, R., and Rivoli, P. (1989). Seasonal and day-of-the-week effects in four emerging stock markets. *Financial Review* 24:541–50.

Ariel, R. A. (1987). A monthly effect in stock returns. *Journal of Financial Economics* 18:161–74.

Baillie, R. T., and DeGennaro, R. P. (1990). Stock returns and volatility. *Journal of Financial and Quantitative Analysis* 25:203–14.

Balaban, E., Bayar, A., and Kan, Ö. B. (2001). Stock returns, seasonality and asymmetric conditional volatility in world equity markets. *Applied Economics Letter* 8:263–68.

Bildik, R. (2004). Are calendar anomalies still alive? Evidence from Istanbul Stock Exchange. Available at SSRN: http://ssrn.com/abstract=598904.

Corhay, A., and Rad, A. T. (1994). Expected returns and volatility in European stock markets. *International Review of Economics and Finance* 3:45–56.

Cross, F. (1973). The behaviour of stock prices on Fridays and Mondays. *Financial Analysts Journal* 29:67–69.

Demirer, R., and Karan, M. B. (2002). An investigation of the day-of-the-week effect on stock returns in Turkey. *Emerging Markets Finance and Trade* 38:47–77.

French, K. R. (1980). Stock returns and weekend effect. *Journal of Financial Economics* 8:55–70.

French, K. R., Schwert, G. W., and Stambaugh, R. (1987). Expected stock returns and volatility. *Journal of Financial Economics* 19:3–29.

Gibbons, M. R., and Hess, P. (1981). Day of the week effects and assets returns. *Journal of Business* 54:579–96.

Glosten, L. R., Jagannathan, R., and Runkle, D. E. (1993). On the relation between the expected value and volatility of the nominal excess return on stocks. *Journal of Finance* 48:1779–801.

Gültekin, M. N., and Gültekin, N. B. (1983). Stock market seasonality: International evidence. *Journal of Financial Economics* 12:469–81.

Jaffe, J., and Westerfield, R. (1985). The week-end effect in common stock returns: The international evidence. *Journal of Finance* 40:453–61.

Jones, C. P., Pearce, D. K., and Wilson, J. W. (1987). Can tax-loss selling explain the January effect? A note. *Journal of Finance* 42:453–61.

Karan, M. B., and Uygur, A. (2001). İstanbul Menkul Kıymetler Borsası'nda Haftanın Günleri ve Ocak Ayı Etkilerinin Firma Büyüklüğü Açısından Değerlendirilmesi. *Ankara Üniversitesi SBF Dergisi* 56:103–15.

Kato, K., and Schallheim, J. S. (1985). Seasonal and size anomalies in the Japanese stock market. *Journal of Financial and Quantitative Analysis* 20:107–18.

Keim, D. B. (1983). Size related anomalies and stock return seasonalities. *Journal of Financial Economics* 12:13–22.

Keim, D. B., and Stambaugh, R. F. (1984). A further investigation of the weekend effect in stock returns. *Journal of Finance* 39:819–83.

Lakonishok, J., and Levi, M. (1982). Weekend effects on stock returns: A note. *Journal of Finance* 37:883–89.

Lakonishok, J., and Smidt, S. (1984). Volume and turn-of-the-year behavior. *Journal of Financial Economics* 13:435–55.

Lakonishok, J., and Smidt, S. (1988). Are seasonal anomalies real? A ninety-year perspective. *Review of Financial Studies* 1:403–25.

Nelson, D. B. (1991). Conditional heteroscedasticty in assets returns: A new approach. *Econometrica* 59:347–70.

Oğuzsoy, C. B., and Güven, S. (2003). Stock returns and the day-of-the-week effect in Istanbul Stock Exchange. *Applied Economics* 35:959–71.

Seyyed, F. J., Abraham, A., and Al-Haji, M. (2005). Seasonality in stock returns and volatility: The Ramadan effect. *Research in International Business and Finance* 19:374–83.

Theodossiou, P., and Lee, U. (1995). Relationship between volatility and expected returns across international stock markets. *Journal of Business Finance and Accounting* 22:289–300.

CHAPTER 27

Are Macroeconomic Variables Important for the Stock Market Volatility? Evidence from the Istanbul Stock Exchange

M. Nihat Solakoglu, Nazmi Demir, and Mehmet Orhan

CONTENTS

27.1   INTRODUCTION

The volatility of stock market returns is important for many reasons. For a risk-averse investor, an increase in the return volatility indicates a riskier environment with a possible consequence of lower investment. This change, as discussed by Schwert (1989), should influence capital investment decisions, consumption decisions, and even some other business cycle variables. There is empirical evidence suggesting that the stock market volatility increases at downturns of business cycles (for instance, see Errunza and Hogan, 1998; Hamilton and Lin, 1996). Moreover, the source of the return volatility reveals whether the changes in volatility are caused by temporary factors (perhaps, because of irrational investor behavior) or by the underlying fundamental factors. If the large part of return volatility is explained by these fundamentals, this should lower investor concerns on irrational behavior (Liljeblom and Stenius, 1997; Binder and Merges, 2001).[*]

Nevertheless, there have not been many studies investigating the relationship between stock market volatility and underlying fundamentals. While Schwert (1989) finds weak evidence for the relationship between stock market volatility and macroeconomic variables,[†] Liljeblom and Stenius (1997) report a statistically significant impact of several macroeconomic variables on stock market volatility, though only a small percentage of the volatility is reported to be explained by macroeconomic fundamentals. On the other hand, Abugri (2008), for Latin American markets, shows that stock market volatility is influenced consistently and significantly by global factors, but the impact of local factors is not consistent among markets.

For the Australian market, Kearney and Daly (1998) conclude that return volatility is influenced by macroeconomic variables, and in particular volatility of inflation and interest rates becomes important. Similarly, Binder and Merges (2001) show that economic factors explain a significant percentage of the variation in stock market volatility. In an intertemporal asset pricing model, Rodriguez et al. (2002) show that observed volatility of asset returns can be predicted by real fundamentals. For a cross section of countries, Diebold and Yılmaz (2007) find a positive relationship between stock market volatility and volatility of fundamental variables. In addition, Pierdzioch et al. (2008) find a link between business cycle variables and stock market volatility. Moreover, different from other mentioned studies, they show that this link is independent of whether revised or real-time macroeconomic data are utilized.

---

[*] It can be shown theoretically that volatility of real economic factors impacts stock market volatility through their effect on stock prices, which reflects claims on future profits of the corporations. In a discounted present value model, a change in the volatility of real economic factors will cause a change in the discount rate and the volatility of future expected cash flows (see Schwert, 1989; Liljeblom and Stenius, 1997).

[†] However, he finds stronger evidence when the causal relationship runs from financial volatility to macroeconomic volatility.

This study also investigates the relationship between stock market volatility and macroeconomic variables using a low-frequency monthly data set for an emerging market, Turkey. The Turkish capital markets have, to some extent, been harmonized with the EU acquis as well as other international standards, through the introduction of new regulations, intended to restore confidence particularly for financial investors* (SPO, 2006). Since several studies find that business cycles impact stock market volatility, we try to identify breaks or volatility shifts in data. However, rather than using our own judgment or experience, where both are subjective, we prefer to employ the ICSS algorithm introduced by Inclan and Tiao (1994). It should be kept in mind that the Turkish capital market is still at its early years, with characteristic low saving ratios, less developed capital market culture, and limited variety of financial instruments. However, its integration with international markets seems to have improved considerably judging from the fact that the share of foreign investors in the ISE has exceeded two-thirds of the total market value of all stocks traded. Hence, to take into account the effect of foreign investors' behavior, we also consider the volatility of the S&P500 index.

The rest of the chapter is organized as follows: In Section 27.2, we discuss stock market volatility and volatility estimation along with data. The ICSS algoritm and our estimation methodology will be discussed in Section 27.3. Results will be presented in Section 27.4. Our concluding remarks and suggestions for further research are left for the last section.

27.2 VOLATILITY ESTIMATION

Although we investigate the link between return volatility and macroeconomic variables, the return volatility is not directly observable, and hence it needs to be proxied by alternative measures. A common measure used in this particular literature is the standard deviation of returns in a certain time period. Some other measures are the sum of the squared daily returns, a measure based on GARCH(1,1), and a measure based on daily high and low prices (for example, see Schwert, 1989; Binder and Merges, 2001; Liljeblom and Stenius, 1997).

---

* The new legislations intended to bring about investors' confidence consisted of the establishment of the Investors' Protection Fund, the transition to the registry system in stocks and bonds, the establishment of Turkish Derivative Exchange (TurkDEX), and the implementation of International Financial Reporting Standards all for transparency and corporate governance whose stocks are traded in the Istanbul Stock Exchange.

TABLE 27.1   Measures of Exchange Rate Risk

Formula	Measure	Acronym
(1) $h_t = \left\{ \left[\sum_{j=t-12}^{t} \left(r_j - \mu^b \right)^2 \right] / 11 \right\}$	Rolling variance of monthly returns (12 months)	V1
(2) $h_t = \mu + \alpha\, e_{t-1}^2 + \beta\, h_{t-1}$	GARCH(1,1)	V2
(3) $h_t = \left(\sum_{j=1}^{t} w_{jt}\, r_j^2 \right) / (t-1)$	Recursive variance estimate with $w_t = 1, \forall t$	V3
(4) $h_t = \sum_{j=1}^{T} w_{jt}\, r_j^2$	Nonparametric estimate of conditional variance	V4
(5) $h_t = (1-\lambda) \sum_{j=0}^{\infty} \lambda^{j-1}(r_{t-j} - \mu)^2$	Exponentially weighted moving average ($\lambda = 0.94$)	V5
(6) $h_t = \dfrac{1}{T} \sum_{t=1}^{T} \min((r_t - \mu), 0)^2$	Downside risk (semivariance)	V6

Note: The variable r represents monthly return, h represents the volatility estimate, and μ is the mean of the return series.

To test the robustness of the findings, some studies employ several measures of volatility (e.g., Binder and Merges, 2001; Liljeblom and Stenius, 1997). We follow a similar approach and select several volatility measures. In choosing the measures, we depend on this particular literature as well as the literature investigating the predictive power of alternative volatility measures (see West and Cho, 1995; Jorion, 1995; Pagan and Schwert, 1990; Brooks, 2002). Table 27.1 presents the measures of volatility used in this study. The market portfolio we select is the ISE100 index, which is a value-weighted index compromising 100 companies listed at the Istanbul Stock Exchange. The source of the data is the official web page of the Istanbul Stock Exchange.[*]

First and second measures are commonly known and do not require explanations.[†] For the other measures, though, a brief explanation could be useful. The third measure, a recursive variance estimate, gives the sample variance at time t. In calculating the sample variance, the starting date does not move together with the end date. In other words, the sample window does not move as t moves, as in moving average measures. As a

---

[*] http://www.imkb.gov.tr.
[†] ARCH-GARCH type measures are introduced by Engle (1982) and Bollerslev (1986).

result, this measure estimates the unconditional variance at each time t, but when observed for the entire sample, it is a measure of conditional variance at time t.

For volatility measure 4 (V4) in Table 27.1, we used a nonparametric estimator as defined by the following equation:

$$\hat{\sigma}_t^2 = \sum_{j=1}^{T} w_{jt}\, r_j^2, \quad \sum_{j=1}^{T} w_{jt} = 1$$

The weights, w_{jt}, are chosen to depend on information sets I_t and I_j in such a way that, if I_t and I_j are far apart, w_{jt} is close to zero. This makes the estimate equivalent to the sample variance of $\hat{r}_t^2$ using only the observations that are close to I_t. The weights are calculated by the following formula:

$$w_{jt} = K_{jt} / \sum_{k=1}^{T} K_{kt} \quad \text{and} \quad K_{jt} = (2\pi b^2)^{-0.5} \exp\!\left(-0.5(r_t - r_j)^2 / b^2\right)$$

where K is the Gaussian kernel and has the properties that it is nonzero, integrates to unity, and is symmetric. The bandwidths, b, are set to $\hat{\sigma}\, T^{-1/5}$, where $\hat{\sigma}$ is the sample standard deviation of r_t. As in West and Cho (1995) and Pagan and Schwert (1990), we did not try any other kernel or experiment with different bandwidths and weighting schemes.

The fifth measure is the exponentially weighted moving average measure with the decay factor set equal to 0.94, as recommended by RiskMetrics (Brooks, 2002). The last measure of volatility is concerned only with a decline in the value of a portfolio for an investor. Hence, the relevant measure of risk, one can argue, should be the downside risk. Hence, as a last measure, a semivariance estimate is included to measure the downside risk.[*]

We estimated all measures of volatility between January 1990 and June 2007. For all measures, the starting value is set equal to 0.1 for January 1990. Table 27.2 provides some descriptive statistics and the bivariate correlations among these six measures.

As it is clear from the table, the measures of volatility all behave as if they have a different data generating process. For instance, volatility measures V1 and V2 (rolling variance estimate and variance prediction from a GARCH(1,1) process) have a negative correlation. Moreover, the highest

---

[*] For an application of downside risk, see Ang et al. (2006).

TABLE 27.2   Descriptive Statistics and Bivariate Correlations

	V1	V2	V3	V4	V5	V6
Mean	0.05447	0.15354	0.07058	0.00048	0.11890	7.50801
Median	0.04519	0.15952	0.06590	0.00000	0.11747	7.91624
Max	0.14932	0.17040	0.12160	0.10000	0.29299	14.26745
Min	0.00416	0.01204	0.05433	0.00000	0.02482	0.05353
Stdev	0.03471	0.01993	0.01430	0.00690	0.04940	2.44170
Skewness	0.83795	−3.61987	1.54956	14.38750	0.61189	−1.08639
Kurtosis	3.05936	20.85892	4.75960	208.00170	3.91714	5.09260
Jargue-Bera	24.6063	3249.3550	374970.0	111.1311	20.4644	79.6248
Probability	0.00001	0.00000	0.00000	0.00000	0.00004	0.00000

Bivariate Correlations of Volatility Measures

	V1	V2	V3	V4	V5	V6
V1	1.00000	−0.55624	0.54960	0.09173	0.61419	−0.13059
V2		1.00000	−0.35263	−0.18751	−0.47792	0.10474
V3			1.00000	0.14392	0.06178	−0.11777
V4				1.00000	−0.02670	−0.21129
V5					1.00000	0.20662
V6						1.00000

positive correlation is 0.614 between V1 and V5, which is far below one expects. Hence, it should not be surprising to observe mixed findings in the related literature. In addition, along with the effect of the volatility measure selected, the country level factors will also be important for the results, as shown by Abugri (2008).

27.3   MODEL SPECIFICATION AND IMPLEMENTATION

Our analysis is performed using monthly data between January 1990 and June 2007. The source for ISE100, as indicated earlier, is Istanbul Stock Exchange. For the macroeconomic variables, we used IMF's International Financial Statistics, Central Bank of Turkey, and Statistical Institute of Turkey.[*] The S&P500 monthly index values are obtained from yahoo.com/finance web page. Following the literature on this area, we select the following macroeconomic variables: industrial production index, M2 definition of money supply, interest rates as measured by the weighted 3-month time deposit rates, exchange rates as defined by the price of the U.S. dollar

---

[*] The web addresses are www.tcmb.gov.tr and www.tuik.gov.tr for the Central Bank and Statistical Institute of Turkey, respectively.

in terms of local currency, and terms of trade as measured by the ratio of unit price index for exports and imports. The S&P500 index is chosen to represent the global factors' impact on ISE100 to capture the change in foreign (and perhaps the domestic) investor behavior.

During the analysis window, both the Turkish economy and the rest of the world have gone through several important economic and financial events, such as the Asian crisis in 1997,[*] the burst of the stock bubble in the United States in 2000, the September 11 terrorist attacks, and particularly for Turkey, the severe financial crisis in the 2000–2001 period, with many banks going into bankruptcy or taken over by the Saving Deposits Insurance Fund. Therefore, we believe the size and the significance of the association between stock market volatility and macro factors can be influenced by the structural breaks. As mentioned earlier, we use the ICSS algorithm introduced by Inclan and Tiao (1994) to detect multiple break-points in a time series by testing for volatility shifts.

To briefly explain the algorithm, lets assume ε_t is the series in question with zero mean and σ^2_t as the unconditional variance. Let us define cumulative sum of squares between time 1 and k as

$$C_k = \sum_{t=1}^{k} \varepsilon_t^2, \text{ where } t=1,\dots,T \text{ and } k=1,\dots,T$$

The centered and normalized cumulative sum of squares until time k is represented by the D_k statistics:

$$D_k = \frac{C_k}{C_T} - \frac{k}{T}, \text{ with } D_0 = D_T = 0$$

If there is no volatility shift in the series, the plot of D_k against k will oscillate around zero. On the other hand, with a volatility shift, we will observe D_k statistics drifting away from zero. The asymptotic critical value of 1.358 can be used to create boundaries to identify the point in time with a volatility shift.[†] Figure 27.1 presents D_k statistics and identified breaks using market portfolio return from the Istanbul Stock Exchange.

---

[*] Also known as the IMF crises.

[†] Critical values are calculated from the distribution of D_k under the the null hypothesis of homogeneous variance. One can use the critical values to obtain upper and lower boundaries to detect volatility shifts. For details on the ICSS algorithm and some uses, see Inclan and Tiao (1994), Ewing and Malik (2005), and Marcelo et al. (2008).

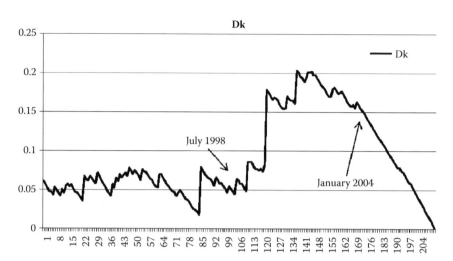

FIGURE 27.1   The D_k statistics and breaks.

The ICSS algorithm identifies two shifts in volatility between January 1990 and June 2007. The first break is determined as July 1998, whereas the second break is determined as January 2004. Hence, our sample period is divided into three subperiods. The second subperiod, between July 1998 and January 2004, represents a volatile environment with several local and global crises influencing the Istanbul Stock Exchange and Turkish economy. The last subperiod, on the other hand, corresponds to a stable environment. The first subperiod is not as stable as the last subperiod, but it is also not as volatile as the second subperiod.

27.4   ESTIMATION RESULTS AND DISCUSSION

We estimated the following model to evaluate the effect of macro variables on stock market volatility. Although we utilize several measures of volatility, we only consider monthly standard deviations of macro variables to estimate fundamental volatilities.[*] In calculating volatilities, monthly percentage changes are used. The relationship between stock market volatility and macro factors is investigated using the following equation:

$$SMV_t = \beta_0 + \beta_1 IR_t + \beta_2 M2_t + \beta_3 TOT_t + \beta_4 INF_t + \beta_4 ER_t$$
$$+ \beta_5 IP_t + \beta_7 SP500_t + \varepsilon_t$$

---

[*] Since we use generated variables as our dependent variables, this approach will lead to inefficient but still unbiased and consistent estimates.

TABLE 27.3   Coefficient Estimates: 1990:M01–2007:M06

Variables	Volatility Measures					
	V1	**V2**	**V3**	**V4**	**V5**	**V6**
Constant	0.0129	0.1617***	0.0535***	0.0030	0.0613***	3.9410***
	(.0083)	(.0072)	(.0045)	(.003)	(.0124)	(.9014)
IR	0.0016***	−0.0007***	−0.0002***	0.00001	0.0033***	−0.0074
	(.0003)	(.0002)	(.0001)	(.)	(.0003)	(.0123)
M2	−0.0007**	0.0002	−0.0008***	−0.0001	−0.0017***	−0.0375
	(.0003)	(.0002)	(.0002)	(.0001)	(.0004)	(.0277)
TOT	−0.0048**	0.0014	−0.0017	−0.0002	−0.0034	−0.3940*
	(.0023)	(.0016)	(.0012)	(.0002)	(.0029)	(.2264)
CPI	0.0182***	−0.0043**	0.0087***	0.0004	0.0149***	−0.2908
	(.0032)	(.0019)	(.0015)	(.0004)	(.0039)	(.3185)
US$	−0.0076***	0.0027***	−0.0019***	−0.0002	−0.0073***	0.2737***
	(.0011)	(.0007)	(.0004)	(.0002)	(.0014)	(.083)
IP	0.0016*	−0.0002	0.0025***	−0.0003	0.0015	0.4306***
	(.0009)	(.0007)	(.0005)	(.0003)	(.0013)	(.1007)
SP500	0.0099***	−0.0027***	0.0015***	−0.00003	0.0124***	0.5123***
	(.001)	(.0006)	(.0004)	(.0001)	(.0015)	(.0892)
F stat	38.74***	6.22***	19.93***	0.276	39.06***	9.78***
Adj R^2	0.5583	0.149	0.388	−0.0248	0.5604	0.2272

Note: Standard errors are provided in parentheses.

***, **, and * represent significance at 1%, 5%, and 10% levels, respectively.

where SMV is a measure of stock market volatility, IR is the interest rate, M2 is the M2 definition of money supply, TOT is a measure of the terms of trade calculated as the ratio of export unit price index to import unit price index, INF stands for inflation based on the consumer price index, ER is the exchange rate as measured by the price of US$ in terms of local currency, and IP is the industrial production index. We use SP500 as an additional variable to account for the impact of foreign investor behaviors on volatility.* Results for the full sample are presented in Table 27.3.†

F-statistics for all models, except for V4 (nonparametric measure of volatility), indicate that the regressions have overall significance. In other words, the stock market volatility is explained sufficiently by the variables

---

* As discussed in Abugri (2008), the selection of variables in the model is criticized for being subjective and it is unavoidable.

† For all estimations, we used White's heteroskedasticity consistent coefficient covariance matrix.

we use in the model. The adjusted R^2 ranges from 15% to 56% for the significant models. Although Table 27.3 reports significant coefficient estimates for macro factors, we should be cautious since the direction of the influence shifts from volatility measure to measure. For example, inflation volatility influences stock market volatility positively under volatility measure V1. On the other hand, under volatility measure V2, the influence is negative. This finding shows that the effect of inflation volatility on stock market volatility is fragile if we use the definition of robustness or fragility of Leamer (1985) and Leamer and Leonard (1983). For M2, TOT, and IP, the effect seems to be robust, independent of the volatility measure chosen. Can fragility be caused by the volatility shifts that exist in the full sample? To answer this question, we report the regression results for three subperiods separately in Tables 27.4 to 27.6.

The results presented in Table 27.4 cover the first subperiod, January 1990 to June 1998. The Turkish economy experienced one major crisis in

TABLE 27.4 Coefficient Estimates: 1990:M01–1998:M06

	Volatility Measures					
Variables	**V1**	**V2**	**V3**	**V4**	**V5**	**V6**
Constant	0.1204***	0.1059***	0.1148***	0.0236	0.0752***	−4.1348
	(.0212)	(.0307)	(.0117)	(.0228)	(.0148)	(2.4926)
IR	0.0051***	−0.0025	0.0023***	0.0013	−0.0051***	−0.5214***
	(.0015)	(.0019)	(.0007)	(.0013)	(.0013)	(.1436)
M2	−0.0183***	0.0061	−0.0138***	−0.0031	0.0139***	0.1115
	(.0038)	(.005)	(.0019)	(.003)	(.0033)	(.3233)
TOT	−0.0111***	0.0081**	−0.0081***	−0.0021	−0.0056*	1.2086***
	(.0035)	(.0033)	(.0014)	(.0021)	(.0032)	(.4047)
CPI	−0.0009	0.0094*	0.0061***	−0.0042	0.0009	2.3715**
	(.0054)	(.0055)	(.0022)	(.0042)	(.0043)	(.9472)
US$	0.0027	−0.0033	0.0017	0.0013	0.0018	−0.2814
	(.0019)	(.0021)	(.0013)	(.0013)	(.0015)	(.3172)
IP	−0.0042***	0.0018	0.0006	−0.0013	0.0026**	0.8161***
	(.0013)	(.0016)	(.0006)	(.0012)	(.0013)	(.1327)
SP500	0.0163***	−0.0061***	0.0059***	0.0017	0.0004	−0.4615**
	(.002)	(.002)	(.0008)	(.0016)	(.0016)	(.2304)
F stat	18.14***	3.52***	46.65***	1.40	5.28***	10.85***
Adj R^2	0.5429	0.1486	0.7598	0.0273	0.2288	0.4056

Note: Standard errors are provided in parentheses.
***, **, *and* * *represent significance at 1%, 5%, and 10% levels, respectively.*

TABLE 27.5   Coefficient Estimates: 1998:M07–2003:M12

Variables	Volatility Measures					
	V1	**V2**	**V3**	**V4**[a]	**V5**	**V6**
Constant	−0.0320	0.1171**	0.0616***	0.0123	0.0970	8.4205***
	(.05)	(.0468)	(.0038)	(.0066)	(.079)	(1.0068)
IR	0.0013*	−0.0008*	0.0001*	0.0001	0.0019**	0.0373**
	(.0008)	(.0005)	(.)	(.0001)	(.0008)	(.0167)
M2	−0.0083	−0.0044	−0.0008	−0.0011	−0.0027	−0.2856
	(.0074)	(.0071)	(.0005)	(.0009)	(.0111)	(.2497)
TOT	0.0132	0.0068	−0.0007	0.0018	0.0178	0.0858
	(.01)	(.0098)	(.0008)	(.0018)	(.0159)	(.2679)
CPI	0.0305***	0.0037	0.0020***	−0.0005	0.0124	−0.1694
	(.0079)	(.008)	(.0006)	(.0012)	(.0102)	(.181)
US$	−0.0083***	0.0022	0.0002	−0.0005	−0.0056**	0.0904*
	(.0024)	(.0018)	(.0001)	(.0002)	(.0025)	(.0526)
IP	0.0098**	0.0030	0.0006*	−0.0005	0.0085	−0.1397
	(.0039)	(.0027)	(.0003)	(.0006)	(.0062)	(.0959)
SP500	0.0062	0.0022	0.0000	0.0004	−0.0040	0.2789
	(.0048)	(.0049)	(.0004)	(.0007)	(.0088)	(.1747)
F stat	11.00***	1.48	13.61***	1.36	8.33***	9.79***
Adj R^2	0.5186	0.0492	0.5758	0.0371	0.4413	0.4864

Note: Standard errors are provided in parentheses.

***, **, and * *represent significance at 1%, 5%, and 10% levels, respectively.*

[a] Both coefficient estimates and standard errors reported are multiplied by 10,000 due to their small size.

1994 after continuous growth rates for almost a decade. Our first period sliced by the ICSS, January 1990 to June 1998, has overall significance again (even at $\alpha = 1\%$) except for volatility measure V4. The macro variables as a set are able to explain volatility sufficiently. Unfortunately, signs continue to shift, indicating fragility of the parameter estimates. The specification with V1 indicates that all macro variables except inflation and exchange rate are significant. Specification of V3 is similar in that this model has the same significant variables. The highest overall belongs to this model: 76% of the variation in volatility is explained by the macro covariates. The specification with the greatest number of significant variables is the regression of V6 on the regressors. The only insignificant variable according to this model is the exchange rate. The poorest model according to overall significance is the regression with the dependent variable of V4.

TABLE 27.6   Coefficient Estimates: 2004:M01–2007:M06

Variables	Volatility Measures					
	V1	**V2**	**V3**	**V4[a]**	**V5**	**V6**
Constant	−0.0617***	0.1931***	0.0384***	0.0056*	−0.1275***	4.6895***
	0.0098	0.0054	0.0028	0.0031	0.0195	0.4259
IR	0.0027***	−0.0009***	0.0005*	0.0004***	0.0034*	0.0711
	0.0004	0.0003	0.0003	0.0001	0.0019	0.0440
M2	0.0004***	−0.0001	0.0000	0.0000	0.0004	0.0056
	0.0001	0.0001	0.0000	0.0000	0.0003	0.0071
TOT	0.0081**	0.0002	0.0030**	0.0001	0.0225**	0.4916**
	0.0035	0.0023	0.0012	0.0012	0.0082	0.1904
CPI	0.0236**	−0.0238***	0.0066**	−0.0006	0.0711***	1.0628***
	0.0091	0.0048	0.0025	0.0029	0.0173	0.3828
US$	−0.0026***	0.0007*	−0.0009***	−0.0005***	−0.0060**	−0.1444**
	0.0007	0.0004	0.0003	0.0002	0.0023	0.0540
IP	0.0034***	−0.0004	0.0004	−0.0001	0.0051***	0.0563
	0.0006	0.0004	0.0002	0.0002	0.0014	0.0347
SP500	0.0136***	−0.0033**	0.0053***	0.0011*	0.0395***	0.8474***
	0.0025	0.0014	0.0008	0.0006	0.0055	0.1266
F stat	23.09***	9.14***	20.89***	2.48**	25.11***	22.16***
Adj R²	0.7904	0.5814	0.7725	0.2015	0.8046	0.7832

Note: Standard errors are provided in parentheses.

***, **, and * represent significance at 1%, 5%, and 10% levels, respectively.

[a] Both coefficient estimates and standard errors reported are multiplied by 10,000 due to their small size.

Table 27.5 reports the regression results for the second subperiod with global and local crises affecting the stock market. The local crisis of the 2000–2001 period decreased the Turkish GDP from \$204 billion to \$148 billion in 2001. This was the greatest crisis of the last half decade. Inflation and exchange rate are the two variables that appeared the most significant in this period. Interestingly, S&P500 does not play any important role as identified by insignificant coefficients under all volatility measures. In addition, compared to Tables 27.3 and 27.4, there is a large decline in the number of significant coefficients. That is, we do not expect to observe economic and financial processes parallel to the theoretical background since this is a transitory period of the economy. The model with V3 has the highest overall significance, whereas the models with V1 and V4 do not even qualify to be significant. This point is important since it indicates that the measure of volatility preferred is substantially influential. Specification with V1 has the highest number of significant macro variables: 4. The interest rate appears to be the significant variable in almost all specifications. It is interesting to note that money supply, terms of trade, and S&P500 do not appear significant in any model.

The last period specified by the ICSS starts in January 2004 and continues to the time where our data set is available. This is the period of political stability, which in turn brought the economic stability. Turkey achieved high growth rates in these years and the economy functioned very well. Results for this period are reported in Table 27.6. This is the period with the highest overall significance, even for the volatility measure 4. The macro variables as a set are able to explain 20% to 80% of the variation in stock market volatility, depending on the volatility measure chosen. Except for V2, higher return volatility in US$ causes a higher return volatility in Turkey. Exchange rate is significant in all specifications, and interest rate and inflation are significant in all but one specification. Macro variables in the models with V1, V3, V5, and V6 explain about 80% of the variation in volatility. If we exclude the measure V2, we notice robust estimates for the significant results, with volatility of M2 growth being the least significant. As a result, we can argue that, under stable economic and political conditions, there should not be a concern that volatility is largely driven by irrational investor behavior.

27.5 CONCLUSION

In this study, we examine the link between macroeconomic factors and stock market volatility for an emerging market. Stock market volatility is proxied by six different measures. To take into account the effect of foreign investor behavior and global factors, the return volatility of S&P500 index is also included as an explanatory variable. Furthermore, the shifts in volatility are identified using the ICSS algorithm leads to three subperiods, with the second subperiod being a severe crisis period and the last one being a stable environment.

Overall, we find that macro factors explain a significant amount of the variability in stock market volatility, lowering the role of irrational investor behavior. However, under unstable environment as defined by the second subperiod, the size and significance of macro factors, along with global factors, decline significantly. As indicated by the last subperiod with stable economic and political environment, a large part of the changes in stock market volatility can be explained by macroeconomic and global factors together.

However, the direction of the influence of macro factors on the stock market volatility is not robust across measures of volatility, though the signs are consistent for most measures. Thus, a test of robustness or fragility, perhaps using extreme bound analysis, can improve our understanding of the link between macro factors and stock market volatility. Moreover, the global factors, as proxied by the return volatility of the S&P500 index, influences return volatility of the ISE100 index significantly. In particular for the last period, this can be due to the existence of foreign investors in the Istanbul Stock Exchange.[*]

All in all, our study presents some empirical evidence that macro variables do have influence on stock market volatility, in particular under a stable economic environment. Also, we show that the size, the significance, and even the direction of the influence are impacted by the choice of the volatility measure.

REFERENCES

Abugri, B. A. (2008). Empirical relationship between macroeconomic volatility and stock returns: Evidence from Latin American markets. *International Review of Financial Analysis* 17:396–410.

Ang, A., Chen, J., and Xing, Y. (2006). Downside risk. *Review of Financial Studies* 19:1191–239.

Binder, J. J., and Merges, M. J. (2001). Stock market volatility and economic factors. *Review of Quantitative Finance and Accounting* 17:5–26.

Bollerslev, T. (1986). Generalized autoregressive conditional heteroskedasticity. *Journal of Econometrics* 31:307–27.

Brooks, C. (2002). *Introductory econometrics for finance.* Cambridge, UK: Cambridge University Press.

Diebold, F. X., and Yılmaz, K. (2007). Macroeconomic volatility and stock market volatility, world-wide. Working paper, Tüsiad-Koç University Economic Research Forum, Istanbul.

Engle, R. F. (1982). Autoregressive conditional heteroscedasticity with estimates of the variance of United Kingdom inflation. *Econometrica* 50:987–1006.

Errunza, V., and Hogan, K. (1998). Macroeconomic determinants of European stock market volatility. *European Financial Management* 4:361–77.

Ewing, B. T., and Malik, F. (2005). Re-examining the asymmetric predictability of conditional variances: The role of sudden changes in variance. *Journal of Banking and Finance* 29:2655–73.

---

[*] Although we do not report the regression results, the inclusion of S&P500 to the regression increases explanatory power significantly. For example, for the entire time period, the adjusted R^2 increases from 42% to 56% under V1. For the three subperiods, the changes are from 27% to 54% (first subperiod), from 51% to 51% (second subperiod), and from 65% to 79% (third subperiod). Thus, except for in the second subperiod, foreign investor behavior and global factors are important in the determination of the stock market volatility of ISE100.

Hamilton, J. D., and Lin, G. (1996). Stock market volatility and the business cycle. *Journal of Applied Econometrics* 11:573–93.

Inclan, C., and Tiao, G. C. (1994). Use of cumulative sums of squares for retrospective detection of changes of variance. *Journal of the American Statistical Association* 89:913–23.

Jorion, P. (1995). Predicting volatility in the foreign exchange market. *Journal of Finance* 50:507–28.

Kearney, C., and Daly, K. (1998). the causes of stock market volatility in Australia. *Applied Financial Economics* 8:597–605.

Leamer, E. E. (1985). Sensitivity analyses would help. *American Economic Review* 75:308–13.

Leamer, E. E., and Leonard, H. (1983). Reporting the fragility of regression estimates. *Review of Economics and Statistics* 65:306–17.

Liljeblom, E., and Stenius, M. (1997). Macroeconomic volatility and stock market volatility: Empirical evidence on Finnish data. *Applied Financial Economics* 7:419–26.

Marcelo, J. L. M., Quiros, J. L. M., and Quiros, M. M. M. (2008). Asymmetric variance and spillover effects: Regime shifts in the Spanish stock market. *Journal of International Financial Markets, Institutions and Money* 18:1–15.

Pagan, A. R., and Schwert, G. W. (1990). Alternative models for conditional stock volatility. *Journal of Econometrics* 45:267–90.

Pierdzioch, C., Döpke, J., and Hartmann, D. (2008). Forecasting stock market volatility with macroeconomic variables in real time. *Journal of Economics and Business* 60:256–76.

Rodriguez, R., Restoy, F., and Pena, J. I. (2002). Can output explain the predictability and volatility of stock returns. *Journal of International Money and Finance* 21:163–82.

Schwert, G. W. (1989). Why does stock market volatility change over time? *Journal of Finance* 44:1115–51.

SPO. (2006). *Ninth development plan 2007–2013 of Turkey.* State Planning Organization, Prime Ministry of Turkey, Ankara.

West, K. D., and Cho, D. (1995). The predictive ability of several models of exchange rate volatility. *Journal of Econometrics* 69:367–91.

CHAPTER 28

Forecasting Default Probability without Accounting Data

Evidence from Russia

Dean Fantazzini

CONTENTS

28.1 INTRODUCTION

Prediction of a firm's bankruptcy risk has been an active research area in finance for the last 40 years. We can identify at least two distinct literatures: The first relies on accounting-based measures as predictor variables. Included in this literature are Altman (1968), Altman et al. (1977), Ohlson (1980), Zmijewski (1984), and Lau (1987), and most recently Shumway (2001), with Altman (1968) and Shumway (2001) being the most popular. We categorize those studies as statistical approaches since they are based on established statistical methods. For example, Altman (1968) uses discriminative analysis; Shumway (2001) uses the hazard model. The second strand of literature predicts bankruptcies using a structural approach, where we assume that we can observe values of a firm's underlying assets and treat its equity either as a standard call option (Merton, 1974; Moody's KMV*—see Crosbie and Bohn, 2001; Vassalou and Xing, 2004) or a barrier option (see Leland, 1994; Brockman and Turtle, 2003; Reisz and Perlich, 2007). The essential difference between a barrier (down-and-out) call option and a standard call option is that the former is a path-dependent derivative while the latter is not. Recently, Hao (2006) performed an extensive out-of-sample forecasting exercise and found that overall the structural approach has better accuracy than the statistical approach, whereas the standard call option has the best accuracy among all models.

However, all the previous approaches rely on the basic assumption that financial statements are reliable: unfortunately, the recent defaults of business giants such as Enron, Parmalat, and Worldcom clearly show how accountancy data can be misleading and far from the true financial situation of a company, even for financial markets subject to strict business regulations and controls such as the American markets.

Given this evidence, we propose here a novel approach for default forecasting that considers stock prices only, and is able to model nonnormalities, too. We justify this proposal, as a vast literature has shown how quoted prices are mostly driven by private information, and therefore should be closer to the true values than accountancy data. The importance of private information in price determination was first highlighted by French and Roll (1986) in a theoretical study on the volatility of stock returns in trading days and nontrading days, as well as in open-market hours versus closed-market hours. They showed evidence that most stock returns volatility is caused by informed traders, whose private information is incorporated into prices when financial markets are open. The increasing availability of high-frequency data has later allowed us to conduct more precise tests on microstructural models. Hausbrouck (1988), Madhavan and Smith (1991), Hasbrouk and Sofianos (1993), and Madhavan and Sofianos (1997), just to name a few, showed the importance of asymmetric private information for stock prices and futures dynamics (see Biais et al. (2005) and Hansbrouck (2007) for recent surveys about market microstructure studies).

---

* KMV LLC was founded in the early 1990s and was the first company to sell default probabilities by using the Merton's structural credit model. KMV LLC was later acquired by Moody's Investors Service in 2002.

Starting from simple financial identities based on the true accounting figures, we show how the null price can be used as a barrier to separate an operative firm from a defaulted one, and to estimate its default probability by using stock prices only. Besides, little effort has been made in financial studies to estimate the variability of computed default probabilities. The classical way is to produce asymptotic standard errors based on the second derivatives of the objective function that is going to be minimized. However, due to the complicated form of the parametric models, this is a cumbersome task, while the first-order Taylor expansion that is used to approximate the expectations can be misleading. In addition, since default probabilities are usually very small numbers, overflow problems may occur.

In this chapter we provide an alternative way for estimating the confidence bands of the default probabilities by utilization of a parametric bootstrap approach. This leads to more flexible inference and overcomes problems related to the calculations involved in asymptotic standard errors. These bands may exhibit asymmetry and also allow for comparison between different computed probabilities.

We compare the KMV-Merton model with our approach by using the historical prices of some of the most traded Russian stocks, and we show how the former model may suffer from numerical instabilities and provide unreasonable results, as already highlighted in Crosbie and Bohn (2001), Bharath and Shumway (2008), and Hao (2006). We then consider five famous American, Italian, and Russian defaulted stocks whose financial statements were found to be irregular and therefore do not allow use of standard structural or statistical approaches. By using our approach and parametric bootstrap methods, we find that these stocks show a default probability statistically higher than 50% already a couple of months in advance of the default event, whereas the KMV-Merton model tends to jump only a couple of days before the default, when it is already too late.

The chapter proceeds as follows. Section 28.2 reviews the KMV-Merton model. Section 28.3 presents our novel approach for estimating default probabilities and shows how to construct bootstrap confidence bands. Section 28.4 discusses an empirical example with American, Italian, and Russian stocks. We conclude in Section 28.5.

28.2 THE KMV-MERTON MODEL: A REVIEW

Merton-type models (or structural models) base the evaluation of firm-related securities on the structural firm variables, i.e., the firm's assets and liabilities values. Those models date from the early 1970s (see Merton, 1970,

1977). The classic papers by both Black and Scholes (1973) and Merton (1974) point out that the liabilities of a corporate firm may be priced as plain vanilla options. Needless to say, the straightforward use of the Black-Scholes valuation formulas requires some basic assumptions on the behavior of assets, no-arbitrage opportunities, and continuous hedging.

A very common assumption in Merton-type models (e.g., see Ingersoll, 1987; Mason and Merton, 1985; Merton 1977) is that the value A_t of the firm follows a geometric Brownian motion:

$$dA_t = \mu_A A_t\, dt + \sigma_A A_t\, dW_t \tag{28.1}$$

where W_t is a Wiener process and the drift and volatility coefficients μ_A and σ_A do not depend on the capital structure of the firm $q_t = B_t/E_t$, i.e., on how the assets value A_t is split into equity value E_t and bonds value B_t. The independence of (μ_A, σ_A) on q_t simply translates the Miller-Modigliani theorem (see Miller and Modigliani, 1958, 1961).

Another basic assumption is that the assets value is exogenous, so it can be treated as the underlying in an option pricing framework. This means that the assets value does not depend on the dynamics of the firm-related securities, and therefore the equity has a residual value. Thus, the equity value satisfies

$$E_t = A_t N(d_1) - e^{-rT} B_T N(d_2) \tag{28.2}$$

where E_t is the market value of the firm's equity, B_T is the face value of the firm debt, r is the risk-free rate, $N(\cdot)$ is the cumulative standard normal distribution function, and d_1 is given by

$$d_1 = \frac{\log(A_t/B_T) + \left(\mu_A + 0.5\sigma_A^2\right)T}{\sigma_A \sqrt{T}} \tag{28.3}$$

while $d_2 = d_1 - \sigma_A \sqrt{T}$. Under the Merton model assumptions the volatility of the equity is

$$\sigma_E = \frac{A_t}{E_t} N(d_1)\sigma_A \tag{28.4}$$

The underlying variable in the Merton model cannot be directly observed. To overcome this problem, the KMV-Merton model (see also Ronn and Verma, 1986; Crosbie and Bohn, 2001) makes use of the two nonlinear formulas, Equations (28.2) and (28.4), and solves them numerically for A_t and σ_A. Once this numerical solution is obtained, the distance to default can be calculated as d_2, while the implied probability of default is

$$\Pr [A_t \le B_T] = N(- d_2) \tag{28.5}$$

28.3 A NEW APPROACH: THE ZERO PRICE PROBABILITY

The previous KMV-Merton model as well as other structural and statistical approaches for default forecasting assume that the accountancy data represent the true picture of the company financial situation. With regard to the KMV-Merton model, the accountancy data used is the book face value of the firm's total liabilities. The usual practice considers for total liabilities the sum of short-term liabilities plus one-half of long-term liabilities. This assumption, which is made by Moody's KMV for North American firms, ensures that the firm's liabilities are not overstated (see also Vassalou and Xing, 2004; Hao, 2006). Even though Vassalou and Xing (2004) state that using different percentages for long-term liabilities is not deemed to alter the main qualitative results, such an approach may not be robust to "window dressing" policies made to improve the financial score of a company or, in the worst case, to financial frauds. Besides, KMV itself admits that "in practice the market leverage moves around far too much for [Equation (28.4)] to provide reasonable results" and particular iterative methods have to be used instead (see Crosbie and Bohn, 2001).

The recent default of the food giant Parmalat in 2003 clearly showed how the debts reported in the certified balanced sheet can represent only a part of the true debt figures. Summarizing this financial story, in February 2003,

> the chief financial officer (CFO) Fausto Tonna unexpectedly announced a new €500 million bond issue. This came as a surprise both to the markets and to the CEO, Calisto Tanzi. Tanzi fired Tonna and replaced him as CFO with Alberto Ferraris. According to an interview he later gave *Time Magazine,* Ferraris was surprised to discover that, though now CFO, he still didn't have access to some of the corporate books, which were being handled by chief accounting officer Luciano Del Soldato. (http://www.wikipedia.org/wiki/Parmalat) [so] Ferraris asked two trusted members of his staff to mount a quiet investigation. After calling around Parmalat's worldwide

operations, they came back with shocking news: a total debt estimate of €14 billion, more than double that on the balance sheet.
(Gumbel, 2004)

The crisis became public in November 2003 when questions were
raised about transactions with the mutual fund Epicurum, a Cayman-
based company linked to Parmalat causing its stock to plummet. Ferraris
resigned less than a week later and was replaced by Del Soldato. In
December, Del Soldato resigned, unable to get cash from the Epicurum
fund, needed to pay debts and make bond payments. Tanzi himself
resigned as chairman and CEO as the 7 billion euros hole was discovered
in Parmalat's accounting records: Parmalat's bank, Bank of America,
later released a document showing €3.95 billion in Bonlat's bank account
as a forgery (in 1999, Parmalat set up a subsidiary in the Cayman Islands
called Bonlat).

> Calisto Tanzi was detained hours after the firm was declared offi-
> cially insolvent and eventually charged with financial fraud and
> money laundering. Among the questionable accounting practices
> used by Parmalat: it sold itself credit linked notes, in effect placing a
> bet on its own credit worthiness in order to conjure up an asset out
> of thin air. (http://www.wikipedia.org/wiki/Parmalat)

For more details, see the full article reported in *Time Magazine* (Gumbel,
2004) and Castri and Benedetto (2006).

The previous short description of the Parmalat scandal clearly highlights that using accounting data to infer the firm's default probability
can be misleading and result in a very poor estimate. In order to avoid
such problems, we propose here a novel approach that uses the null
price as a default barrier to separate an operative firm from a defaulted
one, and to estimate its default probability without resorting to accountancy data.

Let us consider the following two financial identities based on the true
accountancy data at time T:

$$\begin{cases} E_T = A_T - B_T \\ E'_T = A_T = (A_T - B_T) + B_T = E_T + B_T \end{cases}$$

and consider the financial meanings and signs of E_T and E'_T according to
the situation faced by the firm:

TABLE 28.1   Financial Meaning and Signs of E_T and E_T'

	$E_T = A_T - B_T$	$E_T' = A_T$
Operative	Equity belonging to shareholders (+)	Asset value (+)
Defaulted	Loss given default for debtholders (−)	Equity belonging to debtholders (+)

Table 28.1 shows that the quantity E_T is negative when the firm defaults as it represents the loss given default for debtholders, while it is positive when the firm is operative, representing the equity belonging to shareholders instead. A negative value for E_T is a direct consequence of the limited liability now in place in all modern Western legislations. Besides, losses can be theoretically infinite, like profits: just think of the effects of the September 11, 2001, attacks on airline companies or of the mad cow and bird flue diseases on agriculture companies. Therefore, we can resort to probability density functions with negative domain, too.

The main consequence of the previous discussion is that we can estimate the distance to default simply by using E_T, instead of d_2 as in Merton's framework, and the default probability by $\Pr[E_T \leq 0]$, as the firm defaults when E_T is zero or negative. Furthermore, given that $E_T = S \times P_T$, where P_T is the quoted stock price at time T and S is the number of shares, the default probability of a firm can be retrieved by estimating $\Pr[P_T \leq 0]$, that is, by using the *zero-price probability* (ZPP). While the quoted price P_T is a truncated variable that cannot be negative, the quantity E_T has no lower bound, as it has a different financial meaning whether the firm is operative or defaulted: in the former case E_T is computed daily in (electronic) financial markets, whereas in the latter case the loss given default is computed in bankruptcy courts.

Stock prices are usually nonstationary $I(1)$ variables, and it is common to model their dynamics by considering the log-returns, so that the prices are guaranteed to be positive. However, we are interested in finding $\Pr[P_T \leq 0]$, since we have just shown that the null price can be used as a default barrier. A straightforward way to do that is to consider a conditional model for the *differences in prices levels*, $X_t = P_t - P_{t-1}$, instead of differences in log-prices.

An analytical close-form solution for $\Pr[P_T \leq 0]$ is available for a few special and unrealistic cases, such as normally distributed prices with homoskedastic variance. When this is not the case, simulation methods are required. If we are at time t and want to estimate the default probability at time $t + T$, we can use the following general algorithm:

Proposition (ZPP Estimation Algorithm)

Step 1: Consider a generic conditional model for the differences of prices levels $X_t = P_t - P_{t-1}$, without the log-transformation, given the filtration at time t, F_t:

$$\begin{cases} X_t = E[X_t \mid F_t] + \varepsilon_t \\ \varepsilon_t = H_t^{1/2}\eta_t, \quad \eta_t \sim i.i.d.(0,1) \end{cases} \tag{28.6}$$

where $H_t^{1/2}$ is the conditional standard deviation.

Step 2: Simulate a high number N of price trajectories up to time $t + T$, using the estimated time-series model (28.6) at Step 1.

Step 3: The default probability is simply the ratio n/N, where n is the number of times out of N when the price touched or crossed the barrier along the simulated trajectory.

This method entails a number of important benefits:

1. We only need the stock prices.

2. We do not need either any firm's volatility σ_A or the debt face value, like in Merton-style models.

3. We can consider more realistic distributions than the log-normal.

4. We can estimate the default probability for any given time horizon $t + T$.

5. We can screen the default risk daily or even intradaily. The ZPP can therefore be used as a tool for risk management.

6. The ZPP can be used as an early warning system for financial default in general, since it can be estimated with any financial time series.

28.3.1 Bootstrap Confidence Bands

The bootstrap (Efron and Tibshirani, 1993) was first used to investigate the reproducibility of certain features of phylogenetic trees by Felsenstein

(1985). Efron and Tibshirani (1998) later looked up this problem more generally and termed it the problem of regions. They also link this confidence measure, in certain settings, to frequentist *p*-values and Bayesian posterior probabilities. Bootstrap techniques have expanded the ability for statistical inferences in certain circumstances where classical inferential procedures face problems. Bootstrap estimates of the standard errors for the parameters of a model are useful counterparts when the direct calculation of the standard errors (asymptotic or not) is troublesome (see, e.g., Efron and Tibshirani, 1993). In addition, bootstrap standard errors can better reflect small sample properties of the estimates.

To date, little work has been done to quantify the uncertainty around the computed default probabilities, since the main interest has lain in the point estimate. Instead, if we look at other fields of research that deal with computed probabilities, Van der Laan and Bryan (2001) used the parametric bootstrap in a biostatistics work to determine the single-gene probabilities and their distribution. Karlis and Kostaki (2002) used a similar approach for mortality rates in human populations. Zwane and Van der Heijden (2003) presented an algorithm for the parametric bootstrap that can be used in log-linear modeling when there are continuous covariates. In a following paper, Zwane et al. (2004) proposed the parametric bootstrap to construct a confidence interval for multiple-record systems estimators when registrations refer to different but overlapping populations.

Besides, several authors have used the nonparametric bootstrap in log-linear modeling (see, for example, Huggins, 1989; Tilling and Sterne, 1999; Tilling et al., 2001). But as noted by Norris and Pollock (1996), the nonparametric bootstrap results in a variance estimate that is likely to be smaller than the true variance, because it conditions on the data being observed. This is in line with a simulation conducted by Tilling and Sterne (1999), which showed that the nonparametric bootstrap has a coverage consistently lower than the nominal coverage. Similar findings were reported by Zwane and Van der Heijden (2003), too. "Besides, the parametric bootstrap is asymptotically valid under relatively mild conditions compared to those required by the nonparametric bootstrap" (Gine and Zinn, 1990).

In general, the asymptotic validity of the parametric bootstrap requires that the chosen parametric model be correct. However, for the case of the truncated multivariate normal, Van der Laan and Bryan (2001) show that as long as the parametric model places no constraints on the parameters, even when it is incorrect, the parametric bootstrap will still consistently estimate the degenerate limit distribution. Given this literature background, we resort here to parametric bootstrap methods.

The whole procedure to construct confidence bands around the estimated default probabilities is as follows:

Step 1: Draw a T × 1 vector of standardized innovations η_t from the considered marginal density used in Equation (28.6), for example, Student's t.

Step 2: Create an artificial history for the random variable X_t, by replacing all parameters in Equation (28.6) with their estimated counterparts, together with the standardized innovations η_t drawn in the previous step.

Step 3: Estimate the conditional Equation (28.6) using the data from the artificial history.

Step 4: Calculate a bootstrap estimate of the ZPP using the previous estimates performed on the artificial history.

Step 5: Repeat the above four steps for a large number of times n_B, in order to get a numerical approximation to the distribution of the ZPP.

This distribution forms the basis for computing the bootstrap confidence intervals around the default probabilities.

28.4   EMPIRICAL ANALYSIS

We first analyze the daily data of the five most traded Russian stocks: Gazprom, Lukoil, Norilsk Nickel, Sberbank and United Energy. We then consider five well-known American, Italian, and Russian defaulted stocks whose financial statements were found to be irregular and therefore do not allow use of standard approaches.

28.4.1   Russian Not-Defaulted Stocks

We consider both the standard KMV-Merton approach that makes use of the nonlinear Equations (28.2) and (28.4) and our approach described in Section 28.3. In the former case, the liabilities are the sum of short-term liabilities plus one-half of long-term liabilities, as previously discussed. In the latter case, we consider an AR(1)-threshold-GARCH(1,1) model for the differences in prices levels, $X_t = P_t - P_{t-1}$, together with a Student's t distribution to take the leverage effect as well as leptokurtosis in the data into account (for more details, see Glosten et al., 1993):

$$
\begin{cases}
X_t = \mu + \phi_1 X_{t-1} + \varepsilon_t \\
\varepsilon_t = \sqrt{h_t}\, \eta_t, \quad \eta_t \sim i.i.d.(0,1) \\
h_t = \omega + \alpha\varepsilon_{t-1}^2 + \gamma\varepsilon_{t-1}^2 D_{t-1} + \beta h_{t-1}
\end{cases}
\tag{28.7}
$$

where $D_{t-1} = 1$ if $\varepsilon_{t-1} < 0$. We choose such a specification given its past success in modeling financial variables (see Tsay, 2002; Hansen and Lunde, 2005, and references therein). As for the number N of simulated price trajectories to estimate the ZPP, we set $N = 5,000$.

We test for unit roots in the financial variables under scrutiny by using the Dickey-Fuller test with GLS detrending (DF-GLS) by Elliott et al. (1996) and the test by Kwiatkowski et al. (1992), which is based on the null of covariance stationarity rather than integratedness. A careful analysis of the levels and of the first differences of the prices series, reported in Table 28.2, shows that nonstationarity is the main feature of the variables over the observation period 2002–2008.

We then tested the goodness of fit of the AR(1)-T-GARCH(1,1) models employed for the conditional marginal distributions by using Ljung-Box tests on the standardized residuals in levels $\hat{\eta}_t$ and squares $\hat{\eta}_t^2$ to test the null of no autocorrelation in the mean and in the variance, together with the specification tests discussed in Granger et al. (2006). We used the Kolmogorov-Smirnov test for density specification, together with the "hit" test in order to test jointly for the adequacy of the dynamics and the density specifications in the marginal distribution models, where the null hypothesis is that the density model is well specified. The latter test

TABLE 28.2   Unit Root Tests for the Five Russian Nondefaulted Stocks

Stock	DF-GLS		KPSS	
	Levels	First Differences	Levels	First Differences
Gazprom	0.048	−26.170 (**)	4.479 (**)	0.168
Lukoil	−0.615	−43.822 (**)	4.647 (**)	0.049
Norilsk Nickel	0.595	−46.843 (**)	4.144 (**)	0.245
Sberbank	−0.196	−39.150 (**)	4.135 (**)	0.229
United Energy	−0.779	−41.649 (**)	4.162 (**)	0.169

Note: *, significant at the 5% level; **, significant at the 1% level.

TABLE 28.3 p-values for the Specification Tests

Stock	Ljung-Box(25) η_t	Ljung-Box(25) η_t^2	Kolmogorov-Smirnov	Joint Hit Test
Gazprom	0.281	0.066	0.021	0.075
Lukoil	0.334	1.000	0.604	0.032
Norilsk Nickel	0.734	1.000	0.353	0.081
Sberbank	0.632	0.987	0.107	0.056
United Energy	0.854	0.331	0.282	0.074

divides the support of the density into five regions and then applies interval forecast evaluation techniques to each region separately, and then to all regions jointly (for more details, see Granger et al., 2006). For sake of space and interest, we report in Table 28.3 only the p-values of each test. The full set of results is available from the author upon request.

All the marginal models passed the tests (at least) at the 0.01 level, thus highlighting that they are correctly specified. Overall, the previous tables point out that the AR(1)-TGARCH(1,1) model with a Student's t distribution is a proper choice for our financial variables.

To estimate the default probabilities, we consider a 1-year-ahead horizon. The initialization sample considers prices between January 8, 2002 and April 19, 2006, for a total of 1,000 observations: at the j-th iteration, where j goes from April 20, 2006 to April 23, 2008 (for a total of 500 observations), the estimation sample is augmented to include one more observation, and this procedure is iterated until all days have been included in the estimation sample. Figures 28.1 and 28.2 show the default probabilities computed by using the KMV-Merton model and the ZPP, with a range between 0 and 50%, while Figure 28.3 shows two special cases.

The previous figures highlight some elements of sure interest:

1. The default probabilities estimated with the ZPP are usually higher than the ones obtained by using Merton's model. Such a result may be due to both the log-normality assumption in the Merton model and to incorrect financial statements made to "window dress" the financial health of a company. Regarding this issue, it is possible to show that in the Merton model the default probability rises quickly only when the equity-to-debt ratio is very low, everything else kept fixed.

2. The KMV-Merton model shows some numerical instability problems with noisy data and from the jumps in the debt values at book

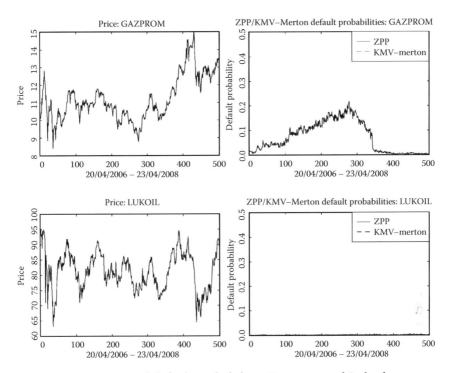

FIGURE 28.1   Estimated default probability: Gazprom and Lukoil.

closure dates at the end of the year: this produces the 1-day peaks
shown in Figure 28.3 (Gazprom and Lukoil).

We remark that Ketz (2003) discusses a wide variety of techniques to hide
debts and financial risk. This explains why the Merton's default probabilities
and firm values are usually underestimated with respect to the ZPP. Besides,
the log-normal is not an appropriate distribution for price dynamics since
it underestimates the tail of the distribution. Increasing volatility and leptokurtosis can also be interpreted as a sign of informed trading (see Biais et al.,
2005; Hansbrouck, 2007): just see the difference in Figure 28.1 before and
after the Gazprom CEO was nominated as candidate for the Russian presi-
dency. Political risk is not accounted for by the KMV-Merton model.

28.4.2 Defaulted Stocks with Irregular Financial Statements
We now consider the last 1,000 trading days of five famous defaulted stocks
whose financial statements were found to be irregular by justice probes and
therefore do not allow use of standard approaches. Instead, we pursue the

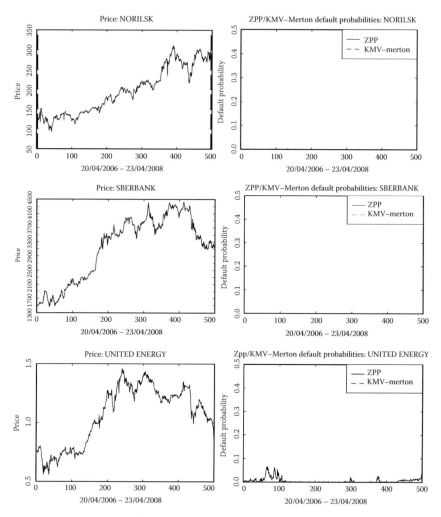

FIGURE 28.2 Estimated default probability: Norilsk, Sberbank, and United Energy.

use of our methodology described in Section 28.3, since a vast literature has shown how quoted prices are mostly driven by private information, and therefore they should be closer to the true values than accountancy data (see the discussion in the introduction). These five stocks are:

1. *Cirio:* September 24, 1999 to July 24, 2003. Second largest default in the European food sector.

2. *Enron:* January 20, 1998 to January 10, 2002. Second largest default in American history.

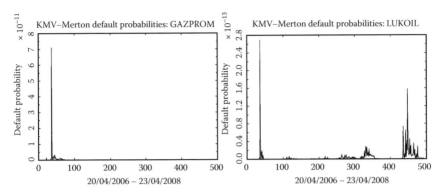

FIGURE 28.3   Problems with KMV-Merton's model: Numerical instability (GAZPROM, LUKOIL).

3. *Parmalat:* February 22, 2000 to December 22, 2003. Largest default in European history.

4. *Worldcom:* July 16, 1998 to July 12, 2002. Largest default in American history.

5. *Yukos:* July 3, 2000 to June 30, 2004. Largest default in Russian history.

Similarly to Section 28.4.1, we employ AR(1)-T-GARCH(1,1) models with a Student's t distribution to model leptokurtosis in the price differences X_t. Tests for unit roots reported in Table 28.4 show the five series are $I(1)$ variables, while the specification tests in Table 28.5 highlight no significant misspecification in the model dynamics and marginal distributions. The latter evidence is important, as it allows us to use parametric bootstrap procedures to construct confidence intervals around the computed default probabilities.

TABLE 28.4   Unit Root Tests for the Five Defaulted Stocks

Stock	DF-GLS		KPSS	
	Levels	First Differences	Levels	First Differences
Cirio	−2.083	−20.583 (**)	0.486 (*)	0.065
Parmalat	−1.372	−12.395 (**)	0.768 (**)	0.045
Enron	−2.620	−38.785 (**)	0.453 (*)	0.022
Worldcom	−1.359	−34.876 (**)	0.833 (**)	0.057
Yukos	−0.238	−28.006 (**)	3.672 (**)	0.021

Note: *, significant at the 5% level; **, significant at the 1% level.

TABLE 28.5   *p*-values for the Specification Tests

Stock	Ljung-Box(25) η_t	Ljung-Box(25) η_t^2	Kolmogorov-Smirnov	Joint Hit Test
Cirio	0.112	0.182	0.018	0.017
Enron	0.953	0.096	0.380	0.837
Parmalat	0.906	0.971	0.091	0.530
Worldcom	0.167	0.758	0.435	0.057
Yukos	0.263	0.405	0.556	0.237

Figures 28.4 and 28.5 show the last 1,000 end-of-day quoted prices before default (500 days for Yukos), together with the computed default probabilities and the 90% confidence bands.

A first interesting thing to note is that the confidence intervals show a certain degree of asymmetry. Moreover, we observe a strong difference between American and Italian stocks. The computed 1-year-ahead default probabilities for Enron and Worldcom are higher than 50% a couple of months in advance of the default event. As for the Italian stocks, instead, it is interesting to observe that already 2 to 3 years in advance the computed probabilities are above 50%. This evidence seems consistent with preliminary justice probes, which highlight that financial distress was already known to the two Italian company managements in the 1990s. Besides, while the large confidence bands suggest we should consider the computed default probabilities with some care (particularly for Parmalat), nevertheless they can also be interpreted as a strong sign of market uncertainty about the future of the company.

As for Yukos, the biggest Russian default so far, the market seemed to price the difficulties regarding this company already a couple of months in advance of the arrest of Yukos' CEO in October 2003.

Overall, the previous figures seem to highlight a different degree of efficiency between the Italian market and the American market in the case of financial frauds. Some episodes concerning the Enron and Parmalat defaults may help to shed some light over this different behavior: the wife of Enron CEO, Lisa Lay, has been accused of selling 500,000 shares of Enron stock totaling $1.2 million on November 28, 2001. Records show that Mrs. Lay placed the sale order sometime between 10:00 and 10:20 a.m. News of Enron's problems, including the millions of dollars in losses it had been hiding, went public about 10:30 that morning, and the stock price soon fell to below $1. Similarly, former Enron executive Paula Rieker was charged

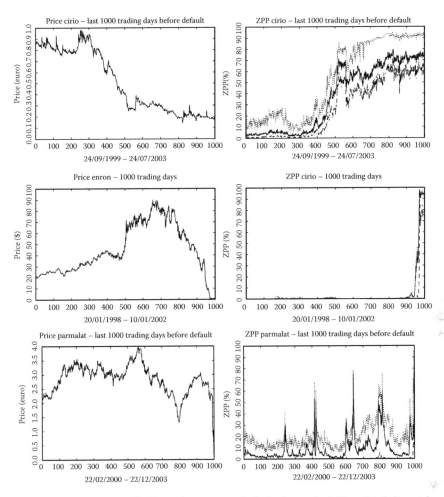

FIGURE 28.4 Prices (left) and estimated default probability (right), with 90% bootstrap confidence bands.

with criminal insider trading. Rieker obtained 18,380 Enron shares for $15.51 a share, and she sold that stock for $49.77 a share in July 2001, a week before the public was told what she already knew about a $102 million loss. She pleaded guilty. As for Parmalat, instead, the financial police found that a sum between 1 and 2 billion euros was distracted from Parmalat bank accounts to tourist societies and other firms belonging to the Tanzi family (yet external to the food group giant) between 1993 and 2003. Therefore, the two defaults seem not of the same kind: while the information that Enron was collapsing was already known to Enron executives for over a year, the

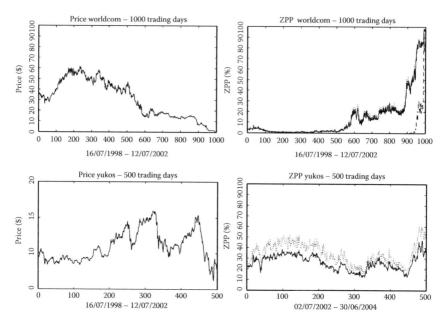

FIGURE 28.5   Prices (left) and estimated default probability (right), with 90% bootstrap confidence bands.

last-minute timeline of their sales seems to show that "they left the ship before it sank." The case for Parmalat is rather different; instead, the management seems to have organized a precise illegal system to drain money out of the company in a systematic way. The ongoing justice probe and the following trial will, hopefully, clarify these points.

Instead, the Yukos case is very interesting because it is the first default of such large dimensions in the Russian Federation, and it highlights a certain degree of market inefficiency, given that the implied default probability was higher than 20% well in advance of the main public events. However, given the peculiarities of this default, some caution should be taken before retrieving any lessons from this case.

28.5   CONCLUSION

In this chapter we described a new methodology to assess the default probability of a quoted firm, together with parametric bootstrap methods to build confidence intervals around the computed probabilities. Our approach needs neither any firm's volatility nor the debt face value, like in Merton-style models. Moreover, we can consider more realistic distributions than the log-normal one. Besides, Merton's default probabilities and firm values may be largely underestimated in the presence of financial fraud or window dressing. Similarly, political risk is not accounted for by the KMV-Merton model, as the case of Gazprom and Yukos clearly highlighted.

The empirical analysis showed that the numerical stability of our approach is much better than Merton's. Furthermore, by using five defaulted stocks whose balance sheets were irregular, we found that the Italian market seemed much less efficient than the American market in case of financial frauds. Particularly, estimated default probabilities in the former market were well above 50% already a couple of years before the defaults, while in the latter markets this phenomenon took place only in the last 100 trading days. As for Russian markets, the ZPP showed that we are able to take political and financial factors into account, delivering much more reliable estimates of the default probability than the KMV-Merton model. However, more future research is needed before drawing any lessons.

An avenue for further research is to perform a back-testing analysis with larger data sets so that we can sharpen the comparison between American, Italian, and Russian markets in terms of efficiency and forecasting performances. Besides, given the strong interdependencies in world financial markets, a multivariate extension of our approach is called for.

REFERENCES

Altman, E. I. (1968). Financial ratios, discriminant analysis and prediction of corporate bankruptcy. *Journal of Finance* 23:589–609.

Altman, E. I., Haldeman, R., and Narayanan, P. (1977). ZETA analysis: A new model to identify bankruptcy risk of corporations. *Journal of Banking and Finance* 1:29–54.

Bharath, S., and Shumway, T. (2008). Forecasting default with the KMV-Merton model. *Review of Financial Studies*, 21(3) 1339–1369.

Biais, B., Glosten, L., and Spatt, C. (2005). Market microstructure: A survey of microfoundations, empirical results and policy implications. *Journal of Financial Markets* 8:217–64.

Black, F., and Scholes, M. (1973). The pricing of options and corporate liabilities. *Journal of Political Economy* 81:637–59.

Brockman, P., and Turtle, H. (2003). A barrier option framework for corporate security valuation. *Journal of Financial Economics* 67:511–29.

Castri, S., and Benedetto, F. (2006). There is something about Parmalat (On directors and gatekeepers). Working paper, Bocconi University, Milan, Italy.

Crosbie, P., and Bohn, J. (2001). Modeling default risk. KMV Technical Document, New York.

Efron, B., and Tibshirani, R. (1993). *An introduction to the bootstrap*. London: Chapman Hall.

Efron, B., and Tibshirani, R. (1998). The problem of regions. *Annals of Statistics* 26:1687–718.

Elliott, G., Rothenberg, T., and Stock, J. (1996). Efficient tests for an autoregressive unit root. *Econometrica* 64:813–36.

Felsenstein, J. (1985). Confidence limits on phylogenies: An approach using the bootstrap. *Evolution* 39:783–91.

French, K. R., and Roll, R. (1986). Stock return variances: The arrival of information and the reaction of traders. *Journal of Financial Economics* 17:5–26.

Gine, E., and Zinn, J. (1990). Bootstrapping general empirical measures. *Annals of Probability* 18:851–69.

Glosten, L. R., Jaganathan, R., and Runkle, D. (1993). On the relation between the expected value and the volatility of the normal excess return on stocks. *Journal of Finance* 48:1779–801.

Granger, C., Patton, A., and Terasvirta, T. (2006). Common factors in conditional distributions for bivariate time series. *Journal of Econometrics* 132:43–57.

Gumbel, P. (2004). How it all went so sour. *Time Magazine*, November 21.

Hansen, P., and Lunde, A. (2005). A forecast comparison of volatility models: Does anything beat a GARCH(1,1)? *Journal of Applied Econometrics* 20:873–89.

Hao, H. (2006). Is the structural approach more accurate than the statistical approach in bankruptcy prediction? Working paper, School of Business, Queen's University, Ontario, Canada.

Hasbrouck, J. (1988). Trades, quotes, inventories and information. *Journal of Financial Economics* 22:229–52.

Hasbrouck, J. (2007). *Empirical market microstructure*. London: Oxford University Press.

Hasbrouck, J., and Sofianos, G. (1993). The trades of market makers: An empirical analysis of NYSE specialists. *Journal of Finance* 48:1565–93.

Huggins, R. (1989). On the statistical analysis of capture experiments. *Biometrika* 76:133–40.

Ingersoll Jr., J. E. (1987). *Theory of financial decision making*. Boston: Rowman & Littlefield.

Karlis, D., and Kostaki, A. (2002). Bootstrap techniques for mortality models. *Biometrical Journal* 44:850–66.

Ketz, E. J. (2003). *Hidden financial risk. Understanding off-balance sheet accounting*. Hoboken, NJ: John Wiley & Sons.

Kwiatkowski, D., Phillips, P., Schmidt, P., and Shin, Y. (1992). Testing the null hypothesis of stationary against the alternative of a unit root. *Journal of Econometrics* 54:159–78.

Lau, A. (1987). A 5-state financial distress prediction model. *Journal of Accounting Research* 25:127–38.

Leland, H. (1994). Corporate-debt value, bond covenants, and optimal capital structure. *Journal of Finance* 49:1213–52.

Madhavan, A., and Smidt, S. (1991). A Bayesian model of intraday specialist pricing. *Journal of Financial Economics* 30:99–134.

Madhavan, A., and Sofianos, G. (1997). An empirical analysis of the NYSE specialist trading. *Journal of Financial Economics* 48:189–210.

Mason, S., and Merton, R. C. (1985). The role of contingent claims analysis in corporate finance. In *Recent advances in corporate finance*, ed. E. Altman and M. Ubrahmanyam. Homewood, IL: R.D. Irwin 7–54.

Merton, R. C. (1974). On the pricing of corporate debt: The risk structure of interest rates. *Journal of Finance* 29:449–70.

Merton, R. C. (1977). On the pricing of contingent claims and the Modigliani-Miller theorem. *Journal of Financial Economics* 15:241–50.

Miller, M. H., and Modigliani, F. (1958). The cost of capital, corporation finance and the theory of investment. *American Economic Review* 48:261–97.

Miller, M. H., and Modigliani, F. (1961). Dividend policy, growth, and the valuation of shares. *Journal of Business* 34:411–33.

Norris, J., and Pollock, K. (1996). Including model uncertainty in estimating variances in multiple capture studies. *Environmental and Ecological Statistics* 3:235–44.

Ohlson, J. A. (1980). Financial ratios and the probabilistic prediction of bankruptcy. *Journal of Accounting Research* 18:109–31.

Reisz, A. S., and Perlich, C. (2007). A market-based framework for bankruptcy prediction. *Journal of Financial Stability* 3:85–131.

Ronn, E. I., and Verma, A. K. (1986). Pricing risk-adjusted deposit insurance: An option-based model. *Journal of Finance* 41:871–95.

Shumway, T. (2001). Forecasting bankruptcy more accurately: A simple hazard model. *Journal of Business* 74:101–24.

Tilling, K., and Sterne, J. (1999). Capture-recapture models including covariate effects. *American Journal of Epidemiology* 149:392–400.

Tilling, K., Sterne, J., and Wolfe, C. (2001). Estimation of incidence of stroke using a capture-recapture model including covariates. *International Journal of Epidemiology* 30:1351–59.

Tsay, R. (2002). *Analysis of financial time series*. Hoboken, NJ: John Wiley & Sons.

Van Der Laan, M., and Bryan, J. (2001). Gene expression analysis with the parametric bootstrap. *Biostatistics* 2:445–61.

Vassalou, M., and Xing, Y. (2004). Default risk in equity returns. *Journal of Finance* 59:831–68.

Zmijewski, M. E. (1984). Methodological issues related to the estimation of financial distress prediction models. *Journal of Accounting Research* 22:59–82.

Zwane, E., and Van der Heijden, P. (2003). Implementing the parametric bootstrap in capture-recapture models with continuous covariates. *Statistics and Probability Letters* 65:121–25.

Zwane, E., Van der Pal-de Bruin, K., and Van der Heijden, P. (2004). The multiple-record systems estimator when registrations refer to different but overlapping populations. *Statistics in Medicine* 23:2267–81.

CHAPTER 29

Recent Assessments on Mean Reversion in the Middle East Stock Markets

Sam Hakim and Simon Neaime

CONTENTS

29.1 INTRODUCTION

There has been a considerable amount of attention devoted to the predictability of stock returns. However, many studies (for example, Campbell and Perron, 1991; Ferson et al., 2003) have cast serious doubt on the predictive power of variables believed to forecast stock returns in long-horizon regressions. Finance practitioners and academics have always questioned the long-run time-series properties of equity prices while paying attention to whether the motion of stock prices can be characterized as a random walk or as a mean reverting process. If the dynamics of stock prices over time are mean reverting, then there exists a tendency for the price level to return to its trend path over time, and investors may be able to forecast future returns by using information on past returns. On the other hand, a random walk process implies that any shock to stock prices is permanent, and that there is no tendency for the price level to revisit a specific trend or path over time. In other words, historical observations become totally irrelevant for the purpose of fore-casting future returns. The random walk property also implies that the volatility of stock prices can grow without a bound in the long run. Aside from being an interest by themselves, these time-series properties have important implications for asset pricing. The evidence of mean rever-sion was first documented for the U.S. market. Using U.S. individual firm-level data, DeBondt and Thaler (1985) first report that past losing stocks over the previous 3–5 years significantly outperform past winning stocks over a 3–5 years holding period. Their results indicate that stock prices do not follow a random walk, but contain a strong mean revert-ing component. Fama and French (1988) also report mean reversion in U.S. equity market using long-horizon regressions, and Poterba and Summers (1988) document evidence of mean reversion using the vari-ance ratio test. Recently, researchers have also tested for mean reversion in equity prices using international data. For example, Richards (1997) reports evidence of long-term winner-loser reversals for equity indexes for sixteen countries. Balvers, Wu and Gilliland (2000) find significant evidence of mean reversion across eighteen developed equity markets and demonstrate that one can exploit the property of mean reversion to predict equity returns using a parametric contrarian investment strategy. Other researchers, however, report conflicting results against mean reversion. For example, Lo and MacKinlay (1988) report some evidence against mean reversion in weekly U.S. data. Kim, Nelson and Startz (1991) show that mean reversion exists only in pre-war U.S. data, while Richardson and Stock (1989) and Richardson (1993) argue that the results from Fama and French (1988) and Poterba and Summers (1988) are not robust because of small-sample biases. (Chaudhuri and Wu, 2003) test for mean reversion for emerging market stock prices and find that the null hypothesis of no mean reversion cannot be rejected in general using the standard unit-root test. They argue, however, that emerging markets may be subject to structural changes, and if a structural break is explicitly taken into account in the regression, mean reversion can be detected in fourteen out of seventeen countries. In the context of MENA countries, Hakim and Neaime (2003) indicate evidence of mean reversion in the more mature stock markets of Egypt, Jordan, Turkey, and Morocco, but the sample period was relatively short (at most 5 years in some countries). More recently, Assaf (2006) uses a rescaled variance statistic to investigate the long memory demonstrated in the stock price series of the same four preceding countries. Despite the limited period (5 years) of his analysis, his findings suggest that a significant long-term memory exists, contrary to the hypothesis of market efficiency.

Chaudiri and Wu (2002) state that "Much of the controversy on the issue of mean reversion arises because the speed of reversion may be slow and standard econometric tests do not have sufficient power to discriminate a mean reversion process from a random walk process. In this chapter, we test for mean reversion in stock price indexes of five emerging markets in the Middle East using monthly data from January 1996 through April 2008. Our results provide useful information from this independent sample, and complement the existing studies on stock market efficiency in the Middle East and North Africa (MENA). Chaudhuri and Wu (2002) state that "To overcome the power deficiency problem, we conduct the test in a panel framework. We pool data of five neighboring countries in the Gulf region and utilize the information on the cross-sectional variations in equity returns to increase the power of the test so that mean reversion can be more easily detected. To further improve estimation efficiency, we estimate the system of equations using the seemingly unrelated regression (SUR) technique. We find that the null hypothesis of a random walk can be rejected in favor of mean reversion for two stock index prices.

The remainder of the chapter is organized as follows. Section 29.2 reviews the historical background of the equity markets in the Gulf Cooperation Council countries of the Middle East. Section 29.3 presents the empirical methodology. Section 29.4 describes the data. The empirical results are reported in Section 29.5, and Section 29.6 concludes the chapter.

29.2 BACKGROUND OF THE GULF COOPERATION COUNCIL STOCK MARKETS

Compared to other emerging markets, the stock markets of the MENA region have enjoyed faster price gains over the past decade. The average annual rise in the stock markets of countries such as Saudi Arabia, Qatar, and Abu Dhabi exceeded 200% during each of the last 10 years. The fastest-growing

MENA stock market since 2001 was Abu Dhabi, where the index rose by a remarkable 237%. Furthermore, this unprecedented price growth has been accompanied by a sharp jump in market capitalization. The fastest capitalization growth in the MENA region has been realized in the stock markets of Gulf Cooperation Council (GCC) countries, where their combined market capitalization increased from $120 billion in 2000 to $1.5 trillion in 2006. Despite this impressive growth, the MENA stock markets are still small in international terms and relative to other emerging markets.

Several factors are behind this rapid growth. The surge in oil revenues of MENA oil-exporting countries has fueled an economic boom that has created many profitable business opportunities for private firms. Companies involved in real estate, banking, and telecommunications have done particularly well over the past 3 years, and these good performances are reflected in their stock prices. Also, Arabs are now showing more interest in regional investment opportunities. Traditionally, Arab investors and financial institutions showed a strong preference for investing in U.S. and West European financial markets, but since the tragedy of September 11, they have increased their holdings of Middle Eastern financial assets. Finally, the recent spate of economic reforms and privatization across the region has created many new investment opportunities for the private sector. Several countries privatized their telecom industries over last 5 years, and these new firms have done well in the stock market. While oil assets are expected to remain fully under government control, Saudi Arabia is privatizing one of its petrochemical companies, which might set a precedent for other oil-exporting countries.

Since 2003, the rapid rise in oil prices has fueled an unprecedented rise in the GCC stock markets. This increase has worried investors that equities were being overvalued and eventually led to a sharp market correction and tremendous price volatility. A sharp decline in petroleum prices could trigger another market correction, but the likelihood of such an event occurring seems small given that the oil futures for the next 10 years are well above $100 per barrel.

29.3 EMPIRICAL METHODOLOGY

We adopt and reproduce the methodology from Chaudhuri and Wu (2003, 580):

> Our primary interest in this study is to test whether stock prices in GCC markets follow random walk or mean-reverting processes.

Let p_t^i be the natural logarithm of country i's stock-price index with dividends reinvested at time t, and $R_t^i = p_t^i - p_{t-1}^i$ be its continuously compounded return. Let T be the sample size. Consider the following process:

$$p_t^i = c^i + k^i p_{t-1}^i + v_t^i \tag{29.1}$$

where c^i is a constant parameter and v_t^i, is a stationary process that is allowed to be serially correlated, $i = 1, 2,\ldots, 5; t = 1, 2,\ldots, T$. If $k^i = 1$, the equity price follows a random walk; while if $k^i < 1$, the equity price is mean reverting.

The most widely used and accepted tests for the random walk hypothesis are the augmented Dickey and Fuller (1979, 1981; ADF) tests and the Phillips and Perron (1988; PP) tests. For the ADF tests, one subtracts p_{t-1}^i from both sides of Equation (29.1) to obtain

$$p_t^i - p_{t-1}^i = R_t^i = c^i + (k^i - 1)p_{t-1}^i + v_t^i \tag{29.2}$$

To conduct the tests, it is common to add lagged terms of the dependent variable and obtain the following equation:

$$R_t^i = c^i + b^i p_{t-1}^i + \sum_{j=1}^{m} \gamma_j^i R_{t-j}^i + v_t^i \tag{29.3}$$

where $b^i \equiv (k^i - 1)$. Equation (29.3) tests for the null hypothesis of a random walk against a mean stationary alternative.

Similar to Chaudhuri and Wu (2003, 580), the m

The m extra regressors R_{t-j}^i are added to eliminate possible nuisance-parameter dependencies in the asymptotic distributions of the test statistics caused by serial correlation in the error terms. For a given sample, if the estimate of k^i is not significantly different from unity, then the null hypothesis of a random walk cannot be rejected. On the other hand, if one finds that $k^i < 1$, then the alternative hypothesis of mean reversion is supported. The PP tests work in a similar way except that the extra regressors R_{t-j}^i are not included in the regressions, but the serial correlation of the residuals is corrected via a non-parametric approach. One significant drawback of the popular ADF and PP tests is that they have low power against the alternative of slow-speed mean reversion in small samples (see Campbell and Perron (1991), Cochrane (1991), and DeJong et al. (1992), among others). Therefore, failure to reject the null hypothesis may not be interpreted as decisive evidence against mean reversion. Because of this inherent problem, researchers have advocated pooling data and testing the hypothesis in a panel framework to gain test power. Our study follows this approach. We pool data of five stock markets to estimate the speed of reversion k^i, $i = 1, 2,...,$ 5. To improve estimation efficiency and gain statistical power, we exploit the information in the cross-country correlation of returns and estimate the system using the seemingly unrelated regression (SUR) technique. The panel-based test for the null hypothesis of no mean reversion ($k = 1$) is based on the estimated coefficients $b \equiv (k - 1)$ from Equation (29.3) and the t-statistic $t_\beta = (k - 1)/s(k^\wedge)$, where $k^\wedge$ is the panel estimate of k from either ordinary least squares (OLS) or SUR, and $s(k^\wedge)$ is the standard error of k. It is well known that under the null hypothesis of $k = 1$, the ADF and the PP statistics do not follow limiting normal distributions. We will therefore generate appropriate critical values for our exact sample size through Monte-Carlo simulations.

We describe the simulation procedure below, which is based on the test of Levin and Lin (1992).

First, we estimate Equation (29.3) using ordinary least squares under the null hypothesis by restricting b^i to zero. Following Rapach (2002), by using the restricted OLS estimate, we simulate a panel series of $T + 100$ observations for R_t^i, c^i, and γ_j^i, random draws from a $N(0, \sigma^{\wedge 2})$, where $\sigma^{\wedge 2}$ is the restricted OLS estimate of σ^2, and setting the initial p_{t-1}^i and R_{t-j}^i to zero. An additional 100 observations are generated but discarded to avoid initial value bias. This process is generated 2,000 times so as to achieve the 2,000 simulated panel series. We calculate and store the t_β statistics of the OLS panel test for each simulated panel and then order the simulated t_β statistics such that the 20th and 200th values are the 1% and 10% critical values, respectively.

A common problem with the preceding OLS-based procedure is it ignores the cross-sectional dependence. One way of addressing the cross-country stock market dependence is by estimating Equation (29.3) using a SUR estimator. The SUR estimator is basically a multivariate generalized least squares, using an estimate of the contemporaneous variance-covariance matrix of the disturbances obtained using the OLS residuals from

Equation (29.1). Following Rapach (2002), we estimate Equation (29.3) for the panel data using SUR and restricting b to zero. We generate the 2,000 simulated panel series of $100 + T$ observations using the restricted SUR parameter estimates of c^i and γ_j^i, random draws from a $N(0, \Sigma^\wedge)$, where $\Sigma^\wedge$ is the restricted SUR estimate of contemporaneous covariance matrix for the disturbances, and the initial p_{t-1}^i and R_{t-j}^i are set to zero. We drop the first 100 observations to yield simulated panel series of T observations. We calculate and store the t_β statistics of the SUR panel test for each simulated panel and then order the simulated t_β statistics such that the 20th and 200th values are the 1% and 10% critical values, respectively.

29.4 THE DATA

The data used in this paper are obtained from the Bloomberg database on the Middle East. The sample period is monthly from January 1996 to April 2008, with 148 observations of the stock-price indexes for the following five countries: Saudi Arabia, Kuwait, the Sultanate of Oman, Qatar, and the United Arab Emirates. These indexes include dividends and capital gains and are end-of-month quotes.

29.5 EMPIRICAL RESULTS

Table 29.1 reports the descriptive statistics of the monthly returns of the five stock markets in the sample. The table also provides the coefficient of variation (COF) for each stock market index, which represents the ratio of the standard deviation to the mean. The COF is a useful statistic for comparing the degrees of variation of each market, even if the means are drastically different from each other. In terms of risk per unit of return, Oman ranks the best and Kuwait is the least attractive.

Most distributions exhibit some degree of skewness but with significant variability in kurtosis. The skewness (S) and kurtosis (K) are computed as follows:

$$S = \frac{\sum_{i=1}^{N}(R_i - \bar{R})^3}{N\sigma^3}$$

and

$$K = \frac{\sum_{i=1}^{N}(R_i - \bar{R})^4}{N\sigma^4}$$

TABLE 29.1   Descriptive Statistics of Monthly Stock Market Returns in Five GCC
Countries

	Saudi Arabia	**Qatar**	**Oman**	**Kuwait**	**Abu Dhabi**
Mean	0.014	0.020	0.009	0.016	0.019
Median	0.015	0.014	0.006	0.016	0.009
Maximum	0.162	0.215	0.209	0.184	0.359
Minimum	−0.268	−0.210	−0.151	−0.129	−0.191
Std. dev.	0.070	0.073	0.061	0.049	0.079
Skewness	−0.756	0.164	0.565	0.182	0.848
Kurtosis	4.7	3.8	4.0	4.0	7.2
Coefficient of variation	0.194	0.277	0.146	0.330	0.243
Jarque-Bera	32.3	3.3	11.6	7.5	67.1
p-value	0	0.189	0.003	0.023	0
No. of monthly obs.	147	112	119	147	79

Note: The sample period is from January 1996 through April 2008 for Saudi Arabia and
Kuwait. Other countries are observed over a shorter period because their stock
markets are relatively new.

where $R_i, \bar{R}$ represent the return in week i and the average return for the
series, respectively. For a normal distribution, S and K are 0 and 3, respectively. Clearly, most markets exhibit substantial departures from normality. We formally tested for normality of the return distributions using the
Jarque-Bera statistic (JB). Under the null hypothesis of normality, JB is
distributed χ^2 with 2 degrees of freedom. JB is defined as

$$JB = \frac{T}{6}\left[S^2 + \frac{1}{4}(K-3)^2 \right]$$

where S and K represent the skewness and kurtosis. With the exception of
Qatar, it appears that returns in all four GCC stock markets are not normal.

Table 29.2 reports the monthly cross-correlation in returns among the
five stock markets. All of the cross-correlations are positive, and indeed

TABLE 29.2   Cross-Correlation of Monthly Stock Market Returns in Five GCC
Countries

	Saudi Arabia	**Qatar**	**Oman**	**Kuwait**	**Abu Dhabi**
Saudi Arabia	1				
Qatar	0.2483	1			
Oman	0.3520	0.2970	1		
Kuwait	0.2775	0.1069	0.3158	1	
Abu Dhabi	0.3727	0.3995	0.4090	0.3241	1

TABLE 29.3   ADF and PP Tests for Random Walk in the GCC Stock Market Prices
Using Mackinnon Critical Values

		Kuwait	**Qatar**	**Saudi Arabia**	**Oman**	**Abu Dhabi**
	Months	148	116	148	121	80
Augmented Dickey-Fuller						
	Lag length = 1	1.79	0.01	−0.96	4.49**	−1.07
	Lag length = 3	1.45	−0.02	−1.05	2.95*	−1
Phillips-Perron						
	Trun. lag = 1	2.32	0.23	−0.86	5.00**	−0.96
	Trun. lag = 3	2.03	0.23	−0.95	4.32**	−1.04
	Mackinnon 1% critical value	−3.48	−3.49	−3.48	−3.49	−3.52
	Mackinnon 10% critical value	−2.58	−2.58	−2.58	−2.58	−2.59

Note: This table reports augmented Dickey-Fuller (ADF) and Phillips-Perron (PP) tests
   for the random walk hypothesis for five stock markets in the GCC. Two lag lengths
   and truncation lags for the ADF and PP tests are selected.
Statistically significant at 10% (*), or 5% (**)

some of them are as high as 41% (e.g., Oman with Abu Dhabi). These
relatively high cross-sectional correlations motivate the use of the SUR
estimation technique. It appears that Abu Dhabi shows the highest correlations, with an average of 37.6% with its neighbors. It is followed by
Oman, Saudi Arabia, Qatar, and then Kuwait.

For the purpose of comparison, we first apply the standard ADF and PP
tests to each country and report the results in Table 29.3. For the tests, two
lag lengths are chosen (1 and 3) because it is known that the test results
may be sensitive to the choice of the lag length. Using the critical values[*]
from Mackinnon (1991), we find that the null hypothesis of random walk
can be rejected in favor of mean reversion at the 1% significance only in
the case of the Omani stock market.

It is well known that the distribution of the test statistics for the ADF and
the PP tests is nonnormal. To address this shortcoming, we compute actual
critical values for the exact sample size using Monte Carlo simulation with
2,000 replications under the null hypothesis of no mean reversion ($k = 1$)
with independent and identically distributed (iid) normal innovations. To
that end, we estimate Equation (29.3) using ordinary least squares under the
null hypothesis by restricting b^i to zero and simulate 2,000 panel series of

---

[*] The critical values vary slightly across countries because the number of observations is not
   identical for all five countries.

observations for R_t^i by using the restricted OLS estimate of c^i and y_j^i, random draws from a $N(0, \sigma^{^2})$ where $\sigma^{^2}$ is the restricted OLS estimate of σ^2, and setting the initial p_{t-1}^i and R_{t-j}^i to zero. The calculated t_β statistics of the OLS panels test yields the 1% and 10% critical values, which we report in Table 29.4.

Next, we further exploit the information on the cross-country correlations of returns and estimate the system of five equations using a SUR setup. The bottom part of Table 29.4 reports the estimation and testing results. The critical values obtained from the Monte Carlo simulation of the SUR model are significantly lower than the ones obtained from the OLS-based simulation because of the increase in estimation efficiency, which improves the test power. The results suggest that in addition to Oman, the null hypothesis of random walk can also be rejected in Kuwait, albeit at the 10% significance. Table 29.4 also reports the speed of reversion and implied half-life for each stock market. The speed is calculated as the parameter $k^{^}$ obtained after estimating the coefficient $b^{^}$ in the SUR model ran on Equation (29.3). The speed of mean reversion is also reported for

TABLE 29.4 Monte Carlo Simulation of Critical Values for the ADF and PP Tests for Random Walk in the GCC Stock Market Prices

		Kuwait	Qatar	Saudi Arabia	Oman	Abu Dhabi
Augmented Dickey-Fuller						
	Lag length = 1	1.79	0.01	−0.96	4.49**	−1.07
	Lag length = 3	1.45	−0.02	−1.05	2.95**	−1
Phillips-Perron						
	Trun. lag = 1	2.32*	0.23	−0.86	5.00**	−0.96
	Trun. lag = 3	2.03	0.23	−0.95	4.32**	−1.04
Monte Carlo Simulation of Critical Values						
	10%—OLS model	14.96	6.74	9.29	11.95	9.56
	1%—OLS model	16.13	7.37	10.09	13.19	10.62
	10%—SUR model	2.22	2.24	2.17	2.19	2.21
	1%—SUR model	2.94	2.92	3.07	2.95	2.92
Implied half-life		7.7	9.2	1.4	5.0	5.2
Speed of mean reversion		0.91	0.93	0.60	0.87	0.88
No. of observations		148	116	148	121	80

Note: This table reports augmented Dickey-Fuller (ADF) and Phillips-Perron (PP) tests for the random walk hypothesis for five stock markets in the GCC. Two lag lengths and truncation lags for the ADF and PP tests are selected.

Statistically significant at 10% (*), or 5% (**)

the unit root countries even though there is no evidence of mean reversion in their stock markets. A transitory deviation reverses to the trend path at a speed that varies between 0.60% and 0.93% per month with a wide range of half-lives that vary between 1.4 and 9.2 months. For Oman and Kuwait, the speeds are 0.91% and 0.87%, respectively, and are quoted per month. The implied half-life is calculated as $\log(1/2)/\log(b^\wedge)$, which translates to 7.7 months for Kuwait and 5 months for Oman. The half-life measures the time taken by the price to revert halfway to its long-term trend if no more shocks arrive. The half-life is a gauge of the speed of the mean reversion process. These half-lives are smaller than in more mature markets (see, for example, Balvers et al., 2000; Hakim and Neaime, 2003), suggesting that the adjustment process in these two countries requires proportionately less time than in developed markets.

29.6 CONCLUSION

A key area of research in the financial economics literature has focused on the mean reversion in stock prices. Regardless of the overabundance of studies since the 1960s, the existing literature has not attained a consensus on whether stock prices follow a unit root process. This information is crucial for investors, for if stock prices follow a random walk, then shocks to prices have a permanent effect. Stock prices will achieve a new equilibrium and potential returns cannot be based on past historical movements in stock prices. This also opens up the possibility that volatility in stock markets will increase in the long run without bound. However, if the prices of stocks are mean reverting, then random shocks to prices will only be temporary. This guarantees that investors may be able to predict future movements in stock prices based on past performance and create new trading strategies to produce abnormal returns.

This chapter considered mean reversion in five Middle Eastern countries' stock price indices by employing panel unit root testing approaches on monthly data over the period 1996–2008. We used three different panel unit root tests: the ADF, PP, and Monte Carlo simulations. It is known that the traditional tests, such as ADF and PP, for a random walk in stock prices do not have enough power when compared to other hypotheses of mean reversion in small samples. Due to the 12 years of data in our study, the power issue is of concern. Consequently, we combine data from five countries and apply a panel-based test. The test utilizes cross-country information and increases the efficiency of estimation. This allows us to recognize a slow mean reverting constituent in equity prices and significantly increases the power. The tests suggest that the stock prices for Saudi Arabia, Kuwait, and Abu Dhabi follow a random walk and are characterized by a unit root, consistent with the efficient market hypothesis. However, there is evidence of mean reversion in Kuwait and Oman. The gain in test power permits us to discard the random walk hypothesis and support for mean reversion at conservative significance levels for Oman. We estimate the half-lives for Oman and Kuwait to be between 5 and 8 months, which are faster than those found in developed markets.

REFERENCES

Assaf, A. (2006). The stochastic volatility in mean model and automation: Evidence from TSE. *Quarterly Review of Economics and Finance* 46:241–53.

Balvers, R., Wu, Y., and Gilliland, E. (2000). Mean reversion across national stock markets and parametric contrarian investment strategies. *Journal of Finance* 55:745–72.

Campbell, J. Y., and Perron, P. (1991). *Pitfalls and opportunities: What macroeconomists should know about unit roots.* National Bureau of Economic Research Macroeconomics Annual. Cambridge, MA: MIT Press.

Chaudhuri, K., and Wu, Y. (2003). Random walk versus breaking trend in stock prices: Evidence from emerging markets. *Journal of Banking and Finance* 27:575–92.

Cochrane, J. H. (1991). A critique of the application of unit root tests. *Journal of Economic Dynamics and Control* 15:275–84.

DeBondt, W., and Thaler, R. (1985). Does the stock market overreact? *Journal of Finance* 40:793–805.

DeJong, D. N., et al. (1992). The power problems of unit root tests in time series with autoregressive errors. *Journal of Econometrics* 53:323–43.

Dickey, D. A., and Fuller, W. (1979). Distribution of the estimators in autoregressive time series with a unit root. *Journal of the American Statistical Association* 74:427–31.

Dickey, D. A., and Fuller, W. (1981). Likelihood ratio statistics for autoregressive time series with a unit root. *Econometrica* 49:1057–82.

Fama, E. F., and French, K. R. (1988). Permanent and temporary components of stock prices. *Journal of Political Economy* 96:246–73.

Ferson, W. E., Sarkissian, S., and Simin, T. T. (2003). Spurious regression in financial economics. *Journal of Finance* 58:1393–413.

Hakim, S., and Neaime, S. (2003). Mean reversion across MENA stock markets: Implications for derivative pricing. *International Journal of Business* 8:345–56.

Kim, M. J., Nelson, C. R., and Startz, R. (1991). Mean reversion of stock prices: A reappraisal of the empirical evidence. *Review of Economic Studies* 58:5151–280.

Lo, A. W., and MacKinlay, A. C. (1988). Stock market prices do not follow random walks: Evidence from a simple specification test. *Review of Financial Studies* 1:41–66.

Mackinnon, J. G. (1991). Critical values for cointegration tests. In *Long-run economic relationships: Readings in cointegration*, ed. R. F. Engle and C. W. J. Granger, chap. 13. Oxford: Oxford University Press.

Phillips, P., and Perron, P. (1988). Testing for a unit root in time series regression. *Biometrika* 75:335–46.

Poterba, J. M., and Summers, L. H. (1988). Mean reversion in stock prices: Evidence and implications. *Journal of Financial Economics* 22:27–59.

Rapach, David E. (2002). The long-run relationship between inflation and real stock prices. *Journal of Macroeconomics* 24:331–51.

Richards, A. J. (1997). Winner-loser reversals in national stock market indexes: Can they be explained? *Journal of Finance* 52:2129–44.

Richardson, M. (1993). Temporary components of stock prices: A skeptic's view. *Journal of Business and Economic Statistics* 11:199–207.

Richardson, M., and Stock, J. H. (1989). Drawing inferences from statistics based on multiyear asset returns. *Journal of Financial Economics* 25:323–48.

CHAPTER 30

Stock Market Volatility and Market Risk in Emerging Markets

Evidence from India

Sumon Kumar Bhaumik, Suchismita Bose, and Rudra Sensarma

CONTENTS

30.1 INTRODUCTION

One of the main pillars of the process of global integration is the rapid rise in the volume of cross-border capital flows since the early 1990s, and an important aspect of this phenomenon is the significant increase in the flow of portfolio investment in emerging market securities. Gross portfolio investment in emerging markets increased from USD 12.57 billion in 1990 to

USD 213.39 billion in 2006.[*] The main beneficiaries of this increase were the emerging Asian economies, whose share of these flows increased from 9.5% to 67.4% over the same period. A significant proportion of this investment was in equities, the proportion being as high as 90% in countries like India.

However, investment in emerging market equities is fraught with risk. To begin with, there is exchange rate risk. In the absence of institutions like clearing houses, counterparty risk can be significant as well. Finally, foreign investors in emerging equity markets are exposed to substantial market risk. There are numerous sources for this market risk. Emerging equity markets are often thin. This, by itself, can add to volatility of stock prices and returns, given the negative relationship between trading volumes and volatility that has been estimated in several contexts (e.g., Pyun et al., 2000), and this volatility can be further increased if trading in these thin markets is dominated by large institutional investors (Gabaix et al., 2006). Lack of adequate liberalization of capital markets might itself be a source of excess volatility in these contexts (Bekaert and Harvey, 1997; Huang and Yang, 2000). In particular, in the absence of appropriate regulations, the market can be destabilized by phenomena such as insider trading (Du and Wei, 2004). It has also been hypothesized that investment by foreign investors might itself increase the volatility of emerging stock markets, but there is as yet no empirical evidence in support of this hypothesis (e.g., Choe et al., 1999).

Despite the rising portfolio investment in emerging markets, and the market risk associated with such investments, however, there have been few attempts, if any, to quantify this risk. In this chapter, we address this lacuna in the empirical literature, by estimating value-at-risk (VaR) of three important (and widely cited) market indices at the Bombay Stock Exchange (BSE). The choice of the Indian equity market is easily justified. Starting in the early 1990s, the Indian capital market witnessed the creation of institutions like screen-based trading, clearing houses, dematerialized transactions, and a market regulator in the form of the Securities and Exchange Board of India (SEBI). Weeklong trading cycles have been abandoned in favor of $T + n$ rolling settlement, derivatives trading on market indices and individual stocks has been introduced, and short selling has been legalized. Finally, as highlighted in Figure 30.1, there has been a significant inflow of foreign portfolio investment in India since 2003–04.

---

[*] Source: International Monetary Fund (http://www.imf.org/externa l/pubs/ft/weo/2007/02/c 1/FIG1_15.csv).

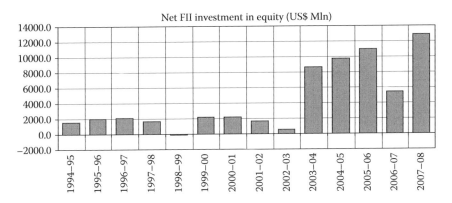

FIGURE 30.1 Foreign portfolio investment in India.

Our results suggest that the 5% VaR of these indices is significantly larger than the average returns, implying that the market risk is significant. There is also evidence to suggest that there was a decline in the market risk (or VaR) after the introduction of derivatives trading at BSE.

The rest of the chapter is structured as follows: In Section 30.2, we discuss the measure of market risk, namely, VaR. The data are discussed in Section 30.3, and the modeling of the data generating process of the returns in Section 30.4. The VaR estimates are reported in Section 30.5, and Section 30.6 concludes.

30.2 VALUE-AT-RISK

In the context of asset prices, the VaR is defined as the estimated loss of value of the asset over a given time period with a very high probability, p. Conversely, any actual loss of value of the asset would exceed the VaR with a very small probability, $1 - p$. A fairly widely used value of p is 0.95. As explained by Füss et al. (2007), under the assumption that the asset returns are normally distributed, VaR of an asset is given by

$$VaR = -(\phi.\sigma - \mu) \qquad (30.1)$$

where ϕ is the density function of the aforementioned distribution, with a standard deviation σ and a mean μ. To recapitulate, σ^2 or variance is a stylized measure of volatility.

In principle, it is possible to measure volatility using simple measures of dispersion such as standard deviation (of returns), and more sophisticated measures would then involve the use of rolling standard deviations such that market risk on any one day t is measured as the dispersion in returns over the previous k days (e.g., Bhaumik and Coondoo, 2003). However, it is now well known that volatility of asset returns is fairly persistent, such that the volatility experienced on a given day is influenced or conditioned by the volatility experienced during the previous days. On account of this observation, presently, stylized modeling of the data generating process of asset returns involves the use of ARCH (autoregressive conditional heteroskedasticity). The ARCH model, first proposed by Engle (1982), is characterized by the following:

$$y_t = \alpha_0 + \sum_{i=1}^{m} \varpi_i x_{t-i} + \sum_{j=1}^{n} \pi_j u_{t-j} + u_t \tag{30.2}$$

$$u_t = \sqrt{h_t}\, v_t \tag{30.3}$$

$$h_t = \beta_0 + \beta_1 u_{t-1}^2 \tag{30.4}$$

where Equation (30.2) suggests that the returns follow a ARMA(m,n) process, h_t is the conditional variance of the error term, and v_t is an *iid* term that has a standard normal distribution with zero mean and a variance of 1. The ARMA characterization of the mean equation is necessitated by the possibility of serial dependence in the series of asset returns.

Bollerslev (1986) extended and generalized Engle's specification by restating Equation (30.4) as follows:

$$h_t = \beta_0 + \beta_1 u_{t-1}^2 + \beta_2 h_{t-1} \tag{30.5}$$

Equations (30.2), (30.3), and (30.5) together constitute the generalized ARCH (or GARCH) model. In our illustration, we have outlined the commonly used ARCH(1) and GARCH(1,1) models. However, in principle, ARCH(x) and GARCH(x,y) models can be of higher orders, i.e., $x > 1$ and $y > 1$. The choice between AR(x) and GARCH(x,y) models can be made on the basis of information criteria. Typically, GARCH(1,1) is found to be a reasonable generalization of higher-order ARCH(x) models.

Once a GARCH model has been estimated, the estimated conditional standard deviation, $\sqrt{h_t}$, can be used to replace σ, the unconditional standard deviation, in Equation (30.1), thereby yielding a more accurate estimation of VaR (Füss et al., 2007). In other words, we have

$$VaR = -\left(\phi.\sqrt{h_t} - \mu_t\right) \qquad (30.6)$$

In this chapter, we use this refined measure of VaR.

30.3 DATA

The data for the analysis were collected from the publicly available archives of Bombay Stock Exchange (BSE), whose thirty-stock market index, the Sensex, is the face of India's capital market to the rest of the world. As highlighted in Panel 2.A of Figure 30.2, trading at BSE has grown steadily since the initiation of the reform of the Indian capital market in the early 1990s, shaking off the impact of the financial crisis in Southeast Asia, the sanctions imposed by much of the industrialized world in response to the nuclear tests of 1998, and a mini-war with Pakistan in 1999. The exponential growth in both dollar-denominated volumes and number of trades was stalled or reversed in 2001–02,[*] in the aftermath of the bursting of the dot-com bubble and 9/11. However, after steady yet unremarkable growth for 4 years, growth in dollar-denominated volumes and trades has once again been exponential since 2005–06. As highlighted in Panel 2.B of the figure, the growth in volumes and trades since the turn of this century has also coincided with an exponential growth in BSE's market capitalization.

In this chapter, we examine the VaR of three market indices at BSE: the aforementioned Sensex, the 100-stock BSE100, and the 200-stock BSE200. Since the average liquidity of the underlying stocks of the three indices differ, with the average liquidity of the Sensex stocks being the highest and that of BSE200 stocks being the lowest, it would allow us to examine the impact of liquidity on VaR. In Panel 3.A of Figure 30.3, we highlight the trends in the three market indices. In keeping with the trend in market capitalization, the BSE indices have risen sharply since April 2004, i.e., since the 2004–05 financial year, and series look correlated. The decline in the indices during the second half of 2007–08 was on account of the widely anticipated slowdown in the global economy that has been brought about by the subprime crisis and high commodity prices. In Panel 3.B, we

---

[*] India's financial year runs from April of one calendar year to March of the following calendar year.

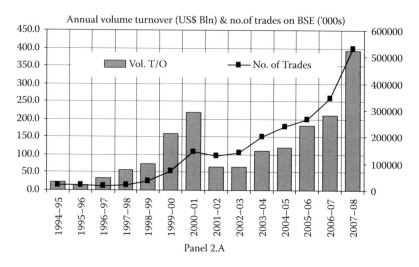

Panel 2.A

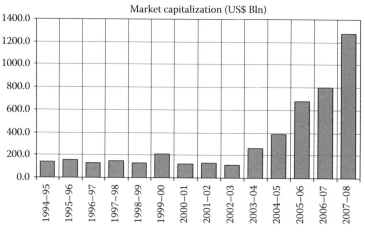

Panel 2.B

FIGURE 30.2   Bombay Stock Exchange.

highlight the returns to the 30-stock Sensex, when returns are computed as the difference in the logarithm of the index in two successive trading days. A visual examination of the returns suggests that, with the exception of 2007–08, volatility of returns has been lower since 2001–02. Since trading in stock index futures was introduced at the BSE in June 2000, a plausible interpretation is that derivatives trading has helped reduce cash market volatility at the exchange.

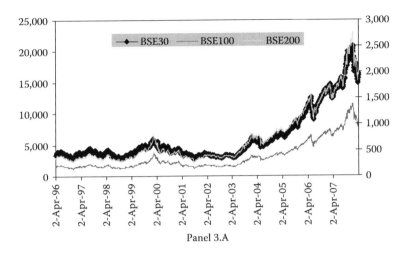

Panel 3.A

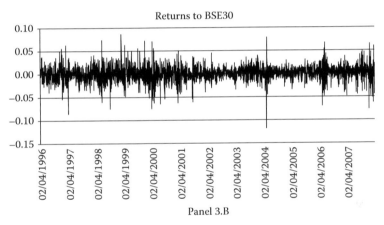

Panel 3.B

FIGURE 30.3   Stock index movements at the Bombay Stock Exchange.

In Table 30.1, we report the descriptive statistics for the data. Given our observation about the likely impact of derivatives trading on market risk at BSE, we treat the introduction of equity derivatives trading at the exchange (i.e., June 2000) as the natural break point in our data, and hence report the summary statistics for returns for the pre- and post-derivatives periods. It can be seen that the mean and, especially, median returns were significantly higher after June 2000, while market risk, as measured by standard deviation, was noticeably lower. This is consistent with our observation based on Panel 3.B of Figure 30.3. The summary statistics also suggest that the distribution of returns in both time periods is leptokurtic and nonnormal.

TABLE 30.1   Descriptive Statistics

	BSE Sensex		BSE100		BSE200	
	Pre-derivatives	Post-derivatives	Pre-derivatives	Post-derivatives	Pre-derivatives	Post-derivatives
Mean	0.023	0.065	0.029	0.069	0.026	0.074
Median	0.030	0.141	0.025	0.199	0.006	0.195
Std. dev.	1.863	1.517	1.846	1.600	1.774	1.589
Skewness	0.030	−0.626	−0.037	−0.754	−0.049	−0.899
Kurtosis	5.031	7.430	5.179	7.814	5.154	8.725
Jarque-Bera	174.678	1733.134	200.940	2081.797	196.707	2945.781
Observations	1015	1963	1015	1963	1015	1963

We have tested for unit roots and signs of autocorrelation in the data, using the augmented Dickey-Fuller (ADF) statistic and the Ljung-Box (Q) statistic, respectively. Estimates of these statistics are reported in Table 30.2; the p-values of the statistics are reported within parentheses. The ADF test statistics suggest that the null hypothesis of unit root can be rejected at the 1% level for returns to all three market indices. Estimates of the Q statistic, on the other hand, indicate the presence of autocorrelation in returns as well as squared returns. This, in turn, implies that the use of ARMA might be appropriate for modeling these returns. We shall discuss this in greater detail in the next section.

30.4   MODELING RETURNS

As mentioned in the previous section, the returns to the three BSE indices chosen for analysis are autocorrelated, and the series do not have unit roots. Hence, an ARMA(m,n) specification can be used to model the underlying data generating process. The usual practice is to fit a number of ARMA models, for different values of m and n, and then to choose one on the basis of information criteria and properties of the estimated

TABLE 30.2   Unit Root and Autocorrelation Tests

	Sensex	BSE100	BSE200
ADF test statistics	−50.825 (0.000)	−49.051 (0.000)	−48.666 (0.000)
Q(2)	16.098 (0.000)	32.954 (0.000)	38.659 (0.000)
Q(5)	20.063 (0.001)	34.996 (0.000)	41.159 (0.000)
$Q^2(2)$	284.71 (0.000)	440.42 (0.000)	528.84 (0.000)
$Q^2(5)$	442.61 (0.000)	640.98 (0.000)	706.14 (0.000)

TABLE 30.3   ARMA Modeling of the Returns

	SENSEX	BSE100	BSE200
Constant	0.032 (0.558)	0.039 (0.499)	0.035 (0.539)
Derivatives	0.029 (0.674)	0.025 (0.732)	0.033 (0.643)
MA(1)	0.073 (0.000)	0.109 (0.000)	0.120 (0.000)
AIC	3.827	3.874	3.832
SIC	3.833	3.880	3.838
Q(2)	2.089 (0.148)	0.714 (0.398)	1.444 (0.229)
Q(5)	5.694 (0.223)	2.521 (0.641)	3.709 (0.447)
$Q^2(2)$	308.75 (0.000)	467.84 (0.000)	565.86 (0.000)
$Q^2(5)$	470.33 (0.000)	674.09 (0.000)	755.72 (0.000)
ARCH LM(2)	254.00 (0.000)	363.74 (0.000)	431.74 (0.000)
ARCH LM(5)	296.66 (0.000)	397.22 (0.000)	453.08 (0.000)

residuals. Thereafter, one has to test for ARCH effects that are characteristic of financial time-series data.

In order to capture a specific aspect of the Indian stock market, and on the basis of our casual observation of the returns data highlighted in Figure 30.3 (Panel 3.B), we extend the ARMA specification in Equation (30.2) to include a dummy variable that takes the value unity for all trading days from June 2000. The dummy variable captures the marginal effect of the introduction of derivatives trading at BSE. In other words, our ARMA specification is

$$y_t = \alpha_0 + \sum_{i=1}^{m} \varpi_i x_{t-i} + \sum_{j=1}^{n} \pi_j u_{t-j} + \theta Dummy + u_t \qquad (30.7)$$

The estimates of the ARMA model are reported in Table 30.3; the p-values for the estimated coefficients and test statistics are reported within parentheses. For each series, we report the estimates of only the model that has the best fit, as indicated by Akaike's (AIC) and Schwarz's (SIC) information criteria. Interestingly, MA(1), i.e., the choice of $m = 0$ and $n = 1$, provides the best fit for all three data series. Earlier research on stock returns in India had found that AR(4) fits data on stock returns in India well (Bhaumik and Bose, 2008). Since MA(1) is a parsimonious equivalent of higher-order AR(m) models, our results are consistent with those in the existing research.

The Ljung-Box (or Q) statistics indicate that the residuals of the MA(1) model are not autocorrelated, but that the squared residuals are correlated. Further, the residuals are leptokurtic. These suggest that there are possible ARCH effects that have to be taken into account, and this is verified by the statistics of Lagrange multiplier (LM) tests that reject the null hypothesis of no ARCH. The existence of ARCH effects is consistent with our methodology for estimating VaR, which replaces unconditional volatility of returns with conditional GARCH volatility.

The generalized form of the model that should therefore be estimated is as follows:

$$y_t = \alpha_0 + \sum_{i=1}^{m} \omega_i x_{t-i} + \sum_{j=1}^{n} \pi_j u_{t-j} + \theta Dummy + u_t$$

$$u_t = \sqrt{h_t} v_t$$

$$h_t = \beta_0 + \sum_i \beta_{1i} u_{t-i}^2 + \sum_j \beta_{2j} h_{t-j} + \psi Dummy \qquad (30.8)$$

when β_{1i} capture the ARCH effects and β_{2j} capture the GARCH effects. In keeping with the existing literature (Bhaumik and Bose, 2008), as well as the summary statistics reported in Table 30.1, we posit that structural changes like the introduction of derivatives trading at BSE would affect not only the mean returns but also the (stochastic) volatility of these returns.

Next, one has to choose among ARCH(x) and GARCH(x,y) models, i.e., the appropriate values of x and y, and as before, the choice is made on the basis of information criteria and properties of the residuals. In Table 30.4, we report only the estimates of the model that best fit the data, as indicated by AIC and BIC. It can be seen that GARCH(1,1), which is a parsimonious approximation of higher-order ARCH models, provides the best fit for the data for all three indices, and this is consistent with experiences of other attempts to model financial time series both in the Indian context and elsewhere. The Q and LM statistics for the GARCH(1,1) models indicate that there is no remaining autocorrelation within residuals and that there are no ARCH effects.

An interesting aspect of the results is that while the introduction of derivatives trading has had no impact on the returns themselves, it has significant reduced volatility of the returns to all three market indices. This is consistent with the evidence that suggests that, by and large, trading in futures contracts based on market indices and individual stocks, which constitute the overwhelming majority of the derivatives traded at

TABLE 30.4   GARCH Modeling of Returns

	SENSEX	**BSE100**	**BSE200**
Mean Equation			
Constant	0.099 (0.107)	0.126 (0.035)	0.118 (0.042)
Derivatives	0.044 (0.522)	0.027 (0.688)	0.039 (0.560)
MA(1)	0.102 (0.000)	0.129 (0.000)	0.137 (0.000)
Variance Equation			
Constant	0.317 (0.000)	0.267 (0.000)	0.253 (0.000)
Derivatives	−0.185 (0.000)	−0.128 (0.000)	−0.116 (0.000)
ARCH(1)	0.148 (0.000)	0.170 (0.000)	0.172 (0.000)
GARCH(1)	0.785 (0.000)	0.771 (0.000)	0.769 (0.000)
AIC	3.631	3.641	3.598
SIC	3.645	3.655	3.612
Q(2)	1.091 (0.296)	1.861 (0.173)	1.742 (0.187)
Q(5)	10.395 (0.034)	11.778 (0.019)	13.015 (0.011)
$Q^2(2)$	0.455 (0.500)	1.178 (0.278)	1.178 (0.278)
$Q^2(5)$	2.5614 (0.634)	2.250 (0.690)	2.729 (0.604)
ARCH LM(2)	0.451 (0.799)	1.180 (0.555)	1.180 (0.554)
ARCH LM(5)	2.526 (0.773)	2.254 (0.813)	2.761 (0.737)

BSE, reduces volatility in cash markets (see Gulen and Mayhew, 2001). The implication is that market risk at BSE has been reduced by the introduction of derivatives trading, and provides *ex post* justification for the introduction of derivatives products at Indian stock exchanges.

To summarize, our analysis indicates that all three market indices of BSE, namely, BSE30 (or Sensex), BSE100, and BSE200, have a data generating process that is captured best by a MA(1)-GARCH(1,1) model. The similarity across the indices is not surprising, given the similarity in the movements of the indices over time. Importantly, the GARCH model generates estimates of conditional variance that can be used to compute values of VaR for the market indices over the 12-year period.

30.5 VALUE-AT-RISK

The 5% VaR for the three market indices is generated for each trading day,[*] using Equation (30.7), and from the daily VaR estimates, monthly averages were computed. The monthly averages of VaR and the returns to the

---

[*] The loss on account of decline in stock prices will exceed this VaR with only a 5% probability.

indices are highlighted in Figure 30.4. Two things are immediately obvious: First, the daily values of the 5% VaR are large relative to the daily returns to the indices, bringing into question the nature of the Sharpe's ratio for BSE, and hence the rationale for the significant increase in portfolio flows into India since 2003–04. Second, on average, VaR values were lower for 2001–02 and beyond, relative to the pre-2001 values. Even though derivatives trading commenced at BSE in June 2000, allowing for a gestation period for enhancement of the maturity and depth of the market, and taking into consideration the coefficient estimates of the GARCH model, it is reasonable to conclude that the decline in VaR roughly coincided with the introduction of derivatives trading. Interestingly, the significant increase in inflow of overseas portfolio investment did not add to VaR from after 2003–04, indicating that, contrary to apprehensions in certain quarters, foreign portfolio flows have not destabilized the Indian stock markets by adding to market risk.

In order to further emphasize the reduction in market risk in India since the early years of this decade, we report the distributions of VaR of the pre- and post-derivatives years for all three indices. The distributions are highlighted in Figure 30.5. It is easily seen that the likelihood of the VaR being 4% or less is much higher in the post-derivatives period than in the pre-derivatives period. Indeed, while a significant proportion of the post-derivatives VaR values are of the order of magnitude of 2%, none of the pre-derivatives VaR values are of that magnitude.

Note that even though the VaR values of the market indices have declined since 2001–02, market risk was very high during some of the months. However, each of these episodes of large VaR in the post-2001–02 period can be explained by large shocks, suggesting that the spikes in the VaR values were on account of exogenous factors rather than on account of exchange-specific factors like noise trading. For instance, the spike in March 2001 was caused by events surrounding revelations that unscrupulous brokers were manipulating prices and a media exposure of government corruption in defense deals. The heightened risk in May 2004 was engendered by political instability in India with uncertainty over government formation following inconclusive election results. The sharp rise in volatility in June 2006 was caused by turmoil in international financial markets, especially in other emerging markets, and finally, March 2008 witnessed fears arising out of the collapse of Bear Stearns and the U.S. feds' emergency cut in discount rate. In fact, the VaR values for the market indices have increased considerably since 2007–08, highlighting the impact of

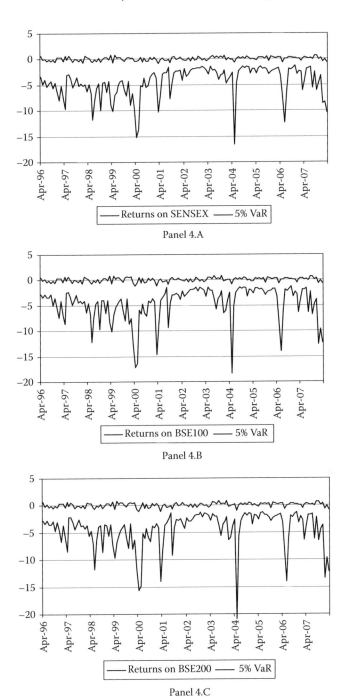

Panel 4.A

Panel 4.B

Panel 4.C

FIGURE 30.4   Estimates of value-at-risk.

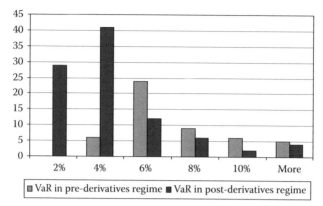

Panel 5.A

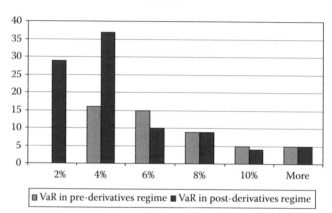

Panel 5.B

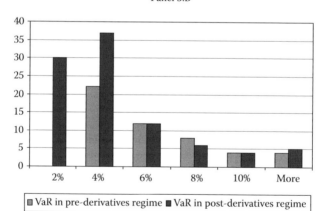

Panel 5.C

FIGURE 30.5 Impact of derivatives trading on value-at-risk.

the global credit crisis and prospects of higher (and sustained) inflation on investors in Indian stocks.

30.6 CONCLUSION

Since the second half of the 1990s, there has been a significant increase in the flow of portfolio capital into emerging markets. However, while the returns from these markets have been attractive, and while the returns often reflect the sound growth potential of the real sectors of these countries, market risk associated with portfolio investment in emerging markets can be significant. Yet, there have been few attempts, if any, to quantify the extent of the market risk. In this chapter, we addressed this lacuna in the literature, using three different market indices from BSE, one of the largest—and certainly the best known—stock exchanges in India, a country that has witnessed very significant inflows of portfolio capital over the past decade.

Our results indicate that while the 5% VaR values of returns for all three indices are skewed toward the lower tail of the distribution, the magnitude of the VaR is quite large in relation to the returns themselves. This brings into question the efficacy of portfolio investment in emerging markets, and raises concerns about the extent to which market risk is factored into decisions to invest in these markets. We also demonstrate that the returns have increased, on average, and the VaR has declined significantly since the introduction of equity derivatives—stock index futures being the most frequently traded instrument—at BSE. In other words, the risk-to-return ratio for equity investment at BSE has improved since the introduction of derivatives trading, and this has obvious implications for financial development in emerging markets that aim to attract greater portfolio investment.

REFERENCES

Bekaert, G., and Harvey, C. R. (1997). Emerging equity market volatility. *Journal of Financial Economics* 43:29–77.

Bhaumik, S. K., and Bose, S. (2008). Impact of derivatives trading on emerging stock markets: Some evidence from India. *Comparative Economic Studies*, forthcoming.

Bhaumik, S. K., and Coondoo, D. (2003). Econometrics of yield spreads in the money market: A note. *Applied Financial Economics* 13:645–53.

Bollerslev, T. (1986). Generalized autoregressive conditional heteroskedasticity. *Journal of Econometrics* 31:307–28.

Choe, H., Kho, B.-C., and Stulz, R. M. (1999). Do foreign investors destabilize stock markets? The Korean experience in 1997. *Journal of Financial Economics* 54:227–64.

Du, J., and Wei, S.-J. (2004). Does insider trading raise market volatility? *Economic Journal* 114:916–42.

Engle, R. F. (1982). Autoregressive conditional heteroskedasticity with estimates of the variance of UK inflation. *Econometrica* 50:987–1008.

Füss, R., Kaiser, D. G., and Adams, Z. (2007). Value at risk, GARCH modelling and the forecasting of hedge fund return volatility. *Journal of Derivatives and Hedge Funds* 13:2–25.

Gabaix, X., Gopikrishnan, P., Plerou, V., and Stanley, H. E. (2006). Institutional investors and stock market volatility. *Quarterly Journal of Economics* 121:461–504.

Gulen, H., and Mayhew, S. (2001). Stock index futures trading and volatility in the international equity markets. *Journal of Futures Markets* 20:661–85.

Huang, B.-N., and Yang, C.-W. (2000). The impact of financial liberalization on stock price volatility in emerging markets. *Journal of Comparative Economics* 28:321–39.

Pyun, C. S., Lee, S. Y., and Nam, K. (2000). Volatility and information flows in emerging equity market: A case of the Korean Stock Exchange. *International Review of Financial Analysis* 9:405–20.

CHAPTER 31

Stock Market Volatility and Political Risk in Latin America

The Case of Terrorism in Colombia

Ignacio Olmeda and Daniel Sotelsek

CONTENTS

31.1 INTRODUCTION

Stock markets are probably the most efficient markets in the world for pricing news, whether intrinsic (e.g., earning announcements of companies) or external (e.g., a climate catastrophe). Several episodes confirm this view; as an example, one might remember that levels of oil prices are hardly anticipated by policy markers, while stock markets are able to predict well in advance a consistent increase in such prices. Keynes (1936) coined the term "animal spirits" as an explanation of wild movements in stock market prices that were not explained by fundamentals. Maybe animal spirits only live in stock markets, and for this reason, only stock markets are able to evaluate how to assign an economic value to news.

In the context of this chapter, virtually every citizen of some (ideal) country would declare a clear rejection of terrorism, and probably also everyone, and with the same determination, would express his conviction that a terrorist act would have immediate negative consequences for the economy (causing, for example, a downturn of the stock market). It is obvious that both opinions are perfectly respectable, but it is also obvious that the second one is a consequence of the first and translates to the real world the ethical biases of the individual. In the context of stock markets, one would argue that the second opinion should at least be tested since animal spirits could determine the irrelevance of such an incident.

In recent times, the empirical investigation in economics and finance is paying more attention to political and social issues that might condition the behavior of markets. For example, some authors (Kurtzman et al., 2004) have demonstrated that *opacity* (defined as the degree to which countries lack clear, accurate, easily discernible and widely accepted practices governing the relationships among businesses, investors, and governments) has a clear impact on the economic activity; for example, they show that every 1-point increase in the opacity index lowers GDP per capita by about US$1,000 and decreases market capitalization by about 1%. Following similar lines of research, and from a macroeconomic viewpoint, some authors (e.g., Alesina and Perroti, 1996) analyze political instability and show that it has a direct effect on the economy, lowering growth and investment rates.

Intuitively, political phenomena such as terrorism would affect more rapidly markets where sentiment can have an immediate effect on supply and demand (for example, a tourist may cancel his reservations in the case that the destination is affected by any kind of terrorist event, affecting prices on the tourism products of the destination, such as hotel rates). The stock market is paradigmatic since investors may take buying and selling decisions in a fast and relatively cheap way so that, again intuitively, one would expect an immediate reaction of the stock market behavior to particular news.

Surprisingly, the analysis on the influence of terrorism on the stock market is quite recent (Karolyi, 2006), particularly since the attacks of 2001, and references in this area are still scarce. Among them, Abadie and Gardeazabal (2003) analyzed the effects of terrorism in Spain and showed that market capitalization is negatively affected by terrorism of ETA.

Guidolin and La Ferrara (2005) analyzed data of internal conflicts for over 30 years and showed that they produce a significant impact both on stock markets indexes and on commodity prices, and Karolyi and Martell (2006) found a negative reaction of stock prices of firms that have suffered some terrorist attack. Nevertheless, as we mention, this area of study is relatively unexplored, and this motivates our study.

In this chapter we analyze one aspect of political fragility in Latin America and its incidence on the stock market. Specifically, we will try to determine how news related to terrorism affects the volatility level of the Colombian stock market. The choice of this market was for several reasons, with two being the most important. First, the Colombian stock market is developed enough to permit conducting empirical analyses; without such a degree of maturity, it would be impossible to determine whether external factors produce a particular reaction, or if it is due to microstructural biases typically present in underdeveloped markets. Second, in the Colombian case terrorism is, unfortunately, a structural problem; it is so frequent and permanent in time that it permits the building of a complete database of terrorist episodes, which can be employed to evaluate empirically their economic consequences under different scenarios.

The rest of the chapter is organized as follows: In Section 31.2 we provide a brief introduction to the phenomenon of terrorism in the Colombian case; we motivate its significance and the economic consequences that have been pointed out in some studies. In Section 31.3 we provide a description of the Colombian stock market, which has experienced tremendous development and which has consistently been positioned, until recent times, as one of the best investment alternatives in the world. Section 31.4 is devoted to empirical analyses. After describing the main patterns of volatility in the Colombian stock market, we estimate multivariate ARCH models to evaluate the effect of news related to terrorism in stock market's volatility. Our main conclusion is that terrorism does not seem to affect the levels of risk in the market. This conclusion is stable along the period of study, which expands along more than 12 years, and after controlling for other factors, such as the level of volatility in the stock markets in the area. A brief recapitulation of the main findings and conclusions close the chapter.

31.2   TERRORISM IN COLOMBIA

Colombia has had the misfortune, for several years, to lead the world ranking in terms of homicides and kidnappings. Terrorism is so deeply imbricated in the Colombian society that it would be impossible to understand the country without a reference to the phenomenon. The Colombian case is particularly complex for several reasons; among them, terrorism is strongly linked to other forms of crime that affects severely the economy. Smuggling, drug trafficking, and money laundering are immediate consequences of terrorism.* Colombian terrorism is also quite idiosyncratic, since it involves several groups with quite different motivations and social support (e.g., the guerrillas versus the paramilitaries).

As we have just mentioned, terrorism in Colombia is, sadly, so frequent and ubiquitous that it would be a futile task to try to produce here a coherent description of the problem. Since the period of extreme violence known as *La Violencia* (1948), where about 200,000 people were killed, Colombia has submerged on the drama of terrorism. It has produced thousands of deaths, and newspapers are full of news related to the phenomenon. As an intuition of the intensity of the problem, the database used in this paper includes more than 500 news items related to terrorism just for the period of study (1996–2008). In Table 31.1 we provide an incomplete chronology of some landmarks in the recent history of Colombian terrorism.

The economic impact of terrorism in Colombia has been analyzed in a number of papers that, from a macroeconomic viewpoint, have shown that terrorism has damaged quite severely the Colombian economy. For example, Cárdenas (2007) shows that crime that, as we mentioned, is closely related to terrorism in the Colombian case, implied a reduction of the output per worker at a rate of about 1% per year during the 1980s. At the microeconomic level, as an example, Pshisva and Suarez (2006) show that terrorist kidnappings that directly target firms have a significant effect on corporate investment. The curse of terrorism in Colombia is, then, clear and reveals one point of relative weakness when compared to the main countries on the LATAM area. In Table 31.2 we show how some factors related to terrorism affect the business climate using estimations of the World Economic Forum (2007). Note that, by far, Colombia shows the worst performance of the countries in the area.

To our knowledge, no study has ever analyzed how such an important idiosyncratic phenomenon affects the Colombian stock market. This is surprising since it is known that the development of the Colombian economy (and especially the development of its small and medium enterprises (SMEs)

---

* The United Nations calculates that about 60% of the cocaine in the world comes from coca leaf cultivated in Colombia.

TABLE 31.1   Brief Chronology of Terrorism Landmarks, 1996–2008

1996	March: Ernesto Samper Pizano (elected president) has been charged of receiving drug cartel money for his election campaign, and he declared that he is willing to leave office before the end of his term, as a way to resolve the political crisis caused by the investigation of the 8,000 process.
1997	July: President Samper's finance campaign treasurer, Guillermo Palomari, confessed to get US$6 millions from the the Cali Cartel as donations for his campaign. October: Two members of the OAS and an official of Human Rights Watch were kidnapped.
1998	March: FARC guerrilla kidnapped two French citizens and an American citizen. July: Andres Pastrana Arango, the elected president, begins peace talks with guerrillas. August: Rebels from de FARC guerrilla attacked an antinarcotics base in Guaviare, killing 40 policemen and kidnapping other 56. October: President Pastrana grants FARC a safe haven the size of Switzerland in the southeast to help move peace talks along. The zone is off-limits to the army.
1999	January: Peace talks formally launched but proceed in stop-start fashion. Pastrana and FARC leader Manuel "Tirofijo" Marulanda meet. February: Two Americans working with the Indians of the Colombian forest were kidnapped and killed by the FARC guerrilla. April: An airplane of Avianca was kidnapped, with some foreign citizens from the United States and Italy. May: ELN rebels kidnapped 160 people in a church in Cali.
2000	August: Congressman Óscar Tulio Lizcano was kidnapped by rebels of the FARC. September: Government freezes talks; alleges FARC harbored hijacker of plane forced to land in safe haven. Later, FARC refuses to resume talks, accuses Pastrana of not stopping paramilitary groups. December: Ex-minister Fernando Araújo was kidnapped by rebels of the FARC guerrilla.
2001	February: The FARC return to peace talks after meeting between Tirofijo and Pastrana. Pastrana extends demilitarized area for 8 months. June: FARC rebels free 359 police and troops in exchange for 14 captured rebels. FARC accused of using safe haven to rearm, prepare attacks, and conduct drug trade. October: Government and FARC sign San Francisco agreement, committing themselves to negotiate cease-fire. Pastrana extends life of safe haven until January 2002. November: The U.S. State Department included the guerrillas of FARC and ELN and the paramilitars of the AUC in its list of terrorist groups.
2002	January: Pastrana accepts FARC cease-fire timetable after a period of extended crisis in the process, and extends safe haven until April. February: Pastrana breaks off three of tortuous peace talks with FARC rebels, says hijacking of aircraft hours earlier is final straw. He orders rebels out of demilitarized zone. Government declares war zone in south after rebels step up attacks. Presidential candidate Ingrid Betancourt and her vice president candidate were kidnapped by rebels of the FARC guerrilla, while visiting the rebels in the safe haven. August: Moments before Alvaro Uribe is sworn in as president, suspected FARC explosions rock Bogota. Twenty people are killed. Days later, Uribe declares state of emergency. September: Ex-minister Consuelo Araújo Noguera is kidnapped by the FARC guerrilla and later killed.

(Continued)

TABLE 31.1   Brief Chronology of Terrorism Landmarks, 1996–2008 (*Continued*)

2003	February: A car bomb exploded in a social club in Bogota. An airplane was crashed by the FARC guerrilla and 3 American contractors were kidnapped. May: Governor Gaviria and Ex-minister Echeverri were killed during a rescue attempt. November: Fighters from right-wing United Self-Defence Forces of Colombia (AUC) begin to disarm. December: The Department of Security captured Wilmar Antonio Marín Cano, chief of an important FARC faction and responsible for most of the kidnappings from the FARC.
2004	January: Simon Trinidad, one of the chief leaders of the FARC guerrillas, was captured by the national army and the national police. February: President Uribe asks the European Parliament to proclaim the guerrillas of ELN terrorist groups. July: Right-wing AUC and government begin peace talks. AUC leaders address Congress. December: Rodrigo Granda, a very important leader of the FARC guerrilla, was captured in Venezuela. Salvatore Mancuso, the paramilitary chief, disarmed 1,400 men and women of his own soldiers, and starts the peace process execution. Venezuela and Colombia break their diplomatic relations, and the frontiers are closed.
2005	January: Bitter 15-day dispute with Venezuela over the capture of a FARC leader on Venezuelan soil. The affair is resolved at talks in Caracas in February. February: Crisis with Venezuela comes to an end, thanks to the intervention and mediation of Cuba, Brazil, and Peru. December: Exploratory peace talks with the second biggest left-wing rebel group, the National Liberation Army (ELN), begin in Cuba.
2006	January: Seven hundred forty-two soldiers of the paramilitary group start getting into the peace process and give up their arms. February: Diplomatic crisis with Ecuador, after some declarations of President Uribe, accusing Ecuador of letting FARC rebels cross the border. August: Carlos Castaño, leader of the paramilitars, was killed by his own peers. September: A list of politicians who are supposed to be linked with the paramilitary is revealed. October: The computer of Jorge 40 reveals some links between the Security Department and the paramilitars.
2007	January: The Ralito Pact, a deal between paramilitars and politicians, is revealed, and many politicians are on it. June: Government releases dozens of jailed FARC guerrillas, in the hope that rebels will reciprocate by releasing hostages. FARC rejects the move, saying it will only free hostages if the government pulls back troops and sets up a demilitarized zone. Twelve deputies held hostage since 2002 die on a rescue attempt. September: In his role as mediator, Venezuelan President Hugo Chavez agrees to invite rebels for talks on hostage release deal. November: Colombia sets deadline of December 31 for President Chavez to reach deal with rebels on prisoner swap. Venezuelan President Hugo Chavez withdraws his country's ambassador to Bogotá in a row over his role in negotiations between the Colombian government and rebel forces. December: FARC announces the release of some hostages, to prove mediation of President Chavez was a good tool to negotiate with the guerrilla.

TABLE 31.1   Brief Chronology of Terrorism Landmarks, 1996–2008 (*Continued*)

2008	January: The FARC releases two high-profile hostages, Clara Rojas and Consuelo Gonzalez, as a result of Mr. Chavez's mediation. Mr. Chavez calls on the U.S. and European governments to stop considering Colombian left-wing rebel groups as terrorists, but Mr. Uribe rejects the idea. February: Four ex-congressmen and women were released by the FARC guerrilla, after 6 years of being held hostage. March: A Colombia cross-border strike into Ecuador kills senior FARC rebel Raul Reyes, sparking a diplomatic crisis. Venezuela and Ecuador cut ties with Colombia and order troops to their borders. Venezuela orders the expulsion of the Colombian ambassador, and announced the cut of diplomatic ties with Colombia, because of the Ecuadorian incident. Nicaragua cuts ties with Colombia because of the Ecuadorian incident. Ivan Rios, another leader of the FARC guerrilla, was killed by the army. Relations between Colombia, Venezuela, and Nicaragua are going back to normal, while the situation with Ecuador is still on the line of crisis. April: The political crisis gets aggravated by the fact that 20% of the Congress is under investigation, of their possible ties with the paramilitary movement. Ecuador continues with the idea of recognizing the political status of the Colombian FARC guerrilla, keeping the relations under crisis.

requires a sound and sustained development of the stock market to channel funds to the productive sector.

31.3   THE COLOMBIAN STOCK MARKET

The Colombian stock market is relatively young, particularly since 2001 a number of important measures were adopted that, jointly with the fantastic dynamism of the Colombian economy in the last few years,[*] led to a significant expansion. Even so, the Colombian stock market is still in its infancy, less than 100 (88) mostly illiquid companies are traded, compared to 400 for Brazil, 368 for Mexico, 239 for Chile, 231 for Peru, and 111 for Argentina.[†] In terms of capitalization Colombia occupies a modest position with 99.923 million US$ (MUS$) after Brazil (1,348,569 MUS$), Mexico (393,568 MUS$), and Chile (215,852 MUS$), and only preceding Peru (68,022 MUS$) and Argentina (55,478 MUS$); moreover, in terms of relative size against the GNP, the Colombian stock market has one of the lowest ratios (40% compared to 120% for Chile), only higher

---

[*] In 2006 Colombia showed an increase of about 6.8% in GNP with a low inflation rate of 4.3% and strong foreign investment.

[†] The stock market data reported here are taken from the last bulletin of the World Federation of Exchanges.

TABLE 31.2 Business Climate and Terrorism

	Business Cost of Terrorism		Business Costs of Crime and Violence		Public Trust of Politicians		Organized Crime	
	Disadvantage	Advantage	Disadvantage	Advantage	Disadvantage	Advantage	Disadvantage	Advantage
Colombia	**129**		112	—	87	—	126	—
Argentina	—	**12**	106	—	125	—	90	—
Brazil	—	**10**	121	—	126	—	125	—
Mexico	53	—	119	—	91	—	**120**	—
Chile	28	—	65	—	28	—	28	—
Peru	97	—	114	—	117	—	107	—
Median	41	—	113	—	104	—	114	—

TABLE 31.3   Performance of the LATAM Markets, 2001–2008

	Colombia	Argentina	Brazil	Chile	Mexico	Peru	
2001	**1.1504**	−0.4624	−0.8711	−0.5877	0.4070	0.4704	
2002	0.9556	−1.2176	−0.9103	−1.3069	−0.6042	1.1002	
2003	2.2979	2.2491	2.4807	3.1042	1.0456	2.6995	
2004	**2.8276**	0.7787	0.8364	1.1794	2.0270	0.1367	Sharpe ratio
2005	**2.6408**	1.6158	1.4217	1.0090	1.8827	1.0119	
2006	0.2768	1.5052	0.7835	1.3026	0.9375	1.1228	
2007	0.3066	−0.4599	1.3940	0.5395	0.2011	1.7907	
2008	**0.9466**	0.6766	0.6906	0.5955	0.2667	0.7754	
2001	**0.2292**	−0.2194	−0.303	−0.1044	0.1041	0.0754	
2002	0.1785	−0.8155	−0.3747	−0.2349	−0.156	0.2145	
2003	0.3975	0.624	0.6295	0.5238	0.191	0.5542	
2004	**0.7231**	0.2314	0.2467	0.2028	0.3558	0.0371	Excess return
2005	**0.6523**	0.4402	0.4026	0.1456	0.3488	0.2167	
2006	0.1207	0.4598	0.2421	0.1978	0.2468	0.3346	
2007	0.0717	−0.1174	0.4879	0.1165	0.0509	0.5543	
2008	0.2643	0.2438	0.2926	0.1823	0.0851	0.2656	
2001	0.0159	0.0992	0.0680	0.0186	0.0354	0.0140	
2002	0.0178	0.3029	0.0798	0.0174	0.0318	0.0220	
2003	**0.0146**	0.0453	0.0343	0.0152	0.0192	0.0221	
2004	0.0344	0.0479	0.0520	0.0167	0.0180	0.0395	Semivariance
2005	0.0339	0.0417	0.0467	0.0123	0.0183	0.0256	
2006	0.1064	0.0496	0.0554	0.0132	0.0372	0.0508	
2007	0.0336	0.0380	0.0713	0.0276	0.0365	0.0582	
2008	**0.0175**	0.0254	0.0378	0.0176	0.0195	0.0246	
2001	**0.0258**	0.0793	0.0701	0.0377	0.0423	0.0331	
2002	**0.0289**	0.1801	0.0665	0.0305	0.0485	0.0447	
2003	0.0292	0.0471	0.0439	0.0245	0.0296	0.0315	
2004	0.0449	0.0545	0.0532	0.0285	0.0286	0.0478	VaR
2005	0.0412	0.0426	0.0475	0.0264	0.0301	0.0355	
2006	0.0917	0.0510	0.0542	0.0252	0.0474	0.0444	
2007	0.0471	0.0406	0.0627	0.0456	0.0453	0.0515	
2008	**0.0366**	0.0535	0.0502	0.0354	0.0500	0.0407	

than Argentina (20%). Domestically, equity represents only about 2% of the total trade, which is a marginal figure compared to the 96% of trading in fixed income. These and other figures suggest that there is still a long way for the development of the Colombian stock market.

In terms of performance, the Colombian stock market is probably one of the most interesting in the world. To give a view of this, in Table 31.3 we show several performance measures (Sharpe's ratio, excess return, semivariance, and 1-day VaR computed at the 99% level[*]) for the main LATAM markets and for the years 2001–2008. We denote boldface when, for a particular year and performance measure, the Colombian stock market dominates the others. Note that about 40% of the time the Colombian stock market has the best performance. It should also be pointed out that, in terms of return, most of the LATAM stock markets (with the exception of Argentina and Chile) show returns along the period between 400% and 600% approximately, making them some of the most profitable in the world.

31.4 EMPIRICAL RESULTS

In our context of analysis, it is important to determine whether terrorism may add extra challenges to the development of the still weak Colombian market. In the short term, terrorism news might cause some shock in both returns and volatilities, while on the mean and long term their influence is less clear and would depend on the constraints that investors of a particular country may face to diversify away this kind of risk.

In this chapter we focus on the effect of terrorism on short-term volatility. This issue is important because to the extent that terrorist episodes are frequent and economically significant, they could provoke risk to reach a level where investors (demanding a higher compensation for risk) and companies (incapable of providing such compensation) are unable to reach equilibrium. We will analyze the nature of this relationship along the recent history of Colombia. The period of study (January 1, 1996–April 31, 2008) is chosen because of the availability of high-quality stock market data as well as compiled records of news related to terrorism.

Since we are interested in this relationship along a relatively long period of time there are some problems with the use of the IGBC (Colombia's stock market general index). For this reason, we employ in our analyses the MSCI index, in U.S. dollars, of Colombia. Since the MSCI indexes are well known, we do not provide a description here.[†] To give a clearer view of our results we analyze each of the years separately. Table 31.4 shows the main characteristics of the series for each of the years in the period of study; as it can be seen, for most of the years the returns are skewed and show evident kurtosis, nonnormality (tested with the Jarque-Bera test (JB)), autocorrelation (Ljung-Box test (LB)), and heteroskedasticity (according to the

---

[*] We employ the 3-month Treasury bills yield as the risk-free rate.
[†] The data employed here were generously provided by MSCI Barra; we refer the reader to their website to find a description of the indexes employed.

TABLE 31.4 Descriptive Statistics

	nobs	mean	std	skew	kurt	JB	p-val	LB(10)	p-val	LB2(10)	p-val
1996	262	0.0002	0.008	-0.026	5.926	93.5	0.000	129.830	0.000	56.020	0.0000
1997	261	0.0012	0.012	0.491	9.044	407.7	0.000	155.590	0.000	116.230	0.0000
1998	261	-0.0023	0.020	1.092	13.200	1183.0	0.000	61.590	0.000	**21.980**	**0.3415**
1999	261	-0.0008	0.019	-0.192	4.628	30.4	0.000	42.190	0.003	86.280	0.0000
2000	260	-0.0020	0.012	-0.401	6.688	154.3	0.000	40.270	0.005	**19.540**	**0.4870**
2001	261	0.0012	0.013	1.144	7.277	255.8	0.000	70.200	0.000	**24.960**	**0.2030**
2002	261	0.0006	0.012	0.007	4.475	23.7	0.001	46.880	0.001	**21.740**	**0.3547**
2003	261	0.0018	0.011	0.230	5.371	63.5	0.000	37.720	0.009	62.900	0.0000
2004	262	0.0031	0.016	-0.102	6.531	136.6	0.000	**28.025**	**0.109**	60.210	0.0000
2005	260	0.0027	0.016	-0.355	3.809	12.6	0.008	36.700	0.012	32.993	0.0338
2006	260	0.0004	0.028	-0.065	11.357	756.8	0.000	43.310	0.002	281.830	0.0000
2007	261	0.0005	0.015	-1.176	9.917	579.7	0.000	**7.931**	**0.992**	**7.408**	**0.9951**
2008	82	0.0008	0.019	-1.075	8.123	105.4	0.000	**14.560**	**0.801**	22.229	0.3289

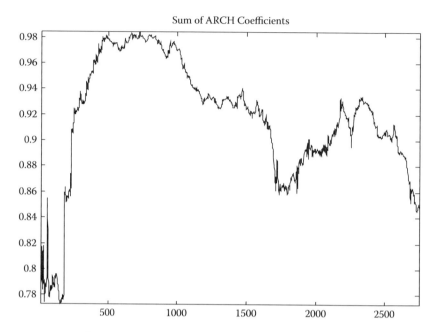

FIGURE 31.1   Decreasing persistence of volatility.

LB test on squared returns (LB2)). In the table we use boldface in the cases where we fail to reject the null.

To analyze the effect of terrorism on market volatility, we built a database of news related to terrorism by reviewing the main Colombian newspapers as well as by using other well-known sources, such as the chronology of terrorist incidents of *Patterns of Global Terrorism* of the U.S. Department of Defense. Obviously our database is quite subjective since, in our view, it would be impossible to exactly determine when news is important enough to be considered. Moreover, in the Colombian case the phenomenon of terrorism is ubiquitous, and in many cases it can be hardly determined whether particular news has some relation to terrorism.

As a previous analysis, we tried to measure the level and persistence of volatility in the Colombian stock market. To analyze the degree of persistence in volatility, in Figure 31.1 we plot the sum of coefficients of the conditional variance of an ARMA(1,1)-GARCH(1,1) model:

$$r_t = \delta + \lambda r_{t-1} + \theta \varepsilon_{t-1} + \varepsilon_t, \varepsilon_t \approx N(0, \sigma_t)$$

$$\sigma_t^2 = \eta + \alpha \sigma_{t-1}^2 + \beta \varepsilon_{t-1}^2$$

(31.1)

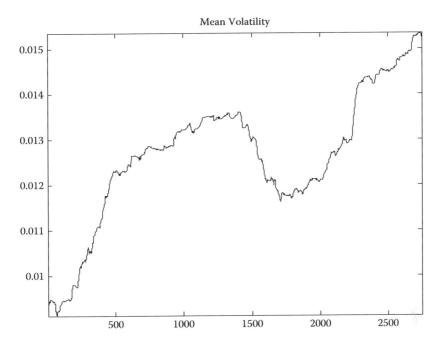

FIGURE 31.2   Increasing levels of volatility.

which is fitted using a rolling window of 5 years along the period of study. Note that after a transient period, the process is close to being integrated in variance ($\alpha + \beta \approx 1$) but then persistence declines monotonically to reach 0.85, a figure similar to the one found in many other stock markets. This reduction of persistence is consistent with an increase in the efficiency of the market, which adjusts more rapidly to the news.

Respecting the level of volatility of the Colombian stock market, we find that it has increased along time. In Figure 31.2 we plot the estimated conditional volatility using Equation (31.1); note that the volatility by the end of the period is 60% higher than at the beginning of the period. As a conclusion of these preliminary analyses we can say that the volatility of the Colombian stock market has increased in recent times, which is consistent with the high returns experienced by the market, but that its level of persistence has decreased, which is consistent with increasing efficiency levels. In our view, we would expect that news specifically related to terrorism would have a decreasing level of importance since other information might be present and spikes in volatility would be rapidly absorbed by the market.

To analyze the effects of news related to terrorism on volatility we used expanded ARCH models that take into account other variables that might

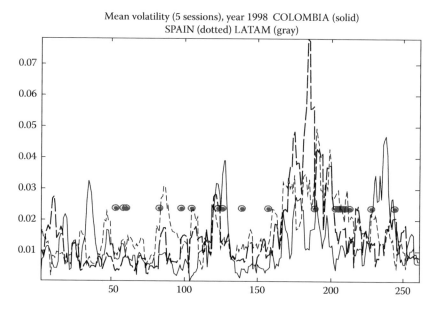

Mean volatility (5 sessions), year 1998 COLOMBIA (solid)
SPAIN (dotted) LATAM (gray)

FIGURE 31.3   Incidence of terrorism news on volatility.

affect volatility. First, since we want to be sure that fluctuations on volatility are exclusively due to news related to terrorism, we have to control
for higher levels of volatility due to general and not intrinsic conditions.
To do so, we build an equally weighted portfolio of the MSCI indexes of
Argentina, Brazil, Chile, Mexico, and Peru and compute the volatility of
this equally weighted portfolio, which is used as a proxy of the general
level of volatility of the area. To control for levels of volatility outside the
LATAM market we consider the Spanish MSCI index. The Spanish stock
market is the natural link of LATAM stock markets to developed markets,
since it is highly integrated with the most important capital markets in
the world.

In Figure 31.3 we show an example of the behavior of returns, volatility,
and news related to terrorism for the year 1998. We plot the volatilities of
the Colombian, LATAM, and Spanish markets computed along the last 5
trading days. In the figure we also plot a grey dot on the mean volatility level
the days that news related to terrorism is found. To the extent that terrorism affects the volatility levels one should expect a sudden spike or a change
in level in volatilities the days following the news. Even though we only
show an example, for all the years the behavior was quite similar to that
shown here. Note that around observation 160 we find such a spike, and

we also find a higher level of volatility after the news. However, the same spike appears in the LATAM as well as in the Spanish case, so that one has to conclude that, in this example, either the news shocked these markets (a conclusion that seems quite implausible) or the higher level on volatility is due to international market instability and not to Colombian terrorism.

Though illustrative, obviously this approach is too simple to reach conclusions. For this reason we employ a more formal procedure to evaluate the incidence of terrorism. First, for each one of the years we estimate a series of ARMA(p,q)-GARCH-M(r,s) models of the form:

$$r_t = \delta + \sum_{i=1}^{p} \lambda_i r_{t-i} + \sum_{j=1}^{q} \theta_j \varepsilon_{t-j} + \varphi T_t + \varepsilon_t, \varepsilon_t \approx N(0, \sigma_t)$$

(31.2)

$$\sigma_t^2 = \eta + \sum_{k=1}^{r} \alpha_k \sigma_{t-k}^2 + \sum_{l=1}^{s} \beta_l \varepsilon_{t-l}^2 + \omega T_t + \gamma_1 \sigma lat_t^2 + \gamma_2 \sigma spa_t^2$$

where T_t is a dummy variable that takes a value equal to 1 when there is news related to terrorism and 0 otherwise, and σlat_t^2 and σspa_t^2 are the volatility levels of the LATAM portfolio and Spain MSCI index, respectively, which are used to control for international instability. Note that if terrorism affects volatility, increasing its level, ω, should be positive and statistically different from zero. Note also that since we compute positive and negative news, the value of φ can be either positive, negative, or zero.

Since we analyze each of the years separately we have to keep the models as parsimonious as possible; for this reason, we select the models by increasing the number of lags until the LB test does not reject the null of no autocorrelation in the residuals and squared residuals of the model, keeping p, q, r, $s \leq 2$. In the case that the model fails to pass the LB test for either the residuals or squared residuals, we employ the most parsimonious model that provides similar results. In most of the cases an AR(1)-GARCH(1,1) model was enough to capture dependence in mean and variance; only in two cases (years 1998 and 2006) were we unable to remove all the structure in the returns and volatilities series.

In Table 31.5 we show the results obtained in each of the estimations as well as the p-value of the corresponding parameters; we denote in boldface the statistics that are significant at a 10% level. Note that the parameter corresponding to the dummy of terrorism is significant at a 10% level only in three cases (years 1999, 2000, and 2003), and it is never significant at

TABLE 31.5 Parameters and Significance of the Models

	δ	λ	θ	φ	η	α	β	ω	γ1	γ2	LB	LB2
1996	**-0.011**	**0.628**	**-0.304**	**0.261**	0.000		**0.984**	0.081	**7.870**	**-6.536**	6.251	12.501
	0.883	0.000	0.009	0.061	1.000		0.000	0.194	0.013	0.010	0.794	0.253
1997	**0.170**	**0.411**		-0.070	0.000	**0.548**	**0.216**	0.138	24.805	6.099	12.950	6.447
	0.095	0.000		0.748	1.000	0.005	0.055	0.710	0.253	0.758	0.226	0.776
1998	**-0.229**	**0.467**		-0.020	0.000	**0.283**	**0.601**	-0.240	17.440	6.676	9.326	**18.776**
	0.088	0.000		0.946	1.000	0.006	0.000	0.442	0.281	0.843	0.501	0.043
1999	0.094	**0.258**		-0.499	0.000		**0.802**	0.430	**69.567**	-22.360	3.225	13.406
	0.495	0.000		0.164	1.000		0.000	0.100	0.000	0.136	0.976	0.202
2000	**-0.147**	**0.176**		0.361	0.004	**0.371**	**0.950**	0.229	**11.263**	-6.060	10.869	8.369
	0.039	0.009		0.310	0.963	0.055	0.000	0.071	0.015	0.339	0.368	0.593
2001	-0.036	**0.291**		0.034	**0.920**		0.218	0.301	-9.604	-14.513	15.440	6.068
	0.765	0.001		0.910	0.011		0.202	0.422	0.645	0.209	0.117	0.810
2002	-0.043	**0.272**		**0.435**	0.333	**0.237**	0.180	0.274	-1.457	22.928	11.256	7.142
	0.646	0.000		0.046	0.392	0.046	0.734	0.301	0.963	0.272	0.338	0.712
2003	**0.194**	**0.249**		-0.292	0.016	**0.150**	**0.752**	**0.276**	13.661	-3.980	7.656	4.951
	0.006	0.000		0.240	0.877	0.033	0.000	0.082	0.209	0.287	0.662	0.894
2004	0.171	**0.207**		**0.576**	0.208	**0.396**	**0.394**	-0.060	23.229	21.411	12.414	6.661
	0.171	0.011		0.042	0.773	0.048	0.006	0.937	0.777	0.892	0.258	0.757
2005	**0.388**	**0.247**		-0.342	0.070		**0.982**	-0.017	**-17.256**	17.915	13.250	11.130
	0.007	0.002		0.298	0.440		0.000	0.898	0.011	0.335	0.210	0.347
2006	0.010	0.202		0.023	0.008	0.940	0.000	-0.007	2.004	-0.780	13.280	**17.550**
	0.444	0.230		0.417	0.573	0.187	1.000	0.466	0.115	0.488	0.208	0.080
2007	**0.203**			-0.253	0.000		0.101	-0.034	70.137	80.453	3.276	12.849
	0.090			0.111	1.000		0.883	0.918	0.372	0.294	0.974	0.232
2008	0.419			-0.247	0.000		**0.985**	-0.062	22.235	-16.369	7.243	11.438
	0.207			0.526	1.000		0.000	0.615	0.125	0.190	0.702	0.116

TABLE 31.6   News and Stock Market Return, 1996

Date	News	Return
Jan. 19	An American citizen is kidnapped.	−0.34%
Feb. 6	Three foreign engineers and a Colombian citizen are kidnapped by the ELN.	−0.12%
Feb. 16	One American citizen is kidnapped by the ELN.	−0.06%
March 25– April 13	Samper's crisis	0.94%
July 14	An Italian citizen is kidnapped.	−0.02%
Aug. 9	An Italian citizen is kidnapped.	0.54%
Aug. 14	Two Brazilian engineers are kidnapped.	−0.76%
Aug. 21	An Italian citizen is kidnapped.	0.82%
Oct. 28	A convoy is attacked and two engineers are kidnapped.	0.54%
Dec. 11	An American geologist is kidnapped.	0.21%
	Sum	**1.76%**

a 5% level. These results seem to indicate that stock market volatility is not affected by news related to terrorism. An explanation of this fact is that investors consider these acts normal and not structural in the sense that they do not have a clear impact on the economy or permit changed expectations of future terrorism activity. Our results are consistent with other authors (e.g., Zussman and Zussman, 2006) who show that the stock market does not react to terrorism-related news when it does not imply either an increase or a decrease in terrorism levels.

An interesting conclusion is that news related to terrorism does not even seem to affect returns since the corresponding coefficient of the mean equation is never significant. This lack of relationship is surprising in some cases: for example, in 1996 (Table 31.6), all the news of the year cannot be interpreted as positive since it does not indicate the resolution of a particular conflict. Overall, returns on the days that the news impacted the market were positive; this explains the positive coefficient in the regression (significant at the 10% level). Similar unexpected results have been found by other authors, such as Pagano and Strother (2007), who report that the S&P500 index responds positively to an increase in the threat of a terrorist attack.

31.5   CONCLUSION

We have analyzed the effect of news related to terrorism on the levels of volatility of the Colombian stock market. Our results suggest that this kind of news does not seem to affect the risk level faced by investors.

A possible explanation is that terrorism in Colombia, though dramatic, is considered a variable that does not condition economic activity. On the other hand, one could argue that terrorism news or acts are so common that they do not seem to affect the sentiment of investors.

Our results must be taken with extreme caution since a more sound methodology for categorizing news related to terrorism is needed. Specifically, it would be interesting to discern between positive and negative news, since they might have different effects of volatility. Moreover, even among negative news it would be interesting to differentiate between those news items that are too specific to impact markets as a whole (for example, the kidnapping of a worker of a specific foreign company) and those that might interfere with the normal economic activity by eroding confidence in the institutional system (such as the detention of politicians linked to terrorist groups). These and other extensions of the present work are under current investigation.

ACKNOWLEDGMENTS

The authors thank the invaluable research assistance of Javier García and of Diana Arteaga, who helped to develop the database and provided useful comments.

REFERENCES

Abadie, A., and J. Gardeazabal. (2003). The economic costs of conflict: A case study of the Basque country. *American Economic Review* 93:113–32.

Alesina, A., and R. Perotti. (1996). Income distribution, political instability and investment. *European Economic Review* 40:1203–28.

Cárdenas, M. (2007). Economic growth in Colombia: A reversal of "fortune"? Working paper, Fedesarrollo, Bogotá, Colombia.

Guidolin, M., and E. La Ferrara. (2005). The economic effects of violent conflict: Evidence from asset market reactions. Working paper, Federal Reserve Bank of St. Louis, MO.

Karolyi, G. A. (2006). Shock markets: What do we know about terrorism and the financial markets? *Canadian Investments Review*, Summer, 9–15.

Karolyi, G. A., and R. Martell. (2006). Terrorism and the stock market. Working paper, Ohio State University, Columbus.

Keynes, J. M. (1936). *The general theory of employment, interest, and money.* London: MacMillan.

Kurtzman, J., G. Yago, and T. Phumiwasana. (2004). The global costs of opacity. *MIT Sloan Management Review* 46:38–44.

Pagano, M., and T. S. Strother. (2007). Systematic risk and the perceived threat of terrorist attacks: The advisory system that cried wolf. 20th Australasian Finance & Banking Conference. Available at http://papers.ssrn.com/sol3/papers.cfm?abstract_id=1016901.

Pshisva, R., and G. A. Suarez. (2006). Captive markets: The impact of kidnappings on corporate investment in Colombia. Working paper, Federal Reserve Board, Washington, DC.

World Economic Forum. (2007). The global competitiveness report 2007–2008. Available at http://www.gcr.weforum.org/.

Zussman, A., and N. Zussman. (2006). Assassinations: Evaluating the effectiveness of an Israeli counterterrorism policy using stock market data. *Journal of Economic Perspectives* 20:193–206.

Index

Printed and bound by CPI Group (UK) Ltd, Croydon, CR0 4YY

23/10/2024

01778369-0001